CALCULUS WITH APPLICATIONS TO BUSINESS AND LIFE SCIENCES
Second Edition

CALCULUS WITH APPLICATIONS TO BUSINESS AND LIFE SCIENCES
Second Edition

Abe Mizrahi
Indiana University Northwest
Michael Sullivan
Chicago State University

John Wiley & Sons
New York • Chichester • Brisbane • Toronto • Singapore

Dedicated to our children:
Tami, Laura
Katy, Mike, Danny, Colleen

Copyright © 1976, 1984 by John Wiley & Sons, Inc.

All rights reserved. Published simultaneously in Canada.

Reproduction or translation of any part of
this work beyond that permitted by Sections
107 and 108 of the 1976 United States Copyright
Act without the permission of the copyright
owner is unlawful. Requests for permission
or further information should be addressed to
the Permissions Department, John Wiley & Sons.

Library of Congress Cataloging in Publication Data:

Mizrahi, Abe.
   Calculus with applications to business and life
sciences.

   Includes index.
   1. Calculus.   I. Sullivan, Michael, 1942-
II. Title.
QA303.M688   1983   515   83-16946
ISBN 0-471-05484-4

Printed in the United States of America

10 9 8 7 6 5 4 3 2 1

Preface

First and foremost, this book is student-oriented. It does not dwell on formalities, but appeals to intuition. Abstraction and sophisticated mathematical theory are minimized without sacrificing an understanding of the underlying mathematical concepts. As much as possible, problems are introduced through real-life situations; the mathematics needed to handle similar situations is then developed. This motivation is highlighted by references to current applications in the social, business, and life sciences. In addition, questions from Certified Public Accountant (CPA) Examinations, Certificate in Management Accounting (CMA) Examinations, and Society of Actuaries Examinations have been reproduced at the end of appropriate chapters to contribute further to the relevance of the mathematics in this book. Textual explanations are precise, brief, and to the point, and they are always accompanied by illustrative examples.

- The textual material is interspersed with more than 230 illustrative examples. Almost every new idea is followed by one or two examples demonstrating how related problems can be solved. Often an example is furnished to motivate the need for a difficult concept before a precise definition is given. (See, for instance, page 100.) When possible, an outline of steps to follow in solving a problem is given. (See pages 18, 102, 166, and 261.)
- There are over 1700 exercises, most of which are keyed to the illustrative examples. The exercises vary in nature and range from drill exercises to those that challenge the superior student. Most exercise sets contain real-world applications of the material. (See pages 130, 162, and 182.) At the end of some chapters, true-false and fill-in-blank questions are provided to aid in the review process. (See pages 88, 135, and 220.)
- Completely worked-out solutions, with illustrations, to odd-numbered problems are given in the back of the book. Completely worked-out solutions to even-numbered problems are available in a supplement to the text. Review exercises are given at the end of each chapter. At the end of most chapters, actual questions from recent CPA Exams, CMA Exams, and Society of Actuaries Exams are reproduced. (See, for example, pages 191, 222, and 325.)
- Every chapter contains word problems as applications, including many real-world applications. (See pages 108, 236, and 268.)
- A second color is used in the text and art to highlight important facts. Over 175 figures illustrate many concepts.
- Flowcharts are utilized whenever possible to outline procedures. See, for instance, pages 149 and 258.
- A solutions manual to even-numbered problems will be available.

This is truly a collaborative effort, and the order of authorship signifies alphabetical precedence. We assume equal responsibility for the book's strengths and weaknesses, and welcome comments and suggestions for its improvement.

Abe Mizrahi
Michael Sullivan

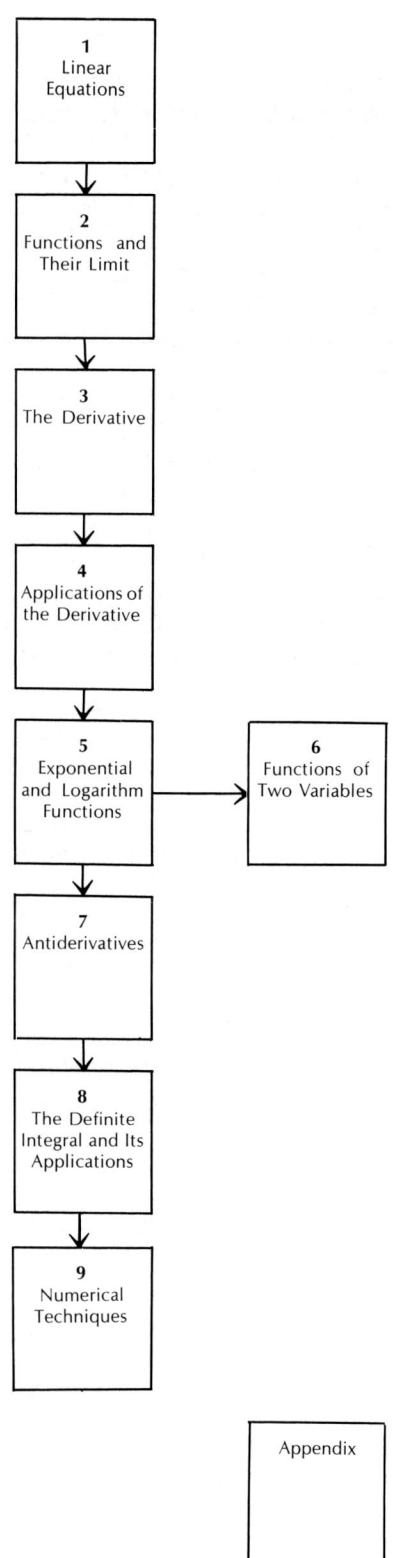

ORGANIZATION

Chapter 1 is the "foundation" or review chapter and covers topics such as rectangular coordinates and linear equations, including applications of linear equations.

Chapter 2 begins with a review of some of the tools required for the study of calculus, and continues with functions and the idea of a limit.

Chapters 3 and 4 introduce the basics of differential calculus and its applications. Chapter 5 covers the exponential and logarithm functions. Chapter 6, which may be postponed, introduces the idea of partial derivatives. Chapters 7 and 8 cover the basic ideas of integral calculus, including necessary applications. Chapter 9 discusses some numerical techniques of calculus.

The diagram on page vi illustrates the division of the book.

The book is suitable for use in a 3 or 4 semester hour course.

Worked-out solutions to approximately 850 odd-numbered problems are given in the back of the book. Worked-out solutions to approximately 850 even-numbered problems are available in a *Solutions Manual*. Problems with asterisks are to be considered more challenging.

At the end of each chapter, a review of important terms, additional problems, and actual questions from recent CPA, CMA, and Society of Actuaries Exams are included. Finally, the text contains all tables required in the exercises.

ACKNOWLEDGMENTS

We thank the many students at Indiana University Northwest and Chicago State University for their comments and criticisms. We also thank these reviewers for their suggestions:

Professor Jerome E. Kaufmann, Western Illinois University;

Dr. Dennis Bertholf, Oklahoma State University;

Professor William W. Hokman, University of Akron;

Professor Koy M. Bass, Jr., San Antonio College;

Professor Martin M. Kotler, Pace University;

Professor William E. Conway, University of Arizona;

Professor J. A. Nordstrom, University of Houston;

Professor Ladnor Geissinger, University of North Carolina;

Professor Arthur Schwartz, University of Michigan;

Professor Edith W. Ainsworth, University of Alabama;

and Professor Garret Etgen, University of Houston.

We especially thank Leroy Peterson and Henry Wyzinski, both at Indiana University Northwest, who checked the accuracy of solutions to all of the exercises.

We are grateful to the following organizations who permitted the reproduction of actual questions from previous professional examinations:

American Institute of Certified Public Accountants (CPA Exams)
Educational Testing Service (Actuary Exams)
Institute of Management Accounting of the National Association of Accountants (CMA Exams)

We are indebted to the people at Wiley, whose talent played a significant part in the publication of this book. We particularly thank Bob Pirtle, mathematics editor, for his patience and help.

A. M.
M. S.

New Features of the Second Edition

*Chapter 1 has been streamlined by moving some of the review material to an appendix while redoing topics appropriate to calculus.

*Chapters 2 and 3 have been combined. New material in graphing, intercepts and symmetry, and the chain rule, have been added, while the hard concepts of infinite limits and limits at infinity have been postponed to Chapter 4.

*Chapter 5 treats exponential and logarithm functions in a more streamlined, practical way, dealing with calculus earlier.

*Chapter 7 was reorganized so that topics contained there can be covered more easily. A new section on differential equations with applications has been added.

*Chapter 8 now contains a section on Riemann sums, so that two approaches to the integral are contained in the same chapter. New sections on applications to averaging and to probability have been added.

*A new chapter on numerical techniques has been added.

*A new section on linear regression has been added to the chapter on functions of two variables. This chapter now appears earlier.

*Every chapter has been rewritten to improve clarity and eliminate formality.

*All exercise sets have been redone.

*Many new examples have been added.

*Many new illustrations have been added.

*Actual questions from CPA, CMA, and Actuary Exams appear at the end of appropriate chapters.

*Where appropriate, a set of true/false and fill-in-the-blank questions appear in the chapter review.

Contents

1 LINEAR EQUATIONS — 1

1. Introduction — 2
2. The Straight Line — 7
3. Parallel and Intersecting Lines — 15
*4. Applications — 23
Chapter Review — 31
    Mathematical Questions from CPA and CMA Exams — 33

2 FUNCTIONS AND THEIR LIMIT — 37

1. Preliminaries — 38
2. Functions — 46
3. Some Useful Functions — 58
4. The Idea of a Limit — 67
5. Algebraic Techniques for Evaluating Limits — 75
6. Continuous Functions — 80
Chapter Review — 87
    Mathematical Question from Actuary Exam — 90

3 THE DERIVATIVE — 91

1. Average Rate of Change — 92
2. The Derivative — 100
3. Two Additional Interpretations of the Derivative — 105
4. Some Derivative Formulas — 109
5. General Formulas for Differentiation — 112
6. The Power Rule; The Chain Rule — 118
7. Higher-Order Derivatives — 125
8. Functions Not Differentiable at c — 130
Chapter Review — 135

4 APPLICATIONS OF THE DERIVATIVE — 139

1. Relative Maxima and Relative Minima — 140
2. Absolute Maximum and Absolute Minimum — 151
3. Applied Problems — 155
4. Concavity — 164
5. Asymptotes — 170
6. Marginal Analysis — 176
*7. Models — 182
    Model 1: Maximizing Tax Revenue — 182
    Model 2: Optimal Trade-In Time — 184
    Model 3: Minimizing Inventory Cost — 185
    Model 4: The Response of the Body to a Drug — 186
Chapter Review — 187
    Mathematical Questions from CPA Exams — 191

5 THE EXPONENTIAL AND LOGARITHM FUNCTIONS — 193

1. The Exponential Function — 194
2. The Derivative of the Exponential Function — 200
3. The Logarithm Function and Its Derivative — 208
4. Logistic Curves — 217
Chapter Review — 220
    Mathematical Questions from Actuary Exams — 222

6 FUNCTIONS OF TWO VARIABLES — 223

1. Introduction — 224
2. Partial Derivatives — 228
3. Maxima and Minima — 232
4. Lagrange Multipliers — 237
5. Least Squares — 240
Chapter Review — 247

7 ANTIDERIVATIVES — 249

1. Antiderivatives; the Indefinite Integral — 250
2. Integration by Substitution — 255
3. Integration by Parts — 260
4. Differential Equations — 263
*5. Application to Marginal Analysis — 269
6. Table of Integrals — 273
Chapter Review — 275

8 THE DEFINITE INTEGRAL AND ITS APPLICATIONS — 277

1. The Definite Integral — 278
2. Application in Geometry: Area under a Graph — 284
3. Applications in Business — 295
4. The Definite Integral as the Limit of a Sum — 300
5. Average Value of a Function — 305
6. Application to Probability — 308
*7. Models — 315
   Model 1: The Learning Curve — 315
   Model 2: Consumer's Surplus — 316
   Model 3: Maximizing Profit Over Time — 319
   Model 4: Average Rate Measure of Synchrony — 321
   Chapter Review — 323
      Mathematical Questions from Actuary Exams — 325

9 NUMERICAL TECHNIQUES — 327

1. Newton's Method of Solving Equations — 328
2. Trapezoidal Rule — 330

APPENDIX: A REVIEW — 335

1. Introduction — 336
2. Exponents — 336
3. Multiplication and Division of Polynomials — 339
4. Factoring — 341
5. Least Common Denominator — 343
6. Geometry Formulas — 344
7. Absolute Value — 347

TABLES — 351

SOLUTIONS — 357

INDEX — 437

*May be omitted without loss of continuity.

CALCULUS
WITH APPLICATIONS TO BUSINESS AND LIFE SCIENCES
Second Edition

1
Linear Equations

1. Introduction
2. The Straight Line
3. Parallel and Intersecting Lines
*4. Applications
Chapter Review
    Mathematical Questions from CPA and CMA Exams

*This section may be omitted without loss of continuity.

1. Introduction

Coordinates

Origin

Real numbers can be represented geometrically on a horizontal line. We begin by selecting an arbitrary point O, called the *origin,* and associate it with the real number 0. We then establish a scale by marking off line segments of equal length (units) on each side of 0. By agreeing that the positive direction is to the right of 0 and the negative direction is to the left of 0, we can successively associate the integers 1, 2, 3,... with each mark to the right of 0 and the integers -1, -2, -3,... with each mark to the left of 0. See Figure 1.

Figure 1

By subdividing these segments, we can locate rational numbers such as $\frac{1}{2}$ and $-\frac{3}{2}$. The irrational numbers are located by geometric construction (as in the case of $\sqrt{2}$) or by other means. In this way, every point P on the line is associated with a unique real number x, called the *coordinate of P* (see Figure 2). Coordinates establish an ordering for the real numbers; that is, if a and b are coordinates of two points P and Q, respectively, then $a < b$ means that P lies to the left of Q on the line.

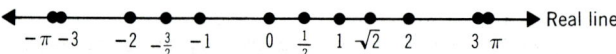

Figure 2

Variables

Equation

Solution

A *variable* is a symbol (usually a letter x, y, etc.) used to represent any real number. An *equation* is a statement involving one or more variables and an "equals" sign ($=$). To *solve* an equation means to find all possible numbers that the variables can assume to make the statement true. The set of all such numbers is called the *solution*. Two equations with the same solution are called *equivalent*.

Example 1 Solve the equation: $3x - 8 = x + 4$

Solution We use the properties of real numbers and proceed as follows:

$$3x - 8 = x + 4$$
$$2x - 8 = 4 \qquad \text{Subtract } x \text{ from both sides}$$
$$2x = 12 \qquad \text{Add 8 to both sides}$$
$$x = 6 \qquad \text{Divide each side by 2}$$

The solution is $x = 6$.

1. INTRODUCTION   3

Example 2   Solve each equation.

(a) $x^2 + x - 12 = 0$ (b) $\dfrac{x^2 - 4}{x - 3} = 0$

Solution   (a) We factor the left side, obtaining

$$(x - 3)(x + 4) = 0$$

The product law states that either the first factor or the second factor must equal zero. Thus,

$$x - 3 = 0 \quad \text{or} \quad x + 4 = 0$$
$$x = 3 \quad \text{or} \quad x = -4$$

The solutions to the equation are $x = 3$ or $x = -4$.

(b) First we observe that $x \neq 3$ since $x = 3$ leads to division by zero, which is not possible. Since a ratio will equal zero only when its numerator equals zero, we set

$$x^2 - 4 = 0$$
$$(x - 2)(x + 2) = 0 \qquad \text{Factor}$$
$$x - 2 = 0 \quad \text{or} \quad x + 2 = 0 \qquad \text{Set each factor equal to zero}$$
$$x = 2 \quad \text{or} \quad x = -2$$

The solutions to the equation are $x = 2$ or $x = -2$. ■

Inequality   An *inequality* is a statement involving one or more variables and one of the inequality symbols ($<, \leq, >, \geq$). To *solve* an inequality means to find all possible numbers that the variables can assume to make the statement true. The set of all such numbers is called the *solution*. Two inequalities with the same solution are called *equivalent*.

To find the solution of an inequality, we apply the laws for inequalities.

Example 3   Solve the inequality: $x + 2 \leq 3x - 5$

Solution $x + 2 \leq 3x - 5$

$-2x \leq -7$ Subtract 2 and then $3x$ from both sides

$x \geq \dfrac{7}{2}$ Multiply by $-\tfrac{1}{2}$ and remember to reverse the inequality because we multiplied by a negative number

The solution is the set of all real numbers to the right of $\tfrac{7}{2}$, including $\tfrac{7}{2}$ (see Figure 3).

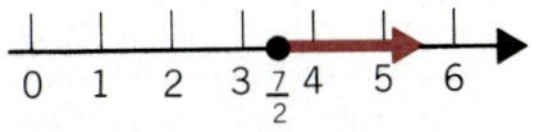

Figure 3

In graphing the solution to an inequality, our practice will be to use a filled-in circle (●) if the number is included (such as $\frac{7}{2}$ in Figure 3) and an open circle (○) if the number is to be excluded.

Rectangular Coordinates

Consider two lines, one horizontal and the other vertical. Call the horizontal line the *x-axis* and the vertical line the *y-axis*.

- x-axis
- y-axis
- Ordered Pair

Any point *P* in the plane formed by the x-axis and y-axis can then be located by using an *ordered pair* of real numbers. Let *x* denote the signed distance of *P* from the y-axis (signed in the sense that if *P* is to the right of the y-axis, then $x > 0$ and if *P* is to the left of the y-axis, then $x < 0$); and let *y* denote the signed distance of *P* from the x-axis. The ordered pair (x, y), the *coordinates of P*, then gives us enough information to locate the point *P*. We can assign ordered pairs of real numbers to every point *P*, as shown in Figure 4.

- Coordinates

If (x, y) are the coordinates of a point *P*, then *x* is called the *abscissa* of *P* and *y* is the *ordinate* of *P*. For example, the coordinates of the origin *O* are $(0, 0)$. The abscissa of any point on the y-axis is 0; the ordinate of any point on the x-axis is 0.

- Abscissa
- Ordinate

The coordinate system described here is a *rectangular* or *cartesian coordinate system* and divides the plane into four sections called *quadrants* (see Figure 5). In quadrant I, both the abscissa *x* and the ordinate *y* of all points are positive; in quadrant II, $x < 0$ and $y > 0$; in quadrant III, both *x* and *y* are negative; and in quadrant IV, $x > 0$ and $y < 0$.

- Rectangular Coordinate System
- Quadrants

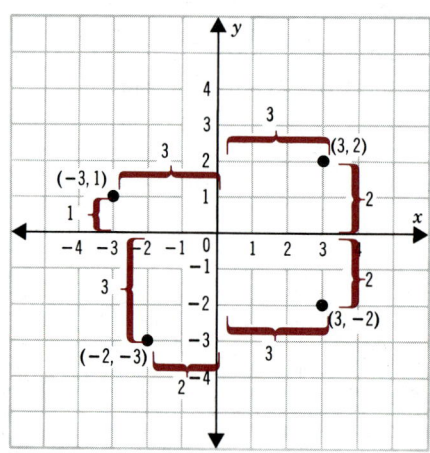

Figure 4

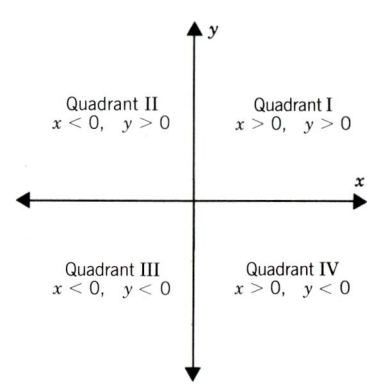

Figure 5

RENÉ DESCARTES (1596–1650), after whom the cartesian coordinate system was named, was born in La Haye, France. After spending several years participating in various wars, he published a treatise introducing analytic geometry. In addition to being a mathematician, he was a respected philosopher and theologian.

Graphs

When a specified relationship between *x* and *y* is given, its *graph* consists of the set of points (*x*, *y*) in the plane that obey the given relationship. To draw a graph, plot a sufficient number of points to see a pattern.

Example 4 Graph the set of points (*x*, *y*) given by the equation

$$y = x$$

Solution We wish to locate all points (*x*, *y*) for which the abscissa *x* and ordinate *y* are equal. Some of these points are

$$(0, 0), (0.1, 0.1), (1, 1), (1.5, 1.5), (3, 3), (-3, -3), (-0.2, -0.2), (8, 8)$$

The graph is given in Figure 6. ■

Example 5 Graph the set of points (*x*, *y*) given by the equation

$$y = 2x + 5$$

Solution We want to find all points (*x*, *y*) for which the ordinate *y* equals twice the abscissa *x* plus 5. To locate some of these points (and thus to get an idea of the pattern of the graph), let us *assign* some numbers *x* and find corresponding values for *y*. Thus:

$$\begin{aligned} &\text{If} \quad x = 0, &&y = 2 \cdot 0 + 5 = 5 \\ &\text{If} \quad x = 1, &&y = 2 \cdot 1 + 5 = 7 \\ &\text{If} \quad x = -5, &&y = 2 \cdot (-5) + 5 = -5 \\ &\text{If} \quad x = 10 &&y = 2 \cdot 10 + 5 = 25 \end{aligned}$$

We form the points (*x*, *y*), namely (0, 5), (1, 7), (−5, −5), and (10, 25). By connecting these points, we obtain the graph. See Figure 7. ■

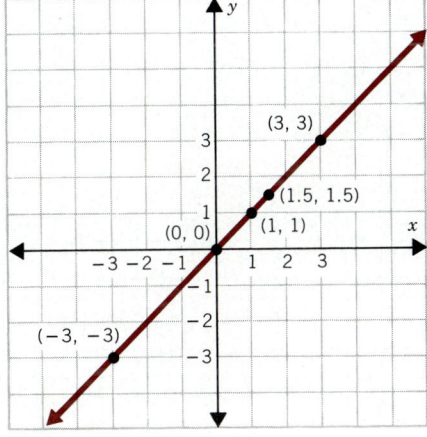

Figure 6

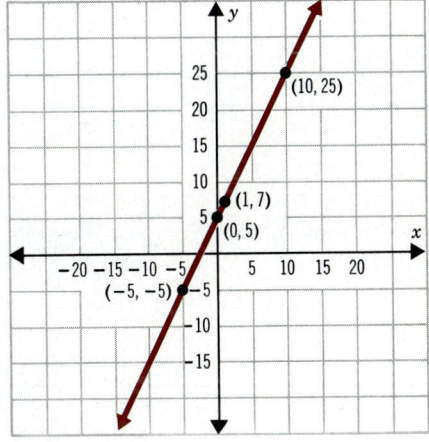

Figure 7

6 CH. 1 LINEAR EQUATIONS

Exercise 1
Solutions to Odd-Numbered Problems begin on page 357.

In Problems 1-26 find the solution x of each equation.

1. $2x + 5 = 7$
2. $x + 6 = 2$
3. $6 - x = 0$
4. $6 + x = 0$
5. $3(2 - x) = 9$
6. $5(x + 1) = 10$
7. $\dfrac{4x}{3} + \dfrac{x}{3} = 5$
8. $\dfrac{2x}{5} + \dfrac{x}{5} = 9$
9. $\dfrac{3x - 5}{x - 3} = 1$
10. $\dfrac{2x + 1}{x - 1} = 3$
11. $x^2 - x - 12 = 0$
12. $x^2 + 7x = 0$
13. $x^2 - 5x + 6 = 0$
14. $x^2 - x - 6 = 0$
15. $x^2 - x = 0$
16. $8x^2 - 2 = 0$
17. $2x^2 + 3x - 2 = 0$
18. $3x^2 - 11x - 4 = 0$
19. $\dfrac{x^2 - 9}{x + 5} = 0$
20. $\dfrac{x^2 - 16}{x - 1} = 0$
21. $\dfrac{x^2 - 10x + 6}{x^2 + x - 5} = 0$
22. $\dfrac{3x^2 + 12x + 12}{6x - 5} = 0$
23. $\dfrac{1}{x - 1} + \dfrac{2}{x + 2} = 0$
24. $\dfrac{2}{x - 3} - \dfrac{5}{x + 4} = 0$
25. $\dfrac{9}{x} - x = 6$
26. $\dfrac{4}{x} - x = 4$

In Problems 27-34 find the solution.

27. $3x + 5 \leq 2$
28. $14x - 21x + 16 \leq 3x - 2$
29. $3x + 5 \geq 2$
30. $4 - 5x \geq 3$
31. $-3x + 5 \leq 2$
32. $8 - 2x \leq 5x - 6$
33. $6x - 3 \geq 8x + 5$
34. $-3x \leq 2x + 5$

35. Find the coordinates of each point in Figure 8.
36. Locate the points $(3, -2)$, $(-2, 3)$, $(5, 0)$, $(-3, -4)$, and $(0, 8)$ using Figure 8 as a background.

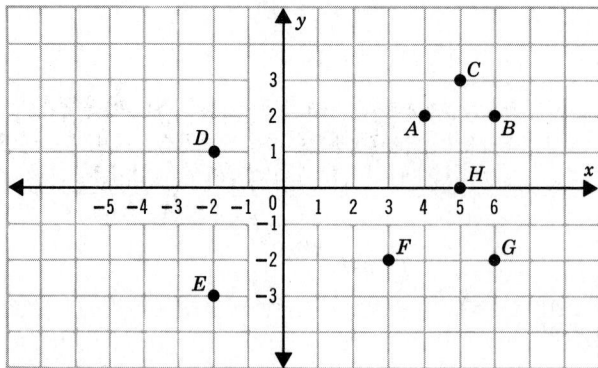

Figure 8

2. THE STRAIGHT LINE

In Problems 37–40 use Figure 8, where $A = (a_1, a_2)$, $B = (b_1, b_2)$, $C = (c_1, c_2)$, and $F = (f_1, f_2)$, to compute each quantity.

37. $\dfrac{f_2 - a_2}{f_1 - a_1}$ 38. $\dfrac{f_2 - c_2}{f_1 - c_1}$ 39. $\dfrac{a_2 - c_2}{a_1 - c_1}$ 40. $\dfrac{a_2 - b_2}{a_1 - b_1}$

In Problems 41–44 copy the tables at the right and fill in the missing values of the given equations. Use these points to graph each equation.

41. $y = x - 3$

x	0	3	2	−2	4	−4
y	−3	0	−1	−5	1	−7

42. $y = -3x + 3$

x	0		2	−2	4	−4
y		0				

43. $2x - y = 6$

x	0	3	2	−2	4	−4
y	−6	0	−2	−10	2	−14

44. $x + 3y = 9$

x	0		2	−2	4	−4
y		0				

In Problems 45–54 graph the set of points (x, y) that obey the given equation.

45. $y = 3x$
46. $y = 4x$
47. $y = 2x - 3$
48. $y = 2x + 1$
49. $y = 0$
50. $x = 0$
51. $y = -2x - 3$
52. $y = -2x - 4$
53. $3x + 2y + 6 = 0$
54. $2x - 3y + 12 = 0$

55. Graph the equations in Problems 51 and 52 on the same coordinate system. Do you notice anything?

2. The Straight Line

Linear Equation In this section, we study a certain type of equation, the *linear equation*, and its graph, the *straight line*. We begin with the result from plane geometry that there is one and only one line L containing two distinct points P and Q. See Figure 9.

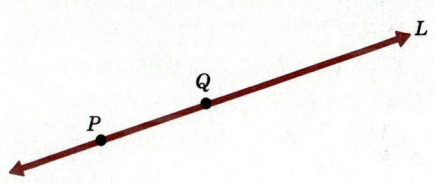

Figure 9

If P and Q are each represented by ordered pairs of real numbers, the following definition can be given:

Slope of a Line Let P and Q be two distinct points with coordinates (x_1, y_1) and (x_2, y_2), respectively. The *slope* m of the line L containing P and Q is defined by the formula*

$$m = \frac{y_2 - y_1}{x_2 - x_1} \qquad \text{if} \quad x_1 \neq x_2$$

If $x_1 = x_2$, the slope m of L is *undefined* (since this results in division by zero) and L is a *vertical line*.

We can also write the slope m of a nonvertical line as†

$$m = \frac{\text{change in } y}{\text{change in } x} = \frac{\Delta y}{\Delta x}$$

That is, the slope m of a nonvertical line L is the ratio of the change in the ordinates from P to Q to the change in the abscissas from P to Q. In other words, the slope m of a line L equals the "rise over run" of the line. See Figure 10.

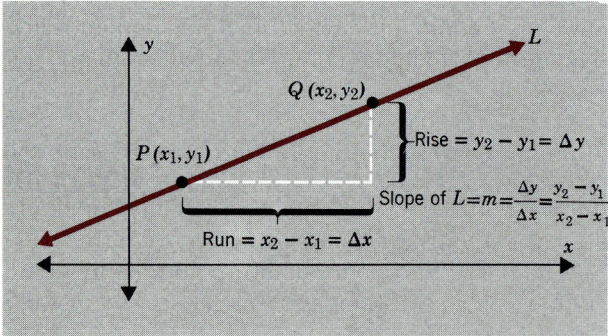

Figure 10

*The following argument, involving similar triangles, shows that the slope of a line L is the same no matter what two distinct points are used: Let L be a nonvertical line joining P and Q and let X and Y be any other two distinct points on L. Construct the triangles depicted in the figure. Since triangle PQA is similar to triangle XYB (why?), it follows that the lengths of the corresponding sides are in proportion. That is, $|AQ|/|BY| = |AP|/|BX|$ or $|AQ|/|AP| = |BY|/|BX|$. But the slope m of L is $|AQ|/|AP|$, and by the foregoing equality, we see that $m = |BY|/|BX|$. In other words, since X and Y are *any* two points, the slope m of a line L is the same no matter what points on L are used to compute m.

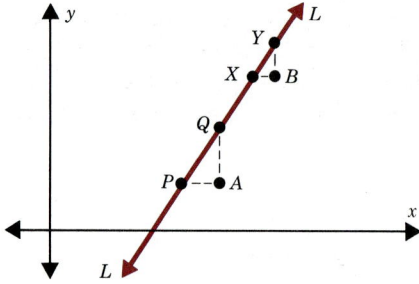

†The symbols Δx and Δy equal the change in x and the change in y, respectively, and do not mean the product of Δ and x or Δ and y.

Since

$$\frac{y_2 - y_1}{x_2 - x_1} = \frac{y_1 - y_2}{x_1 - x_2}$$

the result is the same whether the changes are computed from P to Q or from Q to P.

Example 1 The slope m of the line joining the points $(1, 2)$ and $(5, -3)$ may be computed as

$$m = \frac{-3 - 2}{5 - 1} = \frac{-5}{4} \quad \text{or as} \quad m = \frac{2 - (-3)}{1 - 5} = \frac{5}{-4} = \frac{-5}{4}$$ ∎

To get a better idea of the meaning of the slope of a line L, consider the following example.

Example 2 Compute the slopes of the lines L_1, L_2, L_3, and L_4 containing the following pairs of points. Graph each line.

$$\begin{array}{lll} L_1: & P = (2, 3) & Q_1 = (-1, -2) \\ L_2: & P = (2, 3) & Q_2 = (3, -1) \\ L_3: & P = (2, 3) & Q_3 = (5, 3) \\ L_4: & P = (2, 3) & Q_4 = (2, 5) \end{array}$$

Solution Let m_1, m_2, m_3, and m_4 denote the slopes of the lines L_1, L_2, L_3, and L_4, respectively. Then

$$m_1 = \frac{-2 - 3}{-1 - 2} = \frac{-5}{-3} = \frac{5}{3} \qquad \text{A rise of 5 over a run of 3}$$

$$m_2 = \frac{-1 - 3}{3 - 2} = \frac{-4}{1} = -4$$

$$m_3 = \frac{3 - 3}{5 - 2} = \frac{0}{3} = 0$$

m_4 is undefined

The graphs of these lines are given in Figure 11 on page 10. ∎

As Figure 11 indicates, when the slope m of a line is positive, the line *slants upward* from left to right (L_1); when the slope m is negative, the line *slants downward* from left to right (L_2); when the slope $m = 0$, the line is horizontal (L_3); and when the slope m is undefined, the line is vertical (L_4). Figure 12 on page 10 illustrates the slopes of several lines. Note the pattern.

Equations of Lines

A vertical line is given by the equation

$$x = a$$

where a **is a given real number.**

10   CH. 1   LINEAR EQUATIONS

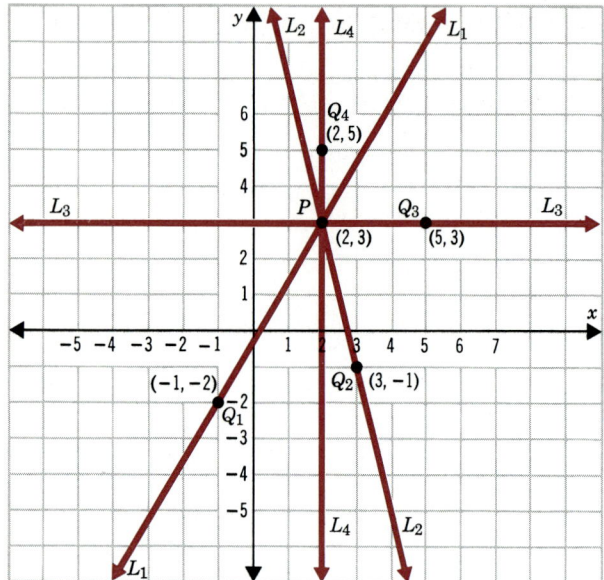

Figure 11

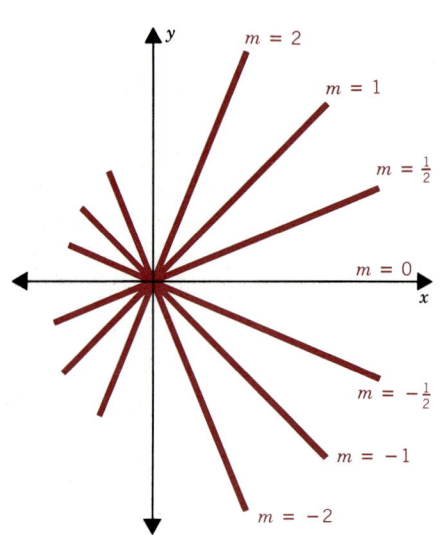

Figure 12

Example 3   The graph of the equation $x = 3$ is a vertical line (see Figure 13).

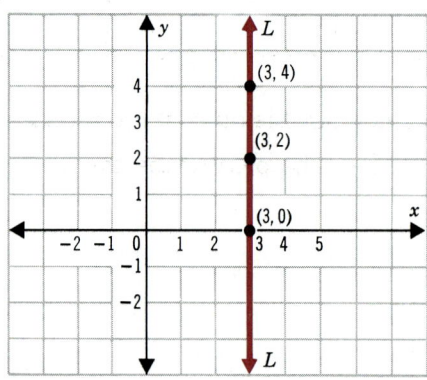

Figure 13

Now let L be a nonvertical line with slope m and containing (x_1, y_1). For (x, y) any other point on L, we have

$$m = \frac{y - y_1}{x - x_1} \quad \text{or} \quad y - y_1 = m(x - x_1)$$

Point-Slope Form   **An equation of a nonvertical line of slope m that passes through the point (x_1, y_1) is**

$$y - y_1 = m(x - x_1)$$

Example 4 An equation of the line with slope 4 and passing through the point (1, 2) is
$$y - 2 = 4(x - 1)$$
$$y = 4x - 2$$
See Figure 14.

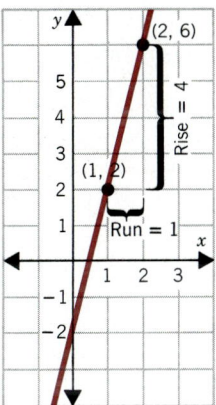

Figure 14

Example 5 Find an equation of the line L passing through the points $(2, 3)$ and $(-4, 5)$. Graph the line L.

Solution Since two points are given, we first compute the slope of the line:
$$m = \frac{5-3}{-4-2} = \frac{2}{-6} = \frac{-1}{3}$$

Using the point (2, 3) (we could use the other point instead, if we wished), and the fact that the slope $m = \frac{-1}{3}$, the point-slope equation of the line is
$$y - 3 = \frac{-1}{3}(x - 2)$$

See Figure 15 for the graph.

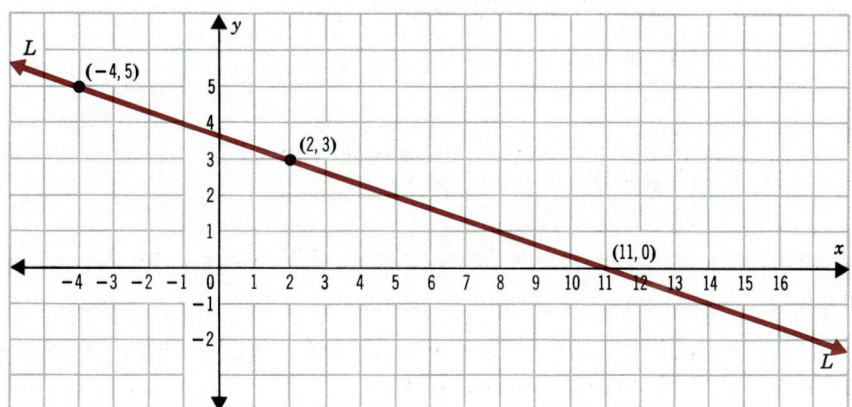

Figure 15

Another form of the equation of the line graphed in Example 5 can be obtained by multiplying both sides by 3 and collecting terms.

$$3(y - 3) = 3(-1/3)(x - 2) \quad \text{Multiply by 3}$$
$$3y - 9 = -1(x - 2)$$
$$3y - 9 = -x + 2$$
$$x + 3y - 11 = 0$$

This last form is referred to as the *general form,* because every line has an equation that can be written this way.

General Form **The equation of a line L is in *general form* when it is written as**

$$Ax + By + C = 0$$

where A, B, C are three real numbers with either $A \neq 0$ or $B \neq 0$.

Intercepts **The points at which the graph of a line L crosses the axes are called *intercepts*. The *x-intercept* is the abscissa of the point at which the line crosses the *x*-axis, and the *y-intercept* is the ordinate of the point at which the line crosses the *y*-axis.**

For example, the line L in Figure 16 has *x*-intercept 3 and *y*-intercept -4. The intercepts are $(3, 0)$ and $(0, -4)$.

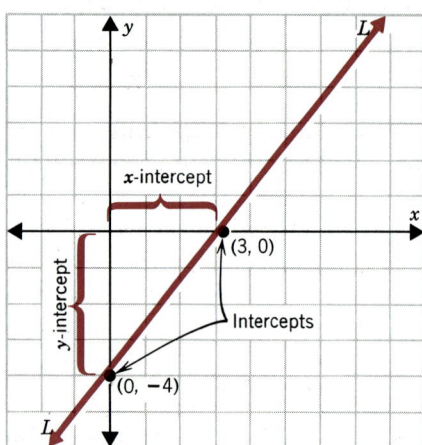

Figure 16

Example 6 Find the intercepts of the line $2x + 3y - 6 = 0$. Graph this line.

Solution To find the point at which the graph crosses the *x*-axis—that is, to find the *x*-intercept—we need to find the number *x* for which $y = 0$. Thus, we set $y = 0$ to get

$$2x + 3(0) - 6 = 0$$
$$2x - 6 = 0$$
$$x = 3$$

The x-intercept is 3. To find the y-intercept, we set $x = 0$ and solve for y:

$$2(0) + 3y - 6 = 0$$
$$3y - 6 = 0$$
$$y = 2$$

The y-intercept is 2.

We now know two points on the line: $(3, 0)$ and $(0, 2)$. See Figure 17 for the graph.

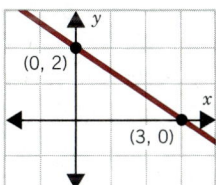

Figure 17

Another useful equation of a line is obtained when the slope m and y-intercept b are known. Since in this event we know both the slope m of the line and a point $(0, b)$ on the line, we may use the point–slope form to obtain the following equation:

$$y - b = m(x - 0) \quad \text{or} \quad y = mx + b$$

Slope-Intercept Form   **An equation of a line L with slope m and y-intercept b is**

$$y = mx + b$$

When the equation of a line is written in slope–intercept form, it is easy to find the slope m and y-intercept b of the line. For example, suppose the equation of a line is

$$y = -2x + 3$$

Compare it to $y = mx + b$:

$$y = -2x + 3$$
$$\uparrow \quad \uparrow$$
$$y = mx + b$$

The slope of this line is -2 and its y-intercept is 3. Let's look at another example.

Example 7   Find the slope m and y-intercept b of the line L given by $2x + 4y - 8 = 0$. Graph the line.

Solution   To obtain the slope and y-intercept, we transform the equation to its slope–intercept form. Thus, we need to solve for y:

$$2x + 4y - 8 = 0$$
$$4y = -2x + 8$$
$$y = \left(\frac{-1}{2}\right)x + 2$$

The coefficient of x, $-\frac{1}{2}$, is the slope, and the y-intercept is 2. To graph this line, we need two points. Normally, the easiest points to locate are the intercepts. Since the y-intercept is 2, we know one intercept is $(0, 2)$. To obtain the x-intercept, set $y = 0$ and solve for x. When $y = 0$, we have

$$2x - 8 = 0$$
$$x = 4$$

Thus, the intercepts are $(4, 0)$ and $(0, 2)$, as shown in Figure 18.

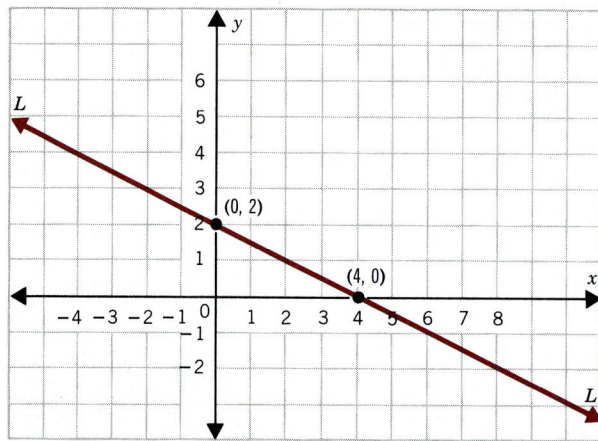

Figure 18

Exercise 2
Solutions to Odd-Numbered Problems begin on page 359.

In Problems 1–6 find the slope m of the line joining the given pair of points.

1. $P = (2, 3)$ $Q = (0, 1)$
2. $P = (1, 1)$ $Q = (5, -6)$
3. $P = (-3, 0)$ $Q = (-5, -4)$
4. $P = (4, -3)$ $Q = (0, 0)$
5. $P = (0.1, 0.3)$ $Q = (1.5, 4.0)$
6. $P = (-3, -2)$ $Q = (6, -2)$

In Problems 7–18 find a general equation for the line having the given properties.

7. Slope = 2; passing through $(-2, 3)$
8. Slope = 3; passing through $(4, -3)$
9. Slope = $-\frac{2}{3}$; passing through $(1, -1)$
10. Slope = $\frac{1}{2}$; passing through $(3, 1)$
11. Passing through $(1, 3)$ and $(-1, 2)$
12. Passing through $(-3, 4)$ and $(2, 5)$
13. Slope = -3; y-intercept = 3

3. PARALLEL AND INTERSECTING LINES

14. Slope = −2;   y-intercept = −2
15. x-intercept = 2;   y-intercept = −1
16. x-intercept = −4;   y-intercept = 4
17. Slope undefined;   passing through (1, 4)
18. Slope undefined;   passing through (2, 1)

In Problems 19–24 find the slope and y-intercept of the given line. Graph each line.

19. $3x - 2y = 6$
20. $4x + y = 2$
21. $x + 2y = 4$
22. $-x - y = 4$
23. $x = 4$
24. $y = 3$

In Problems 25–28 find an equation of the line.

25.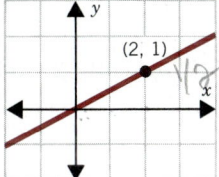

26.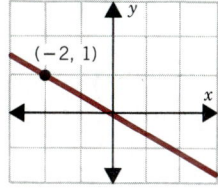

27.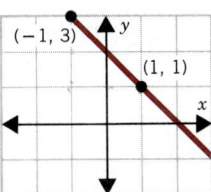

28.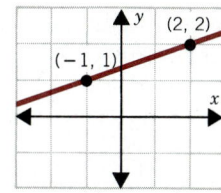

29. *Temperature Conversion.* The relationship between Celsius (°C) and Fahrenheit (°F) for measuring temperature is linear. Find an equation relating °C and °F if 0°C corresponds to 32°F and 100°C corresponds to 212°F. Use the equation to find the Celsius measure of 70°F.

3. Parallel and Intersecting Lines

Let *L* and *M* be two lines. Exactly one of the following three relationships must hold for the two lines *L* and *M*:

1. All the points on *L* are the same as the points on *M*.
2. *L* and *M* have no points in common.
3. *L* and *M* have exactly one point in common.

Identical Lines   If the first relationship holds, the lines *L* and *M* are called *identical lines*. In this case, their slopes and their intercepts will be the same.

When two lines (in a plane) have no points in common, they are said to be **Parallel Lines** *parallel*. Look at Figure 19 on page 16.

16 CH. 1 LINEAR EQUATIONS

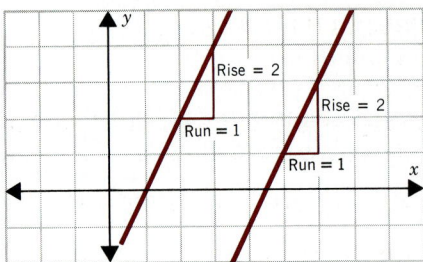

Figure 19

For two lines that are parallel, equal runs result in equal rises. This suggests the following:

Two nonvertical lines are parallel if and only if their slopes are equal.

Example 1 Show that the lines given by the equations below are parallel:

$$L: \quad 2x + 3y - 6 = 0 \qquad M: \quad 4x + 6y = 0$$

Solution To see if these lines have equal slopes, we put each equation into slope–intercept form:

$$
\begin{array}{ll}
L: \quad 2x + 3y - 6 = 0 & M: \quad 4x + 6y = 0 \\
 3y = -2x + 6 & 6y = -4x \\
 y = \dfrac{-2}{3}x + 2 & y = \dfrac{-2}{3}x \\
 \text{Slope} = -2/3 & \text{Slope} = -2/3
\end{array}
$$

Since each has slope $-\frac{2}{3}$, the lines are parallel. See Figure 20.

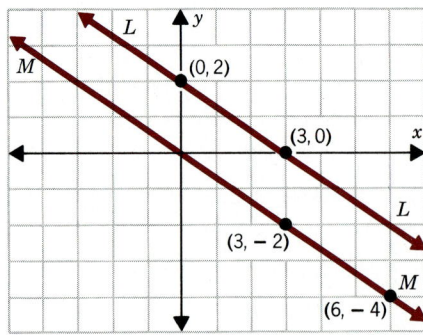

Figure 20

- **Intersecting Lines**

If two lines L and M have exactly one point in common, then L and M are said to *intersect,* and the common point is called the *point of intersection.*

3. PARALLEL AND INTERSECTING LINES   17

Example 2   Find the point of intersection of the two lines:
$$L: \ x + y - 5 = 0 \qquad M: \ 2x + y - 6 = 0$$

Solution   Let the coordinates of the point of intersection of L and M be (x_0, y_0). Since (x_0, y_0) is on both L and M, we must have
$$x_0 + y_0 - 5 = 0 \quad \text{and} \quad 2x_0 + y_0 - 6 = 0$$
Solving for y_0 in each equation, we get
$$y_0 = 5 - x_0 \qquad\qquad y_0 = 6 - 2x_0$$
Setting these equal, we obtain
$$5 - x_0 = 6 - 2x_0$$
$$x_0 = 1$$
Since $x_0 = 1$, then $y_0 = 5 - x_0 = 4$. Thus, the point of intersection of L and M is $(1, 4)$. To check this result, we verify that $(1, 4)$ is on both L and M:
$$\begin{array}{ll} x + y - 5 = 0 & 2x + y - 6 = 0 \\ 1 + 4 - 5 = 0 & 2 \cdot 1 + 4 - 6 = 0 \end{array}$$
This verifies that $(1, 4)$ is the point of intersection. The graphs of the lines are given in Figure 21. ∎

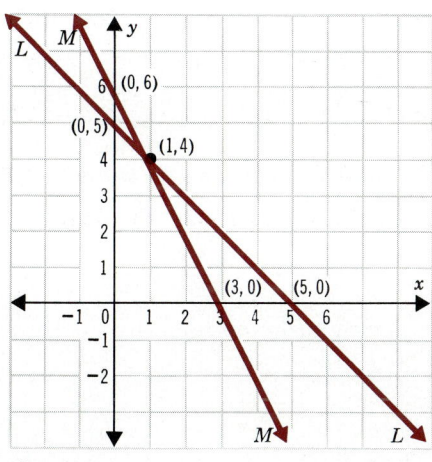

Figure 21

Systems of Equations

System of Two Linear Equations in Two Unknowns   A *system of two linear equations in two unknowns* x and y is of the form
$$\begin{array}{ll} L: & Ax + By + C = 0 \\ M: & A_1 x + B_1 y + C_1 = 0 \end{array}$$
where A, B, C, A_1, B_1, C_1 are real numbers.

18 CH. 1 LINEAR EQUATIONS

Solution of a System of Equations

A *solution* (x, y) of such a system is an ordered pair of real numbers that satisfies both equations. Finding the solution (x, y) of a system of two linear equations in two unknowns is the same as finding the point P of intersection of the two lines determined by the equations. Of course, if the equations of a system represent parallel lines, there will be no solution; if the equations represent identical lines, any point P on the line will be a solution of the system of equations (in this case, there is an infinite number of solutions).

Thus, a system of two linear equations in two unknowns will have:

1. Exactly one solution—the lines intersect
2. No solution—the lines are parallel
3. Infinitely many solutions—the lines are identical

Example 3 Determine whether the following systems of equations have one solution, no solution, or infinitely many solutions:

(a) $x + y + 2 = 0$
     $4x + 4y + 8 = 0$

(b) $x + y + 2 = 0$
     $2x + 2y + 14 = 0$

(c) $x + y + 2 = 0$
     $2x - y + 4 = 0$

Solution Put each equation in slope-intercept form:

(a) $x + y + 2 = 0$ $4x + 4y + 8 = 0$
     $y = -x - 2$ $4y = -4x - 8$
                               $y = -x - 2$

The lines are identical since they have equal slopes and equal y-intercepts. The system has infinitely many solutions.

(b) $x + y + 2 = 0$ $2x + 2y + 14 = 0$
     $y = -x - 2$ $2y = -2x - 14$
                               $y = -x - 7$

The lines are parallel since they have equal slopes (-1) and different y-intercepts. The system has no solution.

(c) $x + y + 2 = 0$ $2x - y + 4 = 0$
     $y = -x - 2$ $y = 2x + 4$

The lines intersect since they have unequal slopes. The system has exactly one solution. ■

We can solve systems of equations with exactly one solution by using two basic methods:

METHOD 1: Substitution

METHOD 2: Add and subtract.

Substitution

The *substitution method* was used in Example 2. The steps to follow are:

STEP 1: Pick one of the equations and solve for one of the unknowns in terms of the other.

STEP 2: Substitute this expression for the same unknown in the other equation.

3. PARALLEL AND INTERSECTING LINES

STEP 3: Solve this equation.

STEP 4: Use the solution found in Step 3 in either of the original equations to get the value of the other unknown.

Let's look at an example.

Example 4 Use the substitution method to find the solution of the system:
$$2x + y + 6 = 0$$
$$4x - 2y + 4 = 0$$

Solution STEP 1: We choose to solve for y in the first equation since this results in the easiest algebra:
$$2x + y + 6 = 0$$
$$y = -2x - 6$$

STEP 2: We replace y in the second equation by this expression:
$$4x - 2(-2x - 6) + 4 = 0$$

STEP 3: Simplify and solve this equation:
$$4x + 4x + 12 + 4 = 0$$
$$8x + 16 = 0$$
$$x = -2$$

STEP 4: Replace x by -2 in the first equation:
$$2(-2) + y + 6 = 0$$
$$-4 + y + 6 = 0$$
$$y + 2 = 0$$
$$y = -2$$

The solution of the system is $x = -2$, $y = -2$.

Add and Subtract The steps used in the *add and subtract method* are listed below:

STEP 1: Multiply each equation by an appropriate nonzero constant so that one of the unknowns drops out when the equations are added (or subtracted).

STEP 2: Add (or subtract) the equations and solve the resulting equation.

STEP 3: Use the solution found in Step 2 in either of the original equations to get the value of the other unknown.

Let's redo the system in Example 4, using the add and subtract method.

Example 5 Use the add and subtract method to solve the system:

$$2x + y + 6 = 0$$
$$4x - 2y + 4 = 0$$

Solution STEP 1: The add and subtract method eliminates one of the variables by adding (or subtracting) the two equations. If we choose to eliminate y in this case, we multiply the first equation by 2, obtaining the equivalent system

$$4x + 2y + 12 = 0$$
$$4x - 2y + 4 = 0$$

STEP 2: Now, when we add these equations, we obtain

$$8x + 16 = 0$$
$$x = -2$$

STEP 3: To find y, use $x = -2$ in one of the original equations, say the first one:

$$2(-2) + y + 6 = 0$$
$$y = -2$$

Thus, the solution of the system is $x = -2$, $y = -2$. Figure 22 illustrates the two lines. ∎

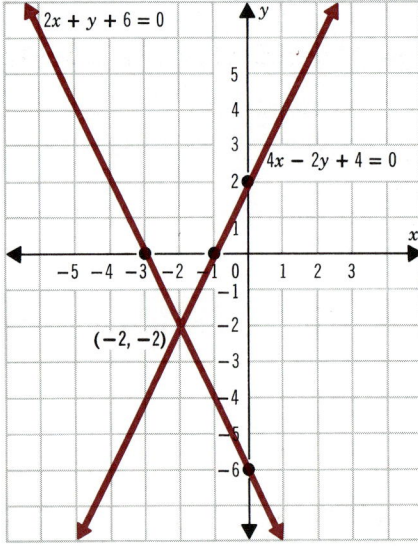

Figure 22

Example 6 Nutt's Nuts, a store that specializes in selling nuts, sells cashews for $1.50 per pound and peanuts for $0.80 per pound. At the end of the month it is found that the peanuts are not selling well. In order to sell 30 pounds of peanuts more quickly, the store manager decides to mix the 30 pounds of peanuts with some cashews and sell the mixture of peanuts and cashews for $1.00 a pound. How many pounds of cashews should be mixed with the peanuts so that the profit remains the same?

Solution There are two unknowns: the number of pounds of cashews (call this x) and the number of pounds of the mixture (call this y). Since we know that the number of pounds of cashews plus 30 pounds of peanuts equals the number of pounds of the mixture, we can write

$$y = x + 30$$

Also, in order to keep profits the same, we must have

$$\begin{pmatrix}\text{Pounds of}\\ \text{cashews}\end{pmatrix} \cdot \begin{pmatrix}\text{Price per}\\ \text{pound}\end{pmatrix} + \begin{pmatrix}\text{Pounds of}\\ \text{peanuts}\end{pmatrix} \cdot \begin{pmatrix}\text{Price per}\\ \text{pound}\end{pmatrix} = \begin{pmatrix}\text{Pounds of}\\ \text{mixture}\end{pmatrix} \cdot \begin{pmatrix}\text{Price per}\\ \text{pound}\end{pmatrix}$$

That is,

$$(1.50)x + (0.80)(30) = (1.00)y$$
$$\tfrac{3}{2}x + 24 = y$$

Thus, we have a system of two linear equations in two unknowns to solve, namely,

$$y = x + 30$$
$$y = \tfrac{3}{2}x + 24$$

Using the substitution method, we find

$$\tfrac{3}{2}x + 24 = x + 30$$
$$\tfrac{1}{2}x = 6$$
$$x = 12$$

The store manager should mix 12 pounds of cashews with 30 pounds of peanuts. See Figure 23.

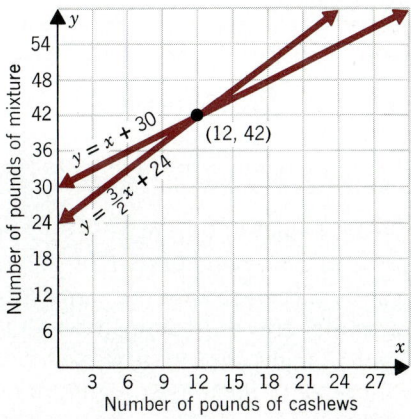

Figure 23

Exercise 3
Solutions to Odd-Numbered Problems begin on page 359.

In Problems 1–4 show that the lines are parallel by finding the slope of each line and showing they are equal.

1. $x + y = 10$
   $3x + 3y = 1$
2. $x - y = 5$
   $-2x + 2y = 7$
3. $2x - 3y + 8 = 0$
   $6x - 9y + 2 = 0$
4. $4x - 2y + 7 = 0$
   $-2x + y + 2 = 0$

In Problems 5–8 find the point of intersection of each pair of lines. Use the substitution method.

5. $x + y - 5 = 0$
   $3x - y - 7 = 0$
6. $2x + y - 7 = 0$
   $x - y + 4 = 0$
7. $3x - 2y + 5 = 0$
   $3x + y - 2 = 0$
8. $4x + y - 6 = 0$
   $4x - 2y = 0$

In Problems 9–14 find the point of intersection of each pair of lines. Use the add and subtract method.

9. $2x - 3y + 4 = 0$
   $3x + 2y - 7 = 0$
10. $3x - 4y - 2 = 0$
    $2x + 5y - 9 = 0$
11. $3x - 4y + 8 = 0$
    $2x + y - 2 = 0$
12. $5x + 2y - 15 = 0$
    $2x - 3y - 6 = 0$
13. $-2x + 3y - 7 = 0$
    $3x + 2y - 9 = 0$
14. $-3x + 4y - 10 = 0$
    $2x - 3y + 7 = 0$

In Problems 15–20 determine whether the system of equations has one solution, no solution, or infinitely many solutions.

15. L: $2x - 3y + 6 = 0$
    M: $4x - 6y + 7 = 0$
16. L: $4x - y + 2 = 0$
    M: $3x + 2y = 0$
17. L: $-2x + 3y + 6 = 0$
    M: $4x - 6y - 12 = 0$
18. L: $2x + 3y - 5 = 0$
    M: $5x - 6y + 1 = 0$
19. L: $3x - 3y + 10 = 0$
    M: $x + y - 2 = 0$
20. L: $2x - 5y - 1 = 0$
    M: $x - 2y - 1 = 0$

21. Find an equation of the line passing through $(1, 2)$ and parallel to $2x - y = 6$.
22. Find an equation of the line passing through $(-1, 3)$ and parallel to $x + y = 4$.
23. *Mixture Problem.* Sweet Delight Candies, Inc. sells boxes of candy consisting of creams and caramels. Each box sells for $4.00 and holds 50 pieces of candy (all pieces are the same size). If the caramels cost $0.05 to produce and the creams cost $0.10 to produce, how many caramels and creams should be in each box for no profit or loss? Would you increase or decrease the number of caramels in order to obtain a profit?
24. *Mixture Problem.* The manager of Nutt's Nuts regularly sells cashews for $1.50 per pound, pecans for $1.80 per pound, and peanuts for $0.80 per pound. How many pounds of cashews and pecans should be mixed with 40 pounds of peanuts to obtain a mixture of 100 pounds that will sell at $1.25 a pound so that the profit or loss is unchanged?
25. *Investment Problem.* Mr. Nicholson has just retired and needs $6000 per year in income to live on. He has $50,000 to invest and

can invest in AA bonds at 15% annual interest or in Savings and Loan Certificates at 7% interest per year. How much money should be invested in each so that he realizes exactly $6000 in income per year?

26. Mr. Nicholson finds after 2 years that because of inflation he now needs $7000 per year to live on. How should he transfer his funds to achieve this amount? (Use the data from Problem 25.)

27. Joan has $1.65 in her piggy bank. She knows she only placed nickels and quarters in the bank and she knows that, in all, she put 13 coins in the bank. Can she find out how many nickels she has without breaking her bank?

28. *Mixture Problem.* A coffee manufacturer wants to market a new blend of coffee that will cost $2.90 per pound by mixing $2.75 per pound coffee and $3 per pound coffee. What amounts of the $2.75 per pound coffee and $3 per pound coffee should be blended to obtain the desired mixture? [*Hint:* Assume the total weight of the desired blend is 100 pounds.]

29. *Mixture Problem.* One solution is 15% acid and another is 5% acid. How many cubic centimeters of each should be mixed to obtain 100 cc of a solution that is 8% acid?

30. *Investment Problem.* A bank loaned $10,000, some at an annual rate of 8% and some at an annual rate of 18%. If the income from these loans was $1000, how much was loaned at 8%? How much at 18%?

31. The Star Theater wants to know whether the majority of its patrons are adults or children. During a week in July, 5200 tickets were sold and the receipts totaled $11,875. The adult admission is $2.75 and the children's admission is $1.50. How many adult patrons were there?

32. After 1 hour of a car ride, 1/3 of the total distance is covered. One hour later, the car is 18 miles past the halfway point. What is the speed of the car (assume it is constant for the entire trip) and what is the total distance to be covered? How long will the trip take? [*Hint:* Distance = Speed • Time.]

4. Applications*

Simple Interest

A knowledge of interest—whether on money borrowed or on money saved—is of ultimate importance today. The old adage "Neither a lender nor a borrower be" is not true in this age of charge accounts and golden passbook savings plans.

Interest

Interest is money paid for the use of money. The total amount of money borrowed (whether by an individual from a bank in the form of a loan or by a

---
*This section may be omitted without loss of continuity.

Principal

Rate of Interest

bank from an individual in the form of a savings account) is called the *principal*. The *rate of interest* is the amount charged for the use of the principal for a given period of time (usually on a yearly or *per annum* basis). The rate of interest is generally expressed as a *percent*.

Simple Interest   *Simple interest* **is interest computed on the principal for the entire period it is borrowed.**

Example 1   If $250 is borrowed for 9 months at a simple interest rate of 8% per annum, what will be the interest charged?

Solution   The actual period the money is borrowed for is 9 months, or 3/4 of a year. Thus, the interest charged will be the product of the principal ($250) times the annual rate of interest (0.08) times the period of time held expressed in years (3/4):

$$\text{Interest charged} = \$(250)(0.08)\left(\frac{3}{4}\right) = \$15$$

In general, if a principal of P dollars is borrowed at a simple interest rate r expressed as a decimal, for a period of t years, the interest I charged is

$$I = Prt$$

The amount A owed at the end of a period of time is the sum of the principal and the interest. That is,

$$A = P + I = P + Prt$$

Example 2   If $500 is borrowed at a simple interest rate of 10% per annum, the amount A due after t years is

$$A = \$500 + \$500(.10)t = \$500 + \$50t$$

Thus, the amount due after 2 years is

$$A = \$500 + \$50(2) = \$600$$

The amount due after 6 months (1/2 year) is

$$A = \$500 + \$50(1/2) = \$525$$

The equation

$$A = 500 + 50t$$

is a linear equation in which A and t are the variables. If we graph this equation using A for the vertical axis and t for the horizontal axis, we can see how the amount A changes over time (see Figure 24). The slope of the line (50) equals the constant annual interest due on the loan.

Break-Even Point

In many businesses, the cost C of production and the number x of items produced can be expressed as a linear equation. Similarly, sometimes the revenue R obtained from sales and the number x of items produced can be expressed as a linear equation. When the cost C of production exceeds the revenue R from sales, the business is operating at a loss; when the revenue R exceeds the cost C, there is a profit; and when the revenue R and the cost C are equal, there is no profit or loss—this is usually referred to as the *break-even point*.

Example 3 Sweet Delight Candies, Inc. has daily fixed costs from salaries and building operations of $300. Each pound of candy produced costs $1 and is sold for $2 per pound. What is the break-even point—that is, how many pounds of candy must be sold daily to guarantee no loss and no profit?

Solution The cost C of production is the fixed cost plus the variable cost of producing x pounds of candy at $1 per pound. Thus,

$$C = \$1 \cdot x + \$300 = x + 300$$

The revenue R realized from the sale of x pounds of candy at $2 per pound is

$$R = \$2 \cdot x = 2x$$

The break-even point is the point where these two lines intersect. Setting $R = C$, we find

$$2x = x + 300$$
$$x = 300$$

That is, 300 pounds of candy must be sold in order to break even.

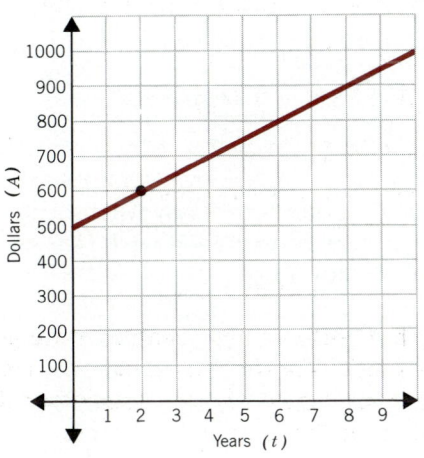

Figure 24

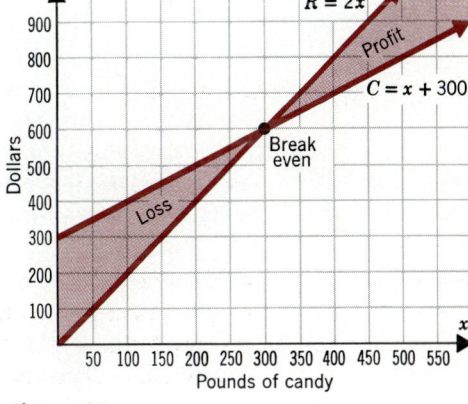

Figure 25

In Figure 25, on page 25, we see a graphical interpretation of the break-even point for Example 3. Note that for $x > 300$, the revenue R always exceeds the cost C so that a profit results. Similarly, for $x < 300$, the cost exceeds the revenue, resulting in a loss.

Example 4
Pricing Candy

After negotiations with employees of Sweet Delight Candies and an increase in the price of chocolate, the daily cost C of production for x pounds of candy is

$$C = \$1.05x + \$330$$

Solution

(a) If each pound of candy is sold for $2, how many pounds must be sold daily to break even?
(b) If the selling price is increased to $2.25 per pound, what is the break-even point?
(c) If it is known that at least 325 pounds of candy can be sold daily, what price should be charged per pound to guarantee no loss?

Solution

(a) If each pound is sold for $2, the revenue R from sales is

$$R = \$2x$$

where x represents the number of pounds sold. When we set $R = C$, we find that the break-even point is the solution of

$$2x = 1.05x + 330$$
$$0.95x = 330$$
$$x = \frac{33{,}000}{95} = 347.37$$

Thus, if 347 pounds of candy are sold, a loss is incurred; if 348 pounds are sold, a profit results.

(b) If the selling price is increased to $2.25 per pound, the revenue R from sales is

$$R = \$2.25x$$

The break-even point is the solution of

$$2.25x = 1.05x + 330$$
$$1.2x = 330$$
$$x = \frac{3300}{12} = 275$$

With the new selling price, the break-even point is 275 pounds.

(c) If we know that at least 325 pounds of candy will be sold daily, the price per pound p needed to guarantee no loss (that is, to guarantee at worst a break-even point) is the solution of

$$325p = (1.05)(325) + 330$$
$$325p = 671.25$$
$$p = \$2.07$$

We should charge at least $2.07 per pound to guarantee no loss, provided at least 325 pounds will be sold.

Example 5 A producer sells items for $0.30 each. If the cost for production is

$$C = \$0.15x + \$105$$

where x is the number of items sold, find the break-even point. If the cost can be changed to

$$C = \$0.12x + \$110$$

would it be advantageous?

Solution The revenue R received is

$$R = \$0.3x$$

The break-even point is the solution of

$$0.3x = 0.15x + 105$$
$$0.15x = 105$$
$$x = 700$$

Thus, for the first cost, the break-even point is 700 items.

To determine the answer to the second part, we find that the break-even point at the new cost is $x = 611.11$. The old break-even point was $x = 700$. Thus, the new cost will require fewer items to be sold in order to break even. Management should probably change over to the new cost. See Figure 26.

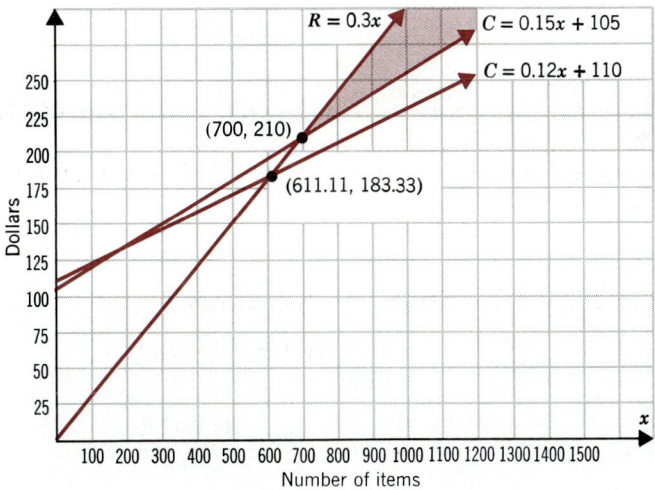

Figure 26

Prediction

Linear equations are sometimes used as predictors of future results. Let's look at an example.

Example 6 In 1981 the cost of an average home was $60,000. One year later the average home sold for $66,000. Assuming this pattern continues, that is, assuming that the increase will remain at $6000 per year, develop a formula for predicting the cost of an average home in 1985. What will it cost in 1990?

Solution We agree to let x represent the year and y represent the cost. We seek a relationship between x and y. Two points on the graph of the equation relating x and y are

$$(1981, 60{,}000) \quad \text{and} \quad (1982, 66{,}000)$$

The assumption that the rate of increase remains constant tells us that the equation relating x and y is linear. The slope of this line is

$$\frac{66{,}000 - 60{,}000}{1982 - 1981} = 6000$$

Using this fact and the point (1981, 60,000), the point-slope form of the equation of the line is

$$y - 60{,}000 = 6000(x - 1981)$$
$$y = 60{,}000 + 6000(x - 1981)$$

For $x = 1985$, we find the cost of an average home to be

$$y = 60{,}000 + 6000(1985 - 1981)$$
$$= 60{,}000 + 6000(4)$$
$$= \$84{,}000$$

For $x = 1990$, we find

$$y = 60{,}000 + 6000(9) = \$114{,}000$$

Figure 27 illustrates the situation.

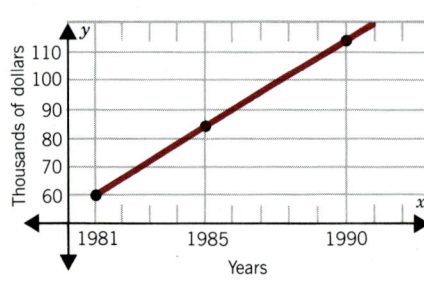
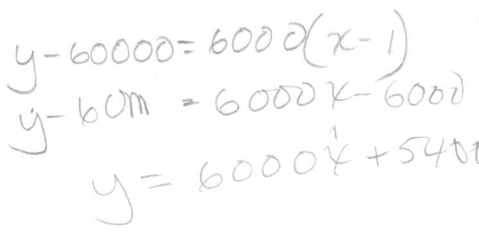

Figure 27

These predictions of future cost are based on the assumption that annual increases remain constant. If this assumption is not accurate, our predictions will be incorrect.

Economics

Supply Equation

Demand Equation

The *supply equation* in economics is used to specify the amount of a particular commodity that sellers have available to offer in the market at various prices. The *demand equation* specifies the amount of a particular commodity that buyers are willing to purchase at various prices.

4. APPLICATIONS 29

Market Price

An increase in price p usually causes an increase in the supply S and a decrease in demand D. On the other hand, a decrease in price brings about a decrease in supply and an increase in demand. The *market price* is defined as the price at which supply and demand are equal.

Example 7
Market Price of Flour

The supply and demand for flour during the period 1920–1935 were estimated as being given by the equations

$$S = 0.8p + 0.5 \qquad D = -0.4p + 1.5$$

where p is measured in dollars and S and D are measured in 50 pound units of flour. Find the market price and graph the supply and demand equations.

Solution

The market price is the point of intersection of the two lines. Thus, the market price p is the solution of

$$0.8p + 0.5 = -0.4p + 1.5$$
$$1.2p = 1$$
$$p = 0.83$$

The graphs are shown in Figure 28. ■

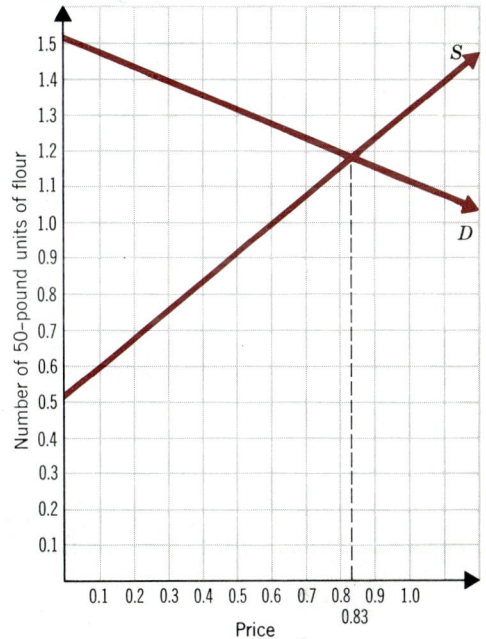

Figure 28

Exercise 5
Solutions to Odd-Numbered Problems begin on page 361.

Simple Interest Problems

1. Suppose you borrow $1000 at a simple interest rate of 18% per annum.
   (a) What is the amount A due after t years? $A = P + Prt$
   (b) How much is due after 6 months? $A = P + Pr \cdot .5$
   (c) How much is due after 1 year? $A = P + Pr \cdot 1$
   (d) How much is due after 2 years? $A = P + Pr \cdot 2$
2. Rework Problem 1 if you borrow $4000 at a simple interest rate of 14%.

Break-Even Problems. In Problems 3–6 find the break-even point for the cost C of production and the revenue R. Graph each result.

3. $C = \$10x + \$600 \qquad R = \$30x$

$10x + 600 = 30x$

4. $C = \$5x + \$200$ $R = \$8x$
5. $C = \$0.2x + \$50$ $R = \$0.3x$
6. $C = \$1800x + \$3000$ $R = \$2500x$

7. A manufacturer produces items at a daily cost of $0.75 per item and sells them for $1 per item. The daily operational overhead is $300. What is the break-even point? Graph your result.

8. If the manufacturer of Problem 7 is able to reduce the cost per item to $0.65, but with a resultant increase to $350 in operational overhead, is it advantageous to do so? Graph your result.

Prediction Problems

9. Suppose the sales of a company are given by

$$S = \$5000x + \$80,000$$

where x is measured in years and $x = 0$ corresponds to the year 1980.
   (a) Find S when $x = 0$.
   (b) Find S when $x = 3$.
   (c) Find predicted sales in 1985, assuming this trend continues.
   (d) Find predicted sales in 1988, assuming this trend continues.

10. Rework Problem 9 if the sales of the company are given by

$$S = \$3000x + \$60,000$$

Economics Problems. In Problems 11–14 find the market price for each pair of supply and demand equations.

11. $S = p + 1$ $D = 3 - p$
12. $S = 2p + 3$ $D = 6 - p$
13. $S = 20p + 500$ $D = 1000 - 30p$
14. $S = 40p + 300$ $D = 1000 - 30p$

15. *Market Price of Sugar.** The supply and demand equations for sugar from 1890 to 1915 were estimated by H. Schulz to be given by

$$S = 0.7p + 0.4 \qquad D = -0.5p + 1.6$$

Find the market price. What quantity of supply is demanded at this market price? Graph both the supply and demand equations. Interpret the point of intersection of the two lines.

16. The market price for a certain product is $5.00 per unit and occurs when 14,000 units are produced. At a price of $1, no units are manufactured and, at a price of $19.00, no units will be purchased. Find the supply and demand equations, assuming they are linear.

*H. Schulz, *Statistical Laws of Demand and Supply with Special Applications to Sugar,* University of Chicago Press, Chicago, 1928.

Chapter Review

Important Terms

system of two linear equations in two unknowns
origin
coordinate
variable
equation
solution
inequality
rectangular coordinates
ordered pair
abscissa
ordinate
quadrants
graphs
linear equation
straight line
slope of a line
vertical line
point-slope form
general form
intercepts
slope-intercept form
identical lines
parallel lines
intersecting lines
x-axis
y-axis
solution of a system of equations
substitution
add and subtract
*simple interest
*principal
*rate of interest
*amount
*break-even point
*prediction
*supply and demand
*market price

True-False Questions (Answers on page 434.)

T F 1. In the slope-intercept equation of a line, $y = mx + b$, m is the slope and b is the x-intercept.
T F 2. The graph of the equation $Ax + By + C = 0$, where A, B, C are real numbers and A, B are not both zero, is a straight line.
T F 3. The y-intercept of the line $2x - 3y + 6 = 0$ is 2.
T F 4. The slope of the line $2x - 4y + 7 = 0$ is $-\frac{1}{2}$.
T F 5. Intersecting lines always have different slopes.

Fill in the Blanks (Answers on page 435.)

1. If (x, y) are rectangular coordinates of a point, the number x is called the _____ and y is called the _____.
2. The points at which the graph of a line crosses the axes are called _____.
3. The slope of a vertical line is _____; the slope of a horizontal line is _____.
4. If a line slants downward as it moves from left to right, its slope will be a _____ number.
5. If two lines have the same slope but different y-intercepts, they are _____.

*From optional sections.

Review Exercises
Solutions to Odd-Numbered Problems begin on page 362.

In Problems 1-6 find the solution x of each equation.

1. $3x + 6 = 2x - 1$
2. $-3x - 2 = 2x + 8$
3. $-2(x + 3) = x + 5$
4. $2x - 3 = -2(x + 2)$
5. $\dfrac{4x - 1}{x + 2} = 5$
6. $\dfrac{3x + 2}{2x - 1} = 1$

In Problems 7-10 find the solution of each inequality.

7. $2x - 1 \leq 5$
8. $8x + 1 \geq 9$
9. $3x + 7 \geq -2x + 2$
10. $-3x + 4 \leq 2x - 1$

In Problems 11-14 graph each linear equation.

11. $y = -2x + 3$
12. $y = 6x - 2$
13. $2y = 3x + 6$
14. $3y = 2x + 6$

In Problems 15-18 find a general equation for the line containing each pair of points.

15. $P = (1, 2)$ $\quad Q = (-3, 4)$
16. $P = (-1, 3)$ $\quad Q = (1, 1)$
17. $P = (0, 0)$ $\quad Q = (-2, 3)$
18. $P = (-2, 3)$ $\quad Q = (0, 0)$

In Problems 19-22 find a general equation for the line.

19. Slope is 2; x-intercept is -1
20. Slope is -1; y-intercept is 1
21. Passing through $(1, 3)$ with slope 1
22. Passing through $(2, -1)$ with slope -2

In Problems 23-26 find the slope and y-intercept of each line. Graph each line.

23. $-9x - 2y + 18 = 0$
24. $-4x - 5y + 20 = 0$
25. $4x + 2y - 9 = 0$
26. $3x + 2y - 8 = 0$

In Problems 27-32 determine whether the system of equations has one solution, no solution, or infinitely many solutions.

27. $3x - 4y + 12 = 0$
    $6x - 8y + 9 = 0$
28. $2x + 3y + 5 = 0$
    $4x + 6y + 10 = 0$
29. $x - y + 2 = 0$
    $3x - 4y + 12 = 0$
30. $2x + 3y - 5 = 0$
    $x + y - 2 = 0$
31. $4x + 6y + 12 = 0$
    $2x + 3y + 6 = 0$
32. $3x - y = 0$
    $6x - 2y + 5 = 0$

33. *Investment Problem.* Mr. and Mrs. Byrd have just retired and find that they need $10,000 per year to live on. Fortunately, they have a nest egg of $90,000 which they can invest in somewhat risky B-rated bonds at 16% interest per year or in a well-known bank at 6% per year. How much money should they invest in each so that they realize exactly $10,000 in income each year?

34. *Mixture Problem.* One solution is 20% acid and another is 12% acid. How many cubic centimeters of each solution should be mixed to obtain 100 cc of a solution which is 15% acid?

35. The annual sales of Motors Inc. for the past 5 years are listed in the table.

Year	Units Sold (in thousands)
1979	3400
1980	3200
1981	3100
1982	2800
1983	2200

(a) Graph this information using the x-axis for years and the y-axis for units sold. (For convenience, use different scales on the axes.)
(b) Draw a line L that passes through two of the points and comes close to passing through the remaining points.
(c) Find the equation of this line L.
(d) Using this equation of the line, what is your estimate for units sold in 1984?

36. *Attendance at a Dance.* A church group is planning a dance in the school auditorium to raise money for its school. The band they will hire charges $500; the advertising costs are estimated at $100; and food is supplied at the rate of $2.00 per person. The church group would like to clear at least $900 after expenses.
(a) Determine how many people need to attend the dance for the group to break even if tickets are sold at $5 each.
(b) Determine how many people need to attend in order to achieve the desired profit if tickets are sold for $5 each.
(c) Answer the above two questions if the tickets are sold for $6 each.

Mathematical Questions From CPA and CMA Exams (Answers on page 433.)

1. CPA Exam—November 1976
The Oliver Company plans to market a new product. Based on its market studies, Oliver estimates that it can sell 5500 units in 1976. The selling price will be $2.00 per unit. Variable costs are estimated to be 40% of the selling price. Fixed costs are estimated to be $6000. What is the break-even point?
(a) 3750 units   (b) 5000 units
(c) 5500 units   (d) 7500 units

2. CPA Exam—November 1976
The Breiden Company sells rodaks for $6.00 per unit. Variable costs are $2.00 per unit. Fixed costs are $37,500. How many rodaks must be sold to realize a profit before income taxes of 15% of sales?
(a) 9375 units   (b) 9740 units
(c) 11,029 units   (d) 12,097 units

3. **CPA Exam—May 1975**
   Given the following notations, what is the break-even sales level in units?

   $$SP = \text{Selling price per unit}$$
   $$FC = \text{Total fixed cost}$$
   $$VC = \text{Variable cost per unit}$$

   (a) $\dfrac{SP}{FC \div VC}$ (b) $\dfrac{FC}{VC \div SP}$ (c) $\dfrac{VC}{SP - FC}$ (d) $\dfrac{FC}{SP - VC}$

4. **CPA Exam—November 1976**
   At a break-even point of 400 units sold, the variable costs were $400 and the fixed costs were $200. What will the 401st unit sold contribute to profit before income taxes?
   (a) $0   (b) $0.50   (c) $1.00   (d) $1.50

Use the following information to answer Problems 5–8:

Akron, Inc. owns 80% of the capital stock of Benson Co. and 70% of the capital stock of Cashin, Inc. Benson Co. owns 15% of the capital stock of Cashin, Inc. Cashin, Inc., in turn, owns 25% of the capital stock of Akron, Inc. These ownership interrelationships are illustrated in the following diagram:

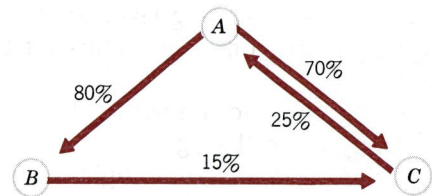

Net income before adjusting for interests in intercompany net income for each corporation follows:

Akron, Inc.	$190,000
Benson Co.	$170,000
Cashin, Inc.	$230,000

The following notations relate to items 5 through 8. Ignore all income tax considerations.

A_e = Akron's consolidated net income; i.e., its net income plus its share of the consolidated net incomes of Benson and Cashin

B_e = Benson's consolidated net income; i.e., its net income plus its share of the consolidated net income of Cashin

C_e = Cashin's consolidated net income; i.e., its net income plus its share of the consolidated income of Akron

5. **CPA Exam—May 1973**
   The equation, in a set of simultaneous equations, which computes A_e is:
   (a) $A_e = .75(190{,}000 + .8B_e + .7C_e)$
   (b) $A_e = 190{,}000 + .8B_e + .7C_e$

(c) $A_e = .75(190,000) + .8(170,000) + .7(230,000)$
(d) $A_e = .75(190,000) + .8B_e + .7C_e$

6. CPA Exam—May 1973
   The equation, in a set of simultaneous equations, which computes B_e is:
   (a) $B_e = 170,000 + .15C_e - .75A_e$
   (b) $B_e = 170,000 + .15C_e$
   (c) $B_e = .2(170,000) + .15(230,000)$
   (d) $B_e = .2(170,000) + .15C_e$

7. CPA Exam—May 1973
   Cashin's minority interest in consolidated net income is:
   (a) $.15(230,000)$ (b) $230,000 + .25A_e$
   (c) $.15(230,000) + .25A_e$ (d) $.15C_e$

8. CPA Exam—May 1973
   Benson's minority interest in consolidated net income is:
   (a) $34,316   (b) $25,500
   (c) $45,755   (d) $30,675

9. CPA Exam—November 1976
   A graph is set up with "depreciation expense" on the vertical axis and "time" on the horizontal axis. Assuming linear relationships, how would the graphs for straight-line and sum-of-the-years'-digits depreciation, respectively, be drawn?
   (a) Vertically and sloping down to the right
   (b) Vertically and sloping up to the right
   (c) Horizontally and sloping down to the right
   (d) Horizontally and sloping up to the right

The following statement applies to items 10 to 12:
In analyzing the relationship of total factory overhead with changes in direct labor hours, the following relationship was found to exist: $Y = \$1000 + \$2X$

10. CMA Exam—December 1973
    The relationship as shown above is:
    (a) Parabolic   (b) Curvilinear
    (c) Linear   (d) Probabilistic
    (e) None of the above

11. CMA Exam—December 1973
    Y in the above equation is an estimate of:
    (a) Total variable costs   (b) Total factory overhead
    (c) Total fixed costs   (d) Total direct labor hours
    (e) None of the above

12. CMA Exam—December 1973
    The $2 in the equation is an estimate of:
    (a) Total fixed costs
    (b) Variable costs per direct labor hour
    (c) Total variable costs
    (d) Fixed costs per direct labor hour
    (e) None of the above

2
Functions and their Limit

1. Preliminaries
2. Functions
3. Some Useful Functions
4. The Idea of a Limit
5. Algebraic Techniques for Evaluating Limits
6. Continuous Functions

Chapter Review
    Mathematical Question from Actuary Exam

1. Preliminaries

Inequalities

In Chapter 1 we solved linear inequalities in one variable. Solving other types of inequalities may be more demanding. Let's look at an example.

Example 1 Solve the inequality: $x^2 + x - 12 > 0$

Solution We factor the left side, obtaining

$$(x - 3)(x + 4) > 0$$

The product of two real numbers is positive either when both factors are positive or when both factors are negative.

Both Positive	or	**Both Negative**
$x - 3 > 0$ and $x + 4 > 0$		$x - 3 < 0$ and $x + 4 < 0$
$x > 3$ and $x > -4$		$x < 3$ and $x < -4$
The numbers x that are greater than 3 and -4 are simply		The numbers x that are less than 3 and -4 are simply
$x > 3$	or	$x < -4$

The solution is $x > 3$ or $x < -4$. See Figure 1.

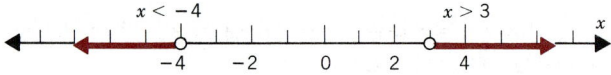

Figure 1

We also can obtain the solution to the inequality of Example 1 by another method. The left-hand side of the inequality is factored so that it becomes $(x - 3)(x + 4) > 0$, as before. We then construct a graph that uses the numbers $x = 3$ and $x = -4$ as cutoff points. See Figure 2. These numbers are the solutions

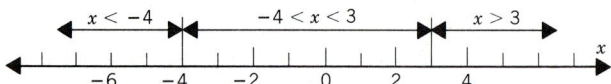

Figure 2

to the equation $x^2 + x - 12 = (x - 3)(x + 4) = 0$, and they separate the line into three parts: $x < -4$, $-4 < x < 3$, and $x > 3$. In the part of the line where $x < -4$, we deduce that both the quantities $(x - 3)$ and $(x + 4)$ are always negative, so their product must always be positive. Therefore, $x < -4$ is a solution of the inequality. In the part of the line where $-4 < x < 3$, we deduce that $(x - 3)$ is always negative and $(x + 4)$ is always positive, so their product is always negative. We conclude that the numbers between -4 and 3 are not solutions of the inequality. In the part of the line where $x > 3$, we deduce that both the quantities $(x - 3)$ and $(x + 4)$ are always positive, so their product is always positive. Hence, numbers greater than 3 are solutions of the inequality. Table 1 summarizes these results.

Table 1

	Sign of $x - 3$	Sign of $x + 4$	Sign of $(x - 3)(x + 4)$	Conclusion
$x < -4$	−	−	+	$x < -4$ is solution
$-4 < x < 3$	−	+	−	$-4 < x < 3$ is not solution
$x > 3$	+	+	+	$x > 3$ is solution

Graphing

In Chapter 1 we graphed equations by connecting points with coordinates that obeyed the equation. Let's look at another example using this technique.

Example 2 Graph the following equations:
(a) $y = x^2$ (b) $x = y^2$

Solution (a) To graph $y = x^2$, we set up a table that provides several points on the graph:

x	0	1	2	3	−1	−2	−3
y	0	1	4	9	1	4	9

In Figure 3 we show these points and, connecting them with a smooth curve, we obtain the graph (a *parabola*).

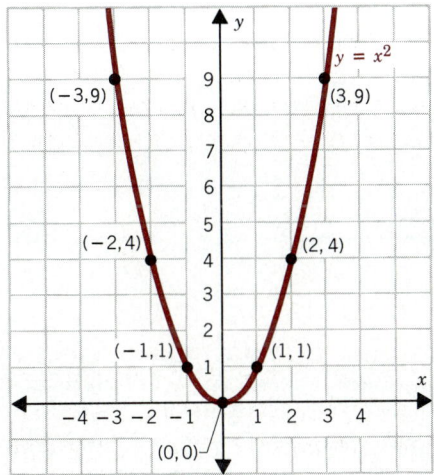

Figure 3

(b) To graph $x = y^2$, we proceed in the same manner:

x	0	1	4	1	4
y	0	1	2	−1	−2

Figure 4 shows the graph of $x = y^2$.

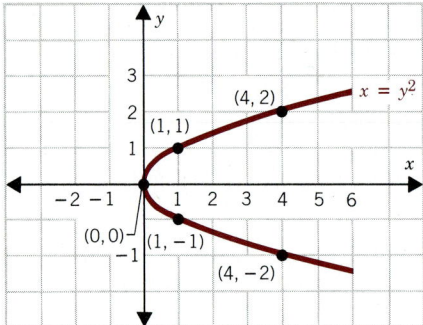

Figure 4

This method of graphing equations by connecting points is inefficient, and will sometimes provide only an incomplete picture. Two useful tools for graphing equations, which you may remember from algebra, are *intercepts* and *symmetry*.

Intercepts

The points at which a graph *intersects* the coordinate axes are called the *intercepts*. The abscissa of a point at which the graph crosses the x-axis is an *x-intercept*, and the ordinate of a point at which the graph crosses the y-axis is a *y-intercept*. See Figure 5. To find the x-intercept(s) of an equation, set $y = 0$ in

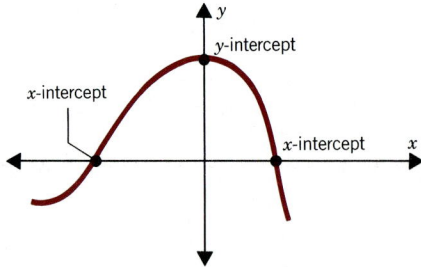

Figure 5

the equation and solve the equation for x. To find the y-intercept(s), set $x = 0$ and solve the equation for y. For example, to find the x-intercept(s) of $y = x^2 − 4$, we set $y = 0$. The resulting equation, $x^2 − 4 = 0$, has two solutions: $x = 2$ and $x = −2$. Thus, the x-intercepts are 2 and −2. The y-intercept, found by setting $x = 0$ in the equation, is $y = −4$. The graph of $y = x^2 − 4$ thus has three intercepts: $(2, 0)$, $(−2, 0)$, and $(0, −4)$. See Figure 6.

1. PRELIMINARIES   41

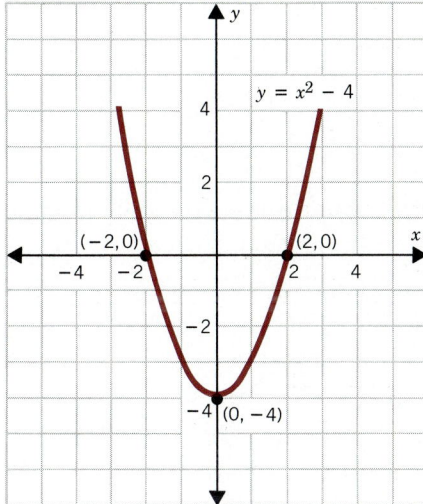

Figure 6

Symmetry

Another useful tool for graphing equations is *symmetry*, particularly symmetry with respect to the *x*-axis, *y*-axis, and origin.

1. *Symmetry with Respect to the x-Axis.*
   For every point (x, y) on a graph, the point $(x, -y)$ is also on the graph.
2. *Symmetry with Respect to the y-Axis.*
   For every point (x, y) on a graph, the point $(-x, y)$ is also on the graph.
3. *Symmetry with Respect to the Origin.*
   For every point (x, y) on a graph, the point $(-x, -y)$ is also on the graph.

Figure 7 illustrates some of the possibilities that can occur.

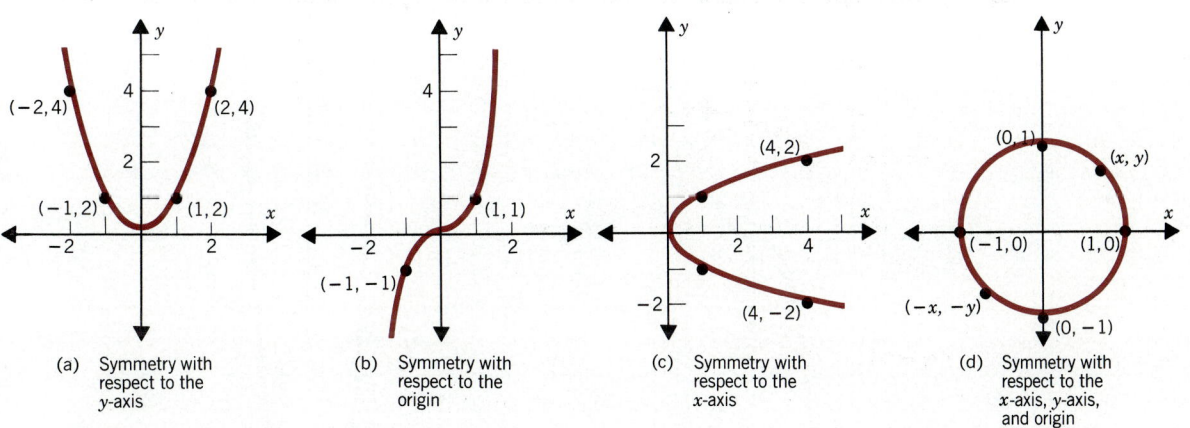

(a) Symmetry with respect to the *y*-axis

(b) Symmetry with respect to the origin

(c) Symmetry with respect to the *x*-axis

(d) Symmetry with respect to the *x*-axis, *y*-axis, and origin

Figure 7

Example 3 Examine the following equations for symmetry with respect to the *x*-axis, *y*-axis, and origin:
(a) $y = x^3$ (b) $y = x^4$

Solution (a) For the graph of the equation $y = x^3$ to be symmetric with respect to the x-axis requires that whenever (x, y) is on the graph, then so is $(x, -y)$. This means that:

$$\text{If} \quad y = x^3, \quad \text{then} \quad -y = x^3.$$

But this is a different equation, so the graph of $y = x^3$ is not symmetric with respect to the x-axis.

For symmetry with respect to the y-axis, we require that:

$$\text{If} \quad y = x^3, \quad \text{then} \quad y = (-x)^3 = -x^3.$$

But this is a different equation, so the graph of $y = x^3$ is not symmetric with respect to the y-axis.

For symmetry with respect to the origin, we require that:

$$\text{If} \quad y = x^3, \quad \text{then} \quad -y = (-x)^3.$$

This is the case since $-y = -(x^3)$ is equivalent to $y = x^3$. So the graph of $y = x^3$ is symmetric with respect to the origin.

(b) Symmetry with respect to the x-axis:

$$\text{If} \quad y = x^4, \quad \text{does} \quad -y = x^4? \quad \text{No.}$$

Symmetry with respect to the y-axis:

$$\text{If} \quad y = x^4, \quad \text{does} \quad y = (-x)^4? \quad \text{Yes.}$$

Symmetry with respect to the origin:

$$\text{If} \quad y = x^4, \quad \text{does} \quad -y = (-x)^4? \quad \text{No.} \quad ∎$$

Figure 8 illustrates the graphs of $y = x^3$ and $y = x^4$ discussed in Example 3.

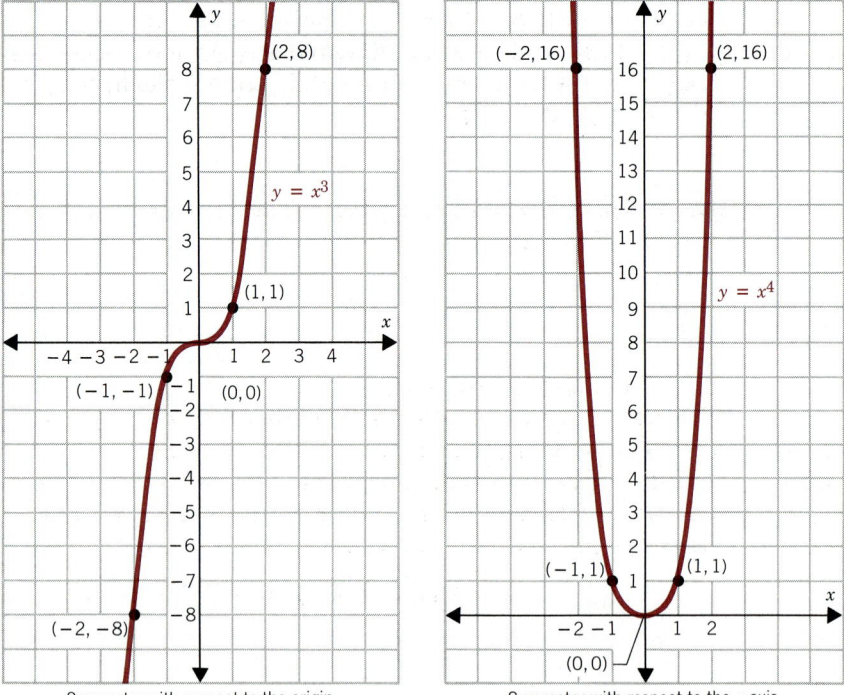

Symmetry with respect to the origin   Symmetry with respect to the y-axis

Figure 8

Sometimes, it is possible to obtain the graph of an equation by a simple *translation*. For example, the graph of the equation $y = x^2 + 1$ may be easily obtained by "lifting" the graph of $y = x^2$ one unit. Some examples are given in Figure 9, where we show translations of some of the graphs given earlier.

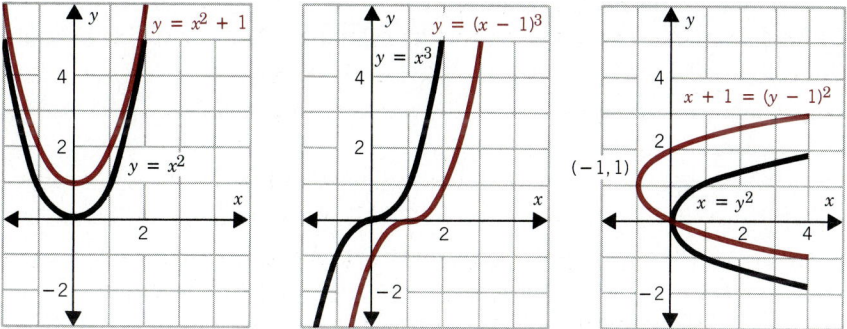

Figure 9

Distance Between Points

Let (x_1, y_1) denote the coordinates of point P_1 and let (x_2, y_2) be the coordinates of point P_2. In moving from P_1 to P_2, the abscissa changes from x_1 to x_2. In calculus we denote this *change in x* by the symbol Δx (read "delta x"). That is, $\Delta x = x_2 - x_1$. Similarly, in moving from P_1 to P_2, the ordinate changes from y_1 to y_2 and this *change in y* is denoted by $\Delta y = y_2 - y_1$. For example, if $P_1 = (5, -2)$ and $P_2 = (4, 7)$, then the change in x from P_1 to P_2 is $\Delta x = 4 - 5 = -1$. The change in y from P_1 to P_2 is $\Delta y = 7 - (-2) = 9$.

If the same scale is used on both the x-axis and the y-axis, then all distances in the plane can be measured using this same scale. In fact, by using the Pythagorean theorem,* we find:

The distance between two points $P_1 = (x_1, y_1)$ and $P_2 = (x_2, y_2)$, which we denote by $|P_1 P_2|$, is

$$|P_1 P_2| = \sqrt{(x_2 - x_1)^2 + (y_2 - y_1)^2} = \sqrt{(\Delta x)^2 + (\Delta y)^2}$$

See Figure 10 on page 44.

Thus, to compute the distance between two points, find the change in their abscissas (Δx), square it, and add this to the square of the change in their ordinates (Δy). The square root of this sum is the distance. For example, to find the distance d between the points $(-2, 5)$ and $(3, 2)$, we compute $\Delta x = 3 - (-2) = 5$ and $\Delta y = 2 - 5 = -3$. Then

$$d = \sqrt{(\Delta x)^2 + (\Delta y)^2} = \sqrt{(5)^2 + (-3)^2} = \sqrt{34}$$

*See the Appendix.

44 CH. 2 FUNCTIONS AND THEIR LIMIT

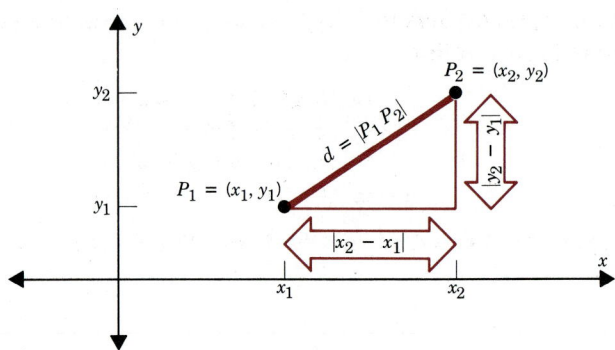

Figure 10

The distance between two points is never a negative number, and the only time the distance between two points can be zero is when the two points are identical.

Notice that it makes no difference whether the distance is computed from P_1 to P_2 or from P_2 to P_1; that is, $|P_1 P_2| = |P_2 P_1|$.

Intervals

Closed Interval Let a and b be two real numbers with $a < b$. A *closed interval* $[a, b]$ is the set of all real numbers x from a to b, inclusive; that is,

$$[a, b] = \{x | a \leq x \leq b\}$$

Open Interval An *open interval* (a, b) consists of all real numbers x between a and b, exclusive of both a and b; that is,

$$(a, b) = \{x | a < x < b\}$$

Half-open (*semi-open*) or *half-closed* (*semi-closed*) intervals are defined by

$$[a, b) = \{x | a \leq x < b\} \qquad (a, b] = \{x | a < x \leq b\}$$

Endpoints In these definitions, a is the *left endpoint* and b is the *right endpoint* of each interval. In Figure 11 an open circle ○ is used to denote the fact that an endpoint

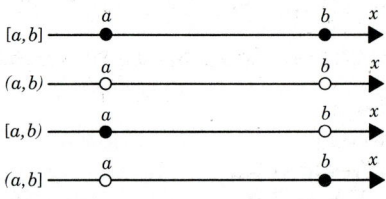

Figure 11

is not included in the interval, while a filled circle ● is used when an endpoint is included in the interval.

For a real number a, the notation $[a, +\infty)$ denotes the set of all real numbers greater than or equal to a; that is,

$$[a, +\infty) = \{x | x \geq a\}$$

The symbol $+\infty$, read "plus infinity," is not a real number; it is just a notational device. Similarly, we define

$$(a, +\infty) = \{x \mid x > a\}$$
$$(-\infty, a] = \{x \mid x \leq a\}$$
$$(-\infty, a) = \{x \mid x < a\}$$
$$(-\infty, +\infty) = \mathbb{R}$$

where $\mathbb{R}$ represents the set of real numbers. See Figure 12 for graphs of these intervals.

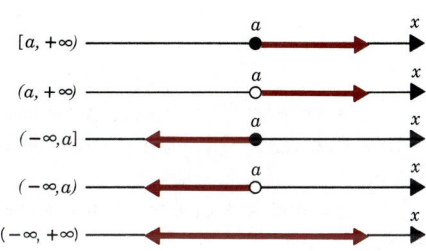

Figure 12

Exercise 1
Solutions to Odd-Numbered Problems begin on page 364.

In Problems 1–12 find the solution of each inequality.

1. $3x + 5 \leq 2$
2. $-3x + 5 \leq 2$
3. $3x + 5 \geq 2$
4. $4 - 5x \geq 3$
5. $6x - 3 \geq 8x + 5$
6. $8 - 2x \leq 5x - 6$
7. $14x - 21x + 16 \leq 3x - 2$
8. $10x - 3x \leq 2x + 5 - 15$
9. $x^2 - 5x + 6 \geq 0$
10. $x^2 + 2x \geq 0$
11. $x^2 + 7x < -12$
12. $x^2 - x < 12$

In Problems 13–20 graph each equation.

13. $y = 3x$
14. $y = -2x$
15. $y = 2x - 3$
16. $y = 2x - 4$
17. $y = 2x^2 + 1$
18. $y = 3x^2 - 6$
19. $y^2 = x + 4$
20. $y^2 = 2x - 4$

In Problems 21–24 find the distance $|P_1 P_2|$ between the points P_1 and P_2.

21. $P_1 = (3, -4); \quad P_2 = (3, 1)$
22. $P_1 = (-1, 0); \quad P_2 = (2, 1)$
23. $P_1 = (-0.6, 2); \quad P_2 = (-0.4, -0.2)$
24. $P_1 = (-5, 1.2); \quad P_2 = (0.6, -0.5)$

In Problems 25–28 use the graph on page 46.

25. Add to the graph to make it symmetric with respect to the x-axis.
26. Add to the graph to make it symmetric with respect to the y-axis.
27. Add to the graph to make it symmetric with respect to the origin.
28. Add to the graph to make it symmetric with respect to the origin, x-axis, and y-axis.

46 CH. 2 FUNCTIONS AND THEIR LIMIT

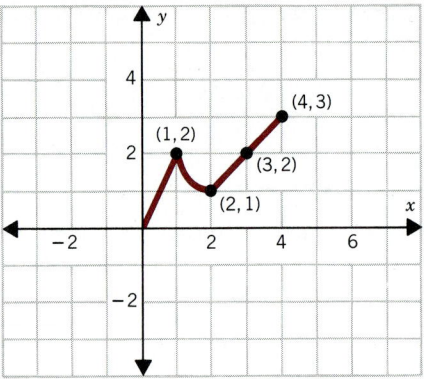

In Problems 29–36 examine each equation for symmetry with respect to the x-axis, y-axis, and origin.

29. $y = 3x^2$
30. $x = 4y^2$
31. $xy = 1$
32. $x^2y = 1$
33. $y = x^3 + 1$
34. $4y^2 = x^2$
35. $y = -x^5 + 3x$
36. $y = (x^2 + 1)^2 - 1$

37. Find the lengths of the medians* of the triangle with vertices at $(0, 0)$, $(0, 6)$, and $(8, 0)$.
38. If two vertices of an equilateral triangle* are $(4, -3)$ and $(0, 0)$, find the third vertex. How many of these triangles are possible?

In Problems 39–42 find the length of each side of the triangle determined by the three points P_1, P_2, and P_3; and state whether the triangle is an isosceles triangle,* a right angle triangle,* neither of these, or both.

39. $P_1 = (2, 1);\quad P_2 = (-4, 1);\quad P_3 = (-4, -3)$
40. $P_1 = (-1, 4);\quad P_2 = (6, 2);\quad P_3 = (4, -5)$
41. $P_1 = (-2, -1);\quad P_2 = (0, 7);\quad P_3 = (3, 2)$
42. $P_1 = (7, 2);\quad P_2 = (-4, 0);\quad P_3 = (4, 6)$

2. Functions

In many applications, a correspondence often exists between two sets of numbers. For example, the revenue R resulting from the sale of x items selling for $10 each is $R = 10x$. If we know how many items have been sold, we can then find the revenue R by using the rule $R = 10x$. This is an example of a *function*.

Very simply, a *function* f is a rule that associates to any given number x a single number $f(x)$, read "f of x." Here, $f(x)$ is the number that results when x is given; $f(x)$ does not mean f times x.

---

*The *medians* of a triangle are the line segments from each vertex to the midpoint of the opposite side. An *equilateral triangle* is one in which all three sides are of equal length. An *isosceles triangle* is one in which at least two of the sides are of equal length. A *right angle triangle* is one in which one of the angles is 90°. The Pythagorean theorem holds for all right triangles.

Example 1 The function f that associates the square of a number to the given number x is
$$f(x) = x^2$$

Example 2 For the function $f(x) = 4x^3 - 2x^2 - 4x + 8$ find:

(a) $f(3)$ (b) $f(-3)$

Solution (a) To find $f(3)$ we substitute 3 for x wherever x appears in the formula for $f(x)$:

$$\begin{aligned}f(3) &= 4(3)^3 - 2(3)^2 - 4(3) + 8 \\ &= 4(27) - 2(9) - 4(3) + 8 \\ &= 108 - 18 - 12 + 8 = 86\end{aligned}$$

(b) The calculation of $f(-3)$ is similar:

$$\begin{aligned}f(-3) &= 4(-3)^3 - 2(-3)^2 - 4(-3) + 8 \\ &= 4(-27) - 2(9) - 4(-3) + 8 \\ &= -108 - 18 + 12 + 8 \\ &= -106\end{aligned}$$

Example 3 For the function $f(x) = \dfrac{6-x}{x^2+5}$ find:

(a) $f(c)$ (b) $f(c+1)$

Solution (a) We substitute c for x wherever x appears in $f(x)$:

$$f(c) = \frac{6-c}{c^2+5}$$

(b) Here, we substitute $c+1$ for x wherever x appears in $f(x)$:

$$f(c+1) = \frac{6-(c+1)}{(c+1)^2+5} = \frac{6-c-1}{c^2+2c+1+5} = \frac{5-c}{c^2+2c+6}$$

We now give a definition of a function.

Function; Domain; Range Let D and R be two given sets of real numbers. A *function* f is a rule that assigns to each number x in D one and only one number $f(x)$ in R. The set D is called the *domain* of the function f, and R is called its *range*.

48 CH. 2 FUNCTIONS AND THEIR LIMIT

Independent Variable
Dependent Variable

The given number *x* is called the *independent variable* and the number associated to it by the function is called the *dependent variable,* since it depends for its value on *x*. We shall usually denote this dependent variable by *y*, and write

$$y = f(x)$$

Functions are often denoted by letters other than *f*, as illustrated below.

Example 4 The cost of eliminating a large part of the pollutants from the atmosphere (or from water) is relatively cheap. However, removing the last traces of pollutants results in a significant increase in cost. A typical relationship between the cost *C*, in thousands of dollars, for removal and the percent *x* of pollutant removed is given by the function

$$C(x) = \frac{3x}{105 - x}$$

The cost of removing 0% of the pollutant is

$$C(0) = 0$$

The cost of removing 50% of the pollutant is

$$C(50) = \frac{150}{55} = 2.727 \text{ thousand dollars}$$

The costs of removing 60% and 70% are

$$C(60) = \frac{180}{45} = 4 \quad \text{and} \quad C(70) = \frac{210}{35} = 6 \text{ thousand dollars}$$

Observe that the cost of removing an additional 10% of the pollutant after 50% had been removed is $1273, while the cost of removing an additional 10% after 60% is removed is $2000. Figure 13 illustrates the graph.

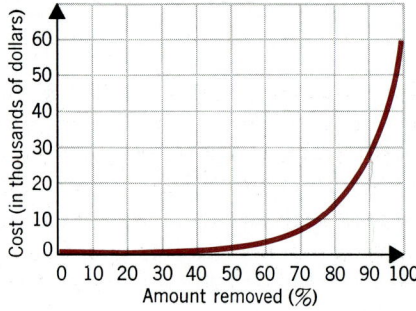

Figure 13

Sometimes it is helpful to visualize a function as an apparatus that manipulates numbers; the domain is the input for the apparatus and the range is the output. We can call such an apparatus an *input-output machine*. See Figure 14.

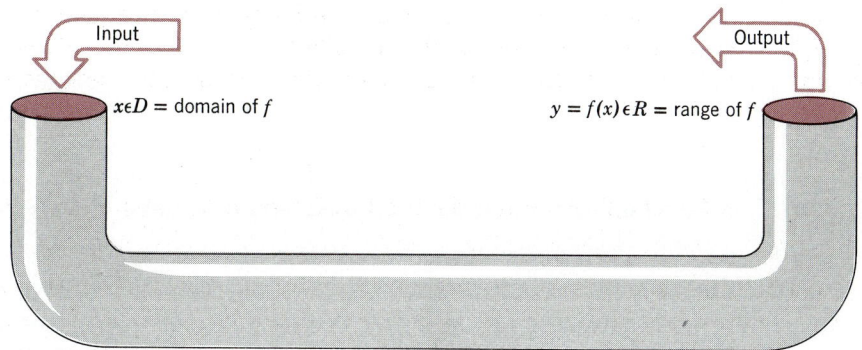

Figure 14

The only restrictions on an input-output machine are:

1. It only accepts numbers from the domain of f; that is, it only accepts numbers for which there is an output.
2. For each input there is exactly one output (which may be repeated for different inputs).

Ordered Pair

We can also consider a function as a set of *ordered pairs* (x, y) in which no different pairs have the same first number. The set of all first numbers is the *domain* and the set of all second numbers is the *range* of the function. Thus, there is associated with each number x in the domain a unique number y in the range.

An advantage of expressing a function (or any correspondence) as a set of ordered pairs is that we can then graph the set of pairs to make a "picture" of the function.

Example 5 Determine whether the correspondences given below are functions.
(a) $y = x^2$ (b) $y^2 = x$

Solution (a) Some of the ordered pairs (x, y) in this set are

$$(2, 2^2) = (2, 4) \quad (0, 0^2) = (0, 0) \quad (-2, (-2)^2) = (-2, 4) \quad (\tfrac{1}{2}, (\tfrac{1}{2})^2) = (\tfrac{1}{2}, \tfrac{1}{4})$$

In this set no two different pairs have the same *first* number (even though there are different pairs that have the same *second* number). This set is the squaring function, which associates with each real number x the value x^2.

(b) The ordered pairs (x, y) for which $y^2 = x$ do not represent a function because there are ordered pairs with the same first number but different second numbers. For example, $(1, 1)$ and $(1, -1)$ are ordered pairs obeying the relationship $y^2 = x$ with the same first number, but different second numbers.

Look back at Figures 3 and 4 on pages 39 and 40 for the graphs of $y = x^2$ and $x = y^2$. Notice that the graph of the function is symmetric with respect to the y-axis, and the graph that does not represent a function is symmetric with respect to the x-axis.

50   CH. 2   FUNCTIONS AND THEIR LIMIT

To summarize, we have determined that a function f associates with real numbers x other real numbers y, and we use the notation $y = f(x)$ to denote the rule that associates x and y. The set of all ordered pairs (x, y), where $y = f(x)$ is the ordinate and x is the abscissa, is called the *graph* of the function f.

Regardless of whether a function is described by a formula, by some rule, or by other means, it will always have a graph. However, not every collection of points is the graph of a function. In fact, a graph provides a visual technique for determining whether a collection of ordered pairs is a function.

If any vertical line intersects the graph in more than one point, the graph is not that of a function.

Example 6   Which of the graphs shown in Figure 15 are graphs of functions?

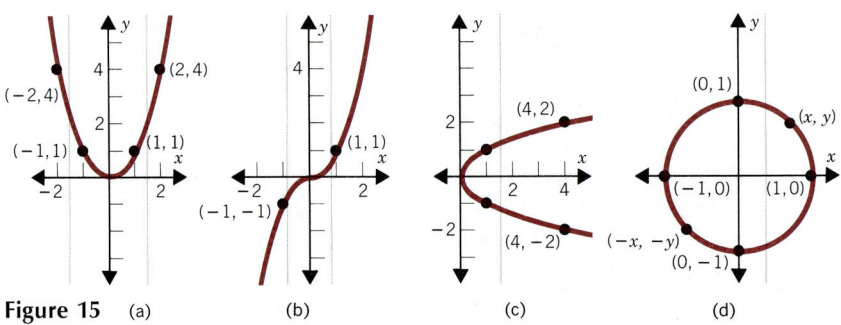

Figure 15   (a)   (b)   (c)   (d)

Solution   The graphs in Figure 15(a) and (b) are graphs of functions. The graphs in Figure 15(c) and (d) are not graphs of functions, since some vertical lines intersect the graphs in more than one point.

Example 7   Let f be a function whose graph is given in Figure 16. The points labeled are on the graph.

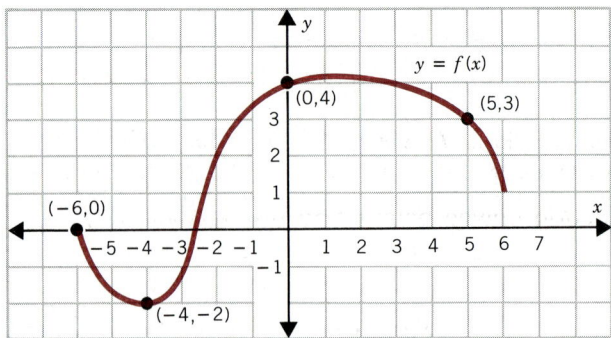

Figure 16

(a)   What is the value of the function when $x = -6$, $x = -4$, $x = 0$, $x = 5$?

(b) Find $f(-6)$, $f(-4)$, $f(0)$, $f(5)$.
(c) What is the domain of f?

Solution (a) Since $(-6, 0)$ is on the graph of f, the ordinate $y = 0$ must be the value of f at the abscissa $x = -6$. That is, $f(-6) = 0$. In a similar way we find that when $x = -4$, then $y = -2$, or $f(-4) = -2$; and when $x = 0$, then $y = 4$ or $f(0) = 4$; and when $x = 5$, then $y = 3$ or $f(5) = 3$.
(b) To find $f(-6)$ we look at the graph and locate the ordinate y, where $x = -6$. From Figure 16, we see that $(-6, 0)$ is on the graph of f. Thus $f(-6) = 0$. In a similar manner we find that $f(-4) = -2$, $f(0) = 4$, and $f(5) = 3$.
(c) To determine the domain of f, we notice that the points on the graph of f all have abscissas between -6 and 6 inclusive, and for each number x between -6 and 6 there is a point $(x, f(x))$ on the graph. Thus, the domain of f is the interval $-6 \leq x \leq 6$.

We now give an example of a function that is given by more than one rule.

Example 8 Graph the function:
$$f(x) = \begin{cases} x & \text{if } x \geq 0 \\ -x & \text{if } x < 0 \end{cases}$$

Solution If $x \geq 0$, then f is represented by the line $y = x$ (slope 1); when $x < 0$, then f is represented by the line $y = -x$ (slope -1). The graph of f is given in Figure 17.

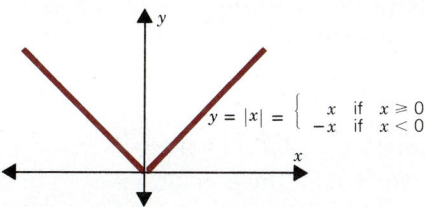

Figure 17

This function is called the *absolute value function*,* and is written as
$$f(x) = |x|$$

Example 9 For the function f given below, find $f(0)$, $f(1)$ and $f(2)$.
$$f(x) = \begin{cases} x/2 & \text{if } -1 \leq x < 1 \\ 2 & \text{if } x = 1 \\ x + \tfrac{1}{2} & \text{if } x > 1 \end{cases}$$

Graph the function f.

*Refer to the Appendix for a review of some of the properties of absolute value.

Solution To find $f(0)$, we observe that when $x = 0$, the rule for f is given by $f(x) = x/2$. So,
$$f(0) = \tfrac{0}{2} = 0$$
When $x = 1$, the rule for f is $f(x) = 2$. Thus,
$$f(1) = 2$$
When $x = 2$, the rule for f is $f(x) = x + \tfrac{1}{2}$. So
$$f(2) = 2 + \tfrac{1}{2} = 2.5$$
The domain of f is all real numbers $x \geq -1$. Its graph is given in Figure 18.

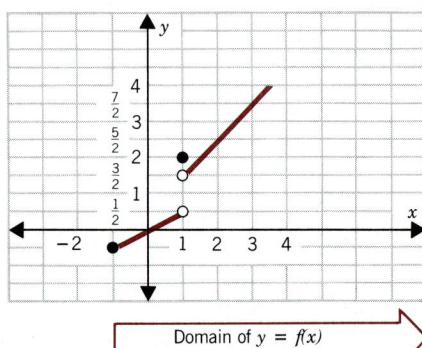

Figure 18

We use a filled circle ● to indicate that at $x = 1$, the value of f is $f(1) = 2$; we use an open circle ○ to illustrate that the function does not assume either of the values $\tfrac{1}{2}$ or $\tfrac{3}{2}$ at $x = 1$. ■

In this book we will sometimes give directions such as "Graph the function $f(x) = x^2$." We actually mean "Graph the equation $y = x^2$, in which the numbers x are restricted to the numbers in the domain of f." But how can we graph a function if its domain is not specified? The answer is simply this: When the domain of a function is *not* specified but a rule of association is known, then we automatically assume that the domain is the largest set of real numbers for which the rule *makes sense* (or more precisely, for which we can compute $f(x)$ as a real number). For example, the operation of squaring can be performed on *any* real number x. Therefore, to associate x with x^2 makes sense for *every* real number x, and thus the domain of $y = f(x) = x^2$ is the set $\mathbb{R}$ of *all* real numbers.

What is the domain of $f(x) = 1/x$? We can divide any nonzero real number into 1. Hence, it makes sense to associate x with $1/x$ as long as $x \neq 0$. The domain of $y = f(x) = 1/x$ is therefore the set of nonzero real numbers; that is, all real numbers x except $x = 0$.

Example 10 Find the domain of the *square root function*:
$$y = f(x) = \sqrt{x}$$
Graph this function.

Solution To find the domain D of f, we ask the question: "What are the numbers x for which we can compute $\sqrt{x}$?" Now, we know it is impossible (in the universe of real numbers) to find the square root of a negative number. Thus, we can only compute $\sqrt{x}$ if $x \geq 0$, and the domain is the set of nonnegative real numbers. See Figure 19.

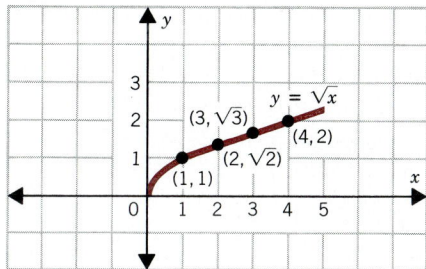

Figure 19

Example 11 Find the domain of the function

$$y = f(x) = 20x + 100$$

Graph this function.

Solution For what real numbers x is it possible to compute $20x + 100$? That is: When can we add 100 to twenty times a number? The answer is always. Thus, the domain of f is the set of real numbers. The graph of this function consists of all points (x, y) for which $y = 20x + 100$. See Figure 20.

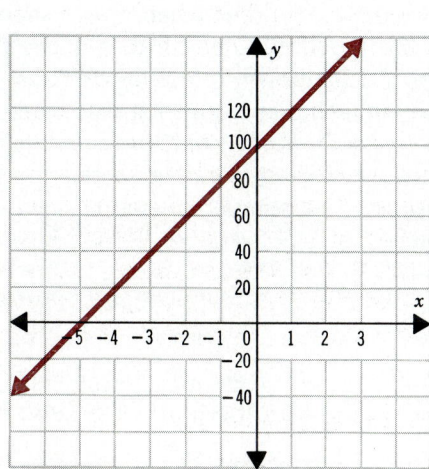

Figure 20

54 CH. 2 FUNCTIONS AND THEIR LIMIT

Example 12 Suppose the function f of Example 11 represents the total expense of production in a factory in which overhead is $100 and the cost for each item manufactured is $20. In this situation, x represents the number of items produced. What is the domain of f? What is the graph of f?

Solution Since the variable x now represents the number of items produced and since it is only meaningful to use nonnegative integers to represent the number of items produced, the domain of this function must be the set

$$\{0, 1, 2, 3, \ldots\}$$

Now, as the variable x takes on the numbers $0, 1, 2, 3, \ldots$, the variable y assumes the values

$$f(0) = 20(0) + 100 = 100$$
$$f(1) = 20(1) + 100 = 120$$
$$f(2) = 20(2) + 100 = 140$$

and so forth. The graph is given in Figure 21.

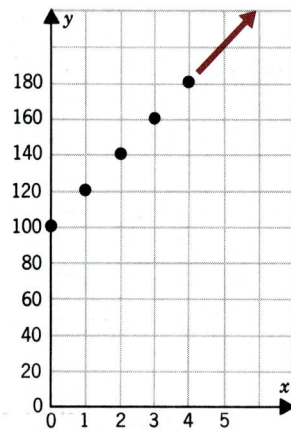

Figure 21

In our study of functions, there are times when practical considerations alter the domain and range of the function. Observe that in the two situations described in Examples 11 and 12, the domain of the first function is $\mathbb{R}$, whereas the domain of the second function is the set of nonnegative integers. The domain of the function in Example 11 is referred to as *continuous*, while the domain of the function in Example 12 is called *discrete*. Time is another example of a variable that is measured continuously; the number of people in a given age bracket must be measured discretely.

The next example illustrates how functions may be constructed.

Example 13 Suppose that a record company sells x record albums each day at $5 apiece.
Constructing a Then the revenue each day is $5 times the number of albums sold:
Profit Function

$$\text{Revenue} = 5x$$

Next, suppose that the cost of making and selling albums is $4 per album plus a fixed daily operating cost (for heat, rent, insurance, etc.) of $400. Then the total cost per day is

$$\text{Cost} = 4x + 400$$

The profit per day is revenue minus cost, or

$$5x - (4x + 400) = x - 400$$

If the letter P denotes the profit function, then $P(x)$ denotes the actual profit when x albums are made and sold:

$$\text{Profit} = P(x) = x - 400$$

Now, what is the profit if 600 albums are made and sold? 400 albums? 300 albums? We can use the formula constructed above:

$$P(600) = 600 - 400 = 200$$
$$P(400) = 400 - 400 = 0$$
$$P(300) = 300 - 400 = -100$$

So, when 600 albums are sold, there is $200 profit; corresponding to 400 albums is no profit or loss; and corresponding to 300 albums is a loss of $100. ∎

Exercise 2

Solutions to Odd-Numbered Problems begin on page 365.

1. A function f is given by

    $$f(x) = 5x + 2$$

    This function takes a number x (input), multiplies it by 5, and then adds 2. Complete the table.

Inputs	Outputs
7	37
1	7
0	2
−4	−18
−2	−8

2. A function f is given by

    $$f(x) = -3x + 2$$

    This function takes a number x (input), multiplies it by −3, and then adds 2. Complete the table.

Inputs	Outputs
3	
1	
0	
−4	
−2	

3. For the function $f(x) = 3x - 2$, find:
    (a) $f(3)$ (b) $f(-2)$ (c) $f(0)$
    (d) $f(x + 2)$ (e) $f(x + h)$ (f) $f(1/x)$

4. For the function $f(x) = x^2 - x$, find:
    (a) $f(1)$ (b) $f(0)$ (c) $f(x + h)$
    (d) $f(-2)$ (e) $f(x + 4)$ (f) $f(1/x)$

In Problems 5-12 use the graph of the function f given below.

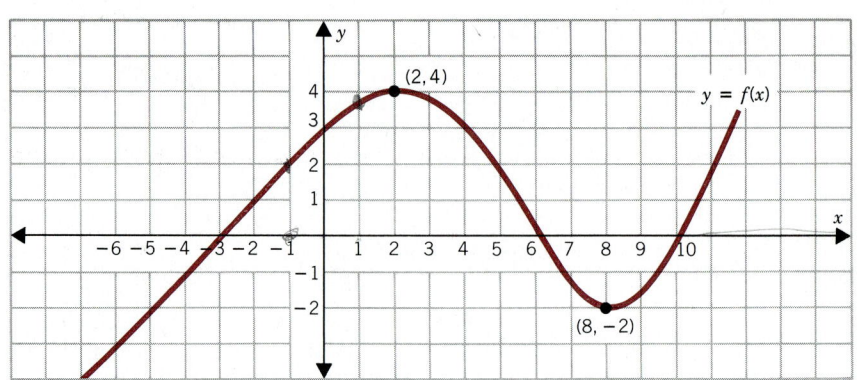

5. Find $f(0)$ and $f(2)$.
6. Find $f(8)$ and $f(-3)$.
7. Is $f(-4)$ positive or negative?
8. Is $f(8)$ positive or negative?
9. For what numbers x is $f(x) = 0$?
10. For what numbers x is $f(x) > 0$?
11. Is $f(-1) > f(1)$?
12. Is $f(2) > f(6)$?

In Problems 13-18 decide which graphs are graphs of functions.

13. 14. 15.

16. 17. 18.

In Problems 19-30, determine whether each given correspondence is a function.

19. $y = x^2 + 2x + 1$
20. $y = x^3 - 3x$
21. $y = \frac{2}{x}$
22. $y = \frac{3}{x} - 4$
23. $y^2 = 1 - x^2$
24. $y = \pm\sqrt{1 - 2x}$
25. $x^2 + y = 1$
26. $x + y^2 = 1$
27. $x^2 y^2 = 5$
28. $x^2 y = 4$
29. $y = (x - 2)^2$
30. $y = \sqrt{x^2}$

In Problems 31-40 find the domain of the function f.

31. $f(x) = 3x + 5$
32. $f(x) = x^2 + 1$
33. $f(x) = \sqrt{x - 1}$
34. $f(x) = \sqrt{2x + 5}$

35. $f(x) = \sqrt{x^2 + 4}$
36. $f(x) = \sqrt{x^2 - 4}$
37. $f(x) = \dfrac{2x}{x-2}$
38. $f(x) = \dfrac{x^2}{x^2 - 4}$
39. $f(x) = \sqrt{3/x}$
40. $f(x) = \dfrac{3x^2}{x^4 + 1}$

In Problems 41–48 find the domain of the function f and graph each function.

41. $f(x) = \begin{cases} 2x & \text{if } x \neq 0 \\ 0 & \text{if } x = 0 \end{cases}$

42. $f(x) = \begin{cases} 3x & \text{if } x \neq 0 \\ 4 & \text{if } x = 0 \end{cases}$

43. $f(x) = \begin{cases} 3x + 2 & \text{if } x \neq 1 \\ 1 & \text{if } x = 1 \end{cases}$

44. $f(x) = \begin{cases} 2x - 1 & \text{if } x \neq 2 \\ 3 & \text{if } x = 2 \end{cases}$

45. $f(x) = \begin{cases} 2x - 3 & \text{if } x < 0 \\ x - 3 & \text{if } 0 \leq x < 5 \end{cases}$

46. $f(x) = \begin{cases} 1 & \text{if } x \leq 2 \\ -1 & \text{if } 2 < x \end{cases}$

47. $f(x) = \begin{cases} 4x + 5 & \text{if } -2 \leq x < 0 \\ 4 & \text{if } x = 0 \\ 2x & \text{if } x > 0 \end{cases}$

48. $f(x) = \begin{cases} 4 - x & \text{if } x \leq 0 \\ x - 2 & \text{if } 0 < x \end{cases}$

49. Is the point $(3, 14)$ on the graph of the function given below?
$$f(x) = \dfrac{x + \frac{1}{2}}{x - 6}$$

50. Is the point $(1, \frac{3}{5})$ on the graph of the function given below?
$$f(x) = \dfrac{x^2 + 2}{x + 4}$$

51. Is the point $(-1, \frac{3}{2})$ on the graph of the function given below?
$$f(x) = \dfrac{3x^2}{x^4 + 1}$$

52. Is the point $(\frac{1}{2}, -\frac{2}{3})$ on the graph of the function given below?
$$f(x) = \dfrac{2x}{x - 2}$$

53. For the function in Problem 41 find:
    (a) $f(-1)$ (b) $f(-\frac{1}{2})$ (c) $f(\frac{1}{2})$ (d) $f(4)$
54. For the function in Problem 42 find:
    (a) $f(-1)$ (b) $f(0)$ (c) $f(2)$ (d) $f(3)$
55. *Falling Rocks on Jupiter.* If a rock falls from a height of 20 meters on the planet Jupiter, then its height H after x seconds is approximately
$$H(x) = 20 - 13x^2$$
    (a) What is the height of the rock when $x = 1$ second? When $x = 1.1$ seconds? When $x = 1.2$ seconds? When $x = 1.3$ seconds?
    (b) When does the rock strike the surface of the planet?

56. *Falling Rocks on Earth.* If a rock falls from a height of 20 meters here on the earth, the height H after x seconds is approximately

$$H(x) = 20 - 4.9x^2$$

Use this function to answer all parts of Problem 55.

57. The cost C, in thousands of dollars, for removal of pollution from a certain lake is

$$C(x) = \frac{5x}{110 - x}$$

where x is the percent of pollutant removed. Find the cost of removing the following percents of pollutant:
(a) 10% (b) 30% (c) 50% (d) 70% (e) 90%
Graph the function.

58. *Constructing a Profit Function.* Each day, a magazine distributor sells x copies of a certain magazine for $1.25 per copy. The cost to the distributor of each magazine is $0.50 per copy, and fixed costs for overhead, distribution, and so on, are $150 per day.
Find the daily profit function.
(a) What is the profit if 100 copies are sold?
(b) What is the profit if 200 copies are sold?
(c) What is the profit if 300 copies are sold?
(d) What is the profit if 500 copies are sold?

59. Rework Problem 58 if the magazine sells for $1.50 per copy and the cost of each magazine to the distributor is $0.60 per copy. Assume fixed costs remain the same.

60. An airplane crosses the Atlantic Ocean (3000 miles) with an airspeed of 500 miles per hour. The cost C (in dollars) per person is

$$C(x) = 100 + \frac{x}{10} + \frac{36000}{x}$$

where x is the groundspeed (airspeed $\pm$ wind).
(a) What is the cost per passenger for quiescent conditions (no wind)?
(b) What is the cost per passenger with a headwind of 50 miles per hour?
(c) What is the cost per passenger with a tailwind of 100 miles per hour?
(d) What is the cost per passenger with a headwind of 100 miles per hour?

3. Some Useful Functions

Many business situations lead to functions that can be easily classified. In this section, we give definitions of these functions, determine their domains, and graph them.

3. SOME USEFUL FUNCTIONS   59

The first important function we discuss is one we have already encountered—the *linear function,* or the function with a graph that is a nonvertical straight line. (see Chapter 1)

Linear Function   A *linear function* **is a function of the form**
$$y = f(x) = mx + b$$
in which m **and** b **are real constants.**

It is clear that the linear function has as its domain the set $\mathbb{R}$ of real numbers. The graph of the linear function is a straight line with slope m and y-intercept b. Another function we need for later use is the *quadratic function*.

Quadratic Function   A *quadratic function* **is a function of the form**
$$y = f(x) = ax^2 + bx + c$$
in which $a \neq 0, b, c$ **are all real constants.**

Quadratic functions occur quite frequently. For example, if the equation that relates the number x of units sold and the price p per unit is linear, then the revenue derived from selling x units at the price p is a quadratic function. Figure 22 illustrates the graph of such a revenue function.

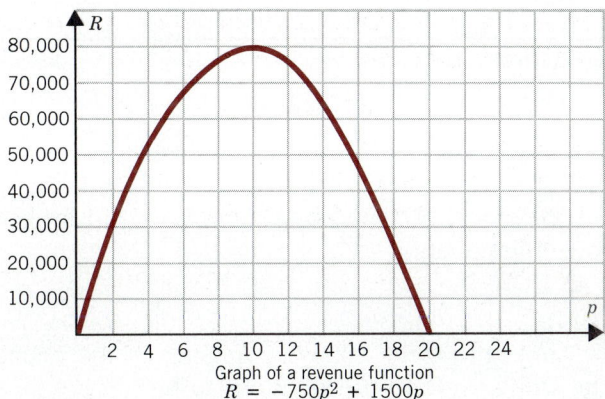

Graph of a revenue function
$R = -750p^2 + 1500p$

Figure 22

When an object is propelled upward at an angle, its path is the graph of some quadratic function. See Figure 23 for an illustration.

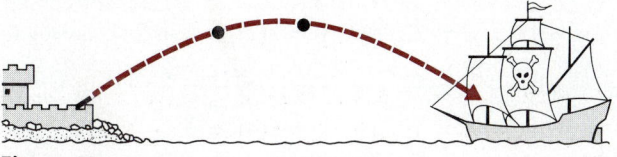

Figure 23

No matter what number is assigned to x, it is possible to compute $f(x) = ax^2 + bx + c$. That is, the domain of the quadratic function is the set $\mathbb{R}$ of real numbers. The x-intercepts of a quadratic function are of particular importance. The x-intercepts are the points at which the graph crosses the x-axis; that is, the points at which $y = 0$. If such points exist, they satisfy the equation

$$ax^2 + bx + c = 0$$

Quadratic Formula If the quantity $b^2 - 4ac \geq 0$, this equation has two solutions given by the *quadratic formula*:

$$x_1 = \frac{-b + \sqrt{b^2 - 4ac}}{2a} \quad \text{and} \quad x_2 = \frac{-b - \sqrt{b^2 - 4ac}}{2a}$$

Discriminant The quantity $b^2 - 4ac$ is called the *discriminant* of a quadratic equation, and its value gives us the following information about the solutions of the equation $ax^2 + bx + c = 0$:

1. $b^2 - 4ac > 0$ **indicates two distinct real solutions. In this case, the graph of** $f(x) = ax^2 + bx + c$ **crosses the x-axis at two points.**

2. $b^2 - 4ac = 0$ **indicates two equal real solutions. In this case, the graph of** $f(x) = ax^2 + bx + c$ **touches the x-axis at one point.**

3. $b^2 - 4ac < 0$ **indicates no real number solution. In this case, the graph of** $f(x) = ax^2 + bx + c$ **does not cross the x-axis.**

Example 1 Graph the quadratic function

$$y = f(x) = x^2 + x - 2$$

Find the y-intercept and the x-intercepts, if they exist.

Solution The y-intercept is -2, obtained by setting $x = 0$. To find the x-intercepts, we need to solve the equation

$$x^2 + x - 2 = 0$$

Since $b^2 - 4ac = 1 - 4(1)(-2) = 1 + 8 = 9 > 0$, the equation has two solutions. By factoring, we find the solutions to be

$$x = -2 \quad \text{and} \quad x = 1$$

(We also could have used the quadratic formula to find the solutions.) Thus, the graph of $f(x) = x^2 + x - 2$ crosses the x-axis at $(-2, 0)$ and at $(1, 0)$.

To find other points on the graph, we set x equal to several numbers:

if $x = 2$, $y = f(2) = 4 + 2 - 2 = 4$
if $x = -1$, $y = f(-1) = (-1)^2 + (-1) - 2 = -2$
if $x = 4$, $y = f(4) = 4^2 + 4 - 2 = 18$
if $x = -4$, $y = f(-4) = (-4)^2 - 4 - 2 = 10$

This selection of points, when connected by a smooth curve, should indicate the general nature of the graph. See Figure 24.

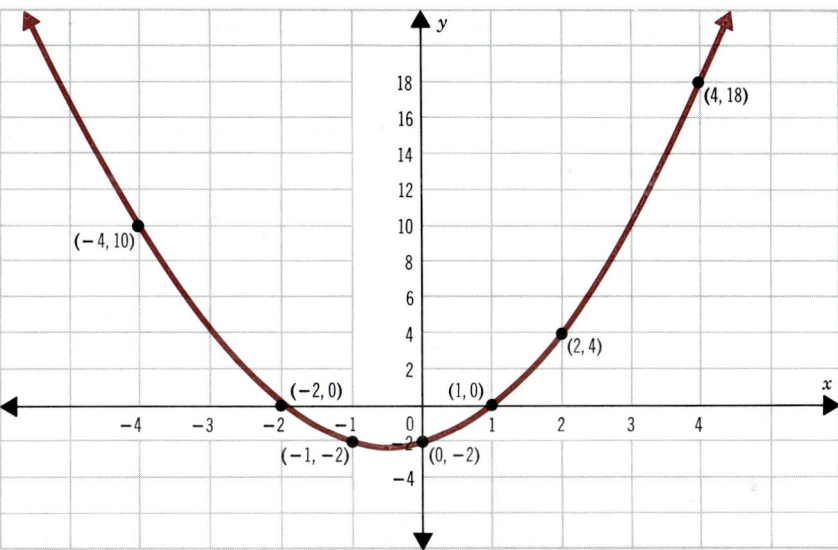

Figure 24

Parabola
Vertex

The graph of a quadratic function is called a *parabola*. **The minimum or maximum point of a parabola is called the** *vertex*.

To analyze the quadratic function more accurately, we proceed to complete the square. First,

$$y = a\left(x^2 + \frac{b}{a}x + \frac{c}{a}\right)$$

Adding and subtracting $b^2/4a^2$, we obtain

$$y = a\left[\left(x^2 + \frac{b}{a}x + \frac{b^2}{4a^2}\right) + \left(\frac{c}{a} - \frac{b^2}{4a^2}\right)\right]$$
$$= a\left[\left(x + \frac{b}{2a}\right)^2 + \left(\frac{4ac - b^2}{4a^2}\right)\right]$$
$$= a\left(x + \frac{b}{2a}\right)^2 + \frac{4ac - b^2}{4a}$$

Now, the value of y depends on the number x in the term $(x + b/2a)^2$. Since this term is either positive or zero, the smallest value of this term occurs when $x = -b/2a$. Thus,

The point on the parabola for which $x = -b/2a$ **is the vertex of the parabola. When** $a > 0$, **the vertex is a** *minimum point* **and the** *graph opens upward*. **When** $a < 0$, **the vertex is a** *maximum point* **and the** *graph opens downward*. See Figure 25 on page 62.

62 FUNCTIONS AND THEIR LIMIT

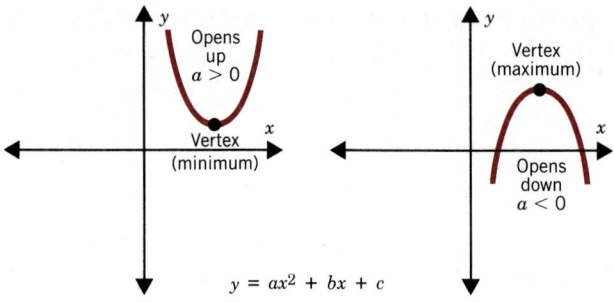

Figure 25

Example 2 Graph the function
$$y = f(x) = -2x^2 - 5x + 3$$
Find its vertex and the x-intercepts, if they exist.

Solution Here we have $a = -2 < 0$ so that the vertex is a maximum point and the graph opens downward. To find the vertex, we remember that it occurs for $x = -b/2a$. Since $a = -2$ and $b = -5$, the abscissa of the vertex is $x = -\frac{5}{4}$. The ordinate of the vertex is

$$y = f\left(\frac{-5}{4}\right) = -2\left(\frac{-5}{4}\right)^2 - 5\left(\frac{-5}{4}\right) + 3$$

$$= -2\left(\frac{25}{16}\right) + \frac{25}{4} + 3 = \frac{-25 + 50 + 24}{8} = \frac{49}{8}$$

Next, since the discriminant $b^2 - 4ac = 25 + 24 = 49$ is positive, the graph crosses the x-axis at two points, which we locate by using the quadratic formula. They are

$$x_1 = \frac{5 + 7}{-4} = -3 \quad \text{and} \quad x_2 = \frac{5 - 7}{-4} = \frac{1}{2}$$

With these three points, we can now graph the parabola. See Figure 26.

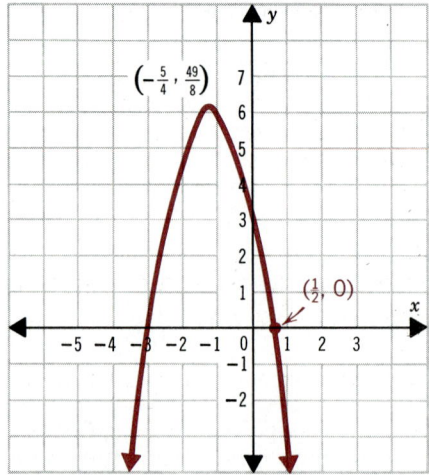

Figure 26

Example 3 Suppose the manufacturer of an electric coffeemaker has found that when x of them are sold at a price of p dollars per unit, the revenue R (in dollars) is

$$R = -750p^2 + 15000p$$

Suppose also that the cost C (in dollars) of producing these coffeemakers is

$$C = -1000p + 69000$$

where p is the price at which each one is sold. What is the lowest price each coffeemaker can be sold at to insure a profit to the manufacturer?

Solution Let's first graph the revenue and cost on the same set of axes. See Figure 27. The manufacturer makes a profit as long as revenue exceeds costs. The break even points occur where revenue equals cost. To find these we set

$$R = C$$
$$-750p^2 + 15000p = -1000p + 69000$$
$$-750p^2 + 16000p - 69000 = 0$$
$$3p^2 - 64p + 276 = 0 \quad \text{Divide by } -250$$
$$(p - 6)(3p - 46) = 0 \quad \text{Factor or use quadratic}$$
$$p = 6 \quad \text{or} \quad p = 46/3 \approx 15.33 \quad \text{formula (we factored)}$$

Thus, the manufacturer should price the coffeemaker no lower than $6.00. Notice that the price should also be no higher than $15.33, since a price higher than this results in a loss to the manufacturers.*

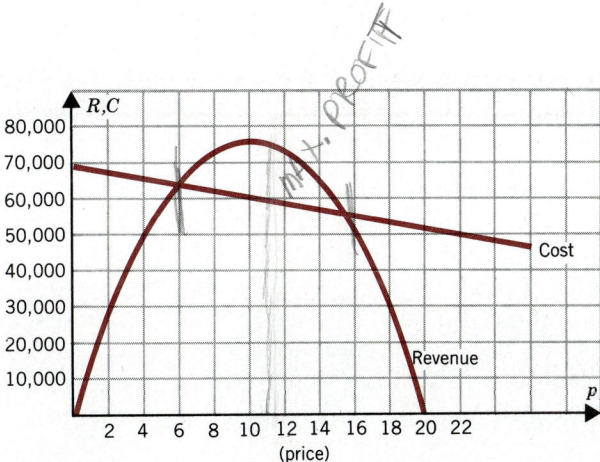

Figure 27

Both the linear function and the quadratic function are examples of a larger class of functions called *polynomial functions*.

Polynomial Function A *polynomial function* **is a function of the form**

$$y = f(x) = a_n x^n + a_{n-1} x^{n-1} + \cdots + a_2 x^2 + a_1 x + a_0$$

*The situation described here is typical when the price p per unit and number x of units sold are related by a linear equation.

64 CH. 2 FUNCTIONS AND THEIR LIMIT

where $a_n \neq 0, a_{n-1}, \ldots, a_2, a_1, a_0$ **are real constants,** $n \geq 0$ **is a nonnegative integer, and** x **is the independent variable.**

For a polynomial function, the real constants $a_n, a_{n-1}, \ldots, a_1, a_0$ are called *coefficients* and $a_n \neq 0$ is called the *leading coefficient*. The exponent n is called the *degree of the polynomial*. Thus, a linear function $f(x) = ax + b, a \neq 0$, is a polynomial of degree 1 and a quadratic function $f(x) = ax^2 + bx + c, a \neq 0$, is a polynomial of degree 2.

• Degree of a Polynomial

The domain of a polynomial function is the set $\mathbb{R}$ of real numbers. For a polynomial function, each exponent of the independent variable x must be a nonnegative integer. Thus, $f(x) = \sqrt{x} + 1$ and $g(x) = \frac{1}{x}$ are not polynomial functions.

We already know how to graph polynomial functions of degree 1 (lines) and of degree 2 (parabolas). Obtaining the graphs of most polynomials of degree 3 or higher is generally not easy with the tools we now have available. We would have to locate several well-chosen points on the graphs (this is easier said than done) and then hope that, by connecting them with a smooth curve, we would obtain an accurate picture. This tedious method is imprecise, and we will soon show how the power of calculus can be used to get accurate graphs without requiring a random selection of many points.

Rational Function   A *rational function* **is a function of the form**

$$R(x) = \frac{P(x)}{Q(x)} = \frac{a_n x^n + \cdots + a_1 x + a_0}{b_m x^m + \cdots + b_1 x + b_0}$$

where P **is a polynomial function of degree** n **and** Q **is a polynomial function of degree** m.

To find the domain of a rational function, remember that the only time we will not be able to compute a value for $R(x)$ is when the x chosen gives a 0 in the denominator. Thus, the domain of R is $\{x \mid Q(x) \neq 0\}$.

The graphs of most rational functions, like those of most polynomial functions, require the use of calculus. We postpone a discussion of their graphs to Chapter 4. As we continue in this book, other types of functions will be encountered, classified, and discussed. For most of them, calculus will not only be useful, but necessary, to obtain a complete description.

Power Function   A *power function* **is a function of the form**

$$y = f(x) = x^r$$

where r **is some real number.**

We have already encountered power functions. The functions $f(x) = x$ and $f(x) = x^2$ are power functions. So is the square root function $f(x) = \sqrt{x} = x^{1/2}$. The function $f(x) = 1/x = x^{-1}$ is a power function.

The domain of a power function $f(x) = x^r$ depends on the number r. Some of the possibilities are listed in the table at the top of page 65.

3. SOME USEFUL FUNCTIONS

	$f(x) = x^r$	Domain
r is a positive integer	x^2, x^3, etc.	All real numbers
r a negative integer	$1/x, 1/x^2$, etc.	All real numbers except 0
$r = 1/n, n > 0$, even integer	$x^{1/2}, x^{1/4}, x^{1/6}$, etc.	All nonnegative real numbers
$r = 1/n, n > 0$ odd integer	$x^{1/3}, x^{1/5}$, etc.	All real numbers

Example 4 Evaluate each power function at the given number x.
(a) $f(x) = x^2$, $x = -4$
(b) $f(x) = x^{1/2}$, $x = 25$
(c) $f(x) = x^{-1/3}$, $x = -8$
(d) $f(x) = x^{2/3}$, $x = -1$

Solution
(a) $f(-4) = (-4)^2 = 16$
(b) $f(25) = (25)^{1/2} = \sqrt{25} = 5$
(c) $f(-8) = (-8)^{-1/3} = 1/(-8)^{1/3} = 1/\sqrt[3]{-8} = 1/-2$
(d) $f(-1) = (-1)^{2/3} = (\sqrt[3]{-1})^2 = (-1)^2 = 1$

Exercise 3
Solutions to Odd-Numbered Problems begin on page 367.

In Problems 1–10, evaluate each power function at the indicated number.

1. $f(x) = x^{1/3}$, $x = 8$
2. $f(x) = x^{1/4}$, $x = 16$
3. $f(x) = x^{1/3}$, $x = \frac{1}{27}$
4. $f(x) = x^{3/2}$, $x = 9$
5. $f(x) = x^{2/3}$, $x = 27$
6. $f(x) = x^{1/3}$, $x = -\frac{1}{27}$
7. $f(x) = x^{-1/3}$, $x = 27$
8. $f(x) = x^{-1/2}$, $x = \frac{1}{100}$
9. $f(x) = x^{-2/3}$, $x = \frac{1}{8}$
10. $f(x) = x^{-1/2}$, $x = \frac{1}{4}$

In Problems 11–22 determine whether the given quadratic function opens upward or downward. Find the vertex, the y-intercept, and the x-intercepts, if they exist. Graph each function.

11. $y = f(x) = 2x^2 + x - 3$
12. $y = f(x) = -2x^2 - x + 3$
13. $y = f(x) = x^2 - 4$
14. $y = f(x) = x^2 + 4x + 4$
15. $y = f(x) = x^2 + 1$
16. $y = f(x) = -3x^2 + 5x + 2$
17. $y = f(x) = -x^2 + 1$
18. $y = f(x) = x^2 + 2x + 1$
19. $y = f(x) = x^2 - 7x + 12$
20. $y = f(x) = x^2 - 10x + 25$
21. $y = f(x) = 4 - x^2$
22. $y = f(x) = x^2 + 4$
23. Find the points of intersection of the graphs of the functions $y = x^2 - 15$ and $y = 2x$.
24. Find the points of intersection of the graphs of the functions $y = x^2 - 20$ and $y = x$.

In Problems 25–37, use the quadratic formula to solve each equation.

25. $x^2 - 2x - 5 = 0$
26. $2x^2 - 3x - 8 = 0$
27. $(\frac{1}{2})x^2 - 3x + 1 = 0$
28. $12x^2 - 2x + 1 = 0$
29. $10x^2 - 5x + 5 = 0$
30. $x^2 - \sqrt{3}x - \frac{1}{4} = 0$

66   CH. 2   FUNCTIONS AND THEIR LIMIT

31. $x + 1/(x - 6) = 2$
32. $x + 1/(x - 3) = 1$
33. $3/x + x = 5$
34. $(x^2 - 5x + 1)/(2x^2 - 6x + 7) = 0$
35. $1 + \dfrac{8}{x} = \dfrac{-15}{x^2}$
36. $x + \dfrac{16}{x} = 10$

37. $\dfrac{3}{1+x} + \dfrac{2}{1-x} = 6$

38. The unit demand function is $x = \frac{1}{2}(18 - p)$, where x is the number of units and p is the price. Let the average cost per unit be $6.
    (a) Find the revenue function R in terms of price p.
    (b) Find the cost function C.
    (c) Find the profit function P.
    (d) Find the price per unit that maximizes the profit function.
    (e) Find the maximum profit.
    (f) Graph R and C.

39. Suppose the manufacturer of a gas clothes dryer has found that when x of them are sold at a price of p dollars per unit the revenue R (in dollars) and the cost C (in dollars) of producing them are:

$$R = -4p^2 + 4000p \qquad C = -400p + 960{,}000$$

Graph R and C on the same set of axes. What is the lowest price and highest price the manufacturer should set for a clothes dryer to insure a profit?

The Algebra of Functions. The example below illustrates the algebraic techniques needed to combine functions by addition, subtraction, multiplication, and division.

Example
Let $f(x) = 4x - 3$, $g(x) = 3x - 2$. Find:
(a) $f(x) + g(x)$,   (b) $f(x) - g(x)$,   (c) $f(x) \cdot g(x)$,   (d) $f(x)/g(x)$

Solution
(a) To compute $f(x) + g(x)$ we add corresponding terms:
$$f(x) + g(x) = (4x - 3) + (3x - 2)$$
$$= 4x - 3 + 3x - 2 = 7x - 5$$

(b) Similarly,
$$f(x) - g(x) = (4x - 3) - (3x - 2)$$
$$= 4x - 3 - 3x + 2 = x - 1$$

(c) To compute $f(x) \cdot g(x)$, we replace $f(x)$ and $g(x)$ by their respective formulas and multiply.
$$f(x) \cdot g(x) = (4x - 3)(3x - 2)$$
$$= 12x^2 - 8x - 9x + 6 = 12x^2 - 17x + 6$$

(d) Similarly,
$$f(x)/g(x) = (4x - 3)/(3x - 2)$$

In Problems 40–49 use the technique of the Algebra of Functions, and the functions, $f(x) = x^2 - 1$, $g(x) = 3x - 1$, $h(x) = 2 - x^2$, $k(x) = x^{3/2}$, $p(x) = x^{-6}$.

40. $f(x) + g(x)$ 41. $g(x) + h(x)$ 42. $f(x) \cdot g(x)$

43. $g(x) \cdot h(x)$ 44. $p(x) \cdot f(x)$ 45. $g(x)/f(x)$
46. $f(x)/h(x)$ 47. $f(x)/k(x)$ 48. $g(x)/p(x)$
49. $k(x)/P(x)$

In Problems 50–53, $f(x)$ and $g(x)$ are given. Express $f(x) + g(x)$ as a rational function. Carry out all multiplications, and simplify, if possible.

50. $f(x) = 2/(x - 2)$, $g(x) = 3/(x + 1)$
51. $f(x) = x/(x - 8)$, $g(x) = -x/(x + 2)$
52. $f(x) = (x + 2)/(x - 5)$, $g(x) = (x - 1)/(x + 7)$
53. $f(x) = (x - 4)/(x + 6)$, $g(x) = (x - 6)/(x + 4)$

4. The Idea of a Limit

With this section we begin the study of calculus, which can be divided into two parts: differential calculus and integral calculus. Our main objective in studying calculus is to gain an understanding of its uses and applications. We will not get too involved with its theory.

The concept of the *limit of a function* is what bridges the gap between the mathematics of algebra and geometry and the mathematics of calculus. Although the *idea* of a limit is somewhat difficult to understand at first, the *evaluation* of limits is fairly easy (after practice).

We begin by asking a question: "What does it mean for a function f to have a limit L as x approaches some fixed number c?" To find an answer, we need to be more precise about f, L, and c. The function f must be defined in an open interval near the number c, but it does not have to be defined at c itself. The limit L is some number. With these restrictions in mind, we introduce the symbolism

(1) $$\lim_{x \to c} f(x) = L$$

which is read as "the limit of $f(x)$ as x approaches c equals the number L." This indicates that f has a limit L as x approaches c. We may describe statement (1) in two ways:

(2) For all x approximately equal to c, but $x \neq c$, the value $f(x)$ is approximately equal to L.

(3) For all x sufficiently close to c, but unequal to c, the value $f(x)$ can be made as close as we please to L.

In Figure 28(a) on page 68 we show the graph of the function f, and observe that as x gets closer to c, the value of f, as measured by its height, gets closer to the number L. This is the key idea behind the notion of a limit. Note that the

68　CH. 2　FUNCTIONS AND THEIR LIMIT

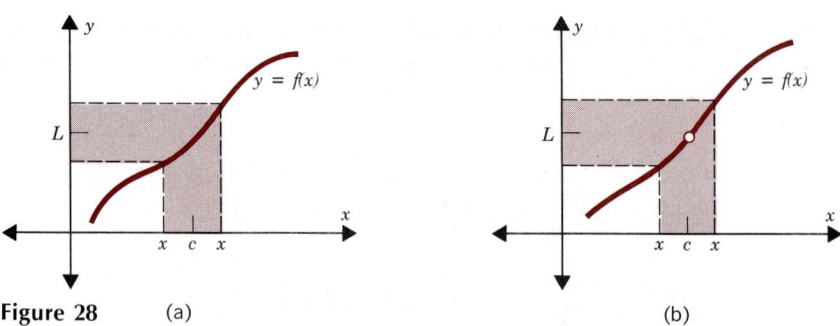

Figure 28　　(a)　　　　　　　　　　　(b)

value of f at c does not matter. As Figure 28(b) illustrates, even though f is not defined at c, it is still true that as x gets closer to c, the value of f gets closer to L. We shall use this idea in our first approach—the graphical approach.

Graphical Approach

Imagine an object and further imagine a solid wall. Suppose we think of the object as being at a fixed distance from the wall and moving halfway to the wall every minute. It is clear that the object is getting "closer and closer" to the wall as time passes. We might even say the object is "approaching" the wall. That is, the distance of the object from the wall is getting "smaller and smaller." In calculus, we say that the distance of the object from the wall approaches zero.

Now suppose we think of this object as a projectile traveling along a path toward the wall. See Figure 29.

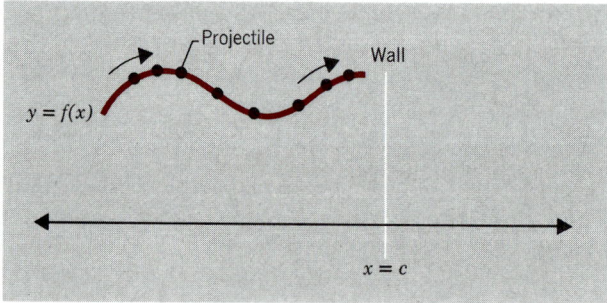

Figure 29

Limit from the Left

It is convenient to denote the path of the projectile by the graph of a function $y = f(x)$ and the wall by a vertical line $x = c$. Based on the motion of the projectile along its path, we want to predict at what height it will strike the wall. This prediction of the height at which the projectile strikes the wall $x = c$, as it proceeds along the path $y = f(x)$, is called the *limit L of f(x) as x approaches c from the left* (since the projectile is moving toward the wall from the left side). We use the notation

$$\lim_{x \to c^-} f(x) = L$$

which is read as "the limit of f(x) as x approaches c from the left equals L," to denote the fact that we predict the projectile hits the wall at the height L. See Figure 30.

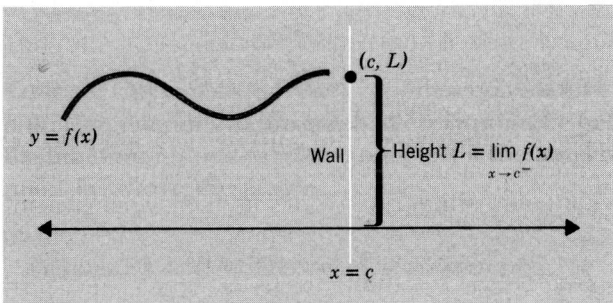

Figure 30

We use the notation $x \to c^-$ with the minus sign ($^-$) to show that x is always a little less than c, that is, something must be "subtracted" from c to yield x. This follows from choosing x to the left of c.

If it is impossible to predict at what height the projectile might strike the wall, then we say the limit of f(x) as x approaches c from the left *does not exist*.

If we envision a projectile approaching the wall (line $x = c$) from the right side, then the prediction of where the projectile strikes the wall is the *limit of f(x) as x approaches c from the right*. We use the notation

Limit from the Right

$$\lim_{x \to c^+} f(x) = R$$

to denote the fact that we predict the projectile will strike the wall at a height R. See Figure 31.

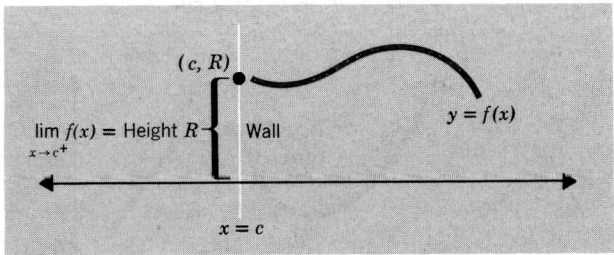

Figure 31

Here, the notation $x \to c^+$ is used because x is always to the right of c, that is, a little larger than c.

If it is impossible to make this prediction, then $\lim_{x \to c^+} f(x)$ does not exist.

Now combine the two situations of a projectile traveling toward the wall from the left and another toward the same wall from the right and think of the union of the two paths as being a function $y = f(x)$. If we predict that the two projectiles strike the wall at exactly the same height (i.e., $R = L$), we say that the *limit of f(x) as x approaches c exists* and equals the common height ($R = L$); that is,

$$\lim_{x \to c} f(x) = R = L$$

If the two projectiles strike the wall at different heights, or if it is impossible to predict where one or both of them strike the wall, then we say that the limit of $f(x)$ as x approaches c does not exist.

Thus, we have the following definition.

Limit **Let $y = f(x)$ be a function and let L and R denote numbers. If**

(a) $\lim_{x \to c^-} f(x) = L$ (b) $\lim_{x \to c^+} f(x) = R$ (c) $L = R$

then the limit of $f(x)$ as x approaches c exists and

$$\lim_{x \to c} f(x) = L = R$$

Notice that three conditions must hold simultaneously in order for the limit of a function at c to exist. If any one of the conditions is not satisfied, the limit does not exist at c.

In Figures 32 and 33 we illustrate two examples of situations involving limits of functions. In Figure 32, $\lim_{x \to c^-} f(x) = L$, $\lim_{x \to c^+} f(x) = R$, and $L = R$, so we have $\lim_{x \to c} f(x) = L$.

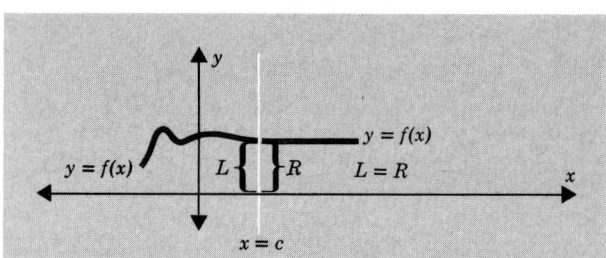

Figure 32

On the other hand, in Figure 33, $\lim_{x \to c^-} f(x) = L$ and $\lim_{x \to c^+} f(x) = R$. Since $L \neq R$, $\lim_{x \to c} f(x)$ does not exist.

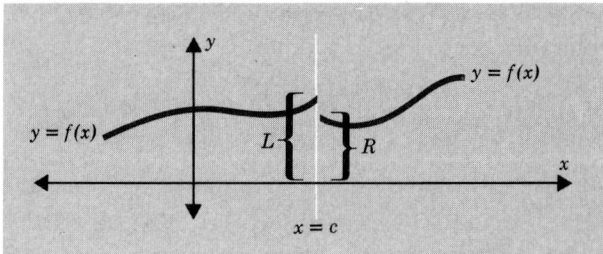

Figure 33

4. THE IDEA OF A LIMIT   71

To summarize, $\lim_{x \to c} f(x) = L = R$ is a statement that describes how a function behaves for numbers x near, but not equal to, c. That is, $\lim_{x \to c} f(x)$ describes the value of $f(x)$ around the number x. Of course, we already know that $f(c)$ equals the value of the function at c, provided the function is defined at c.

Let's look at an example.

Example 1 Use a graphical approach to determine whether the function given below has a limit as x approaches 1.

$$f(x) = \begin{cases} 3x - 1 & \text{if } x \neq 1 \\ 2 & \text{if } x = 1 \end{cases}$$

Solution First, we graph f, as shown in Figure 34. As x approaches 1 from the left, the values of f get closer to 2. We conclude that $\lim_{x \to 1^-} f(x) = 2$. As x approaches 1 from the right, the values of f also get closer to 2, so that $\lim_{x \to 1^+} f(x) = 2$. Since these limits are equal, it follows that $\lim_{x \to 1} f(x) = 2$. ■

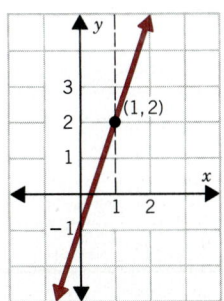

Figure 34

If there is *no single* number that the value of f approaches, we say that f has *no limit as x approaches c* or, more simply, that the *limit does not exist at c*.

Example 2 Use a graphical approach to determine whether the function given below has a limit as x approaches 0.

$$f(x) = \begin{cases} -1 & \text{if } x \leq 0 \\ 1 & \text{if } x > 0 \end{cases}$$

Solution Consult Figure 35. As x approaches 0 from the left (that is, $x < 0$), the value $f(x)$

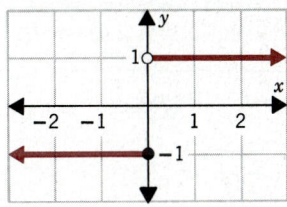

Figure 35

remains at -1. As x approaches 0 from the right (that is, $x > 0$), $f(x)$ remains at 1. Thus, we conclude that

$$\lim_{x \to 0^-} f(x) = -1 \quad \text{and} \quad \lim_{x \to 0^+} f(x) = 1$$

Since these two limits are not equal, we conclude that $\lim_{x \to 0} f(x)$ does not exist.

Now we will study another way of looking at limits.

Numerical Approach
Based on the descriptions (2) and (3) given at the beginning of this section, it is natural to try a numerical approach as an aid in calculating limits.

Example 3 Let $f(x) = x^3$ and $c = 2$. The numerical approach uses a table (and a hand calculator, if available) to guess $\lim_{x \to 2} x^3$.

x	1	1.5	1.6	1.75	1.8	1.9	1.99	1.999	1.9999
$f(x) = x^3$	1	3.375	4.096	5.359	5.832	6.859	7.8806	7.988	**7.9988**

x	3	2.5	2.4	2.25	2.2	2.1	2.01	2.001	2.0001
$f(x) = x^3$	27	15.625	13.824	11.3906	10.648	9.261	8.1206	8.012	**8.0012**

We infer that for x "sufficiently close" to 2, the value of $f(x) = x^3$ can be made "as close as we please" to 8. That is, $\lim_{x \to 2} x^3 = 8$.

Example 4 Use a numerical approach to find: $\lim_{x \to 1} f(x) = \lim_{x \to 1} \dfrac{x^2 - 1}{x - 1}$

Solution As in Example 3, we construct a table. The choices for x, although arbitrary, are selected so that they are close to 1; some are chosen less than 1, and some greater than 1.

x	0	0.5	0.75	0.9	0.99	1.01	1.1	1.25	1.5	2
$f(x) = \dfrac{x^2 - 1}{x - 1}$	1	1.5	1.75	1.9	**1.99**	**2.01**	2.1	2.25	2.5	3

We infer from the table that

$$\lim_{x \to 1} \frac{x^2 - 1}{x - 1} = 2$$

You are right if you make the observation in Example 4 that as long as x is close to 1, but not equal to 1, f can be simplified as $(x^2 - 1)/(x - 1) = x + 1$. Now it is easy to see that for x close to 1, $f(x)$ will be close to $1 + 1 = 2$.

Observe also that the function $f(x) = (x^2 - 1)/(x - 1)$ is not defined at 1, yet its limit at 1 equals 2. This illustrates the following important principle about limits:

The limit L of a function $y = f(x)$ as x approaches the number c does not depend on the value of f at c.

The numerical and graphical approaches to limits described in this section should be viewed as aids to understanding the idea of a limit. Each approach has its drawbacks: the numerical approach may require extensive calculations and the graphical approach requires a knowledge of the graph of the function. In the next section we will discuss a method for finding limits using algebra. This technique is the method we will use in the study of calculus.

Exercise 4
Solutions to Odd-Numbered Problems begin on page 370.

In Problems 1–8 complete each table and evaluate the indicated limits. (A hand calculator will be helpful.)

1.

x	0.9	0.99	0.999	1.001	1.01	1.1
$f(x) = 2x$						

$\lim\limits_{x \to 1^-} f(x) = \underline{}$ $\lim\limits_{x \to 1^+} f(x) = \underline{}$ $\lim\limits_{x \to 1} f(x) = \underline{}$

2.

x	1.9	1.99	1.999	2.001	2.01	2.1
$f(x) = x + 3$						

$\lim\limits_{x \to 2^-} f(x) = \underline{}$ $\lim\limits_{x \to 2^+} f(x) = \underline{}$ $\lim\limits_{x \to 2} f(x) = \underline{}$

3.

x	0.1	0.01	0.001	−0.001	−0.01	−0.1
$f(x) = x^2 + 2$						

$\lim\limits_{x \to 0^-} f(x) = \underline{}$ $\lim\limits_{x \to 0^+} f(x) = \underline{}$ $\lim\limits_{x \to 0} f(x) = \underline{}$

4.

x	−1.1	−1.01	−1.001	−0.999	−0.99	−0.9
$f(x) = x^2 - 2$						

$\lim\limits_{x \to -1^-} f(x) = \underline{}$ $\lim\limits_{x \to -1^+} f(x) = \underline{}$ $\lim\limits_{x \to -1} f(x) = \underline{}$

5.

x	1.9	1.99	1.999	2.001	2.01	2.1
$f(x) = \dfrac{x^2 - 4}{x - 2}$						

$\lim\limits_{x \to 2^-} f(x) = $ _____ $\lim\limits_{x \to 2^+} f(x) = $ _____ $\lim\limits_{x \to 2} f(x) = $ _____

6.

x	-1.1	-1.01	-1.001	-0.999	-0.99	-0.9
$f(x) = \dfrac{x^2 - 1}{x + 1}$						

$\lim\limits_{x \to -1^-} f(x) = $ _____ $\lim\limits_{x \to -1^+} f(x) = $ _____ $\lim\limits_{x \to -1} f(x) = $ _____

7.

x	-1.1	-1.01	-1.001	-0.999	-0.99	-0.9
$f(x) = \dfrac{x^3 + 1}{x + 1}$						

$\lim\limits_{x \to -1^-} f(x) = $ _____ $\lim\limits_{x \to -1^+} f(x) = $ _____ $\lim\limits_{x \to -1} f(x) = $ _____

8.

x	0.9	0.99	0.999	1.001	1.01	1.1
$f(x) = \dfrac{x^3 - 1}{x - 1}$						

$\lim\limits_{x \to 1^-} f(x) = $ _____ $\lim\limits_{x \to 1^+} f(x) = $ _____ $\lim\limits_{x \to 1} f(x) = $ _____

In Problems 9–16 determine whether $\lim\limits_{x \to c^-} f(x)$, $\lim\limits_{x \to c^+} f(x)$, and $\lim\limits_{x \to c} f(x)$ exist by graphing the function.

9. $f(x) = \begin{cases} 2x + 5 & \text{if } x \neq 2 \\ 9 & \text{if } x = 2 \end{cases}$ $c = 2$

10. $f(x) = \begin{cases} 2x + 1 & \text{if } x \neq 0 \\ 1 & \text{if } x = 0 \end{cases}$ $c = 0$

11. $f(x) = \begin{cases} 3x - 1 & \text{if } x \neq 1 \\ 4 & \text{if } x = 1 \end{cases}$ $c = 1$

12. $f(x) = \begin{cases} 2x - 1 & \text{if } x \neq 1 \\ 2 & \text{if } x = 1 \end{cases}$ $c = 1$

13. $f(x) = \begin{cases} 3x - 1 & \text{if } x < 1 \\ \text{Not defined} & \text{if } x = 1 \\ 2x & \text{if } x > 1 \end{cases}$ $c = 1$

14. $f(x) = \begin{cases} 3x - 1 & \text{if } x < 1 \\ 2 & \text{if } x = 1 \\ 3x & \text{if } x > 1 \end{cases}$ $c = 1$

15. $f(x) = \begin{cases} x^2 & \text{if } x \leq 0 \\ 2x + 1 & \text{if } x > 0 \end{cases}$ $c = 0$

16. $f(x) = \begin{cases} x^2 & \text{if } x \leq -1 \\ -3x + 2 & \text{if } x > -1 \end{cases}$ $c = -1$

17. For the function
$$f(x) = 2x + 1 \qquad 0 \leq x \leq 1$$
   (a) Determine whether $\lim_{x \to 0^+} f(x)$ exists.
   (b) Determine whether $\lim_{x \to 1^-} f(x)$ exists.
   (c) Can you find $\lim_{x \to 0^-} f(x)$ or $\lim_{x \to 1^+} f(x)$? Why or why not?

5. Algebraic Techniques for Evaluating Limits

In this section we establish some useful algebraic properties of limits. These properties allow us to find limits without using numerical or graphical approaches.

(1) Limit of a Constant **For the constant function $f(x) = A$, we have**
$$\lim_{x \to c} f(x) = \lim_{x \to c} A = A$$
for any number c.

In other words, *the limit of a constant is the constant.*

The graph of the constant function $f(x) = A$ is a horizontal straight line. Clearly, no matter what the value of x is, the height of $f(x)$ is always A. Thus, as we approach c from either the right or the left, we hit the line $x = c$ at a height of A. That is, $\lim_{x \to c} f(x) = A$. See Figure 36.

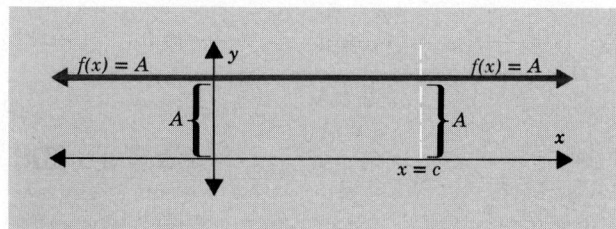

Figure 36

For example,
$$\lim_{x \to 3} 5 = 5$$

76 CH. 2 FUNCTIONS AND THEIR LIMIT

(2) **Limit of $f(x) = x$** For the function $f(x) = x$, we have

$$\lim_{x \to c} f(x) = \lim_{x \to c} x = c$$

for any number c.

The graph of the function $f(x) = x$ is a straight line with slope 1, passing through the origin. See Figure 37. No matter what number c is chosen, as we approach c from either the left or the right, the function approaches the line $x = c$ at the height of c. That is,

$$\lim_{x \to c} f(x) = \lim_{x \to c} x = c$$

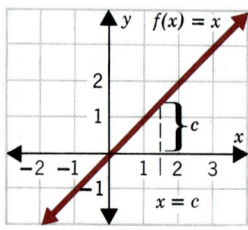

Figure 37

For example,

$$\lim_{x \to 4} x = 4$$

The next four properties are needed for evaluating limits of functions that are more complicated than those just described. We state these properties and give some examples to provide a working knowledge of their many uses.

(3) **Limit of a Sum** Let f and g be two functions whose limits as x approaches c exist; that is, suppose $\lim_{x \to c} f(x)$ and $\lim_{x \to c} g(x)$ are both known real numbers. Then the limit of the function $f + g$ also exists as x approaches c, and

$$\lim_{x \to c} [f(x) + g(x)] = \lim_{x \to c} f(x) + \lim_{x \to c} g(x)$$

Thus, the limit of the sum of two functions equals the sum of their limits.

Because of the relationship between addition and subtraction, the condition stated in (3) leads us to the result

(4)
$$\lim_{x \to c} [f(x) - g(x)] = \lim_{x \to c} f(x) - \lim_{x \to c} g(x)$$

That is, the limit of the difference of two functions equals the difference of their limits.

Example 1 Find: $\lim_{x \to 3} (2 + x)$

5. ALGEBRAIC TECHNIQUES FOR EVALUATING LIMITS

Solution We want to compute the limit of a function $h(x) = 2 + x$, which is really the sum of two other functions, $f(x) = 2$ and $g(x) = x$. From (1) and (2), we know that

$$\lim_{x \to 3} 2 = 2 \quad \text{and} \quad \lim_{x \to 3} x = 3$$

Then, using (3), we have

$$\lim_{x \to 3}(2 + x) = \underbrace{\lim_{x \to 3} 2 + \lim_{x \to 3} x}_{\text{By (3)}} = 2 + 3 = 5$$
$$\text{By (1), (2)}$$

∎

(5) Limit of a Product **Let f and g be two functions whose limits as x approaches c exist; that is, suppose $\lim_{x \to c} f(x)$ and $\lim_{x \to c} g(x)$ are both known real numbers. Then the limit of the function $f \cdot g$ also exists as x approaches c and**

$$\lim_{x \to c}[f(x) \cdot g(x)] = \left[\lim_{x \to c} f(x)\right] \cdot \left[\lim_{x \to c} g(x)\right]$$

Thus, the limit of the product of two functions equals the product of their limits.

The next example illustrates this result.

Example 2 Find: $\lim_{x \to -2}(3x)$

Solution We want to find the limit of a function $h(x) = 3x$, which is the product of two other functions, $f(x) = 3$ and $g(x) = x$. Therefore,

$$\underbrace{\lim_{x \to -2}(3x)}_{\text{By (5)}} = \left[\lim_{x \to -2} 3\right] \cdot \left[\lim_{x \to -2} x\right] = (3) \cdot (-2) = -6$$
$$\text{By (1), (2)}$$

∎

Example 3 Find: $\lim_{x \to 3}(4x + 5)$

Solution
$$\underbrace{\lim_{x \to 3}(4x + 5)}_{\text{By (3)}} = \lim_{x \to 3}(4x) + \underbrace{\lim_{x \to 3} 5}_{\text{By (5)}} = \left(\lim_{x \to 3} 4\right) \cdot \left(\lim_{x \to 3} x\right) + \lim_{x \to 3} 5$$
$$= \underbrace{(4) \cdot (3) + 5}_{\text{By (1), (2)}} = 12 + 5 = 17$$

∎

Example 4 Find: $\lim_{x \to 4} x^2$

Solution
$$\underbrace{\lim_{x \to 4} x^2}_{\text{By (5)}} = \left(\lim_{x \to 4} x\right) \cdot \underbrace{\left(\lim_{x \to 4} x\right)}_{\text{By (2)}} = (4) \cdot (4) = 16$$

∎

Example 4 leads us to an additional property of limits which is a consequence of (2) and the repeated use of (5).

(6) Limit of $f(x) = x^n$ If $n \geq 1$ is a positive integer, then

$$\lim_{x \to c} x^n = \underbrace{\left(\lim_{x \to c} x\right) \cdot \left(\lim_{x \to c} x\right) \cdot \cdots \cdot \left(\lim_{x \to c} x\right)}_{n \text{ times}} = \underbrace{c \cdot c \cdot \cdots \cdot c}_{n \text{ times}} = c^n$$

The last property of limits we take up involves division of functions.

(7) Limit of a Quotient Let f and g be two functions whose limits as x approaches c exist; that is, suppose $\lim_{x \to c} f(x)$ and $\lim_{x \to c} g(x)$ are both known real numbers. If $\lim_{x \to c} g(x) \neq 0$, then the limit of the function f/g as x approaches c also exists and

$$\lim_{x \to c} \frac{f(x)}{g(x)} = \frac{\lim_{x \to c} f(x)}{\lim_{x \to c} g(x)}$$

Thus, if the limit of the denominator function is not zero, the limit of the quotient of two functions equals the quotient of their limits.

Example 5 Find: $\lim_{x \to 3} \dfrac{2x^2 + 1}{x^3 - 2}$

Solution First, we look at the limit of denominator function $g(x) = x^3 - 2$:

$$\lim_{x \to 3} (x^3 - 2) = \lim_{x \to 3} x^3 - \lim_{x \to 3} 2 = 3^3 - 2 = 25$$

Since the limit of the denominator function is not zero, we proceed to compute the limit of the numerator function. Then,

$$\lim_{x \to 3} (2x^2 + 1) = \lim_{x \to 3} (2x^2) + \lim_{x \to 3} 1 = (2)(9) + 1 = 19$$

Thus,

$$\lim_{x \to 3} \frac{2x^2 + 1}{x^3 - 2} \underset{\text{By (7)}}{=} \frac{\lim_{x \to 3} (2x^2 + 1)}{\lim_{x \to 3} (x^3 - 2)} = \frac{19}{25}$$

The previous examples might lead us to think that the evaluation of limits is simply a matter of substituting the number that x approaches into the function. The next few examples are a reminder that this approach cannot always be used.

Example 6 Find: (a) $\lim_{x \to -2} \dfrac{x^2 + 5x + 6}{x^2 - 4}$ (b) $\lim_{x \to 4} \dfrac{x^2 - x - 12}{x^2 - 4x}$

Solution (a) First, we check the limit of the denominator:

$$\lim_{x \to -2} (x^2 - 4) = \lim_{x \to -2} x^2 - \lim_{x \to -2} 4 - 4 = 0$$

Since the limit of the denominator function is zero, we cannot use (7).

5. ALGEBRAIC TECHNIQUES FOR EVALUATING LIMITS 79

However, this does not mean that the limit does not exist! Instead, we use algebraic techniques and factor:

$$\frac{x^2 + 5x + 6}{x^2 - 4} = \frac{(x+3)(x+2)}{(x-2)(x+2)}$$

Since we are interested only in the limit as x approaches -2, and *not* in the value when x equals -2, the quantity $(x+2)$ is not zero. Hence, we can cancel the $(x+2)$'s and then apply (7):

$$\lim_{x \to -2} \frac{x^2 + 5x + 6}{x^2 - 4} = \lim_{x \to -2} \frac{x+3}{x-2} \underset{\text{By (7)}}{=} \frac{\lim_{x \to -2}(x+3)}{\lim_{x \to -2}(x-2)} = \frac{-2+3}{-2-2} = \frac{1}{-4} = -\frac{1}{4}$$

(b) $\displaystyle\lim_{x \to 4} \frac{x^2 - x - 12}{x^2 - 4x} \underset{\text{Factor}}{=} \lim_{x \to 4} \frac{(x-4)(x+3)}{x(x-4)} = \lim_{x \to 4} \frac{x+3}{x} = \frac{7}{4}$

Example 7 For the function $f(x) = x^3$, find:

(a) $\displaystyle\lim_{x \to 2} \frac{f(x) - f(2)}{x - 2}$ (b) $\displaystyle\lim_{x \to c} \frac{f(x) - f(c)}{x - c}$

Solution (a) $f(x) = x^3$ and $f(2) = 8$

$$\lim_{x \to 2} \frac{f(x) - f(2)}{x - 2} = \lim_{x \to 2} \frac{x^3 - 8}{x - 2} \underset{\text{Factor}}{=} \lim_{x \to 2} \frac{(x-2)(x^2 + 2x + 4)}{x - 2}$$

$$= \lim_{x \to 2} (x^2 + 2x + 4) = 4 + 4 + 4 = 12$$

(b) We choose to simplify this expression first:

$$\frac{f(x) - f(c)}{x - c} = \frac{x^3 - c^3}{x - c} \underset{\text{Factor}}{=} \frac{(x-c)(x^2 + cx + c^2)}{x - c} = x^2 + cx + c^2$$

$$\lim_{x \to c} \frac{f(x) - f(c)}{x - c} = \lim_{x \to c} (x^2 + cx + c^2) = c^2 + c^2 + c^2 = 3c^2$$

1-23 odds

Exercise 5
Solutions to Odd-Numbered Problems begin on page 371.

In Problems 1-20 find the indicated limit.

1. $\displaystyle\lim_{x \to 0} 3$ 2. $\displaystyle\lim_{x \to 1} 4$

3. $\displaystyle\lim_{x \to 0} x$ 4. $\displaystyle\lim_{x \to -1} x$

5. $\displaystyle\lim_{x \to 2} (2x - 1)$ 6. $\displaystyle\lim_{x \to -2} (3x + 2)$

7. $\displaystyle\lim_{x \to 4} (x^2 + x)$ 8. $\displaystyle\lim_{x \to 2} (x^2 - x)$

9. $\lim_{x \to -3}(2x^2 - 1)$

10. $\lim_{x \to -4}(3x^2 + 4)$

11. $\lim_{x \to 2}\dfrac{x^2 + 4}{x + 3}$

12. $\lim_{x \to 3}\dfrac{x}{x^2 + 1}$

13. $\lim_{x \to 2}\dfrac{x^2 - 4}{x - 2}$

14. $\lim_{x \to -1}\dfrac{x^2 - 1}{x + 1}$

15. $\lim_{x \to -4}\dfrac{x^3 + 64}{x + 4}$

16. $\lim_{x \to 3}\dfrac{x^3 - 27}{x - 3}$

17. $\lim_{x \to 1}\dfrac{(1/x) - 1}{x - 1}$

18. $\lim_{x \to 2}\dfrac{(1/x^2) - \frac{1}{4}}{x - 2}$

19. $\lim_{x \to 1}\dfrac{\sqrt{x} - 1}{x - 1}$

20. $\lim_{x \to 4}\dfrac{\sqrt{x} - 2}{x - 4}$

21. If $f(x) = x^2$, find: $\lim_{x \to 4}\dfrac{f(x) - f(4)}{x - 4}$

22. If $f(x) = x^2$, find: $\lim_{x \to 1}\dfrac{f(x) - f(1)}{x - 1}$

23. If $f(x) = 3x^2 + x$, find: $\lim_{x \to 2}\dfrac{f(x) - f(2)}{x - 2}$

24. If $f(x) = 2x^2 + x$, find: $\lim_{x \to 4}\dfrac{f(x) - f(4)}{x - 4}$

In Problems 25–28 assume that $\lim_{x \to c} f(x) = 5$ and $\lim_{x \to c} g(x) = 2$ to find each limit.

25. $\lim_{x \to c}[2f(x)]$

26. $\lim_{x \to c}[f(x) - g(x)]$

27. $\lim_{x \to c} g(x)^3$

28. $\lim_{x \to c}\dfrac{f(x)}{g(x) - f(x)}$

*29. If $n \geq 1$ is a positive integer, find
$$\lim_{x \to c}\dfrac{x^n - c^n}{x - c}$$

Hint: $x^n - c^n = (x - c)(x^{n-1} + cx^{n-2} + \cdots + c^{n-2}x + c^{n-1})$

6. Continuous Functions

We have seen that sometimes $\lim_{x \to c} f(x)$ equals $f(c)$ and sometimes it does not. In fact, sometimes $f(c)$ is not even defined and $\lim_{x \to c} f(x)$ exists. Then, what is the relationship between $\lim_{x \to c} f(x)$ and $f(c)$? For the answer, we look at the possibilities:

(a) $\lim_{x \to c} f(x)$ exists and equals $f(c)$

(b) $\lim_{x \to c} f(x)$ exists and does not equal $f(c)$

(c) $\lim_{x \to c} f(x)$ exists and $f(c)$ is not defined

(d) $\lim_{x \to c} f(x)$ does not exist and $f(c)$ is defined

(e) $\lim_{x \to c} f(x)$ does not exist and $f(c)$ is not defined

These situations are illustrated in Figure 38.

Of the five situations, the "nicest" one is that given in Figure 38(a). There, not only does $\lim_{x \to c} f(x)$ exist, but it is equal to $f(c)$. Functions that have this particular

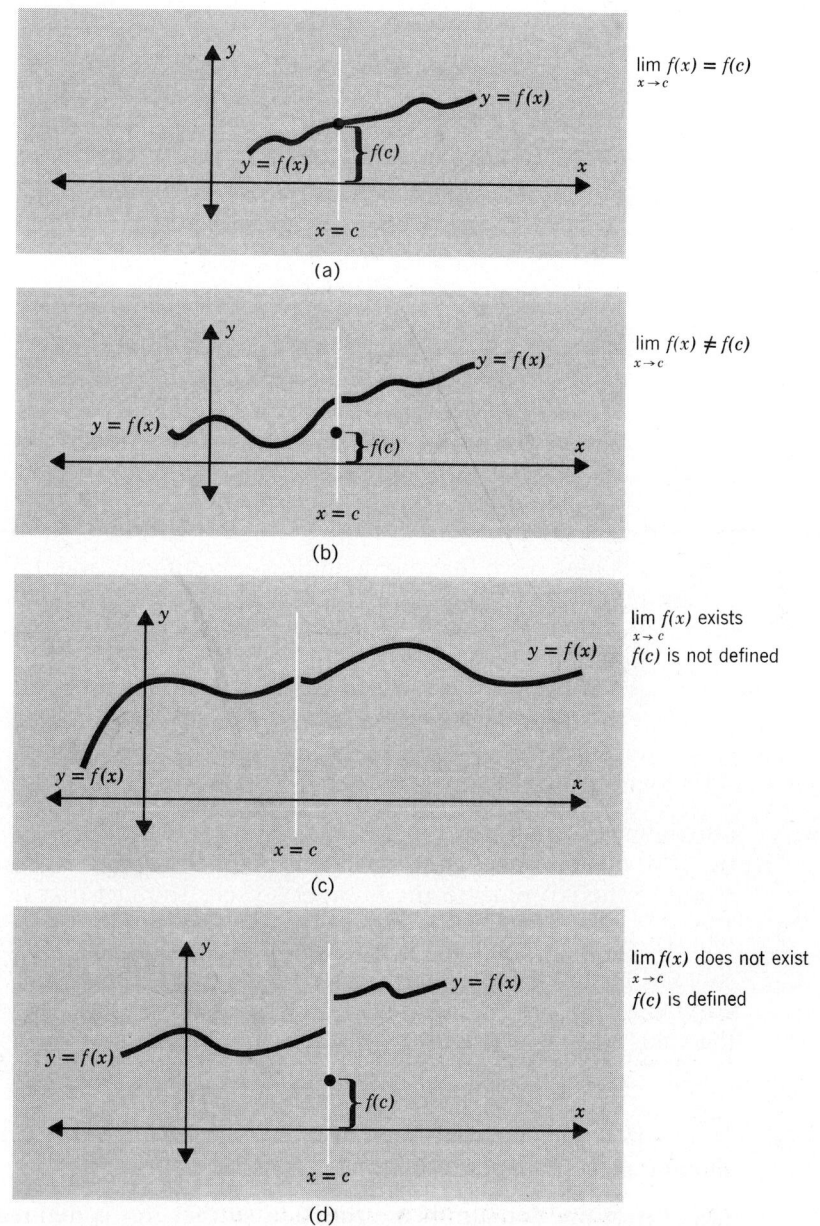

82 CH. 2 FUNCTIONS AND THEIR LIMIT

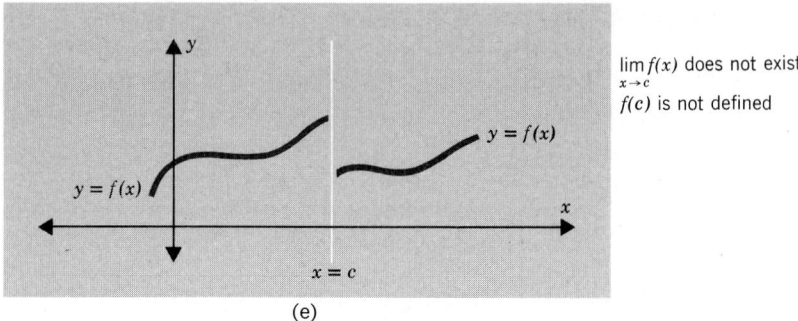

Figure 38

quality are said to be *continuous at c*. This is in agreement with the intuitive notion usually given in elementary courses that states "a function is continuous if its graph can be drawn without lifting the pencil." The functions in Figure 38(b), (c), (d), and (e) are not continuous at c since each has a "break" in the graph at x. This leads us to state the following definition.

(1) **Continuous Function** Let $y = f(x)$ **be a function. If**

$$\lim_{x \to c^-} f(x) = f(c) = \lim_{x \to c^+} f(x)$$

then the function is said to be *continuous at c*. **A function is called** *continuous on an interval* **if it is continuous at every number in the interval.**

At first, it may seem that the definition of continuity and the intuitive concept of continuity have little in common. To see that they are indeed equivalent, let's look at this definition more closely.

The condition for a function to be continuous at c is that

$$\lim_{x \to c^-} f(x) = f(c) = \lim_{x \to c^+} f(x)$$

We interpret $\lim_{x \to c^-} f(x) = f(c)$ to mean that as we travel along the graph of the function approaching c from the left, we expect to hit the line $x = c$ at the height $f(c)$. This means there is no break in the graph as we proceed from numbers near c (but to the left of c) to the number c itself. The fact that $\lim_{x \to c^+} f(x) = f(c)$ says that when we proceed using numbers to the right of c, we are still at a height $f(c)$, so that again there is no jump or break in the graph. Thus, the condition for continuity is that the values of the function for x near c, should be very close to the value of the function at c.

(2) To summarize, a function is continuous at c provided that three conditions are simultaneously satisfied:

(a) c is in the domain of the function so that $f(c)$ is defined

(b) $\lim_{x \to c} f(x)$ **is a real number**

(c) $\lim_{x \to c} f(x) = f(c)$

Discontinuous Function If any one of these conditions is not obeyed, then the function is said to be *discontinuous* at c.

Example 1 (a) The function $f(x) = 3x^3 - 5x + 4$ is continuous at 1, since
$$\lim_{x \to 1} f(x) = \lim_{x \to 1} (3x^3 - 5x + 4) = 2 \quad \text{and} \quad f(1) = 2$$

(b) The function
$$f(x) = \frac{x^2 + 2}{x - 3}$$

is discontinuous at 3, since $f(3)$ is not defined.

Example 2 Discuss the continuity of the function below at 3,
$$f(x) = \begin{cases} \dfrac{x^2 - 9}{x - 3} & \text{if } x \neq 3 \\ 6 & \text{if } x = 3 \end{cases}$$

Solution The function f is defined at 3, since $f(3) = 6$. Also,
$$\lim_{x \to 3^+} \frac{x^2 - 9}{x - 3} = \lim_{x \to 3^+} \frac{(x + 3)(x - 3)}{x - 3} = \lim_{x \to 3^+} (x + 3) = 6$$
$$\lim_{x \to 3^-} \frac{x^2 - 9}{x - 3} = \lim_{x \to 3^-} \frac{(x + 3)(x - 3)}{x - 3} = \lim_{x \to 3^-} (x + 3) = 6$$

Therefore, $\lim_{x \to 3} f(x) = f(3) = 6$. Thus, the three conditions in (2) are satisfied, and the function is continuous at 3. See Figure 39(a).

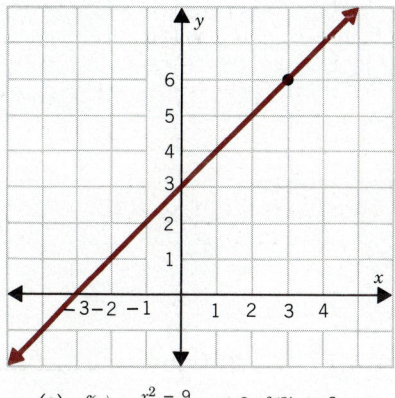

(a) $f(x) = \frac{x^2 - 9}{x - 3}, x \neq 3; f(3) = 6$

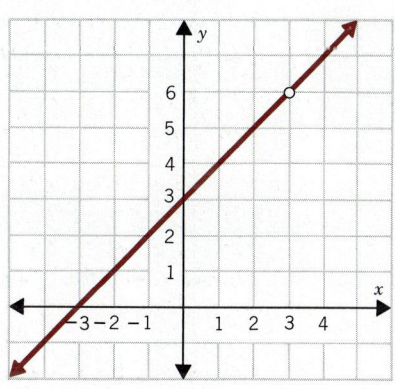

(b) $g(x) = \frac{x^2 - 9}{x - 3}, x \neq 3$

Figure 39

84 CH. 2 FUNCTIONS AND THEIR LIMIT

Consider the function

$$g(x) = \frac{x^2 - 9}{x - 3} \qquad x \neq 3$$

whose graph appears in Figure 39(b) (page 83). It is not continuous at 3, since condition (a) of (2) is not satisfied.

Example 3 The 1982 United States postage rates for first class mail weighing 8 ounces or less are listed in the table. Graph the function describing this rate structure and determine points of discontinuity, if any.

Number of Ounces	≤ 1	≤ 2	≤ 3	≤ 4	≤ 5	≤ 6	≤ 7	≤ 8
Cost in Cents	20	37	54	71	88	105	122	139

Solution For mail weighing up to and including 1 ounce, the charge is 20 cents. For mail weighing over 1 ounce, up to and including 2 ounces, the charge is 37 cents. This pattern continues, as shown in the table. A graph illustrating this function is given in Figure 40.

Notice that the domain of this function is $0 < x \leq 8$, and the range is $\{0.20, 0.37, 0.54, \ldots, 1.39\}$. The function shown in Figure 40 is usually called a *step function*. It is discontinuous at the points 1, 2, 3, 4, 5, 6, 7, and is continuous elsewhere. ∎

• ———————————

Step Function

Of course, just as with the evaluation of limits, the determination of continuity does not rely on graphical techniques. In fact, the definition of continuity makes no mention of the graph of the function. We determine continuity solely based on the expression defining the function.

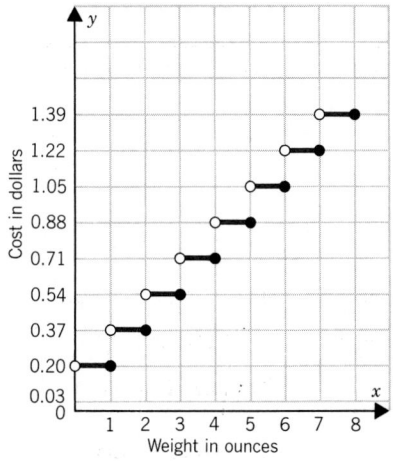

Figure 40

Many of the functions we have already studied are continuous functions. For example:

1. The linear function
$$f(x) = mx + b$$
is continuous for all x.
2. The quadratic function
$$f(x) = ax^2 + bx + c \qquad a \neq 0$$
is continuous for all x.
3. The polynomial function
$$f(x) = a_n x^n + a_{n-1} x^{n-1} + \cdots + a_1 x + a_0 \qquad a_n \neq 0$$
with $n \geq 0$ an integer, is continuous for all x.
4. If c is in the domain of a rational function (that is, the denominator polynomial is not equal to zero at c), then the rational function is continuous at c.

Once a function is known to be continuous at c, then its limit as x approaches c is easy to find—it is $f(c)$. That is,

$$\lim_{x \to c} f(x) = f(c)$$

Example 4 (a) The function $f(x) = x^3 - x^2 + 2$, which is a polynomial of degree 3, is known to be a continuous function. Therefore,

$$\lim_{x \to -2} (x^3 - x^2 + 2) = (-2)^3 - (-2)^2 + 2 = -8 - 4 + 2 = -10$$

(b) The function
$$f(x) = \frac{x^2 + 1}{2x^4 + 15}$$
which is a rational function, is known to be continuous for all x. Therefore,

$$\lim_{x \to 1} \frac{x^2 + 1}{2x^4 + 15} = \frac{1^2 + 1}{2(1^4) + 15} = \frac{2}{17}$$

∎

Exercise 6
Solutions to Odd-Numbered Problems begin on page 372.

In Problems 1–10 compute each limit (see Example 4).

1. $\lim_{x \to 1} (2 - 7x)$
2. $\lim_{x \to 1} (1 - 3x)$
3. $\lim_{x \to 2} (x^4 - 2x + 1)$
4. $\lim_{x \to 5} (x^3 - 4)$

5. $\lim_{x \to 3} (x^2 - 2x + 7)/(2 - x)$
6. $\lim_{x \to 0} -3/(x^2 - 2x + 1)$
7. $\lim_{x \to 1/2} (2 - x^2)/(x^3 - 5x + 1)$
8. $\lim_{x \to 2} [(5x - 2)^2 + 4]/(x^2 - 5x + 1)$
9. $\lim_{x \to -1} (1 - 6x + x^2)/(x^3 - 2x + 1)$
10. $\lim_{x \to 0} [x(x - 1)]/(3x^2 + 2x - 5)$

In Problems 11–18 determine whether the function f is continuous at c.

11. $f(x) = \begin{cases} 2x + 5 & \text{if } x \le 2 \\ 4x + 1 & \text{if } x > 2 \end{cases}$ $c = 2$

12. $f(x) = \begin{cases} 2x + 1 & \text{if } x \le 0 \\ 2x & \text{if } x > 0 \end{cases}$ $c = 0$

13. $f(x) = \begin{cases} 3x - 1 & \text{if } x < 1 \\ 4 & \text{if } x = 1 \\ 2x & \text{if } x > 1 \end{cases}$ $c = 1$

14. $f(x) = \begin{cases} 3x - 1 & \text{if } x < 1 \\ 2 & \text{if } x = 1 \\ 2x & \text{if } x > 1 \end{cases}$ $c = 1$

15. $f(x) = \begin{cases} 3x - 1 & \text{if } x < 1 \\ \text{Not defined} & \text{if } x = 1 \\ 2x & \text{if } x > 1 \end{cases}$ $c = 1$

16. $f(x) = \begin{cases} 3x - 1 & \text{if } x < 1 \\ 2 & \text{if } x = 1 \\ 3x & \text{if } x > 1 \end{cases}$ $c = 1$

17. $f(x) = \begin{cases} x^2 & \text{if } x \le 0 \\ 2x & \text{if } x > 0 \end{cases}$ $c = 0$

18. $f(x) = \begin{cases} x^2 & \text{if } x < -1 \\ 2 & \text{if } x = -1 \\ -3x + 2 & \text{if } x > -1 \end{cases}$ $c = -1$

19. Is the function f defined by $f(x) = \dfrac{x^2 - 4}{x - 2}$

    continuous at 2? If not, can f be defined at 2 to make it continuous?

20. Is the function f defined by $f(x) = \dfrac{x^2 + x - 12}{x - 3}$

    continuous at 3? If not, can f be defined at 3 to make it continuous?

21. *Processing Oil.* An oil refinery has four distillation towers and operates them as they are needed to process available raw materials. Each tower has fixed costs of $300 per week whether operating or not. In addition, each tower, if in operation, will incur fixed costs of $500 per week. The raw material cost is fixed at $0.50 per gallon of refined oil, and each tower can process at most 10,000 gallons of refined oil each week.

(a) Find the cost function $C(x)$; where x is the number of gallons of refined oil for the refinery.
(b) Find the domain of this function and graph it.
(c) Find any points of discontinuity of the function.

22. *Quantity Discounts.* The owner of a grocery store can buy bulk coffee from a particular distributor according to the following price schedule:

$3.00 per pound for 5 pounds or less
$2.50 per pound for more than 5 pounds but less than 10 pounds
$2.25 per pound for 10 or more pounds but less than 20 pounds
$2.00 per pound for 20 or more pounds

(a) Find a cost function $C(x)$, where $C(x)$ represents the cost of buying x pounds.
(b) Graph the cost function.
(c) Find those values of x where the function is discontinuous.

23. *Postage Rates.* The 1982 United States postage rates for third class mail are given in the table. Graph the function describing the cost and determine the points of discontinuity.

Ounces	≤ 1	≤ 2	≤ 3	≤ 4	≤ 6	≤ 8
Cost (Cents)	20	37	54	71	85	95

Chapter Review

Important Terms

inequalities
intercepts
symmetry
distance between points
closed interval
open interval
endpoints
function
domain
range
independent variable
limit
limit from the left
limit from the right

dependent variable
ordered pair
linear function
quadratic function
quadratic formula
discriminant
parabola
vertex of a parabola
polynomial function
degree of a polynomial
rational function
power function
continuous at c
discontinuous
step function

True-False Questions (Answers on page 434.)

T F 1. Vertical lines intersect the graph of a function at most once.
T F 2. The domain of the function $f(x) = \sqrt{x}$ is the set of all real numbers.
T F 3. The quadratic formula can be used to find the x-intercepts of a polynomial of degree 2.
T F 4. For the interval [2, 3], the endpoints 2 and 3 are excluded.
T F 5. The distance between two points is never a negative number.

CH. 2 FUNCTIONS AND THEIR LIMIT

T F 6. The limit of the sum of two functions equals the sum of their limits.
T F 7. The limit of a function f as x approaches c always equals $f(c)$.

T F 8. $\lim\limits_{x \to 4} \dfrac{x^2 - 16}{x - 4} = 8$

T F 9. The function

$$f(x) = \dfrac{x}{x^2 + 4}$$

is continuous for all x.

T F 10. The limit of a quotient of two functions equals the quotient of their limits.

Fill in the Blanks
(Answers on page 435.)

1. The graph of a quadratic function is called a _____.
2. For a function $y = f(x)$, the number x is called the _____ variable; y is called the _____ variable.
3. The maximum point or minimum point of a parabola is called its _____.
4. The graph of a polynomial of degree _____ is a straight line.
5. A _____ function is the ratio of two polynomials.
6. The notation _____ may be described by saying "For x approximately equal to c, but $x \neq c$, the value $f(x)$ is approximately equal to L."
7. If $\lim\limits_{x \to c^-} f(x) = L$ and $\lim\limits_{x \to c^+} f(x) = R$, then $\lim\limits_{x \to c} f(x)$ exists provided _____ R.
8. If there is no single number that the value of f approaches when x is close to c, then $\lim\limits_{x \to c} f(x)$ does _____ _____.
9. When $\lim\limits_{x \to c} f(x) = f(c)$, we say f is _____ at c.
10. $\lim\limits_{x \to c} \dfrac{f(x)}{g(x)} = \dfrac{\lim\limits_{x \to c} f(x)}{\lim\limits_{x \to c} g(x)}$ provided $\lim\limits_{x \to c} f(x)$ and $\lim\limits_{x \to c} g(x)$ each exist and $\lim\limits_{x \to c} g(x)$ _____ 0.

Review Exercises
Solutions to Odd-Numbered Problems begin on page 373.

In Problems 1–6 determine whether each given correspondence is a function.

1. $y + x^2 = 1$
2. $y - 3x + x^2 = 0$
3. $x^2 y = 4$
4. $xy^2 = 9$

5. $y^2(4 - x) = 2$
6. $y(3 - x) = 2$
7. For the function $f(x) = x^2 + 2x$, find:
   (a) $f(0)$ (b) $f(-3)$ (c) $f(2)$ (d) $f(2 + h)$
8. For the function $f(x) = 2x^2 - x$, find:
   (a) $f(0)$ (b) $f(-2)$ (c) $f(3)$ (d) $f(1 + h)$

In Problems 9-14 graph each quadratic function. Label the vertex, the y-intercept, and the x-intercepts, if any.

9. $f(x) = x^2 + 2x$
10. $f(x) = x^2 - 2x$
11. $f(x) = x^2 + 2x - 8$
12. $f(x) = x^2 - 10x + 24$
13. $f(x) = x^2 + 2x + 4$
14. $f(x) = x^2 + 2x + 3$

15. Graph and find the domain of the function
$$f(x) = \begin{cases} x^2 & \text{if } x < 0 \\ 2x + 1 & \text{if } x \geq 0 \end{cases}$$

16. Graph and find the domain of the function
$$f(x) = \begin{cases} 3x - 2 & \text{if } x \leq 0 \\ x^2 + 1 & \text{if } x > 0 \end{cases}$$

In Problems 17-28 find the domain of the function f.

17. $f(x) = x^2 + 8$
18. $f(x) = x^3 - 2$
19. $f(x) = \sqrt{2 - x}$
20. $f(x) = \sqrt{x + 5}$
21. $f(x) = \dfrac{1}{x - 4}$
22. $f(x) = \dfrac{1}{x - 1}$
23. $f(x) = \dfrac{x}{(x - 2)(x + 3)}$
24. $f(x) = \dfrac{x^2}{(x + 3)^2}$
25. $f(x) = \dfrac{2x + 1}{3x^2 - 5x - 2}$
26. $f(x) = \dfrac{x - 3}{2x^2 + x - 1}$
27. $f(x) = \dfrac{x}{\sqrt{x^2 - 1}}$
28. $f(x) = \dfrac{2}{\sqrt{x^2 - 4}}$

29. Find the points of intersection of the graphs of the functions $y = x^2 - 3$ and $y = 2x$.
30. Find the points of intersection of the graphs of the functions $y = 3x^2 + 9$ and $y = 2x^2 - 5x + 3$.

In Problems 31-42 find each limit.

31. $\lim\limits_{x \to 3}(3x - 4)$
32. $\lim\limits_{x \to 2}(2x + 5)$
33. $\lim\limits_{x \to 2} x^2$
34. $\lim\limits_{x \to 1} x^2$
35. $\lim\limits_{x \to -2}(3 - 2x)$
36. $\lim\limits_{x \to 2}(8 - x)$
37. $\lim\limits_{x \to 1} \dfrac{x^2 - 1}{x - 1}$
38. $\lim\limits_{x \to -3} \dfrac{x^2 - 9}{x + 3}$
39. $\lim\limits_{x \to 1} \dfrac{x^3 - 1}{x - 1}$
40. $\lim\limits_{x \to -1} \dfrac{x^3 - 1}{x + 1}$
41. $\lim\limits_{x \to 1} \dfrac{(1/x) - 1}{x - 1}$
42. $\lim\limits_{x \to 2} \dfrac{(1/x) - \frac{1}{2}}{x - 2}$

In Problems 43-46 find the limit given below for each function.
$$\lim_{x \to 4} \frac{f(x) - f(4)}{x - 4}$$

43. $f(x) = 4x$
44. $f(x) = 2x + 1$
45. $f(x) = x^2 + x$
46. $f(x) = x^2 - x$

In Problems 47-52 determine whether the function f is continuous at c.

47. $f(x) = 2x^2 + 5$; $c = 0$
48. $f(x) = 3x^3 + x - 2$; $c = 1$
49. $f(x) = \begin{cases} \dfrac{x^2 - 16}{x - 4} & \text{if } x \neq 4 \\ 8 & \text{if } x = 4 \end{cases}$ $c = 4$

50. $f(x) = \begin{cases} \dfrac{x^2 + x}{x} & \text{if } x \neq 0 \\ 1 & \text{if } x = 0 \end{cases}$ $c = 0$

51. $f(x) = \begin{cases} \dfrac{x^3 - 8}{x - 2} & \text{if } x \neq 2 \\ 4 & \text{if } x = 2 \end{cases}$ $c = 2$

52. $f(x) = \begin{cases} \dfrac{x^3 + 1}{x + 1} & \text{if } x \neq -1 \\ 0 & \text{if } x = -1 \end{cases}$ $c = -1$

Mathematical Question From Actuary Exam (Answer on page 433.)

Actuary Exam Part I

1. $\lim\limits_{x \to 3} \dfrac{x^2 - x - 6}{x^2 - 9} =$

   (a) 0   (b) $\frac{5}{6}$ (c) 1   (d) $\frac{5}{3}$ (e) undefined

3
The Derivative

1. Average Rate of Change
2. The Derivative
3. Two Additional Interpretations of the Derivative
4. Some Derivative Formulas
5. General Formulas for Differentiation
6. The Power Rule; the Chain Rule
7. Higher-Order Derivatives
8. Functions Not Differentiable at c

Chapter Review

CH. 3 THE DERIVATIVE

The first part of this chapter deals with the idea of an average rate of change with interpretations in geometry and physics. Then we discuss the meaning of a derivative and how it is interpreted relative to these same ideas. The last part of the chapter gives formulas that make differentiation of many functions a relatively straightforward procedure.

1. Average Rate of Change

On a Saturday in October, the United States Weather Bureau listed the hourly temperatures in degrees Celsius in Chicago from 3 AM to midnight as shown in Table 1.

Table 1

Temperature, °C	7	7	7	8	7	7	7	7	8	8	9
Time	3 AM	4 AM	5 AM	6 AM	7 AM	8 AM	9 AM	10 AM	11 AM	Noon	1 PM

Temperature, °C	10	10	9	10	10	10	13	13	12	10	9
Time	2 PM	3 PM	4 PM	5 PM	6 PM	7 PM	8 PM	9 PM	10 PM	11 PM	Midnight

Most people would agree that the change in temperature from 3 AM to 8 PM is 6°C. But what is the general formula for computing change? The change of 6°C is obtained by taking the reading at 3 AM (7) and subtracting it from the reading at 8 PM (13). That is,

$$\begin{pmatrix} \text{Change from} \\ \text{3 AM to 8 PM} \end{pmatrix} = \begin{pmatrix} \text{Temperature} \\ \text{at 8 PM} \end{pmatrix} - \begin{pmatrix} \text{Temperature} \\ \text{at 3 AM} \end{pmatrix}$$
$$= \text{Final reading} - \text{Initial reading}$$
$$= 13 - 7 = 6$$

The graph of the data listed in Table 1 is given in Figure 1, using the x-axis for time and the y-axis for temperature. There, we use 13, 14, 15, and so on, to stand for 1 PM, 2 PM, 3 PM, and so on. Also, to aid in visualizing this situation, we have connected the points with lines.

Example 1 What is the change in temperature from 11 AM to 8 PM? What is the corresponding change in time?

Solution To compute the change in temperature, we note that the temperature at 11 AM is 8°C and at 8 PM is 13°C. Thus, the change in temperature from 11 AM to 8 PM is $13 - 8 = 5$°C. The change in time from 11 AM to 8 PM is 9 hours. ∎

When we measure time along the x-axis, as in Figure 1, we can express the change in time from 11 AM to 8 PM as

$$\Delta x = 20 - 11 = 9$$

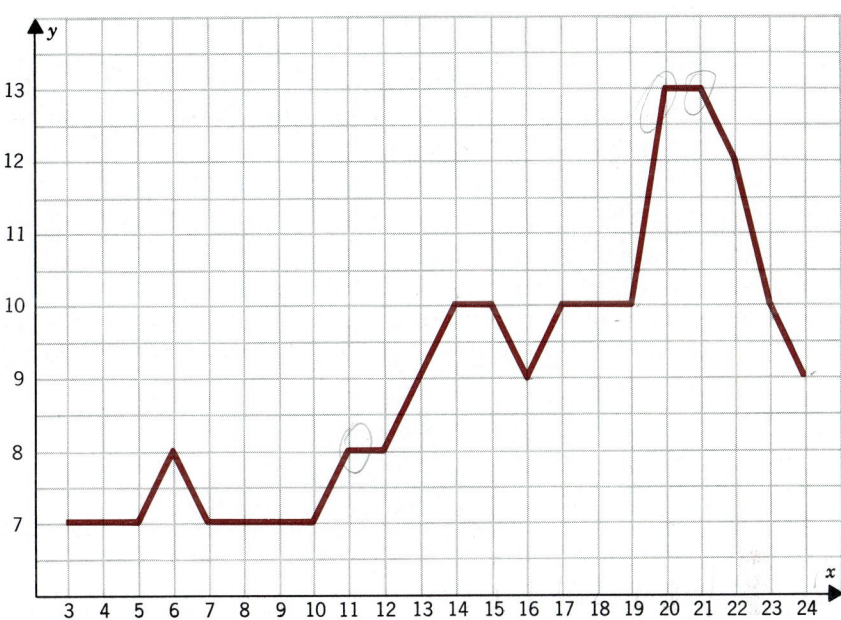

Figure 1

where 20 is the designation for 8 PM. The change in temperature is

$$\Delta y = 13 - 8 = 5$$

The value of Δy can also be zero or negative, as the next example illustrates.

Example 2 For the hourly temperature readings given in Table 1, find the change in temperature from 8 PM to midnight. What is the change from 1 PM to midnight?

Solution The change in temperature from 8 PM to midnight is

$$\Delta y = 9 - 13 = -4$$

The change from 1 PM to midnight is

$$\Delta y = 9 - 9 = 0$$

The next example illustrates how to determine Δx and Δy when the function is given.

Example 3 For the function

$$y = f(x) = 2x - 2$$

find Δx and Δy, if x changes from 0 to 4.

Solution Since the change in x is from 0 to 4, then

$$\Delta x = 4 - 0 = 4$$

94　CH. 3　THE DERIVATIVE

To find the change in y, we need to find the values of y at 0 and at 4. At $x = 0$, the value of y is $f(0) = -2$. At $x = 4$, the value of y is $f(4) = 6$. The change in y is

$$\Delta y = f(4) - f(0) = 6 - (-2) = 8$$

■

Figure 2 illustrates the geometric significance of the changes Δx and Δy for the function given in Example 3.

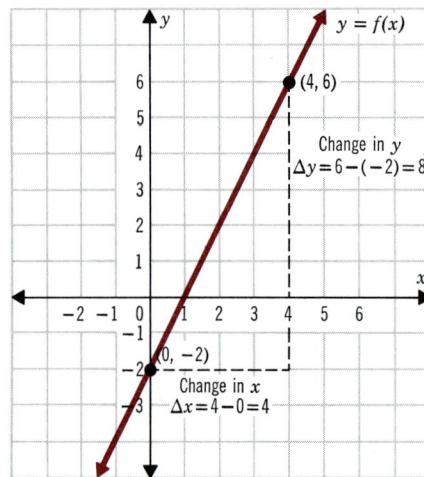

Figure 2

Each of the above examples deals with computing change, but none of them gives us information about the *average rate* at which the change is taking place. There is a significant difference between a change in temperature of 5 degrees over 10 hours and a change of 5 degrees over 1 hour. To measure this information we divide the change in temperature by the change in time. This ratio is called the *average rate of change*—in this case, the average rate of change of temperature with respect to time.

Example 4　For the hourly temperature readings given in Table 1, find the average rate of change of temperature with respect to time from 5 AM to noon.

Solution　We begin by computing the change in temperature from 5 AM to noon, namely,

$$\Delta y = 8 - 7 = 1$$

The corresponding change in time is

$$\Delta x = 12 - 5 = 7$$

The average rate of change is

$$\frac{\Delta y}{\Delta x} = \frac{1}{7} = 0.143$$

■

We interpret the value 0.143 to mean that the change in temperature from 5 AM to noon, *on the average,* amounts to an increase in temperature of 0.143 degree each hour from 5 AM to noon. Of course, in reality the temperature has sometimes remained unchanged and sometimes increased by more than 0.143 degree over each hour of this period. However, the *average* change each hour is 0.143 degree.

Average Rate of Change **For a function** $y = f(x)$**, the** *average rate of change of y with respect to x* **is the ratio of the change in** y **to the change in** x**.**
If the change in x **is from** c **to** d**, then**

$$\text{Average rate of change} = \frac{\Delta y}{\Delta x} = \frac{f(d) - f(c)}{d - c}$$

Example 5 For the function $y = f(x) = x^2$, find the average rate of change as x changes from 1 to 3.

Solution We begin by computing the change in x:

$$\Delta x = 3 - 1 = 2$$

The corresponding change in y is the difference between the values of $f(x) = x^2$ at $x = 3$ and $x = 1$. Thus,

$$\Delta y = f(3) - f(1) = 3^2 - 1^2 = 9 - 1 = 8$$

The average rate of change is

$$\frac{\Delta y}{\Delta x} = \frac{8}{2} = 4$$

We can use the result of Example 5 to get a geometrical interpretation of average rate of change. See Figure 3.

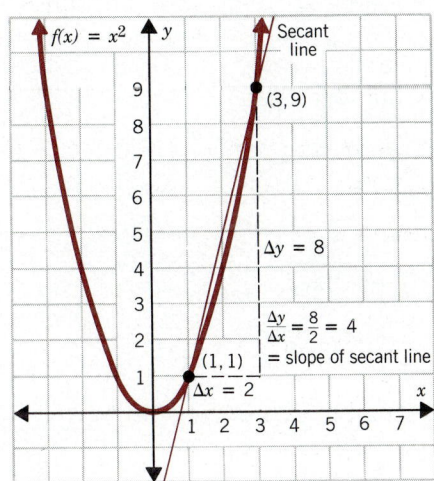

Figure 3

We conclude that:

The *average rate of change* of a function equals the slope of the line joining two points on the graph of the function. This line is called a *secant line*.

The average rate of change also has an important physical interpretation.

Example 6 Mr. Doody and his family leave on a car trip Saturday morning at 5 AM and arrive at their destination at 9 AM. When they began the trip, the car's odometer read 26,700 kilometers and, when they arrived, it read 27,000 kilometers. What was the average velocity for the trip?

Solution Most of us know that average velocity is computed by dividing the distance traveled by the elapsed time. The distance in this case is $27{,}000 - 26{,}700 = 300$ kilometers and the elapsed time is $9 - 5 = 4$ hours. Thus, the average velocity is

$$\text{Average velocity} = \frac{300}{4} = 75 \text{ kilometers per hour}$$

Average Velocity **The *average velocity* is the ratio of the change in distance to the change in time. If s denotes distance and t denotes time, we have**

$$\textbf{average velocity} = \frac{\Delta s}{\Delta t}$$

Difference Quotient The average rate of change $\Delta y / \Delta x$ of a function $y = f(x)$ from c to d is often referred to as a *difference quotient*.

Example 7 Compute the difference quotient $\Delta y / \Delta x$ for the function $f(x) = x^2$, where:
(a) $c = 1$, $d = 2$
(b) $c = 1$, $d = 1.5$
(c) $c = 1$, $x \neq 1$ is any number

Solution (a) For $c = 1$ and $d = 2$, we have

$$\frac{\Delta y}{\Delta x} = \frac{f(d) - f(c)}{d - c} = \frac{4 - 1}{2 - 1} = 3$$

(b) For $c = 1$ and $d = 1.5$, we have

$$\frac{\Delta y}{\Delta x} = \frac{f(d) - f(c)}{d - c} = \frac{2.25 - 1}{1.5 - 1} = 2.5$$

(c) For $c = 1$ and $x \neq 1$ an arbitrary number, we have

$$\frac{\Delta y}{\Delta x} = \frac{f(x) - f(c)}{x - c} = \frac{x^2 - 1}{x - 1} = x + 1$$

1. AVERAGE RATE OF CHANGE 97

Exercise 1
Solutions to Odd-Numbered Problems begin on page 374.

1. For the function
$$f(x) = 5x - 2$$
   (a) Find the change in y from $x = 0$ to $x = 4$.
   (b) Find the change in y from $x = -3$ to $x = 2$.
   (c) Find the average rate of change for part (a).
   (d) Find the average rate of change for part (b).
   (e) Compute the average rate of change from $x = c$ to $x = d$, where $c < d$. What do you conclude? Why? Look at the graph for a hint.

2. For the function
$$f(x) = x^2 - 1$$
   (a) Find the change in y from $x = 1$ to $x = 3$.
   (b) Find the change in y from $x = -1$ to $x = 1$.
   (c) Find the average rate of change in part (a).
   (d) Find the average rate of change in part (b).

3. For the function
$$f(x) = x^3$$
   (a) Find the change in y from $x = -1$ to $x = 1$.
   (b) Find the change in y from $x = 0$ to $x = 4$.
   (c) Find the average rate of change in part (a).
   (d) Find the average rate of change in part (b).

4. For the function
$$f(x) = x^2 - 3x$$
   (a) Find the change in y from $x = -1$ to $x = 1$.
   (b) Find the change in y from $x = 0$ to $x = 2$.
   (c) Find the average rate of change in part (a).
   (d) Find the average rate of change in part (b).

5. The function $s = f(t) = 6t(t + 1)$ relates the distance s in kilometers a car travels in time t (in hours). Compute the car's average velocity from t_0 to t_1 for:
   (a) $t_0 = 2$, $t_1 = 3$ (b) $t_0 = 2$, $t_1 = 2.5$

6. Follow the directions in Problem 5 for:
   (a) $t_0 = 2$, $t_1 = 2.1$ (b) $t_0 = 2$, $t_1 = 2.01$

For the functions in Problems 7–16 compute the difference quotient
$$\frac{\Delta y}{\Delta x} = \frac{f(x) - f(1)}{x - 1} \qquad x \neq 1 \text{ is any number}$$

Simplify your answer.

7. $f(x) = 3x - 5$ 8. $f(x) = 2x + 3$
9. $f(x) = 3x^2 - 2$ 10. $f(x) = 2x^2 + 5$

11. $f(x) = x - x^2$
12. $f(x) = x^2 - x$
13. $f(x) = x^2 + 2x - 2$
14. $f(x) = 2x^2 - x$
15. $f(x) = 3x^2 - 2x + 5$
16. $f(x) = 5x^2 + x - 2$

In Problems 17–22 find the slope of the line from $(0, f(0))$ to an arbitrary point $P = (x, y) = (x, f(x))$.

17. $f(x) = 3x - 2$
18. $f(x) = 3x + 8$
19. $f(x) = x^2 + 4$
20. $f(x) = x^2 - 4$
21. $f(x) = x^3 + 2$
22. $f(x) = x^3$

23. Suppose the function $s = f(t) = 16t^2$ relates the distance s (in feet) an object travels in time t (in seconds). Compute the average velocity, $\Delta s/\Delta t$, from $t = 3$ to:
    (a) $t = 3.5$ (b) $t = 3.1$

24. Follow the directions in Problem 23 if the change in time is from $t = 3$ to:
    (a) $t = 3.01$ (b) $t = 3.0001$

25. The distance a man can walk in time t obeys the formula

$$s = \sqrt{t}$$

where s is measured in kilometers and t is measured in hours. What is his average velocity from $t = 0$ to $t = 16$ hours? What is his average velocity from $t = 1$ to $t = 4$ hours? From $t = 1$ to $t = 2$ hours?

26. **Corn Production.** Consider the corn production data given in the table.

Investment, x (Dollars)	Corn Output, y (Bushels)
700	12,700
1000	13,100
1200	14,700
1400	17,100
1600	19,000
2000	22,100
2200	21,700

Graph the data. Find the average rate of increase from $700 to $2200. What is the average rate of increase from $1000 to $1400? What is the average rate of increase from $1000 to $1200?

27. **Corn Yield.** By adding fertilizer to the soil, increased crop yields can sometimes be obtained. The table lists the corn yield obtained per acre by adding certain amounts of fertilizer.

Fertilizer (Pounds)	Corn Yield/Acre (Bushels)
0	45
10	46.2
20	48
30	48.2
40	47.6
50	47.3

Graph the data, using the x-axis to represent the amount of fertilizer added. What is the average rate of increase in yield from 0 to 50 pounds? What is the average rate of increase in yield from 10 to 40 pounds? From 10 to 30 pounds?

28. *Amino Acids.* A protein disintegrates into amino acids according to the formula

$$M = \frac{28}{t+2}$$

where M, the mass of the protein, is measured in grams and t is time measured in hours. Find the average reaction rate from $t = 0$ to $t = 2$ hours. Interpret your answer.

29. *Glucose Conversion.* In a metabolic experiment, the mass M of glucose decreases according to the formula

$$M = 4.5 - 0.03t^2$$

where M is measured in grams and t is time measured in hours. Find the average reaction rate from $t = 0$ to $t = 2$. Interpret your answer.

*30. *Estimating Zeros.* A number r is said to be a *zero* of the function $y = f(x)$ if $f(r) = 0$. Geometrically, this is a point where the function crosses the x-axis. If a continuous function changes sign on an interval, then it must cross the x-axis at least once on that interval. This result, which is intuitively obvious, can be used to estimate the zeros of functions.

If at $x = a$, $f(a) < 0$ and if at $x = b$, $f(b) > 0$, then there is a zero r of the function in the interval (a, b). See the illustration. To estimate r, we construct the line joining the two points $(a, f(a))$ and $(b, f(b))$. The place where this line crosses the x-axis will be our estimate of the zero r of the function. See the illustration.

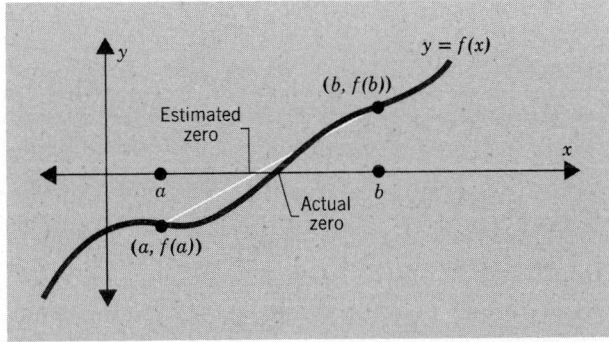

Using this result, estimate the zeros of the following functions between the values given:
(a) $f(x) = x^2 - 5$; $a = 2$, $b = 3$
(b) $f(x) = x^2 + \frac{3}{2}x - 1$; $a = 0$, $b = 1$
(c) $f(x) = x^3 + x^2 - 2x - 2$; $a = 1$, $b = 2$

2. The Derivative

We saw in Section 1 that the average rate of change of a function can be interpreted as the average velocity when the function used relates the distance traveled to the time t. Let's look at this interpretation more closely now.

The function $s = 16t^2$ can be used to measure the distance s (in feet) an object dropped from a tall building has traveled after a time t (in seconds). When $0 \leq t \leq 6$, we calculate the average velocity of the object over the entire time interval of 6 seconds to be

$$\frac{\Delta s}{\Delta t} = \frac{f(6) - f(0)}{6 - 0} = \frac{16(6)^2 - 16(0)^2}{6} = 96 \text{ feet per second}$$

This average accurately describes the velocity of the object over the 6 second interval, but it gives no information at all about the actual velocity at any particular instant of time. We now want to find the velocity of the object at a particular instant.

Instantaneous Velocity

To see how this might be done, we will seek the exact velocity of the object at the instant when $t = 3$. This velocity is called the *instantaneous velocity at 3*.

So far, we have no *mathematical* method for finding instantaneous velocities. However, we can *estimate* the instantaneous velocity at $t = 3$ seconds by computing the average velocity for some intervals of time beginning at $t = 3$. For example, let's compute the average velocity for the 1 second interval beginning at $t = 3$ and ending at $t = 4$. The distance of the object from the starting position is

$$s = f(3) = 16(9) = 144 \text{ feet}$$

At $t = 4$,

$$s = f(4) = 16(16) = 256 \text{ feet}$$

Thus, over the 1 second interval from $t = 3$ to $t = 4$,

$$\text{Average velocity} = \frac{\Delta s}{\Delta t} = \frac{f(4) - f(3)}{4 - 3} = \frac{256 - 144}{1} = 112 \text{ feet per second}$$

The average velocity for the smaller intervals of time $\Delta t = 0.5, 0.1, 0.01, 0.0001$ may be found similarly (see Problems 23 and 24 in Exercise 1); and Table 2 gives

Table 2

Start	End, t	Δt	Average Velocity $\frac{\Delta s}{\Delta t} = \frac{f(t) - f(3)}{t - 3}$ (Feet per Second)
3	4	1	$\frac{\Delta s}{\Delta t} = \frac{f(4) - f(3)}{1} = \frac{16(16) - 16(9)}{1} = 112$
3	3.5	0.5	$\frac{\Delta s}{\Delta t} = \frac{f(3.5) - f(3)}{0.5} = \frac{16(3.5)^2 - 16(9)}{0.5} = 104$
3	3.1	0.1	$\frac{\Delta s}{\Delta t} = \frac{f(3.1) - f(3)}{0.1} = \frac{16(3.1)^2 - 16(9)}{0.1} = 97.6$
3	3.01	0.01	$\frac{\Delta s}{\Delta t} = \frac{f(3.01) - f(3)}{0.01} = \frac{16(3.01)^2 - 16(9)}{0.01} = 96.16$
3	3.0001	0.0001	$\frac{\Delta s}{\Delta t} = \frac{f(3.0001) - f(3)}{0.0001} = \frac{16(3.0001)^2 - 16(9)}{0.0001} = 96.0016$

the five estimates obtained for the instantaneous velocity. We can see from the table that for this example, the larger the time interval Δt, the larger the average velocity $\Delta s/\Delta t$. The most accurate estimates for instantaneous velocity will correspond to very small time intervals Δt. For example, over the interval $\Delta t = 0.0001$ second, we would not expect the velocity of the object to change very much. Thus, the average velocity of 96.0016 feet per second during the very short time interval $\Delta t = 0.0001$ should be very close to the instantaneous velocity at $t = 3$.

But what is the exact or instantaneous velocity at $t = 3$? It must be close to 96.0016, but is it 96.0016? Or is it 96.0001? Or what?

To obtain the precise answer, we first use some algebra. Specifically, we find the average velocity for the object in the time interval that begins at 3 and ends at t, where $t \neq 3$ is close to 3:

$$f(3) = 16(9) = 144 \qquad f(t) = 16t^2$$

Thus,

$$\text{Average velocity} = \frac{\Delta s}{\Delta t} = \frac{f(t) - f(3)}{t - 3} = \frac{16t^2 - 144}{t - 3} = \frac{16(t^2 - 9)}{t - 3}$$

$$= \frac{16(t - 3)(t + 3)}{t - 3}$$

Since $t \neq 3$ we may cancel $(t - 3)$ in the numerator and denominator to get

$$\frac{\Delta s}{\Delta t} = 16(t + 3)$$

We are now at the important step in this procedure. As t gets closer and closer to 3, but not equal to 3, the values of $\Delta s/\Delta t = 16(t + 3)$ get closer to 96. This is apparent since we can make $\Delta s/\Delta t = 16(t + 3)$ as close as we please to 96 by taking t sufficiently close to 3.

Using the terminology we introduced in Chapter 2, 96 is called the *limit* of the average velocity $\Delta s/\Delta t$ as Δt approaches 0, or equivalently, as $t \to 3$. In symbols, we write

$$\lim_{\Delta t \to 0} \frac{\Delta s}{\Delta t} = \lim_{t \to 3} \frac{f(t) - f(3)}{t - 3} = \lim_{t \to 3} [16(t + 3)] = 96$$

Intuition tells us that the limit 96 feet per second is what we mean by the (instantaneous) velocity of the object at time $t = 3$ seconds.

(1) **Instantaneous Velocity** In general, if $s = f(t)$ is a function that describes the distance s a particle travels in time t, the *(instantaneous) velocity v* of the particle at time t_0 is defined as the limit of the average rate of change $\Delta s/\Delta t$ as Δt approaches 0. Specifically, the velocity v at time t_0 is

(2) $$v = \lim_{\Delta t \to 0} \frac{\Delta s}{\Delta t} = \lim_{t \to t_0} \frac{f(t) - f(t_0)}{t - t_0}$$

provided this limit exists.

The limit in (2) has an important generalization:

(3) Derivative of a Function Let $y = f(x)$ be a function and let c be in the domain of f. The *derivative of f at c*, denoted by $f'(c)$, and read "f prime of c," is the number

(4) $$f'(c) = \lim_{x \to c} \frac{f(x) - f(c)}{x - c}$$

provided this limit exists.

The calculation of the derivative of a function f is easily done by following the four steps outlined below:

(5) Four-Step Method

STEP 1: Find $f(c)$.

STEP 2: Subtract $f(c)$ from $f(x)$ to get Δy.

STEP 3: Divide Δy by Δx to get

$$\frac{\Delta y}{\Delta x} = \frac{f(x) - f(c)}{x - c}$$

STEP 4: Find the limit (if it exists) of the expression found in Step 3 as $\Delta x \to 0$ (or equivalently, as $x \to c$):

$$\lim_{x \to c} \frac{f(x) - f(c)}{x - c}$$

Example 1 Use the four-step method to find $f'(2)$ if $f(x) = x^2 + 2x$.

Solution STEP 1: $f(2) = 2^2 + 2(2) = 8$

STEP 2: $\Delta y = f(x) - f(2) = x^2 + 2x - 8$

STEP 3: $\dfrac{\Delta y}{\Delta x} = \dfrac{f(x) - f(2)}{x - 2} = \dfrac{x^2 + 2x - 8}{x - 2}$

STEP 4: $\lim\limits_{x \to 2} \dfrac{f(x) - f(2)}{x - 2} = \lim\limits_{x \to 2} \dfrac{x^2 + 2x - 8}{x - 2} \underset{\text{Factor}}{=} \lim\limits_{x \to 2} \dfrac{(x - 2)(x + 4)}{x - 2} = 6$

Hence, $f'(2) = 6$.

Example 2 Find the derivative of $f(x) = x^3$ at 2.

Solution We go directly to Step 3:

$$\frac{\Delta y}{\Delta x} = \frac{f(x) - f(2)}{x - 2} = \frac{x^3 - 8}{x - 2}$$

$$f'(2) = \lim_{x \to 2} \frac{f(x) - f(2)}{x - 2} = \lim_{x \to 2} \frac{x^3 - 8}{x - 2} = \lim_{x \to 2} \frac{(x - 2)(x^2 + 2x + 4)}{x - 2} = 12$$

2. THE DERIVATIVE

In Example 2 we calculated the derivative of $f(x) = x^3$ at 2. It is often just as easy to find the derivative at an arbitrary number c, as the next example illustrates.

Example 3 Find the derivative of $f(x) = x^3$ at c.

Solution We begin at Step 3 of the four-step method and simplify:

$$\frac{\Delta y}{\Delta x} = \frac{f(x) - f(c)}{x - c} = \frac{x^3 - c^3}{x - c} = \frac{(x-c)(x^2 + cx + c^2)}{x - c} = x^2 + cx + c^2$$

$$f'(c) = \lim_{x \to c} \frac{f(x) - f(c)}{x - c} = \lim_{x \to c}(x^2 + cx + c^2) = 3c^2$$

■

Notice that for $c = 2$, $f'(c) = 3c^2 = (3)(2^2) = 12$, which agrees with the answer found in Example 2.

Since the limit

$$f'(c) = \lim_{x \to c} \frac{f(x) - f(c)}{x - c} = 3c^2$$

exists for any choice of c, it is convenient to replace c by x, so that we can write $f'(x) = 3x^2$. In general, the derivative $f'(x)$ of a function f at x is itself a function, since it gives a rule for associating a number x with a number $f'(x)$. This function f' is called the *derived function of f* or the *derivative of f*. We also say that f is "differentiable". Thus, "differentiate f" means the same as "find the derivative of f".

When it is required to calculate the derivative of a function f, it is often advantageous to use the following alternate formula for the derivative instead of (4):

(6)
$$f'(x) = \lim_{\Delta x \to 0} \frac{f(x + \Delta x) - f(x)}{\Delta x}$$

We justify this alternate formula by setting $\Delta x = x - c$ in (4). Then

$$\Delta y = f(x) - f(c) = f(c + \Delta x) - f(c)$$

$$f'(c) = \lim_{\Delta x \to 0} \frac{\Delta y}{\Delta x} = \lim_{\Delta x \to 0} \frac{f(c + \Delta x) - f(c)}{\Delta x}$$

Replacing c by x gives the alternate formula

$$f'(x) = \lim_{\Delta x \to 0} \frac{f(x + \Delta x) - f(x)}{\Delta x}$$

Example 4 Use (6) to find the derivative of $f(x) = x^2 + 2x$.

Solution We compute

$$f'(x) = \lim_{\Delta x \to 0} \frac{\Delta y}{\Delta x} = \lim_{\Delta x \to 0} \frac{f(x + \Delta x) - f(x)}{\Delta x}$$

$$= \lim_{\Delta x \to 0} \frac{[(x + \Delta x)^2 + 2(x + \Delta x)] - (x^2 + 2x)}{\Delta x}$$

$$= \lim_{\Delta x \to 0} \frac{[x^2 + 2x\Delta x + (\Delta x)^2 + 2x + 2\Delta x] - x^2 - 2x}{\Delta x}$$

$$= \lim_{\Delta x \to 0} \frac{2x\Delta x + (\Delta x)^2 + 2\Delta x}{\Delta x}$$

Factoring Δx from the numerator and canceling it with the Δx in the denominator, we obtain

$$f'(x) = \lim_{\Delta x \to 0} \frac{\Delta x(2x + \Delta x + 2)}{\Delta x} = \lim_{\Delta x \to 0} (2x + \Delta x + 2) = 2x + 2$$

Thus, $f'(x) = 2x + 2$.

Exercise 2
Solutions to Odd-Numbered Problems begin on page 375.

In Problems 1–8 find the derivative of each function at the given number by using (4).

1. $f(x) = 2x + 3$, at 1
2. $f(x) = 3x - 5$, at 2
3. $f(x) = x^2 - 2$, at 0
4. $f(x) = 2x^2 + 4$, at 1
5. $f(x) = 3x^2 + x + 5$, at -1
6. $f(x) = 2x^2 - 4x - 7$, at -1
7. $f(x) = 1/x$, at 4
8. $f(x) = 1/x$, at 2

In Problems 9–20 find the derivative of each function by using (6).

9. $f(x) = 2x + 3$
10. $f(x) = 3x - 5$
11. $f(x) = x^2 - 2$
12. $f(x) = 2x^2 + 4$
13. $f(x) = 3x^2 + x + 5$
14. $f(x) = 2x^2 - x - 7$
15. $f(x) = 5$
16. $f(x) = -2$
17. $f(x) = 1/x$
18. $f(x) = 4/x$
19. $f(x) = mx + b$
20. $f(x) = ax^2 + bx + c$

21. The distance s (in meters) that a particle moves in time t (in seconds) is given by $s = f(t) = 3t^2 + 4t$. Find the velocity at $t = 0$. At $t = 2$. At any time t.

22. The distance s (in meters) that a particle moves in time t (in seconds) is $s = f(t) = t^2 - 4t$. Find the velocity at $t = 0$. At $t = 3$. At any time t.

*23. A ball is thrown upward. Let the height in feet of the ball be given by $s(t) = 100t - 16t^2$, where t is the time elapsed in seconds. What is the velocity when $t = 0$, $t = 1$, and $t = 4$? At what time does the ball strike the ground? At what time does the ball reach its highest point? (At this time the ball should be "stationary" so that its velocity is 0.)

*24. A rock is dropped from a height of 88.2 meters and falls toward earth in a straight line. In t seconds the rock falls $9.8t^2$ meters.
    (a) How long does it take for the rock to hit the ground?
    (b) What is the average velocity of the rock during the time it is falling?
    (c) What is the average velocity of the rock for the first 3 seconds?
    (d) What is the velocity of the rock when it hits the ground?

3. Two Additional Interpretations of the Derivative

Geometric Interpretation

Let's examine what the derivative of a function $y = f(x)$ means geometrically. Since the derivative is the limit of an average rate of change, we begin by recalling that the average rate of change $\Delta y/\Delta x$ equals the slope of the secant line joining two points on the graph of f. If we fix the initial point at $(c, f(c))$ and let the terminal point be at $(x, f(x))$, with $\Delta x \neq 0$, the slope of this secant line is

$$m_{\text{sec}} = \frac{\Delta y}{\Delta x} = \frac{f(x) - f(c)}{x - c}$$

See Figure 4.

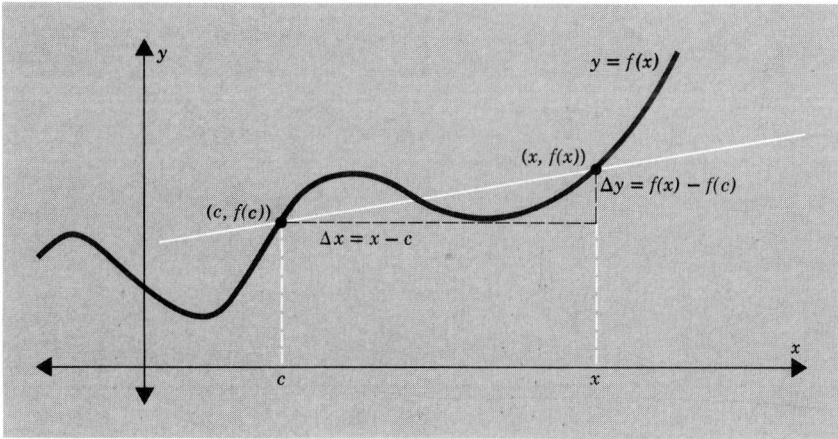

Figure 4

Now look at Figure 5, where we show the point $(x, f(x))$ approaching $(c, f(c))$ as x approaches c. The secant lines L_1, L_2, and L_3 shown in Figure 5 tend to a limiting position—*the tangent line to the graph of f at c.*

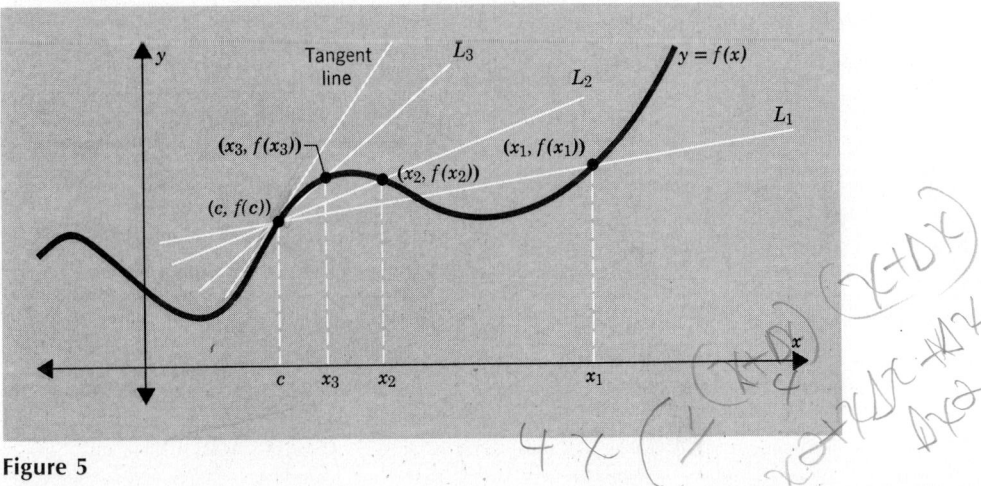

Figure 5

But what is the slope of this tangent line? We know it passes through the point $(c, f(c))$. We also know that the slope m_{sec} of each secant line joining the points $(c, f(c))$ and $(x, f(x))$ is

$$m_{sec} = \frac{f(x) - f(c)}{x - c}$$

As x gets closer to c the secant line approaches a limiting position—the tangent line. So we may expect the limiting value of the slopes of the secant lines to equal the slope m_{tan} of the tangent line. That is,

$$m_{tan} = \lim_{x \to c} \frac{f(x) - f(c)}{x - c}$$

This limit, if it exists, is the derivative of f at c.

(1) Slope of Tangent Line We define the *tangent line* to the graph of f at c to be the line passing through the point $(c, f(c))$ and having the slope

$$f'(c) = \lim_{x \to c} \frac{f(x) - f(c)}{x - c}$$

provided this limit exists.

Equation of Tangent Line

Using the point-slope form of the equation of a line, we find that an *equation of the tangent line* to the graph of f at $(c, f(c))$ is

(2)
$$y - f(c) = f'(c)(x - c)$$

Example 1 Find the slope of the tangent line to the graph of $f(x) = 2x^2$ at $(1, 2)$. What is its equation? Graph the function and show its tangent line.

Solution
$$f'(1) = \lim_{x \to 1} \frac{f(x) - f(1)}{x - 1} = \lim_{x \to 1} \frac{2x^2 - 2}{x - 1} = \lim_{x \to 1} \frac{2(x - 1)(x + 1)}{x - 1} = \lim_{x \to 1} [2(x + 1)] = 4$$

Using (2), we find an equation of this tangent line to be
$$y - 2 = 4(x - 1)$$
$$y = 4x - 2$$

The graphs of $y = 2x^2$ and its tangent line at $(1, 2)$ are shown in Figure 6.

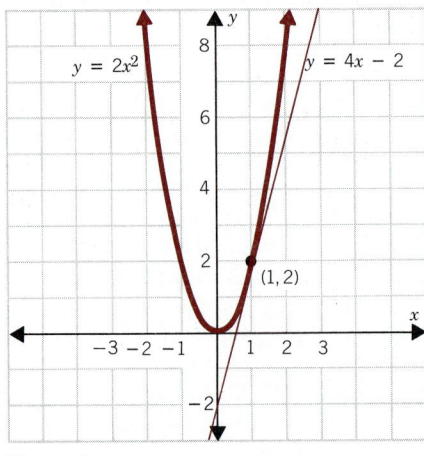

Figure 6

Rates of Change

Our third interpretation of the derivative involves rates of change. Let x and y denote two physical quantities that are related by the function $y = f(x)$. We have defined the ratio $\Delta y/\Delta x$ of the corresponding changes in these quantities as the *average rate of change of y with respect to x over the interval Δx*. The limit of the ratio $\Delta y/\Delta x$ as Δx approaches 0, if it exists, is called the *(instantaneous) rate of change of y with respect to x*. Thus,

(3) $$f'(x) = \lim_{\Delta x \to 0} \frac{\Delta y}{\Delta x} = \text{(Instantaneous) rate of change of } y \text{ with respect to } x$$

In other words, the expression "rate of change of a function" means the derivative of the function.

Example 2 Show that the rate of change of the area of a circle with respect to its radius is equal to its circumference.

Solution The area A of a circle of radius R is $A = \pi R^2$. Thus, the rate of change of area with respect to radius is

$$A'(R) = \lim_{\Delta R \to 0} \frac{\Delta A}{\Delta R} = \lim_{\Delta R \to 0} \frac{\pi(R + \Delta R)^2 - \pi R^2}{\Delta R}$$

$$= \lim_{\Delta R \to 0} \frac{\pi[R^2 + 2R(\Delta R) + (\Delta R)^2] - \pi R^2}{\Delta R}$$

$$= \lim_{\Delta R \to 0} \frac{2\pi R(\Delta R) + \pi(\Delta R)^2}{\Delta R}$$

$$= \lim_{\Delta R \to 0} \frac{\pi \Delta R[2R + \Delta R]}{\Delta R}$$

$$= \lim_{\Delta R \to 0} \pi[2R + \Delta R] = 2\pi R = \text{Circumference}$$

Example 3 In a metabolic experiment, the mass M of glucose decreases according to the formula $M(t) = 4.5 - 0.03t^2$, where M is measured in grams and t is time (in hours). Find the reaction rate at 1 hour.

Solution The reaction rate at $t = 1$ is $M'(1)$. Thus,

$$M'(1) = \lim_{t \to 1} \frac{M(t) - M(1)}{t - 1} = \lim_{t \to 1} \frac{(4.5 - 0.03t^2) - (4.5 - 0.03)}{t - 1}$$

$$= \lim_{t \to 1} \frac{(-0.03)(t^2 - 1)}{t - 1} = \lim_{t \to 1} \frac{(-0.03)(t + 1)(t - 1)}{t - 1}$$

$$= (-0.03)(2) = -0.06$$

The reaction rate at $t = 1$ is -0.06; that is, the mass M at $t = 1$ is decreasing at the rate of 0.06 gram per hour.

Exercise 3

Solutions to Odd-Numbered Problems begin on page 376.

In Problems 1-6 find an equation for the tangent line to the graph of each function at the indicated point. Graph each function and show this tangent line.

1. $f(x) = x^2$, at $(3, 9)$
2. $f(x) = x^2$, at $(-1, 1)$
3. $f(x) = x^2 + 2x + 1$, at $(1, 4)$
4. $f(x) = x^3 + 1$, at $(1, 2)$
5. $f(x) = 1/x$, at $(1, 1)$
6. $f(x) = \sqrt{x}$, at $(4, 2)$

7. Does the tangent line to the graph of $y = x^2$ at $(1, 1)$ pass through the point $(2, 5)$?
8. Does the tangent line to the graph of $y = x^3$ at $(1, 1)$ pass through the point $(2, 5)$?
9. A dive bomber is flying from right to left along the graph of $y = x^2$. When a rocket bomb is released, it follows a path that is approximately along the tangent line. Where should the pilot release the bomb if the target is at $(1, 0)$?
10. Answer the question in Problem 9 if the plane is flying from right to left along the graph of $y = x^3$.
11. A circle of radius R has area $A = \pi R^2$ and circumference $C = 2\pi R$. If the radius changes from R to $(R + \Delta R)$, find the:
    (a) Change in area
    (b) Change in circumference
    (c) Average rate of change of area with respect to radius
    (d) Average rate of change of circumference with respect to radius
    (e) Rate of change of circumference with respect to radius
12. *Respiration Rate.* A human being's respiration rate R (in breaths per minute) is given by $R = -10.35 + 0.59p$, where p is the partial pressure of carbon dioxide in the lungs. Find the rate of change in respiration when $p = 50$.
13. *Physiology.* Physiologists have discovered that it is possible to approximate the surface area of an animal once its mass is known. The relationship between surface area A (in square meters) and mass m (in kilograms) is $A = km^{2/3}$, where the constant k depends on the type of animal. Some values of k are listed below.

Bird	.1	Man	.11
Cat	.1	Monkey	.118
Cow	.09	Moose	.09
Small Dog	.112	Rabbit	.0975
Large Dog	.101	Rat	.091
Guinea Pig	.09	Sheep (sheared)	.084
Horse	.1	Swine	.09

(a) Find the surface area of a person whose mass is 64 kilograms.
(b) Find the rate of change of surface area with respect to mass for a small dog.

Hint: The derivative of A is $2/3 \, km^{-1/3}$.

4. Some Derivative Formulas

The technique for computing the derivative of a function, as introduced in Section 3, is a rather long, tedious process. It requires calculating the difference quotient $\Delta y/\Delta x$ of a function $y = f(x)$, and then finding the limit of $\Delta y/\Delta x$ as $\Delta x \to 0$. Fortunately, for most of the functions we will study, it is possible to use formulas for finding the derivatives. In this section we give a formula for the derivative of the function $f(x) = x^r$ (r any real number). In the next section, some general formulas for finding derivatives are discussed.

We begin by considering the constant function $f(x) = A$, where A is a real number. The derivative is obtained by following the four steps listed on page 102.

The difference quotient of $f(x) = A$ is

$$\frac{\Delta y}{\Delta x} = \frac{f(x) - f(c)}{x - c} = \frac{A - A}{x - c} = \frac{0}{x - c} = 0$$

$$\lim_{\Delta x \to 0} \frac{\Delta y}{\Delta x} = \lim_{\Delta x \to 0} 0 = 0$$

(1) **Derivative of a Constant Function** For the constant function $f(x) = A$, the derivative is $f'(x) = 0$. In other words, the derivative of a constant is zero.

Prime Notation Besides the *prime notation* f' we have used so far, there are several other ways to denote the derivative of a function $y = f(x)$. The most common ones are

$$y' \quad \text{and} \quad \frac{dy}{dx}$$

Leibniz Notation The notation dy/dx, often referred to as the *Leibniz notation*, may be interpreted as

$$\frac{dy}{dx} = \frac{d}{dx}(y) = \frac{d}{dx}f(x)$$

where d/dx is an instruction to compute the derivative (with respect to the independent variable x) of the function $y = f(x)$.

In terms of the Leibniz notation, the result (1) may also be stated as

(2)
$$\frac{d}{dx}A = 0$$

where A is a constant. For example,

$$\frac{d}{dx}6 = 0$$

In subsequent work with derivatives we shall use the prime notation or the Leibniz notation, or sometimes a mixture of the two, depending on which is more convenient.

We now investigate the derivative of the function $f(x) = x^n$ at c for some positive integers n to see if a pattern appears.

For $n = 1$,
$$\lim_{x \to c} \frac{f(x) - f(c)}{x - c} = \lim_{x \to c} \frac{x - c}{x - c} = \lim_{x \to c} 1 = 1$$

For $n = 2$,
$$\lim_{x \to c} \frac{f(x) - f(c)}{x - c} = \lim_{x \to c} \frac{x^2 - c^2}{x - c} = \lim_{x \to c} (x + c) = 2c$$

For $n = 3$,
$$\lim_{x \to c} \frac{f(x) - f(c)}{x - c} = \lim_{x \to c} \frac{x^3 - c^3}{x - c} = \lim_{x \to c}(x^2 + cx + c^2) = 3c^2$$

If we replace c by x in these results, we have

$$\frac{d}{dx} x = 1 \qquad \frac{d}{dx} x^2 = 2x \qquad \frac{d}{dx} x^3 = 3x^2$$

From this pattern we conjecture the following:

(3)   **Derivative of x^n**   **For the function $f(x) = x^n$, n a positive integer, the derivative is $f'(x) = nx^{n-1}$. That is,**

(4)
$$\frac{d}{dx} x^n = nx^{n-1}$$

We can prove this result without too much trouble, as follows:

$$\lim_{x \to c} \frac{f(x) - f(c)}{x - c} = \lim_{x \to c} \frac{x^n - c^n}{x - c} = \lim_{x \to c} [x^{n-1} + cx^{n-2} + \cdots + c^{n-1}]$$
$$= \underbrace{c^{n-1} + c^{n-1} + \cdots + c^{n-1}}_{n \text{ terms}} = nc^{n-1}$$

In words, formula (4) states that:

The derivative of x raised to the power n is n times x raised to the power $n - 1$.

Example 1   (a) $\dfrac{d}{dx} x^6 = 6x^5$ (b) $\dfrac{d}{dx} x^{13} = 13x^{12}$

(c) $\dfrac{d}{dx} x^{65} = 65x^{64}$ (d) $\dfrac{d}{dx} x = 1$ ∎

So far, formula (4) has only been shown to be valid when n is a positive integer. However, it turns out that this formula is true if the exponent is any real number. That is,

(5)
$$\frac{d}{dx} x^r = rx^{r-1} \qquad \text{for any real number } r$$

Example 2 (a) $\frac{d}{dx}x^{4/3} = \frac{4}{3}x^{1/3}$ (b) $\frac{d}{dx}x^{\sqrt{2}} = \sqrt{2}\,x^{\sqrt{2}-1}$

(c) $\frac{d}{dx}\sqrt{x} = \frac{d}{dx}x^{1/2} = \frac{1}{2}x^{-1/2} = \frac{1}{2\sqrt{x}}$ (d) $\frac{d}{dx}\frac{1}{x} = \frac{d}{dx}x^{-1} = -1x^{-2} = \frac{-1}{x^2}$

■

Example 3 Find $f'(4)$ if $f(x) = x^{3/2}$.

Solution
$$f'(x) = \tfrac{3}{2}x^{(3/2)-1} = \tfrac{3}{2}x^{1/2} = \tfrac{3}{2}\sqrt{x}$$
$$f'(4) = \tfrac{3}{2}\sqrt{4} = \tfrac{3}{2}(2) = 3$$

■

Formula (5) allows us to compute some derivatives with ease. However, do not forget that a derivative is, in actuality, the limit of a difference quotient.

Exercise 4

Solutions to Odd-Numbered Problems begin on page 377.

In Problems 1–16 find the derivative of each function.
1. $f(x) = 2$ 2. $f(x) = -5$ 3. $f(x) = x^4$ 4. $f(x) = x^2$
5. $f(x) = x^5$ 6. $f(x) = x^8$ 7. $f(x) = x^{-2}$ 8. $f(x) = x^{-1}$
9. $f(x) = x^{2/3}$ 10. $f(x) = x^{1/3}$ 11. $f(x) = 1/x^3$ 12. $f(x) = 1/x^2$
13. $f(x) = x^{1.3}$ 14. $f(x) = x^{2.1}$ 15. $f(x) = x^{\sqrt{5}}$ 16. $f(x) = x^{\sqrt{3}}$

In Problems 17–22 find the slope of the tangent line at the point indicated.

17. $f(x) = x^4$, at $(1, 1)$ 18. $f(x) = x^4$, at $(2, 16)$
19. $f(x) = \sqrt{x}$, at $(4, 2)$ 20. $f(x) = 1/x^2$, at $(3, \tfrac{1}{9})$
21. $f(x) = 1/\sqrt[3]{x}$, at $(-8, -\tfrac{1}{2})$ 22. $f(x) = \sqrt[3]{x}$, at $(-8, -2)$

In Problems 23–28, find the indicated derivative

23. $\frac{d}{dx}(x^5)$ 24. $\frac{d}{dx}(x^{-5})$ 25. $\frac{d}{dx}(x^{-1/3})$

26. $\frac{d}{dx}(x^{1/3})$ 27. $\frac{d}{dx}(\sqrt[3]{x})$ 28. $\frac{d}{dx}(\sqrt[3]{x^2})$

29. Use $f(x) = 1/x$ and the Four-Step Method (page 102) to show that $f'(2) = -\tfrac{1}{4}$.
30. Use $f(x) = \sqrt{x}$ and the Four-Step Method (page 102) to show that $f'(4) = \tfrac{1}{4}$.
31. A young child travels s feet down a slide in t seconds, where $s = t^{3/2}$. What is the child's instantaneous velocity after 1 second? If the slide is 8 feet long, with what velocity does the child strike the ground?

5. General Formulas for Differentiation

In Section 4 we stated a formula for finding the derivative of x^r (for any real r). However, many of the functions actually encountered in business consist not of the function $f(x) = x^r$, but of various combinations of it. In this section we will learn how to find the derivatives of functions that are sums, products, and quotients of the function $f(x) = x^r$.

For the purpose of this section, we assume that all functions mentioned actually have derivatives.

Our first result tells how to find the derivative of a constant times a function.

(1) Derivative of a Constant Times a Function The derivative of a constant times a function equals the constant times the derivative of the function. That is, if C is a constant and f is a differentiable function, then

$$\text{(2)} \qquad \frac{d}{dx}[Cf(x)] = C\frac{d}{dx}f(x)$$

Example 1 (a) $\frac{d}{dx}(10x^3) = 10\frac{d}{dx}x^3 = 10(3x^2) = 30x^2$

(b) $\frac{d}{dx}(-\frac{1}{3}x^6) = -\frac{1}{3}\frac{d}{dx}x^6 = -\frac{1}{3}(6x^5) = -2x^5$ ∎

Often, a complicated-looking function is really just the sum of two simple functions. The next result may be used to find the derivatives of such functions.

(3) Derivative of a Sum The derivative of the sum of two differentiable functions equals the sum of their derivatives. That is,

$$\text{(4)} \qquad \frac{d}{dx}[f(x) + g(x)] = \frac{d}{dx}f(x) + \frac{d}{dx}g(x)$$

We verify (4) as follows. To compute

$$\frac{d}{dx}[f(x) + g(x)]$$

we need to find the limit of the difference quotient of $f(x) + g(x)$. We use the alternate form of the derivative. [see equation (6) on page 103]:

$$\frac{d}{dx}[f(x) + g(x)] = \lim_{\Delta x \to 0} \frac{[f(x + \Delta x) + g(x + \Delta x)] - [f(x) + g(x)]}{\Delta x}$$

$$= \lim_{\Delta x \to 0} \frac{[f(x + \Delta x) - f(x)] + [g(x + \Delta x) - g(x)]}{\Delta x}$$

$$= \lim_{\Delta x \to 0}\left[\frac{f(x + \Delta x) - f(x)}{\Delta x} + \frac{g(x + \Delta x) - g(x)}{\Delta x}\right]$$

5. GENERAL FORMULAS FOR DIFFERENTIATION 113

$$= \lim_{\Delta x \to 0}\left[\frac{f(x + \Delta x) - f(x)}{\Delta x}\right] + \lim_{\Delta x \to 0}\left[\frac{g(x + \Delta x) - g(x)}{\Delta x}\right]$$

$$= \frac{d}{dx}f(x) + \frac{d}{dx}g(x)$$

Example 2 Find the derivative of $f(x) = x^2 + \sqrt{x}$.

Solution The function f is the sum of the two functions x^2 and $\sqrt{x}$. Thus,

$$\frac{d}{dx}f(x) = \frac{d}{dx}(x^2 + \sqrt{x}) = \frac{d}{dx}x^2 + \frac{d}{dx}\sqrt{x}$$

$$= 2x + \frac{1}{2\sqrt{x}}$$

■

The result above extends to sums and differences of more than two functions.

(5) Derivative of a Difference The derivative of the difference of two differentiable functions equals the difference of their derivatives. That is,

(6)
$$\frac{d}{dx}[f(x) - g(x)] = \frac{d}{dx}f(x) - \frac{d}{dx}g(x)$$

The proof of (6) is similar to that of (4) and is left as an exercise. The results (4) and (6) may be combined. Let's look at an example.

Example 3 Find the derivative of $f(x) = 6x^4 - 3x^2 + 10x - 8$.

Solution
$$f'(x) = \frac{d}{dx}(6x^4 - 3x^2 + 10x - 8)$$

$$= \frac{d}{dx}(6x^4) - \frac{d}{dx}(3x^2) + \frac{d}{dx}(10x) - \frac{d}{dx}8$$

$$= 24x^3 - 6x + 10$$

■

The above examples demonstrate how to find the derivative of a function that is the sum or difference of two or more functions whose derivatives are known. The next result indicates how to find the derivative of a function that is the product of two functions whose derivatives are known.

(7) Derivative of a Product The derivative of the product of two differentiable functions equals the first function times the derivative of the second plus the second function times the derivative of the first. That is,

(8) $$\frac{d}{dx}[f(x)g(x)] = f(x)\frac{d}{dx}g(x) + g(x)\frac{d}{dx}f(x)$$

Observe that, unlike the situation with limits, the derivative of a product does not equal the product of the derivatives.

Example 4 Find the derivative of $F(x) = (x^2 + 2x - 5)(x^3 - 1)$.

Solution The function F is the product of the two polynomial functions $f(x) = x^2 + 2x - 5$ and $g(x) = x^3 - 1$ so that by (8), we have

$$F'(x) = (x^2 + 2x - 5)\left[\frac{d}{dx}(x^3 - 1)\right] + (x^3 - 1)\left[\frac{d}{dx}(x^2 + 2x - 5)\right]$$
$$= (x^2 + 2x - 5)(3x^2) + (x^3 - 1)(2x + 2)$$
$$= 5x^4 + 8x^3 - 15x^2 - 2x - 2$$

Now that you know the rule for the derivative of a product, be careful not to use it unnecessarily. When one of the factors is a constant, you should use (2). For example, it is easier to work

$$\frac{d}{dx}[5(x^2 + 1)] = 5\frac{d}{dx}(x^2 + 1) = (5)(2x) = 10x$$

than it is to work

$$\frac{d}{dx}[5(x^2 + 1)] = 5\left[\frac{d}{dx}(x^2 + 1)\right] + (x^2 + 1)\left[\frac{d}{dx}5\right] = (5)(2x) + (x^2 + 1)(0) = 10x$$

The next formula provides a means for finding the derivative of the quotient of two functions.

(9) **Derivative of a Quotient** If f and $g \neq 0$ are two differentiable functions, then

(10) $$\frac{d}{dx}\left[\frac{f(x)}{g(x)}\right] = \frac{g(x)\frac{d}{dx}f(x) - f(x)\frac{d}{dx}g(x)}{[g(x)]^2}$$

Example 5 Find the derivative of: $F(x) = \dfrac{x^2 + 1}{x - 3}$

Solution Here, the function F is the quotient of $f(x) = x^2 + 1$ and $g(x) = x - 3$. Thus, we use (10) to get

$$\frac{d}{dx}\left(\frac{x^2 + 1}{x - 3}\right) = \frac{(x - 3)\frac{d}{dx}(x^2 + 1) - (x^2 + 1)\frac{d}{dx}(x - 3)}{(x - 3)^2}$$

5. GENERAL FORMULAS FOR DIFFERENTIATION

$$= \frac{(x-3)(2x) - (x^2+1)(1)}{(x-3)^2}$$

$$= \frac{2x^2 - 6x - x^2 - 1}{(x-3)^2} = \frac{x^2 - 6x - 1}{(x-3)^2}$$

A change in the symbol used for the independent variable does not affect the formula. For example,

$$\frac{d}{dt} t^{-2} = -2t^{-3} = -\frac{2}{t^3} \qquad \frac{d}{ds} \frac{3}{s^4} = 3(-4)s^{-5} = -\frac{12}{s^5}$$

In fact, each of the derivative formulas given thus far can be written without reference to the independent variable of the function. If f and g are differentiable functions, we have the following formulas:

Derivative of a Sum	$(f + g)' = f' + g'$
Derivative of a Difference	$(f - g)' = f' - g'$
Derivative of a Product	$(f \cdot g)' = f \cdot g' + g \cdot f'$
Derivative of a Quotient	$\left(\dfrac{f}{g}\right)' = \dfrac{gf' - fg'}{g^2}$

Exercise 5
Solutions to Odd-Numbered Problems begin on page 378.

In Problems 1–38 find the derivative of the function f by using the formulas of this section.

1. $f(x) = 3x^{15}$
2. $f(x) = \frac{1}{3}x^{12}$
3. $f(x) = 3x + 2$
4. $f(x) = 5x - \frac{1}{2}$
5. $f(x) = x^2 + 3x - 4$
6. $f(x) = 4x^4 + 2x^2 - 2$
7. $f(x) = 8x^5 - 5x + 1$
8. $f(x) = 9x^3 - 2x^2 + 4x + 4$
9. $f(x) = \frac{1}{3}x^4 - 3x + \frac{3}{2}$
10. $f(x) = -3x^4 - 2x^3$
11. $f(x) = \pi x^3 + \frac{3}{2}x^2$
12. $f(x) = 4 - \pi x^2$
13. $f(x) = \frac{1}{3}(x^5 - 8)$
14. $f(x) = \dfrac{x^3 + 2}{5}$
15. $f(x) = ax^2 + bx + c$
16. $f(x) = ax^3 + bx^2 + cx + d$
17. $f(x) = (3x^2 - 5)(2x + 1)$
18. $f(x) = (3x - 2)(4x + 5)$
19. $f(t) = (2t^5 - t)(t^3 + 1)$
20. $f(u) = (u^4 - 3u^2 + 1)(u^2 - u + 2)$
21. $f(t) = t^{-3}$
22. $f(u) = u^{-4}$
23. $f(x) = \dfrac{10}{x^4} + \dfrac{3}{x^2}$
24. $f(x) = \dfrac{2}{x^5} - \dfrac{3}{x^3}$
25. $f(s) = \dfrac{2s}{s+1}$
26. $f(z) = \dfrac{z+1}{2z}$
27. $f(x) = \dfrac{4x^2 - 2}{3x + 4}$
28. $f(x) = \dfrac{-3x^3 - 1}{2x^2 + 1}$

29. $f(t) = 3t + \dfrac{1}{3t}$

30. $f(u) = 4u - \dfrac{1}{4u}$

31. $f(u) = \dfrac{1 - 2u}{1 + 2u}$

32. $f(w) = \dfrac{1 - w^2}{1 + w^2}$

33. $f(x) = 3x^3 - \dfrac{1}{3x^2}$

34. $f(x) = x^5 - \dfrac{5}{x^5}$

35. $f(t) = \dfrac{1}{t} - \dfrac{1}{t^2} + \dfrac{1}{t^3}$

36. $f(v) = \left(\dfrac{1-v}{v}\right)(1-v^2)$

37. $f(w) = \dfrac{1}{w^3 - 1}$

38. $f(v) = \dfrac{1}{v^2 + 5}$

In Problems 39–42 find the slope of the tangent line to the graph of the function f at the indicated point. What is an equation of the tangent line?

39. $f(x) = x^3 + 3x - 1$, at $(0, -1)$
40. $f(x) = x^4 + 2x - 1$, at $(1, 2)$
41. $f(x) = \dfrac{x^3}{x + 1}$, at $(1, \tfrac{1}{2})$
42. $f(x) = \dfrac{x^2}{x - 1}$, at $(-1, -\tfrac{1}{2})$

In Problems 43–48 find those x, if any, at which the graph of the function f has a horizontal tangent line—that is, at which $f'(x) = 0$.

43. $f(x) = 3x^2 - 12x + 4$
44. $f(x) = x^2 + 4x - 3$
45. $f(x) = x^3 - 3x + 2$
46. $f(x) = x^4 - 4x^3$
47. $f(x) = \dfrac{x^2}{x + 1}$
48. $f(x) = \dfrac{x^2 + 1}{x}$

49. If $y = x^2(3x - 2)$, find y' by:
   (a) Using the derivative of a product formula
   (b) Multiplying the two factors first and then differentiating
   (c) Compare the answers from parts (a) and (b).

50. If $y = (x^2 + 2)(x - 1)$, find y' by:
   (a) Using the derivative of a product formula
   (b) Multiplying the two factors first and then differentiating
   (c) Compare the answers from parts (a) and (b).

In Problems 51–56 find the indicated derivative.

51. $\dfrac{d}{dx}(\sqrt{3}x + \tfrac{1}{2})$

52. $\dfrac{d}{dx}\left(\dfrac{2x^4 - 5}{8}\right)$

53. $\dfrac{dA}{dR}$ if $A = \pi R^2$

54. $\dfrac{dC}{dR}$ if $C = 2\pi R$

55. $\dfrac{dV}{dR}$ if $V = \tfrac{4}{3}\pi R^3$

56. $\dfrac{dP}{dT}$ if $P = 0.2T$

57. In t seconds, the position of an object is a distance of s meters from the origin, where $s = t^3 - t + 1$. Find the velocity at $t = 0$. At $t = 5$.

58. In t seconds, the position of an object is a distance of s meters from the origin, where $s = t^4 - t^3 + t$. Find the velocity at $t = 0$. At $t = 1$.

5. GENERAL FORMULAS FOR DIFFERENTIATION   117

59. *Satisfaction and Reward.* The relationship† between satisfaction S and *total reward r* has been found to be

$$S(r) = \frac{ar}{g - r}$$

where $g \geq 0$ is the predetermined goal level and $a > 0$ is the perceived justice per unit of reward. Show that the instantaneous rate of change of satisfaction with respect to reward is inversely proportional to the square of the difference between the personal goal of the individual and the amount of reward received.

60. *Work Output.* The relationship‡ between the amount $A(t)$ of work output and the elapsed time t, $t \geq 0$, was found through empirical means to be

$$A(t) = a_3 t^3 + a_2 t^2 + a_1 t + a_0$$

where a_0, a_1, a_2, a_3 are constants. Find the instantaneous rate of change of work output at time t.

61. A large container is being filled with water. After t hours there are $8t - 4t^{1/2}$ liters of water in the container. At what rate is the water filling the container (in liters per hour) when $t = 4$?

62. *Intensity of Illumination.* The intensity of illumination I on a surface is inversely proportional to the square of the distance r from the surface to the source of light. If the intensity is 1000 units when the distance is 1 meter, find the rate of change of the intensity with respect to the distance when the distance is 10 meters.

*63. Prove that if f and g are differentiable functions and if $F(x) = f(x) - g(x)$, then $F'(x) = f'(x) - g'(x)$.

*64. Prove that if $g \neq 0$ is a differentiable function, then

$$\frac{d}{dx}\left[\frac{1}{g(x)}\right] = \frac{-g'(x)}{[g(x)]^2}$$

*65. Prove that if $f, g,$ and h are differentiable functions, then

$$\frac{d}{dx}[f(x)g(x)h(x)] = f(x)g(x)h'(x) + f(x)h(x)g'(x) + h(x)g(x)f'(x)$$

From this, deduce that

$$\frac{d}{dx}[f(x)]^3 = 3[f(x)]^2 f'(x)$$

Hint: Use (8) twice.

In Problems 66–68 use the result of Problem 65 to find dy/dx.

66. $y = (x^2 + 1)(x - 1)(x + 5)$
67. $y = (x - 1)(x^2 + 5)(x^3 - 1)$
68. $y = (x^4 + 1)^3$

†R. Carzo and J. N. Yanouzas, *Formal Organization: A Systems Approach*, Richard D. Irwin Press, Homewood, Ill., 1967.
‡M. R. Neifeld and A. T. Poffenberger, "A Mathematical Analysis of Curves," *Journal of General Psychology*, **1** (1928), pp. 448–456.

6. The Power Rule; the Chain Rule

In Section 5, the rules for differentiating functions that could be represented as the sum, product, or quotient of the function x^r were given. At this stage you may be tempted to think that you can find the derivative of any function made up of such functions. But this is not necessarily the case; for example, look at the function $f(x) = \sqrt{2x + 3}$. We need a further rule or technique to handle this type of function.

We begin with an example to illustrate the procedure.

Example 1 Consider the function: $y = (x^2 + 1)^3$

Solution Here, y is a function of x that can be written as a combination of two simpler functions. If we set

$$y = [g(x)]^3 \quad \text{and} \quad g(x) = x^2 + 1$$

then the substitution of $g(x) = x^2 + 1$ in the formula for y gives

$$y = (x^2 + 1)^3$$

■

Other examples are

$y = (x^2 + 1)^{1/2}$	if	$y = [g(x)]^{1/2}$	and	$g(x) = x^2 + 1$
$y = (x^2 + 1)^{1/3}$	if	$y = [g(x)]^{1/3}$	and	$g(x) = x^2 + 1$
$y = (x^3 - x^2 + 1)^{-3/4}$	if	$y = [g(x)]^{-3/4}$	and	$g(x) = x^3 - x^2 + 1$

To get some idea of how to find the derivative of functions of the form $[g(x)]^n$, with n an integer, let's see what happens when $n = 2$, $n = 3$, and $n = 4$.

If $n = 2$,

$$\frac{d}{dx}[g(x)]^2 = \frac{d}{dx}[g(x)g(x)] = g'(x)g(x) + g(x)g'(x) = 2g(x)g'(x)$$

If $n = 3$,

$$\frac{d}{dx}[g(x)]^3 = \frac{d}{dx}\{[g(x)]^2 g(x)\} = [g(x)]^2 g'(x) + g(x)\left\{\frac{d}{dx}[g(x)]^2\right\}$$

$$= [g(x)]^2 g'(x) + g(x)[2g(x)g'(x)] = 3[g(x)]^2 g'(x)$$

If $n = 4$,

$$\frac{d}{dx}[g(x)]^4 = \frac{d}{dx}\{[g(x)]^3 g(x)\} = [g(x)]^3 g'(x) + g(x)\left\{\frac{d}{dx}[g(x)]^3\right\}$$

$$= [g(x)]^3 g'(x) + g(x)\{3[g(x)]^2 g'(x)\} = 4[g(x)]^3 g'(x)$$

Let's summarize what we've found:

$$\frac{d}{dx}[g(x)]^2 = 2g(x)g'(x) \qquad \frac{d}{dx}[g(x)]^3 = 3[g(x)]^2 g'(x) \qquad \frac{d}{dx}[g(x)]^4 = 4[g(x)]^3 g'(x)$$

6. THE POWER RULE; THE CHAIN RULE

Based on these results, we conjecture the following formula, which is called the *power rule*:

(1) **Power Rule** **If g is a differentiable function and r is any real number, then**

(2) $$\frac{d}{dx}[g(x)]^r = r[g(x)]^{r-1}g'(x)$$

Notice the similarity between the power rule and the formula

$$\frac{d}{dx}x^r = rx^{r-1}$$

The main difference between these formulas is the factor $g'(x)$.

Example 2 Find the derivative of the function: $f(x) = (x^2 + 1)^3$

Solution We could, of course, expand the right-hand side and proceed according to techniques discussed earlier. However, the significance of the power rule is that it enables us to find derivatives of functions like this without resorting to tedious (and sometimes impossible) computation.

The function $f(x) = (x^2 + 1)^3$ is the function $g(x) = x^2 + 1$ raised to the power 3. By the power rule,

$$\frac{d}{dx}f(x) = \frac{d}{dx}(x^2 + 1)^3 = 3(x^2 + 1)^2 \frac{d}{dx}(x^2 + 1)$$

$$= 3(x^2 + 1)^2(2x) = 6x(x^2 + 1)^2 \qquad \blacksquare$$

Example 3 Additional examples of the power rule are:

(a) $f(x) = (3 - x^3)^{-5}$
$f'(x) = (-5)(3 - x^3)^{-6}(-3x^2) = 15x^2(3 - x^3)^{-6}$

(b) $f(x) = (2x + 3)^{3/2}$
$f'(x) = (\frac{3}{2})(2x + 3)^{1/2}(2) = 3(2x + 3)^{1/2}$

(c) $f(x) = \sqrt[3]{(x^3 - 3x^2 + 1)}$
$f'(x) = (\frac{1}{3})(x^3 - 3x^2 + 1)^{-2/3}(3x^2 - 6x) = (x^2 - 2x)(x^3 - 3x^2 + 1)^{-2/3}$ $\blacksquare$

Often, we must use at least one other rule along with the power rule to differentiate a function. Here are two examples.

Example 4 Find the derivative of the function: $f(x) = x(x^2 + 1)^3$

Solution The function f is the product of x and $(x^2 + 1)^3$. We begin by using the rule for differentiating a product. That is,

$$f'(x) = x\frac{d}{dx}(x^2 + 1)^3 + (x^2 + 1)^3 \frac{d}{dx}x$$

120 CH. 3 THE DERIVATIVE

We continue by using the power rule:
$$f'(x) = (x)(3)(x^2 + 1)^2(2x) + (x^2 + 1)^3$$
$$= (x^2 + 1)^2[6x^2 + (x^2 + 1)] = (x^2 + 1)^2(7x^2 + 1)$$

Example 5 Find the derivative of the function: $f(x) = \left(\dfrac{3x + 2}{4x^2 - 5}\right)^5$

Solution Here, f is the quotient $(3x + 2)/(4x^2 - 5)$ raised to the power 5. Thus, we begin by using the power rule and then use the rule for differentiating a quotient:

$$f'(x) \underset{\uparrow}{=} (5)\left(\dfrac{3x + 2}{4x^2 - 5}\right)^4 \left[\dfrac{d}{dx}\left(\dfrac{3x + 2}{4x^2 - 5}\right)\right]$$
Apply power rule

$$\underset{\uparrow}{=} (5)\left(\dfrac{3x + 2}{4x^2 - 5}\right)^4 \left[\dfrac{(4x^2 - 5)(3) - (3x + 2)(8x)}{(4x^2 - 5)^2}\right]$$
Apply quotient rule

$$= \dfrac{5(3x + 2)^4(-12x^2 - 16x - 15)}{(4x^2 - 5)^6}$$

Example 6
Distance Between Moving Objects

A destroyer is traveling due north on a straight course at a constant velocity of 15 kilometers per hour. At 2 PM the destroyer is 50 kilometers due south of a tanker that is moving west at 5 kilometers per hour. At what velocity are the ships approaching each other 2 hours later? At 6 PM, are the ships approaching each other or receding from each other? When are the ships closest?

Solution First we construct Figure 7, which shows the relative position of the destroyer and the steamer at an arbitrary time t after 2 PM.

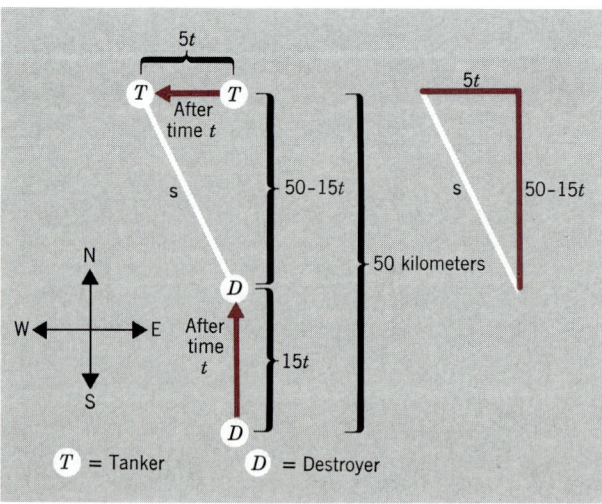

Figure 7

From the Pythagorean theorem, the distance s separating the two ships at time t is
$$s(t) = \sqrt{(5t)^2 + (50 - 15t)^2}$$
$$= \sqrt{250t^2 - 1500t + 2500}$$

Note that when $t = 0$, the ships are 50 kilometers apart as stipulated. The velocity at which the ships are approaching 2 hours later is given by $s'(2)$. Using the power rule, we have

$$s'(t) = \frac{1}{2}[250t^2 - 1500t + 2500]^{-1/2}(500t - 1500)$$
$$= \frac{250(t-3)}{\sqrt{250t^2 - 1500t + 2500}} = \frac{50(t-3)}{\sqrt{10t^2 - 60t + 100}}$$

When $t = 2$,
$$s'(2) = \frac{-50}{\sqrt{40 - 120 + 100}} = \frac{-50}{2\sqrt{5}} = -5\sqrt{5}$$

The fact that $s'(2) < 0$ means the ships are approaching at the rate of $5\sqrt{5}$ kilometers per hour.

At 6 PM (when $t = 4$), we have
$$s'(4) = \frac{50}{\sqrt{160 - 240 + 100}} = \frac{50}{2\sqrt{5}} = 5\sqrt{5}$$

Thus, the ships are moving apart at the rate of $5\sqrt{5}$ kilometers per hour at 6 PM.

Since for $t < 3$, $s'(t) < 0$ (the ships are approaching) and for $t > 3$, $s'(t) > 0$ (the ships are receding), it follows that at $t = 3$ the ships are closest.

The Chain Rule*

The Power Rule, which was just presented, is a special case of a more general, and more powerful rule, called the *Chain Rule*. This rule enables us to find the derivative of a function that is the combination of two simpler functions. For example, we can write each of the following functions as combinations of simpler functions:

$$y = (5x + 1)^3 \qquad y = u^3, \quad u = 5x + 1$$
$$y = \sqrt{x^2 + 1} \qquad y = \sqrt{u}, \quad u = x^2 + 1$$
$$y = \sqrt[3]{2x + 5} \qquad y = \sqrt[3]{u}, \quad u = 2x + 5$$

In the above examples, after substitution, y becomes a function of x. To find the derivative of this function y of x, use the Chain Rule.

Chain Rule Suppose y is a differentiable function of u and u is a differentiable function of x. That is, suppose $y = f(u)$ and $u = g(x)$. Then, after substitution, y is a function of x. The Chain Rule states that the derivative of y with respect to

---
*This section may be omitted without loss of continuity.

x is the derivative of y with respect to u times the derivative of u with respect to x. That is,

(3) $$\frac{dy}{dx} = \frac{dy}{du}\frac{du}{dx}$$

Example 7 Find the derivative of y with respect to x if
$$y = f(u) = u^2 - 4 \quad \text{and} \quad u = g(x) = x^3$$

Solution We first find $\frac{dy}{du} = 2u$ and $\frac{du}{dx} = 3x^2$. Then by the Chain Rule (3) we have

$$\frac{dy}{dx} = \frac{dy}{du}\frac{du}{dx} = (2u)(3x^2) \underset{\underset{u=x^3}{\uparrow}}{=} (2x^3)(3x^2) = 6x^5$$

The solution is easy to check since if $y = f(u) = u^2 - 4$ and $u = g(x) = x^3$, then, after substitution,

$$y = (x^3)^2 - 4 = x^6 - 4$$

so that

$$\frac{dy}{dx} = 6x^5$$

Although the second approach seems shorter and easier (that is, to substitute first, then differentiate), the advantage of the Chain Rule will become apparent as the functions we deal with become more complicated.

Example 8 Find the derivative of $y = (4x^3 - 6x^2 + 5x - 2)^4$ using:
(a) The Power Rule
(b) The Chain Rule

Solution (a) Power Rule:

$$\frac{dy}{dx} = 4(4x^3 - 6x^2 + 5x - 2)^3 \left[\frac{d}{dx}(4x^3 - 6x^2 + 5x - 2)\right]$$
$$= 4(4x^3 - 6x^2 + 5x - 2)^3(12x^2 - 12x + 5)$$

(b) Chain Rule: Here, $y = u^4$ and $u = 4x^3 - 6x^2 + 5x - 2$. Thus,

$$\frac{dy}{dx} = \frac{dy}{du}\frac{du}{dx} = (4u^3)(12x^2 - 12x + 5) \underset{\underset{u = 4x^3 - 6x^2 + 5x - 2}{\uparrow}}{=} 4(x^3 - 6x^2 + 5x - 2)^3(12x^2 - 12x + 5)$$

As this example illustrates, whenever a problem involves finding the derivative of a function raised to a power, either the Power Rule or Chain Rule may be used. However, the Power Rule, as a special case of the Chain Rule, will not always work—especially when exponential and logarithmic functions are involved (Chapter 5).

6. THE POWER RULE; THE CHAIN RULE

Partial Proof of Chain Rule

To find the derivative dy/dx when $y = f(u)$ and $u = g(x)$ we use the four-step rule (page 102). Thus, we look at the difference quotient $\Delta y/\Delta x$. If $\Delta u \neq 0$,* we can write

$$\frac{\Delta y}{\Delta x} = \frac{\Delta y}{\Delta u} \cdot \frac{\Delta u}{\Delta x}$$

Thus,†

$$\frac{dy}{dx} = \lim_{\Delta x \to 0} \frac{\Delta y}{\Delta x} = \lim_{\Delta x \to 0}\left(\frac{\Delta y}{\Delta u} \cdot \frac{\Delta u}{\Delta x}\right) = \lim_{\Delta x \to 0} \frac{\Delta y}{\Delta u} \cdot \lim_{\Delta x \to 0} \frac{\Delta u}{\Delta x}$$

$$= \left(\lim_{\Delta u \to 0} \frac{\Delta y}{\Delta u}\right) \cdot \left(\frac{du}{dx}\right) = \frac{dy}{du} \frac{du}{dx}$$

Proof of Power Rule Using Chain Rule

Let $y = u^r$ and $u = g(x)$. Then $y = [g(x)]^r$ and

$$\frac{dy}{dx} = \frac{dy}{du} \frac{du}{dx} = (ru^{r-1})(g'(x)) = r[g(x)]^{r-1} g'(x)$$

Exercise 6
Solutions to Odd-Numbered Problems begin on page 379.

In Problems 1-24 find the derivative of the function f using the power rule.

1. $f(x) = (3x + 5)^2$
2. $f(x) = (2x - 5)^3$
3. $f(x) = (6x - 5)^{-3}$
4. $f(x) = (4x + 1)^{-2}$
5. $f(x) = (x^2 + 5)^4$
6. $f(x) = (x^3 - 2)^5$
7. $f(t) = (t^5 - t^2 + t)^7$
8. $f(u) = (u^4 - u^2 + u - 1)^6$
9. $f(x) = \left(x - \frac{1}{x}\right)^3$
10. $f(x) = \left(x + \frac{1}{x}\right)^3$
11. $f(z) = \frac{(3z - 1)^3}{z + 1}$
12. $f(w) = \frac{(w + 1)^4}{w}$
13. $f(x) = \frac{(x^2 + 1)^2}{x^3 - 1}$
14. $f(x) = \frac{(x^3 + 2)^3}{x^2 - 1}$
15. $f(x) = \sqrt{x^2 + 1}$
16. $f(x) = \sqrt{x^2 - 4}$
17. $f(x) = \sqrt[3]{3x - 1}$
18. $f(x) = \sqrt[3]{5x + 2}$
19. $f(x) = x\sqrt{x^2 + 1}$
20. $f(x) = x\sqrt{x^2 - 1}$
21. $f(x) = x^2\sqrt{3x + 1}$
22. $f(x) = x^2\sqrt{4x - 1}$

*The case for $\Delta u = 0$ is more complicated. Interested readers should consult books in calculus.
†Since $u = g(x)$ is continuous, when $\Delta x \to 0$, then $\Delta u \to 0$. For, as $\Delta x \to 0$, we see that

$$\lim_{\Delta x \to 0} \Delta u = \lim_{\Delta x \to 0}\left(\frac{\Delta u}{\Delta x} \cdot \Delta x\right) = g'(x) \cdot 0 = 0$$

23. $f(x) = \left(\dfrac{3x-1}{3x+1}\right)^2$ 24. $f(x) = \left(\dfrac{2x+1}{2x-1}\right)^3$

In Problems 25-34, find dy/dx using the Chain Rule.*

*25. $y = f(u) = u^5, \quad u = g(x) = x^3 + 1$
*26. $y = f(u) = u^3, \quad u = g(x) = 2x + 5$
*27. $y = f(u) = \dfrac{u}{u+1}, \quad u = g(x) = x^2 + 1$
*28. $y = f(u) = \dfrac{u-1}{u}, \quad u = g(x) = x^2 - 1$
*29. $y = f(u) = (u+1)^2, \quad u = g(x) = \dfrac{1}{x}$
*30. $y = f(u) = (u^2 - 1)^3, \quad u = g(x) = \dfrac{1}{x+2}$
*31. $y = f(u) = (u^3 - 1)^5, \quad u = g(x) = x^{-2}$
*32. $y = f(u) = (u^2 + 4)^4, \quad u = g(x) = x^{-2}$
*33. $y = f(u) = \dfrac{1}{u^3 + 1}, \quad u = g(x) = \dfrac{1}{x^3 + 1}$
*34. $y = f(u) = \dfrac{4}{u^2 - 4}, \quad u = g(x) = \dfrac{3}{x^2 + 1}$

*35. Find the derivative y' of $y = (x^3 + 1)^2$ by:
   (a) Using the Chain Rule
   (b) Using the Power Rule
   (c) Expanding and then differentiating
   (d) Compare the answers from parts (a)-(c).
*36. Follow the directions in Problem 35 for the function $y = (x^2 - 2)^3$.
37. If $f(x) = x\sqrt{1-x^2}$, find the numbers x at which $f'(x) = 0$. Are there any numbers x for which $f'(x)$ does not exist?
38. Follow the directions of Problem 37 for $f(x) = x^2\sqrt{4-x}$.
39. *Distance Between Moving Objects.* At 6 PM one ship, traveling 8 kilometers per hour in an easterly direction, is 52 kilometers due west of a second ship that is moving at a velocity of 12 kilometers per hour due north. At what rate are the ships approaching each other at 7 PM? Are they approaching or receding from each other at 10 PM? At what time are they closest?
40. *Distance Between Moving Objects.* A submarine, submerged at 200 feet, is moving along at the rate of 15 feet per second and passes under a stationary destroyer. How fast is the distance from the submarine to the destroyer changing after 1 minute?
41. *Pollution.* The amount of pollution in a certain lake is found to be

$$A(t) = (t^{1/4} + 3)^3$$

where t is measured in years and $A(t)$ is measured in appropriate units. What is the instantaneous rate of change of the amount of

---
*From optional section.

pollution? At what rate is the amount of pollution changing after 16 years?

42. *Distance Between Moving Objects.* At noon, a carrier is 10 kilometers due west of a destroyer. The carrier is traveling east at 2 kilometers per hour and the destroyer is going south at 4 kilometers per hour. When are the ships closest?

7. Higher-Order Derivatives

Earlier, we concluded that the derivative of a function $y = f(x)$ is also a function called the derivative function $f'(x)$.

For example, if

$$f(x) = 6x^3 - 3x^2 + 2x - 5$$

then

$$f'(x) = 18x^2 - 6x + 2$$

Second Derivative

The derivative of the function $f'(x)$ is called the *second derivative of f* and is denoted by $f''(x)$. For the example above,

$$f''(x) = \frac{d}{dx} f'(x) = 36x - 6$$

By continuing in this fashion, we can find the third derivative $f'''(x)$, the fourth derivative $f^{(4)}(x)$, and so on, provided that these derivatives exist.*

For example, the first, second, and third derivatives of the function

$$f(x) = x^4 + 3x^3 - 2x^2 + 5x - 6$$

are

$$f'(x) = 4x^3 + 9x^2 - 4x + 5$$

$$f''(x) = \frac{d}{dx} f'(x) = 12x^2 + 18x - 4$$

$$f'''(x) = \frac{d}{dx} f''(x) = 24x + 18$$

For the above example, observe that $f^{(4)}(x) = 24$ and all derivatives of order 5 or more equal 0.

*The symbols $f'(x)$, $f''(x)$, and so on for higher-order derivatives have several parallel notations. If $y = f(x)$, we may write

$$y' = f'(x) = \frac{dy}{dx} = \frac{d}{dx} f(x)$$

$$y'' = f''(x) = \frac{d^2y}{dx^2} = \frac{d^2}{dx^2} f(x)$$

$$y''' = f'''(x) = \frac{d^3y}{dx^3} = \frac{d^3}{dx^3} f(x)$$

$$\vdots$$

$$y^{(n)} = f^{(n)}(x) = \frac{d^n y}{dx^n} = \frac{d^n}{dx^n} f(x)$$

The result obtained in this example can be generalized:
For a polynomial function f of degree n, we have

$$f(x) = a_n x^n + a_{n-1} x^{n-1} + \cdots + a_1 x + a_0$$
$$f'(x) = n a_n x^{n-1} + (n-1) a_{n-1} x^{n-2} + \cdots + a_1$$

Thus, the first derivative of a polynomial function of degree n is a polynomial function of degree $n - 1$. By continuing the differentiation process, it follows that the nth-order derivative of f is

$$f^{(n)}(x) = n(n-1)(n-2) \cdot \cdots \cdot (3)(2)(1) a_n$$

a polynomial of degree 0—a constant. Therefore, all derivatives of order greater than n will equal 0.

In some applications it is important to find both the first and second derivatives of a function and to solve for those x that make these derivatives equal 0.

Example 1 For $f(x) = 4x^3 - 12x^2 + 2$, find those x, if any, at which $f'(x) = 0$. For what numbers x will $f''(x) = 0$?

Solution $f'(x) = 12x^2 - 24x = 12x(x - 2) = 0$ when $x = 0$ or $x = 2$
$f''(x) = 24x - 24 = 24(x - 1) = 0$ when $x = 1$

Example 2 For the function $f(x) = \sqrt{x^2 + 4}$, find those numbers x at which $f'(x) = 0$ and $f''(x) = 0$.

Solution Use the power rule with $y = \sqrt{g(x)}$ and $g(x) = x^2 + 4$. Then,

$$f'(x) = \frac{1}{2\sqrt{x^2 + 4}} (2x) = \frac{x}{\sqrt{x^2 + 4}}$$

We see that $f'(x) = 0$ if

$$\frac{x}{\sqrt{x^2 + 4}} = 0 \quad \text{or} \quad x = 0$$

Next, by using the rule for differentiating a quotient, we get

$$f''(x) = \frac{(1)\sqrt{x^2 + 4} - (x)\dfrac{2x}{2\sqrt{x^2 + 4}}}{x^2 + 4}$$

$$= \frac{x^2 + 4 - x^2}{(x^2 + 4)^{3/2}} = \frac{4}{(x^2 + 4)^{3/2}}$$

Notice that $f''(x)$ is never zero; in fact, $f''(x) > 0$ for all x.

Acceleration

We conclude this section with an interpretation of the second derivative that is used in physics. Suppose the position of a particle at time t is a distance s from the origin, where s is given as a function of t, say, as $s = f(t)$. Then the first

7. HIGHER-ORDER DERIVATIVES

derivative ds/dt is the velocity v of the particle. That is,

$$v = \frac{ds}{dt}$$

Acceleration The *acceleration a* of this particle is defined as the rate of change of velocity with respect to time. That is,

$$a = \frac{dv}{dt} = \frac{d}{dt}v = \frac{d}{dt}\left(\frac{ds}{dt}\right) = \frac{d^2s}{dt^2}$$

In other words, acceleration is the second derivative of the function $s = f(t)$ with respect to time.

Example 3 A ball is thrown vertically upward with an initial velocity of 19.6 meters per second. The distance s (in meters) of the ball above the ground is $s = -4.9t^2 + 19.6t$, where t is the number of seconds elapsed from the moment that the ball is thrown.
(a) What is the velocity of the ball at the end of 1 second?
(b) When will the ball reach its highest point?
(c) What is the maximum height the ball reaches?
(d) What is the acceleration of the ball at any time t?
(e) How long is the ball in the air?
(f) What is the velocity of the ball upon impact?
(g) What is the total distance traveled by the ball?

Solution (a)
$$v = \frac{ds}{dt} = -9.8t + 19.6$$

At $t = 1$, $v = 9.8$ meters per second.

(b) The ball will reach its highest point when $v = 0$.

$$v = -9.8t + 19.6 = 0 \quad \text{when } t = 2 \text{ seconds}$$

(c) At $t = 2$, $s = -4.9(4) + 19.6(2) = 19.6$ meters.

(d)
$$a = \frac{d^2s}{dt^2} = -9.8 \text{ meters per second per second}$$

(e) We can answer this question in two ways. First, since it takes 2 seconds for the ball to reach its maximum height, it follows that it will take another 2 seconds to reach the ground, for a total time of 4 seconds in the air. The second way is to set $s = 0$ and solve for t:

$$-4.9t^2 + 19.6t = 0$$

$$t = 0 \quad \text{or} \quad t = \frac{19.6}{4.9} = 4$$

The ball is at ground level when $t = 0$ and when $t = 4$.

(f) Upon impact, $t = 4$. Hence, when $t = 4$,

$$v = (-9.8)(4) + 19.6 = -19.6 \text{ meters per second}$$

The minus sign here indicates that the direction of the velocity is downward.

(g) The total distance traveled is

$$\text{Distance up} + \text{Distance down} = 19.6 + 19.6 = 39.2 \text{ meters}$$

See Figure 8 for an illustration.

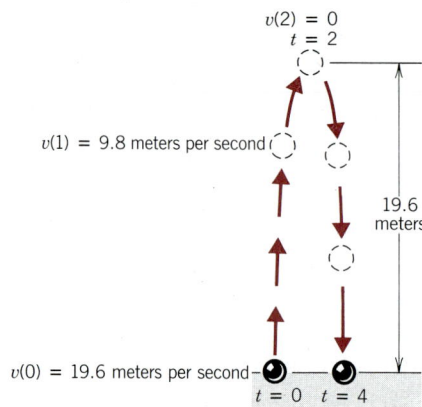

Figure 8

For Example 3, the acceleration of the ball is constant. This is approximately true for all falling bodies provided air resistance is ignored. In fact, the constant is even the same for all falling bodies, as Galileo (1564–1642) discovered in the sixteenth century. We can use calculus to see this. Galileo found by experimentation that all falling bodies obey the law that the distance they fall when they are dropped is proportional to the square of the time t it takes to fall that distance. Of great importance is the fact that the constant of proportionality c is the same for all bodies. Thus, Galileo's law states that the distance s a body falls in time t is given by

$$s = -ct^2$$

The reason for the minus sign is that the body is falling and we have chosen our coordinate system so that the positive s direction is up, along the vertical axis.

The velocity v of this freely falling body is

$$v = \frac{ds}{dt} = -2ct$$

and its acceleration a is

$$a = \frac{dv}{dt} = \frac{d^2s}{dt^2} = -2c$$

Thus, the acceleration of a freely falling body is a constant. Usually, we denote this constant by $-g$ so that

$$a = -g$$

Acceleration of Gravity The number g is called the *acceleration of gravity*. For our planet, g may be approximated by 32 feet per second per second or 980 centimeters per second per second.* On the planet Jupiter, $g \approx 2600$ centimeters per second per second, and on our moon, $g \approx 160$ centimeters per second per second.

Exercise 7
Solutions to Odd-Numbered Problems begin on page 380.

In Problems 1-16 find f' and f'' for the function f.

1. $f(x) = 2x + 5$
2. $f(x) = 3x + 2$
3. $f(x) = 3x^2 + x - 2$
4. $f(x) = -5x^2 - 3x$
5. $f(x) = x + \dfrac{1}{x}$
6. $f(x) = x - \dfrac{1}{x}$
7. $f(t) = \dfrac{t}{t+1}$
8. $f(u) = \dfrac{u+1}{u}$
9. $f(x) = \dfrac{x^2}{x+1}$
10. $f(x) = \dfrac{x^3}{x-1}$
11. $f(w) = (w^2 + 3)^3$
12. $f(v) = (v^3 + 1)^2$
13. $f(x) = \sqrt{x}$
14. $f(x) = \dfrac{1}{\sqrt{x}}$
15. $f(z) = (3z^2 + 1)^{3/2}$
16. $f(t) = (t^3 - 1)^{2/3}$

In Problems 17-22 find the indicated derivative.

17. $f^{(4)}(x)$, if $f(x) = x^3 - 3x^2 + 2x - 5$
18. $f^{(5)}(x)$, if $f(x) = 4x^3 + x^2 - 1$
19. $\dfrac{d^{20}}{dx^{20}}(8x^{19} - 2x^{14} + 2x^5)$
20. $\dfrac{d^{14}}{dx^{14}}(x^{13} - 2x^{10} + 5x^3 - 1)$
21. $\dfrac{d^8}{dx^8}(\tfrac{1}{8}x^8 - \tfrac{1}{7}x^7 + x^5 - x^3)$
22. $\dfrac{d^6}{dx^6}(x^6 + 5x^5 - 2x + 4)$

In Problems 23-26 find the velocity v and acceleration a of an object whose position s at time t is given.

23. $s = 16t^2 + 20t$
24. $s = 16t^2 + 10t + 1$
25. $s = 4.9t^2 + 4t + 4$
26. $s = 4.9t^2 + 5t$

*27. Find the nth-order derivative of: $f(x) = (2x + 3)^n$

*28. Find the nth-order derivative of: $f(x) = \dfrac{1}{3x - 4}$

*29. Find the second derivative of $f(x) = x^2 g(x)$, where g' and g'' exist.
*30. Find the second derivative of $f(x) = g(x)/x$, where g' and g'' exist.

---

*The earth, as you know, is not perfectly round; it bulges slightly at the equator. But neither is it perfectly oval, and its mass is not distributed uniformly. As a result, the acceleration of any freely falling body varies slightly from these constants.

31. *Freely Falling Body.* A ball is thrown vertically upward with an initial velocity of 80 feet per second. The distance s (in feet) of the ball from the ground after t seconds is $s = 6 + 80t - 16t^2$.
   (a) What is the velocity of the ball after 2 seconds?
   (b) When will the ball reach its highest point?
   (c) What is the maximum height the ball reaches?
   (d) What is the acceleration of the ball at any time t?
   (e) How long is the ball in the air?
   (f) What is the velocity of the ball upon impact?
   (g) What is the total distance traveled by the ball?
32. *Freely Falling Body.* An object is propelled vertically upward with an initial velocity of 39.2 meters per second. The distance s (in meters) of the object from the ground after t seconds is $s = -4.9t^2 + 39.2t$.
   (a) What is the velocity of the object at any time t?
   (b) When will the object reach its highest point?
   (c) What is the maximum height?
   (d) What is the acceleration of the object at any time t?
   (e) How long is the object in the air?
   (f) What is the velocity of the object upon impact?
   (g) What is the total distance traveled by the object?
33. A bullet is fired horizontally into a bale of paper. The distance s (in meters) the bullet travels in the bale of paper in t seconds is given by $s = 8 - (2 - t)^3$ for $0 \leq t \leq 2$. Find the velocity of the bullet after 1 second. Find the acceleration of the bullet at any time t.

8. Functions Not Differentiable at c

So far we have been totally concerned with functions f that are differentiable at a number c. In this section we look at some functions that are not differentiable at c—that is, functions f for which

(1)
$$\lim_{x \to c} \frac{f(x) - f(c)}{x - c}$$

does not exist. Two* of the most common ways that the limit above may fail to exist are:

1. When f has a corner at c (see Figure 9)

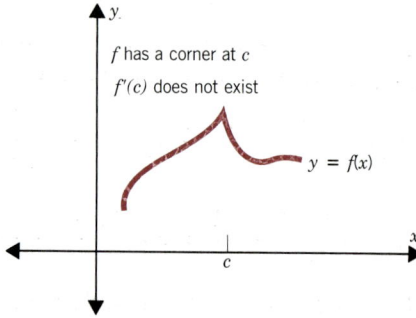

Figure 9

*There are other ways, but they will be discussed later.

8. FUNCTIONS NOT DIFFERENTIABLE AT c

2. When *f is not continuous at c* (see Figure 10)

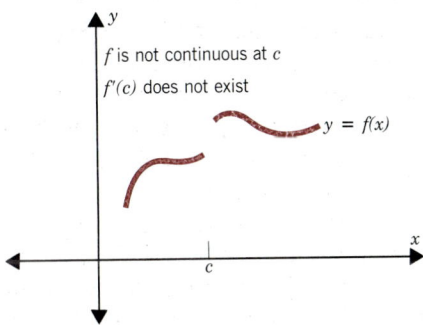

Figure 10

The first of these situations can be described most easily in terms of one-sided derivatives.

The derivative of a function *f* at *c* is given by

$$f'(c) = \lim_{x \to c} \frac{f(x) - f(c)}{x - c}$$

A criterion for this limit to exist is that the one-sided limits

$$\lim_{x \to c^-} \frac{f(x) - f(c)}{x - c} \quad \text{and} \quad \lim_{x \to c^+} \frac{f(x) - f(c)}{x - c}$$

One-Side Derivatives

exist and are equal. These limits, when they exist, are referred to as the *left derivative of f at c* and the *right derivative of f at c*, respectively. Collectively, these are called *one-sided derivatives of f at c*.

Example 1 For the function below, determine whether $f'(1)$ exists.

$$f(x) = \begin{cases} -2x^2 + 4 & \text{if } x < 1 \\ x^2 + 1 & \text{if } x \geq 1 \end{cases}$$

Solution We need to find the limit

$$\lim_{x \to 1} \frac{f(x) - f(1)}{x - 1} = \lim_{x \to 1} \frac{f(x) - 2}{x - 1}$$

But we face a difficulty. If $x < 1$, then $f(x) = -2x^2 + 4$; if $x \geq 1$, then $f(x) = x^2 + 1$. Consequently, we calculate the one-sided derivatives of *f* at 1:

$$\lim_{x \to 1^-} \frac{f(x) - f(1)}{x - 1} = \lim_{x \to 1^-} \frac{(-2x^2 + 4) - 2}{x - 1} = \lim_{x \to 1^-} \frac{-2(x^2 - 1)}{x - 1} = -4$$

$$\lim_{x \to 1^+} \frac{f(x) - f(1)}{x - 1} = \lim_{x \to 1^+} \frac{(x^2 + 1) - 2}{x - 1} = \lim_{x \to 1^+} \frac{(x - 1)(x + 1)}{x - 1} = 2$$

The left derivative of *f* at 1 is -4 and the right derivative of *f* at 1 is 2. Thus, $f'(1)$ does not exist.

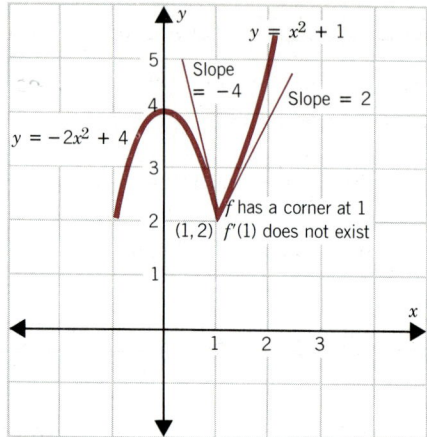

Figure 11

Figure 11 illustrates the graph of the function f in Example 1. At 1, where the derivative does not exist (and hence there is no tangent line), the graph of f has a *corner*. We usually say that the graph of f is not *smooth* at a corner.

Functions defined by more than one rule may be smooth where the split occurs. Let's look at an example.

Example 2 For the function below, determine whether $f'(1)$ exists.

$$f(x) = \begin{cases} 2x^2 + 1 & \text{if } x \leq 1 \\ 4x - 1 & \text{if } x > 1 \end{cases}$$

Solution Since the rule for f changes at 1, we calculate the one-sided derivatives of f at 1.

$$\lim_{x \to 1^-} \frac{f(x) - f(1)}{x - 1} = \lim_{x \to 1^-} \frac{(2x^2 + 1) - (3)}{x - 1} = \lim_{x \to 1^-} \frac{2(x^2 - 1)}{x - 1} = \lim_{x \to 1^-} [2(x + 1)] = 4$$

$$\lim_{x \to 1^+} \frac{f(x) - f(1)}{x - 1} = \lim_{x \to 1^+} \frac{(4x - 1) - (3)}{x - 1} = \lim_{x \to 1^+} \frac{4(x - 1)}{x - 1} = 4$$

Thus, $f'(1) = 4$. The graph of f is smooth at 1; that is, the graph of f has a tangent line at 1 and its slope is $f'(1) = 4$. See Figure 12.

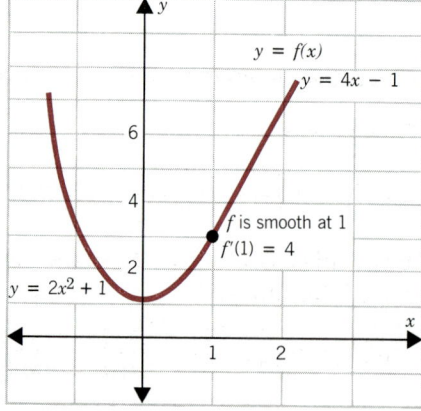

Figure 12

Example 3 Consider the function $f(x) = |x|$. Does it have a tangent line at 0?

Solution In order for $f(x) = |x|$ to have a tangent line at 0, it must have a derivative at 0. Thus, we need to examine the limit

$$\lim_{x \to 0} \frac{|x| - |0|}{x - 0} = \lim_{x \to 0} \frac{|x|}{x}$$

and determine whether it exists. Since for $x < 0$, $|x| = -x$ and for $x > 0$, $|x| = x$, we find that

$$\lim_{x \to 0^-} \frac{|x|}{x} = \lim_{x \to 0^-} \frac{-x}{x} = -1 \quad \text{and} \quad \lim_{x \to 0^+} \frac{|x|}{x} = \lim_{x \to 0^+} \frac{x}{x} = 1$$

Thus, $\lim_{x \to 0} \frac{|x|}{x}$ does not exist, so that $f(x) = |x|$ has no derivative at 0. This means $f(x) = |x|$ has no tangent line at 0.

Figure 13 shows the graph of this function at the point where it has no derivative. Notice the corner at 0.

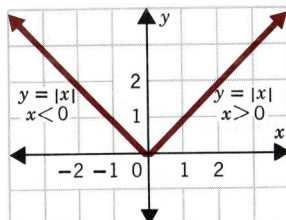

Figure 13

Derivatives and Continuity

The next result states a relationship between a continuous function and a differentiable function.

If a function f has a derivative at c, then it is continuous at c.

Proof
To show that f is continuous at c, we need to verify that $\lim_{x \to c} f(x) = f(c)$. We begin by observing that $x \neq c$, so we have

$$f(x) - f(c) = \left[\frac{f(x) - f(c)}{x - c} \right] (x - c)$$

Now we take the limit of both sides as $x \to c$ and use the fact that the limit of a product equals the product of the limits:

$$\lim_{x \to c} [f(x) - f(c)] = \lim_{x \to c} \left\{ \left[\frac{f(x) - f(c)}{x - c} \right] (x - c) \right\}$$

$$= \left[\lim_{x \to c} \frac{f(x) - f(c)}{x - c} \right] \left[\lim_{x \to c} (x - c) \right]$$

Since f has a derivative at c, we know that

$$\lim_{x \to c} \frac{f(x) - f(c)}{x - c} = f'(c)$$

is a number. Also, since $\lim_{x \to c}(x - c) = 0$, we find

$$\lim_{x \to c}[f(x) - f(c)] = [f'(c)][0] = 0$$

That is, $\lim_{x \to c} f(x) = f(c)$, so f is continuous at c.

Thus, every differentiable function is continuous. Or, to put it another way:

If a function f is not continuous at a number c, then it has no derivative at c.

The converse of this statement is not true. There are functions that are continuous at c and yet have no derivative at c. For example, the absolute value function from Example 3 is continuous at 0, but it is not differentiable at 0.

Exercise 8

Solutions to Odd-Numbered Problems begin on page 381.

Problems 1-4 refer to the graph below.

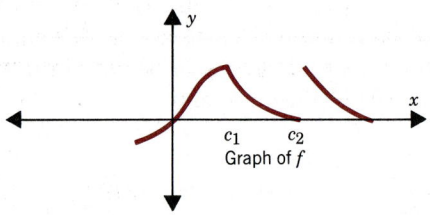

Graph of f

1. Is f continuous at c_1?
2. Is f continuous at c_2?
3. Is f differentiable at c_1?
4. Is f differentiable at c_2?

In Problems 5-12 determine whether the given function f has a derivative at c. Use Example 2 as a guide.

5. $f(x) = \begin{cases} 2x + 3 & \text{if } x < 1 \\ x^2 + 4 & \text{if } x \geq 1 \end{cases} \quad c = 1$

6. $f(x) = \begin{cases} 3 - 4x & \text{if } x < -1 \\ 2x + 9 & \text{if } x \geq -1 \end{cases} \quad c = -1$

7. $f(x) = \begin{cases} -4 + 2x & \text{if } x \leq \frac{1}{2} \\ 4x^2 - 4 & \text{if } x > \frac{1}{2} \end{cases} \quad c = \frac{1}{2}$

8. $f(x) = \begin{cases} 2x^2 + 1 & \text{if } x < -1 \\ -1 - 4x & \text{if } x \geq -1 \end{cases}$ $c = -1$

9. $f(x) = |x^2 - 4|$ $c = 2$

10. $f(x) = |x^2 - 4|$ $c = -2$

11. $f(x) = \begin{cases} 2x^2 + 1 & \text{if } x < -1 \\ 2 + 2x & \text{if } x \geq -1 \end{cases}$ $c = -1$

12. $f(x) = \begin{cases} 5 - 2x & \text{if } x < 2 \\ x^2 & \text{if } x \geq 2 \end{cases}$ $c = 2$

13. For the function

$$f(x) = \begin{cases} x^3 & \text{if } x \leq 0 \\ x^2 & \text{if } x > 0 \end{cases}$$

determine whether:
(a) f is continuous at 0 (b) $f'(0)$ exists

Graph the function and give a geometric interpretation to the answers found in parts (a) and (b).

14. Repeat Problem 13 for the function

$$f(x) = \begin{cases} 2x & \text{if } x \leq 0 \\ x^2 & \text{if } x > 0 \end{cases}$$

15. *Crash Testing an Automobile.* Calculate analytically the velocity (in feet per second) of an automobile if its position is given by

$$s = \begin{cases} t^3 & \text{if } 0 \leq t < 5 \\ 125 & \text{if } 5 \leq t \end{cases}$$

where s is distance (in feet) and t is time (in seconds). If this represents a crash test in which a vehicle is accelerated into a brick wall, find the velocity just before and after impact. Are the formulas quoted accurate during impact?

Chapter Review

Important Terms

average rate of change
difference quotient
slope of secant line
average velocity
instantaneous velocity
derivative of a function
four-step method
slope of tangent line

derivative of a constant function
prime notation
Leibniz notation
composite function
power rule
second derivative
acceleration
one-sided derivative
*chain rule

True-False Questions
(Answers on page 434.)

T F 1. The derivative of a function is the limit of a certain difference quotient.
T F 2. The derivative of a product equals the product of the derivatives.
T F 3. If $f(x) = 1/x$, then $f'(x) = -1/x^2$.

*From optional section.

T F 4. The expression "rate of change of a function" means the derivative of the function.
T F 5. Every function has a derivative at each number in its domain.

Fill in the Blanks
(Answers on page 435.)

1. The derivative of f at c equals the slope of the _____ line to f at c.
2. If $s = f(t)$ denotes the distance of a particle from the origin at time t, then $f'(t)$ is the _____ of the particle.
3. The derivative of $f(x) = (x^2 + 1)^{3/2}$ is obtained using the _____ _____.
4. The rate of change of velocity with respect to time is called _____.
5. The fifth-order derivative of a polynomial of degree 4 equals _____.

Review Exercises
Solutions to Odd-Numbered Problems begin on page 383.

In Problems 1-4 find the average rate of change from $x = 0$ to $x = 1$ for each function.

1. $f(x) = 2x^2 + 1$
2. $f(x) = 2x^3 - 3$
3. $f(x) = x^2 + 2x - 3$
4. $f(x) = 3x^2 - x + 4$

In Problems 5-8 use the definition of the derivative of a function given on page 102 to find the derivative of each function.

5. $f(x) = 2x^2 + 1$, at $x = 0$
6. $f(x) = 2x^3 - 3$, at $x = 1$
7. $f(x) = x^2 + 2x - 3$, at $x = -1$
8. $f(x) = 3x^2 - x + 4$, at $x = 2$

In Problems 9-30 find the derivative of each function.

9. $f(x) = 3x^4 - 2x^2 + 5x - 2$
10. $f(x) = 5x^3 + 2x^2 - 6x - 9$
11. $f(x) = 8\sqrt{x}$
12. $f(x) = 6\sqrt[3]{x}$
13. $g(x) = \dfrac{4}{x}$
14. $h(x) = -\dfrac{2}{x^2}$
15. $f(t) = (t^2 + 1)^3$
16. $f(w) = (w^3 - 3)^2$
17. $f(x) = \dfrac{3}{x} - \dfrac{x}{3}$
18. $f(u) = \dfrac{4}{u} + \dfrac{u}{4}$
19. $f(x) = (3x^2 + x + 1)(4x^3 - x + 2)$
20. $f(x) = (5x^3 - x + 1)(2x^2 + x + 3)$
21. $g(t) = \dfrac{t}{t^2 + 1}$
22. $f(u) = \dfrac{u}{u^2 - 1}$
23. $f(w) = \tfrac{2}{5}w^{5/2} - 2w^{3/2}$
24. $g(u) = \tfrac{3}{5}u^{5/3} + 3u^{4/3}$
25. $f(x) = (3x - 2)^3 + 5(3x - 2)^2$
26. $g(x) = (2x - 3)^4 - 6(2x - 3)^2$
27. $f(x) = \sqrt{x + \sqrt{x}}$
28. $f(x) = \sqrt{x - \sqrt{x}}$
29. $f(t) = t^2\sqrt{t - 1}$
30. $f(u) = u\sqrt{u^3 + 1}$

In Problems 31-36 find $f'(1)$ and $f''(1)$ for each function.

31. $f(x) = 3x^4 - 2x^2 + 4$
32. $f(x) = 6x^3 - 3x^2 + 2$
33. $f(x) = x^{2/3}$
34. $f(x) = x^{1/3}$
35. $f(x) = \dfrac{3}{x}$
36. $f(x) = \dfrac{4}{x}$

In Problems 37-40 find $f'(x)$ and $f''(x)$. Find all numbers x for which $f'(x) = 0$ and calculate $f''(x)$ at these numbers.

37. $f(x) = 2x^3 + 3x^2 - 12x + 6$
38. $f(x) = 2x^3 - 15x^2 + 36x - 2$
39. $f(x) = (x^2 - 1)^{3/2}$
40. $f(x) = (x^2 + 1)^{3/2}$

41. The pressure P of a certain gas is related to its volume V according to the equation $P = 2/V$. Find the rate of change of pressure with respect to its volume.

42. The voltage V produced by a certain thermocouple as a function of the temperature T is given by $V = 4.0T + 0.005T^2$. If the temperature changes slowly, find the rate of change of voltage with respect to temperature when the temperature is 200°C.

43. The ends of a 10-ft beam are supported at different levels. The deflection y of the beam is given by $y = kx^2(x^3 + 450x - 3500)$, where x is the horizontal distance from one end and k is a constant. Determine the expression for the instantaneous rate of change of deflection with respect to x.

44. Determine the sign of the derivative of the function $f(x) = (2x - 1)/(1 - x^2)$ for x equal to $-2, 0, 2$. Is the slope of a tangent line to the graph of f ever negative?

45. Suppose we borrow \$100 and are told that we must pay back an amount, A, where

$$A = 100 + .05T$$

and T is the time, in months, before we pay it back. What is the rate of change of the amount with respect to time? What is this rate called?

46. The heart rate of a resting mammal is inversely proportional to body weight. In particular, if W is the body weight in kilograms and B denotes the number of heart beats per minute, then W and B are related by the equation

$$B = (241)W^{-0.25}$$

Find the rate of change of B with respect to W.

47. The value $v(t)$ of a car t years after its purchase is given by the equation $v(t) = 5000/(1 + t) + 100$ dollars. Find (a) the average rate of depreciation over the first 3 years, (b) the rate at which it depreciates, (c) its rate of depreciation after 1 year, (d) its rate of depreciation after 3 years.

4
Applications of the Derivative

1. Relative Maxima and Relative Minima
2. Absolute Maximum and Absolute Minimum
3. Applied Problems
4. Concavity
5. Asymptotes
6. Marginal Analysis
7. Models
   Model 1: Maximizing Tax Revenue
   Model 2: Optimal Trade-In Time
   Model 3: Minimizing Inventory Cost
   Model 4: The Response of the Body to a Drug

Chapter Review
   Mathematical Questions from CPA Exams

CH. 4 APPLICATIONS OF THE DERIVATIVE

In this chapter we discuss applications of the derivative. The first application is the analysis of the behavior of a function and its graph. The derivative can help us determine the shape of a graph—that is, whether it increases or decreases on an interval, where its high and low points are, and so on.

Additional applications concern models in business and economics that require optimizing certain functions. For example, when a government formulates fiscal policy, it may find that it is not always better to raise taxes to increase revenue. Sometimes higher taxes do not produce higher revenue, as we shall see. Another model involves the question of when the best time might be to replace an appliance, taking into consideration depreciation, resale value, repair costs, and so on.

The use of the derivative is not restricted to these areas since just about every discipline is concerned with some aspect of the theory of optimization. The discussion given here will enable you to understand applications of the derivative when you encounter them in any area of interest.

1. Relative Maxima and Relative Minima

In observing the behavior of a continuous function, such as the one whose graph is shown in Figure 1, we notice, as we proceed from left to right, that portions of the function are increasing and other parts are decreasing. Some points on the graph are higher than the surrounding points, while others are lower than the surrounding points. In this section we discuss techniques for determining when functions increase or decrease and for finding these high and low points. But first we need to introduce some new terminology.

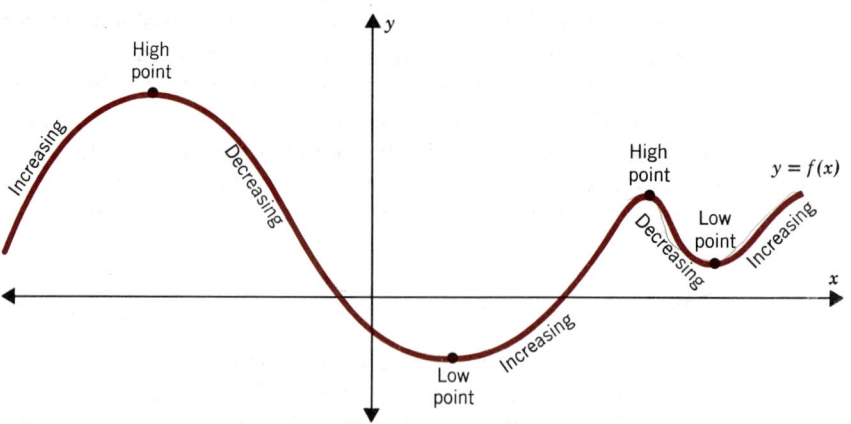

Figure 1

To arrive at definitions for an *increasing function* and a *decreasing function*, we examine the graphs given in Figures 2 and 3. Notice in Figure 2 that when the function is increasing, its height at x_1 is smaller than it is at x_2; that is, $f(x_1) < f(x_2)$. Similarly, in Figure 3, where the function is decreasing, its height at x_1 is larger than it is at x_2; that is, $f(x_1) > f(x_2)$.

1. RELATIVE MAXIMA AND RELATIVE MINIMA 141

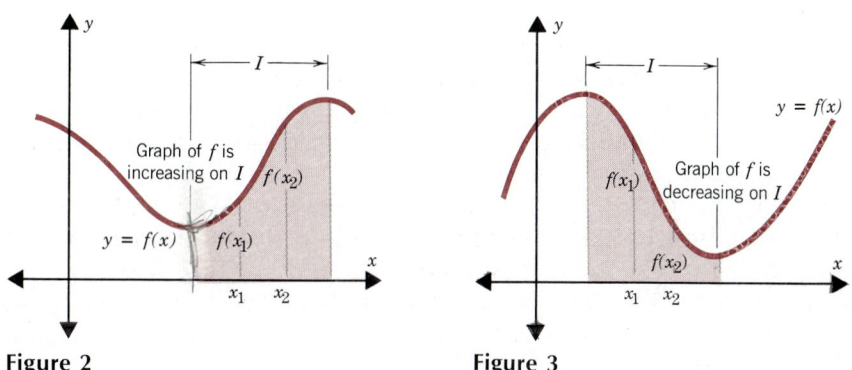

Figure 2   **Figure 3**

Increasing Function; Decreasing Function   **Let $y = f(x)$ be a function defined on an interval I and let (a, b) and (c, d) be subintervals contained in I.**

(a) $y = f(x)$ **is** *increasing* **on (a, b) means that for any choice of $x_1 < x_2$ in (a, b), then $f(x_1) < f(x_2)$.**
(b) $y = f(x)$ **is** *decreasing* **on (c, d) means that for any choice of $x_1 < x_2$ in (c, d), then $f(x_1) > f(x_2)$.**

The terms *increasing* and *decreasing* describe the behavior of the graph as you examine it from left to right on some open interval. Thus, to determine whether a function is, say, increasing on an open interval requires testing *any* choice of numbers x_1 and x_2 in that interval and showing that $f(x_1) < f(x_2)$ when $x_1 < x_2$. Does this mean that every time we want to determine the behavior of a function, we have to go through the tedious process of selecting intervals and numbers in those intervals? The answer is "No" for differentiable functions, since the derivative provides us with a simple straightforward tool. Let's see how.

If we construct tangent lines where a function is increasing, we observe that these tangent lines have something in common—their slopes are positive. Since the first derivative of a function gives the slope of its tangent line at each point, we can use the first derivative to determine where a function is increasing by merely determining where, if at all, its derivative is positive.

Test for Increasing and Decreasing Function   **A differentiable function is:**

(a)   **Increasing on (a, b) if $f'(x) > 0$ throughout (a, b)**
(b)   **Decreasing on (c, d) if $f'(x) < 0$ throughout (c, d).**
  If $f'(x) = 0$, the tangent line to the function is horizontal, and the function is neither increasing nor decreasing at x.

Example 1   For the function $f(x) = x^2 + 1$, find the points, if any, at which the function has a horizontal tangent line. Determine for what numbers x the function is increasing or decreasing. Use this information to graph the function.

Solution

$$f'(x) = 2x$$

Now, $f'(x) = 0$ whenever $x = 0$. Thus, the function has a horizontal tangent line at the point $(0, 1)$. Also, $f'(x) = 2x$ is negative whenever $x < 0$. Thus, for $x < 0$, the function is decreasing. Similarly, when $x > 0$, the function is increasing. Figure 4 illustrates these results.

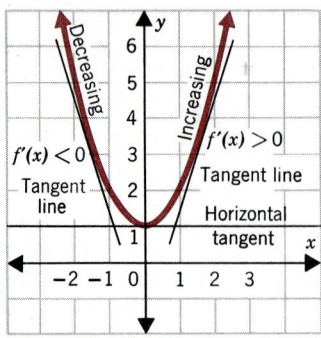

Figure 4

When the derivative of a function is negative, the graph of the function is falling; on the other hand, when the derivative is positive, the graph is rising.

Example 2 Consider the function

$$f(x) = 2x^3 - 9x^2 + 12x - 5$$

Find the points, if any, at which the function has a horizontal tangent line, and determine where the function is increasing and where it is decreasing. Graph the function.

Solution

$$f'(x) = 6x^2 - 18x + 12 = 6(x^2 - 3x + 2)$$

Factoring gives

$$f'(x) = 6(x - 2)(x - 1)$$

Now, $f'(x) = 0$ whenever $x = 2$ or $x = 1$. Thus, the function has horizontal tangent lines at the points $(2, f(2)) = (2, -1)$ and $(1, f(1)) = (1, 0)$. We use these as cutoff points, which partition the real line into three intervals. See Figure 5.

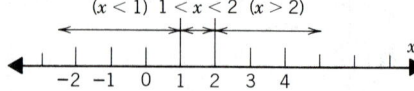

Figure 5

On each of these intervals we calculate the sign of the factors of $f'(x)$, which tells us whether f is increasing or decreasing. See Table 1.

To graph the function, we first plot the points at which it has a horizontal tangent line. See Figure 6. For x between 1 and 2, we know the function is

1. RELATIVE MAXIMA AND RELATIVE MINIMA   143

Table 1

Interval	Sign of $x-1$	Sign of $x-2$	Sign of $f'(x) = 6(x-2)(x-1)$	Behavior
$x < 1$	Negative	Negative	Positive	f is increasing
$1 < x < 2$	Positive	Negative	Negative	f is decreasing
$x > 2$	Positive	Positive	Positive	f is increasing

decreasing and so we sketch this part of the graph showing y decreasing as x moves from 1 to 2. For $x > 2$, the function is increasing, and for $x < 1$, it is increasing, so that as x moves from left to right, the value of $f(x)$ gets larger. The completed sketch is shown in Figure 6.

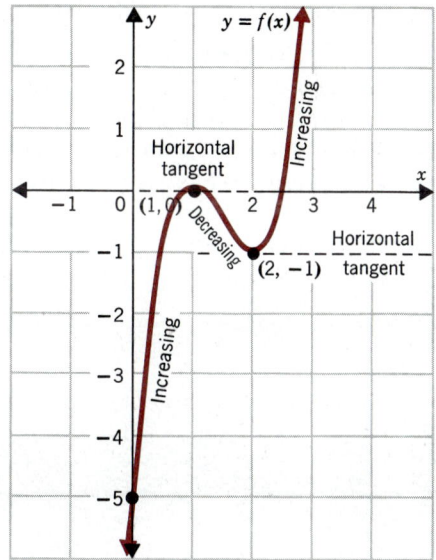

Figure 6

In Figure 6, notice that for $x = 1$, the value of $f(x)$ is larger than it is for any other number x close to 1. In this case, we say the point $(1, 0)$ is a *relative maximum*. The reason we use the word "relative" is that it is a maximum point only around $x = 1$, or in an interval containing $x = 1$, that the value $f(1)$ is larger than it is for other numbers x. As you can see from the graph, the value of $f(x)$ does get larger than $f(1)$ for numbers x "far away" from $x = 1$.

Similarly, for numbers x in a small interval containing $x = 2$, you can see that the value of $f(x)$ is smallest at $x = 2$. That is, the point $(2, -1)$ is a *relative minimum*.

Relative Maximum A point $(c, f(c))$ is called a *relative maximum* of a function $y = f(x)$ if

$$f(c) > f(x)$$

for any choice of $x \neq c$ in an interval containing c. Here, $f(c)$ is called a *relative maximum value* **of the function.**

144 CH. 4 APPLICATIONS OF THE DERIVATIVE

Relative Minimum A point $(c, f(c))$ is called a *relative minimum* of a function $y = f(x)$ if

$$f(c) < f(x)$$

for any choice of $x \neq c$ in an interval containing c. Here, $f(c)$ is called a *relative minimum value* of the function.

To find all the relative maxima and relative minima of a function $y = f(x)$ could be a difficult task. However, *if the function is differentiable,* a fairly straightforward test is available for locating relative maxima and relative minima. It is apparent that every relative maximum and relative minimum of a function $y = f(x)$ must occur at a point at which the tangent line is horizontal, that is, at a number x for which $f'(x) = 0$. For this reason, such numbers x are called *critical numbers*. The corresponding point on the graph is called a *critical point*.

• Critical Number
• Critical Point

But there are critical points that are neither a relative maximum nor a relative minimum. If we examine the ways in which a horizontal tangent line can occur, we get a clue as to how we can ascertain whether the critical point is a relative maximum, a relative minimum, or neither. See Figure 7.

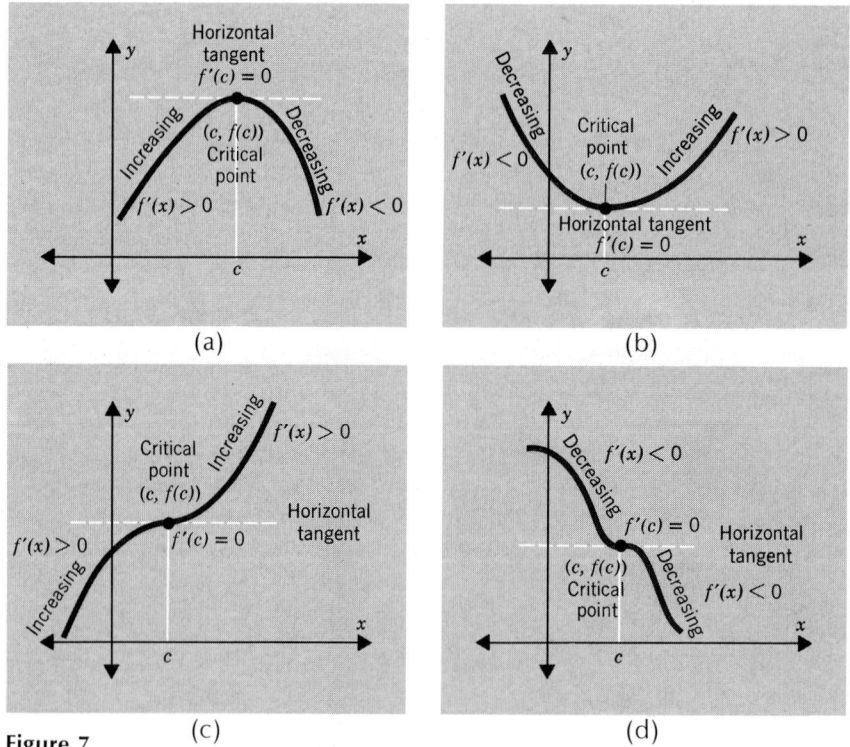

Figure 7

This leads us to formulate the following test for relative maxima and relative minima:

1. RELATIVE MAXIMA AND RELATIVE MINIMA   145

First-Derivative Test   Let $y = f(x)$ be a function that is differentiable on an open interval I. Suppose $(c, f(c))$ is a critical point of the function, so that $f'(c) = 0$.

(a) A critical point $(c, f(c))$ is a relative maximum if the function is increasing to the left of c and decreasing to the right of c. Refer to Figure 7(a).

(b) A critical point $(c, f(c))$ is a relative minimum if the function is decreasing to the left of c and increasing to the right of c. Refer to Figure 7(b).

Figures 7(c) and 7(d) illustrate the situations where the critical point is neither a relative maximum nor a relative minimum.

Example 3   Find all critical points of the function

$$f(x) = 3x^3 - 9x$$

Determine which, if any, are relative maxima or relative minima and graph the function.

Solution   First, find $f'(x)$:

$$f'(x) = 9x^2 - 9 = 9(x^2 - 1) = 9(x - 1)(x + 1)$$

The critical numbers obey $f'(x) = 0$ and are found to be $x = -1$ and $x = 1$. The corresponding critical points are $(-1, 6)$ and $(1, -6)$. The information we have so far is given in Figure 8(a).

To see if $(-1, 6)$ is a relative maximum, a relative minimum, or neither, we must find the sign of $f'(x)$ around $x = -1$. Now, for x just less than -1, we have $f'(x) > 0$ and for x just larger than -1 (but not larger than 1), we have $f'(x) < 0$. This means the function is increasing to the left of $x = -1$ and is decreasing to the right of $x = -1$. Hence, $(-1, 6)$ is a relative maximum.

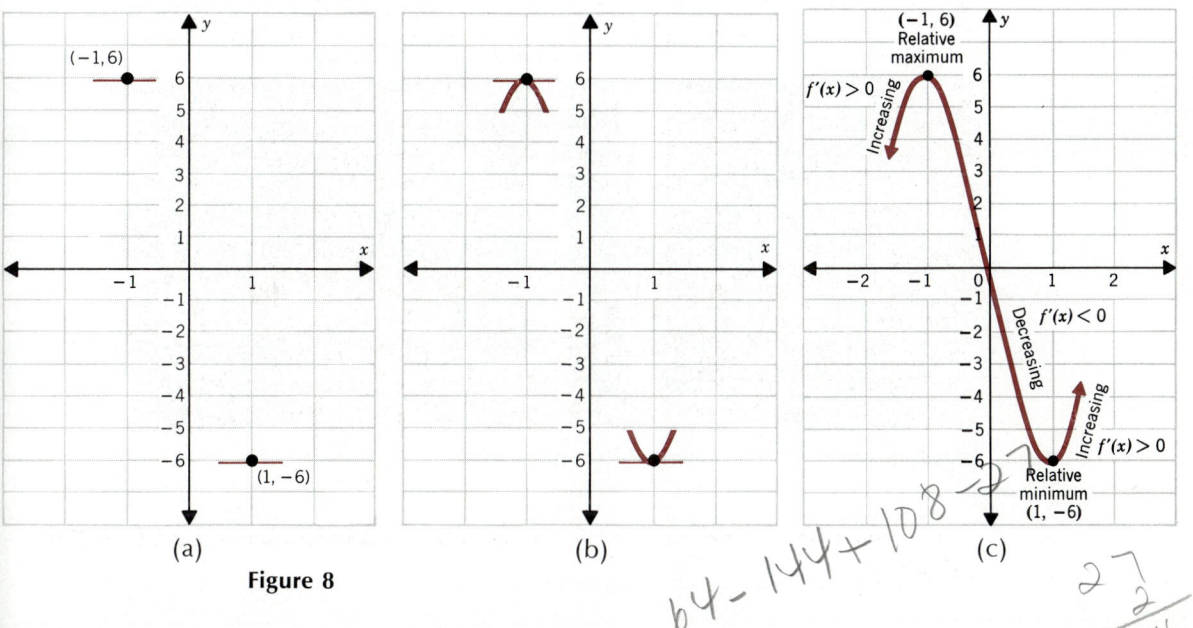

Figure 8

For the critical point $(1, -6)$, we notice that when $x < 1$, $f'(x) < 0$ and when $x > 1$, $f'(x) > 0$. Hence, $(1, -6)$ is a relative minimum. Figure 8(b) illustrates this latest information.

To graph the function, notice that for all $x < -1$, the first derivative is positive so that the function is increasing; and for all $x > 1$, $f'(x)$ is also positive so that the function is increasing.

The behavior of the function for $-1 < x < 1$ remains to be found. Clearly, in the interval $-1 < x < 1$, the function cannot have a relative maximum or a relative minimum since, if it did, a horizontal tangent would occur there, and we know this only happens at $(-1, 6)$ and $(1, -6)$. Also, since the function is continuous* and passes through $(0, 0)$, all we need to do is connect the graph, as Figure 8(c) illustrates. This is a result of the fact that $f'(x) < 0$ for $-1 < x < 1$, which shows that the function is decreasing in that interval.

You may find it useful to employ a table that summarizes the behavior of a function, as we did in Example 2. Table 2 lists the important information for the function in Example 3.

Table 2

Interval	Sign of $x + 1$	Sign of $x - 1$	Sign of $f'(x)$	Behavior
$x < -1$	Negative	Negative	Positive	Increasing
$-1 < x < 1$	Positive	Negative	Negative	Decreasing
$x > 1$	Positive	Positive	Positive	Increasing

Caution
The first-derivative test for locating the relative maxima and relative minima of a function requires that the function be differentiable.

For functions that are not differentiable, the absence of a horizontal tangent does not imply that the function has no relative maxima or relative minima in an interval.

For example, the function $f(x) = |x|$ on the interval $[-1, 1]$ has no horizontal tangent anywhere and yet $(0, 0)$ is an obvious relative minimum. Of course, the reason for this is that $f(x) = |x|$ has no derivative at $x = 0$. See Example 3, page 133.

The next example shows that not every critical point has to be a relative maximum or a relative minimum.

Example 4 Find all critical points of the function
$$f(x) = x^3 + 1$$

---

*In Chapter 3 we noted that differentiability implies continuity.

Determine which, if any, are relative maxima or relative minima and graph the function.

Solution
$$f'(x) = 3x^2$$
Clearly, the only critical number is $x = 0$. The critical point is $(0, 1)$.

Now, for $x < 0$, we have $f'(x) > 0$, so that the function is increasing. For $x > 0$, we have $f'(x) > 0$, so that the function is increasing. Thus, the critical point $(0, 1)$ is neither a relative maximum nor a relative minimum. However, at the point $(0, 1)$, the function has a horizontal tangent. Also, since $f'(x) > 0$ for all $x \neq 0$, we conclude that the function is increasing everywhere, except at $(0, 1)$. See Figure 9.

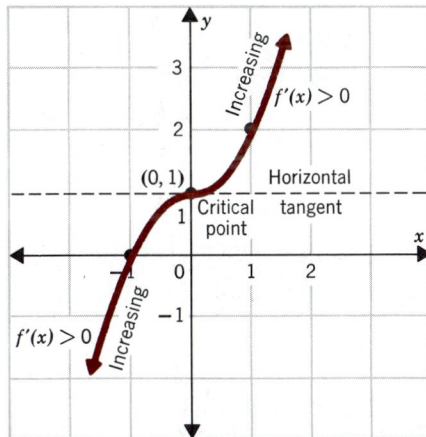

Figure 9

So far, we have discussed how to find relative maxima and relative minima for differentiable functions $y = f(x)$ by applying the first-derivative test. To use this test, we first find all critical points. Each critical point then becomes a candidate for being a relative maximum, a relative minimum, or neither. To decide what the critical point is, requires determining the sign of $f'(x)$ around the critical number. This is sometimes cumbersome. The next test is sometimes easier to use, although it requires that the function have a second derivative.

Second-Derivative Test Let $y = f(x)$ be a function that is differentiable on an open interval I and suppose that the second derivative $f''(x)$ exists on I. Let $(c, f(c))$ be a critical point of the function so that $f'(c) = 0$.

1. If $f''(c) < 0$, then $(c, f(c))$ is a relative maximum.
2. If $f''(c) > 0$, then $(c, f(c))$ is a relative minimum.
3. If $f''(c) = 0$, the test gives no results.

If $f''(c) = 0$, or if the function has no second derivative, then the second-derivative test cannot be used; however, the first-derivative test can still be used.

To see just how much faster the second-derivative test can be, we return to the function of Example 3, where

$$f(x) = 3x^3 - 9x$$
$$f'(x) = 9x^2 - 9 = 9(x-1)(x+1)$$

The critical numbers are $x = 1$ and $x = -1$, and the second derivative is

$$f''(x) = 18x$$

so

$$f''(1) = 18 > 0 \quad \text{and} \quad f''(-1) = -18 < 0$$

Thus, the point $(1, -6)$ is a relative minimum and the point $(-1, 6)$ is a relative maximum.

Example 5 Find all relative maxima and relative minima of the function

$$f(x) = 2x^3 + 3x^2 - 12x + 1$$

Solution First, we find the critical numbers. They obey

$$f'(x) = 6x^2 + 6x - 12 = 0$$
$$x^2 + x - 2 = 0$$
$$(x+2)(x-1) = 0$$
$$x = -2 \qquad x = 1$$

Thus, the critical numbers are -2 and 1.
The second derivative is

$$f''(x) = 12x + 6$$

We now test each critical number:

$$f''(-2) = 12(-2) + 6 = -18 < 0 \qquad f''(1) = 18 > 0$$

Thus, $(-2, 21)$ is a relative maximum, and $(1, -6)$ is a relative minimum. Figure 10 gives a sketch of the graph.

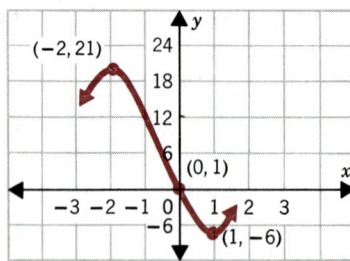

Figure 10

Although the second-derivative test provides a fairly easy way to find relative maxima and relative minima for functions that have second-order derivatives, it gives no result about relative maxima and relative minima when $f''(c) = 0$. However, as we shall see in a later section, this does provide additional information about the behavior of the function.

1. RELATIVE MAXIMA AND RELATIVE MINIMA

The flowchart in Figure 11 may be helpful.

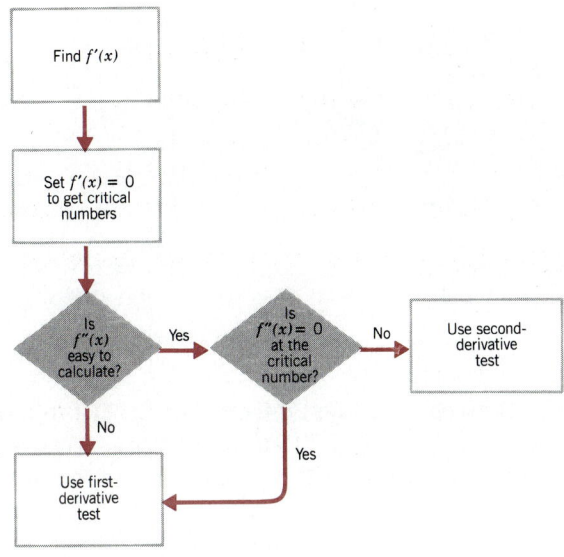

Figure 11

Exercise 1
Solutions to Odd-Numbered Problems begin on page 384.

In Problems 1–6 the derivative of a function $y = f(x)$ is given. Determine where the function $y = f(x)$ is increasing. Determine where the function $y = f(x)$ is decreasing. What are the critical numbers of each function?

1. $f'(x) = x - 1$
2. $f'(x) = x + 3$
3. $f'(x) = x(x - 1)$
4. $f'(x) = x(x + 2)$
5. $f'(x) = x^2(2x - 1)$
6. $f'(x) = x^2(x + 1)$

In Problems 7–14 find the critical numbers of each function. Using the first-derivative test, tell whether each critical point is a relative maximum, a relative minimum, or neither. Sketch the graph.

7. $f(x) = x^2 - 4x + 6$
8. $f(x) = -x^2 + 4x$
9. $f(x) = x^3 - 9x^2 - 27$
10. $f(x) = x^3 - 3x^2 + 1$
11. $f(x) = x^3 - 9x^2 + 27x - 27$
12. $f(x) = 2x^3 - 15x^2 + 36x$
13. $f(x) = x^4 - 2x^2 + 1$
14. $f(x) = x^4 - 8x^2 + 1$

In Problems 15–22 find all critical numbers of each function. Using the second-derivative test, tell whether each critical point is a relative maximum, a relative minimum, or neither. Sketch the graph.

15. $f(x) = x^2 - 8x + 7$
16. $f(x) = -x^2 + 4x + 2$
17. $f(x) = x^3 - 6x^2 + 1$
18. $f(x) = -x^3 + 3x^2$
19. $f(x) = x^3 - 9x^2 - 21x + 2$
20. $f(x) = x^3 - x^2 - x + 2$
21. $f(x) = x^4 - 6x^2 + 1$
22. $f(x) = -3x^4 + 6x^2 + 1$

150   CH. 4   APPLICATIONS OF THE DERIVATIVE

In Problems 23-44 use either the first-derivative test or the second-derivative test to locate the relative maxima and the relative minima of each function.

23. $f(x) = -2x^2 + 4x - 5$
24. $f(x) = -3x^2 - 12x + 2$
25. $f(x) = 2x^3 + 3x^2 + 4$
26. $f(x) = x^3 - 6x^2 + 2$
27. $f(x) = x^3 + 6x^2 + 12x + 1$
28. $f(x) = -x^3 + 3x + 4$
29. $f(x) = 3x^4 - 12x^3 + 5$
30. $f(x) = x^4 - 4x + 2$
31. $f(x) = x^4 + 4x^3 + 6x^2$
32. $f(x) = x^4 + 2x^3 - 3$
33. $f(x) = x^3 + 3x^2 - 9x + 1$
34. $f(x) = 2x^3 + 3x^2 - 36x + 4$
35. $f(x) = x^2(x + 1)$
36. $f(x) = x(x^2 - 1)$
37. $f(x) = x^4 + x^2$
38. $f(x) = x^4 - x^2$
39. $f(x) = x^{2/3} + x^{1/3}$
40. $f(x) = x^{2/3} - x^{1/3}$
41. $f(x) = x^{2/3}(x - 10)$
42. $f(x) = x^{1/3}(x - 4)$
43. $f(x) = x^{2/3}(x^2 - 8)$
44. $f(x) = x^{1/3}(x^2 - 2)$

45. *Distance Between Moving Objects.* At 10 AM a ship moving east at 6 kilometers per hour is 30 kilometers due west of another ship moving north at 12 kilometers per hour. Find the time at which the two ships are closest. What is the minimum distance?

46. *Maximum Profit.* A firm can sell as many units as it can produce at $30 per unit. The total cost to the firm of producing x units is

$$C = \$25 + 2x + 0.01x^2$$

The profit function is given by $30x - C$. Find the number of units the firm should produce to obtain maximum profit. What is the maximum profit?

47. *Transatlantic Crossings.* An airplane crosses the Atlantic Ocean (3000 miles) with an air speed of 500 mph.
    (a) Find the time saved with a 25 mile per hour tailwind.
    (b) Find the time lost with a 50 mile per hour headwind.
    (c) If the cost per person is

$$C(x) = 100 + \frac{x}{10} + \frac{36{,}000}{x}$$

   where x is the ground speed and $C(x)$ is the cost in dollars, what is the cost per passenger when there is no wind?
    (d) What is the cost with a tailwind of 25 miles per hour?
    (e) What is the cost with a headwind of 50 miles per hour?
    (f) What ground speed minimizes the cost?
    (g) What is the minimum cost per passenger?

*48. If $f(x) = ax^2 + bx + c$, $a \neq 0$, prove that the point

$$\left(\frac{-b}{2a}, \frac{4ac - b^2}{4a}\right)$$

is a relative maximum if $a < 0$ and is a relative minimum if $a > 0$.

*49. *Rolle's Theorem.* If a function $y = f(x)$ has the following three properties:

1. It is continuous on the closed interval $[a, b]$

2. It is differentiable on the open interval (a, b)
3. $f(a) = f(b) = 0$

then there is at least one number c, $a < c < b$, at which $f'(c) = 0$. This result is called *Rolle's theorem*.

For each function below, verify the conclusion of Rolle's theorem by finding the number(s) c on the interval indicated.
(a) $f(x) = 2x^2 - 2x$; $[0, 1]$
(b) $f(x) = x^4 - 1$; $[-1, 1]$
(c) $f(x) = x^4 - 2x^2 - 8$; $[-2, 2]$

*50. **Mean Value Theorem.** If $y = f(x)$ is a continuous function on an interval $[a, b]$ and is differentiable on (a, b), there is at least one number in (a, b) at which the slope of the tangent line equals the slope of the line joining $(a, f(a))$ and $(b, f(b))$. That is, there is a number c, $a < c < b$, at which

$$f'(c) = \frac{f(b) - f(a)}{b - a}$$

This result is called the *mean value theorem*. See the illustration.

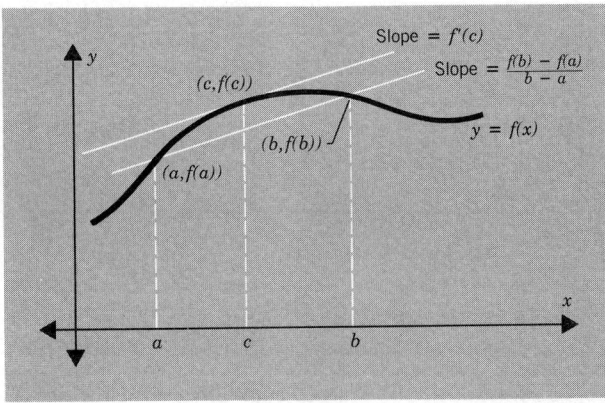

For each function below, verify the conclusion of the mean value theorem by finding the number(s) c on the interval indicated.
(a) $f(x) = x^2$; $[0, 3]$
(b) $f(x) = \frac{1}{x^2}$; $[1, 2]$
(c) $f(x) = x^{3/2}$; $[0, 1]$

2. Absolute Maximum and Absolute Minimum

In the definition of relative maximum (relative minimum), the use of the term "relative" refers to the fact that, at the critical number, the function has the largest (smallest) value in some interval containing the critical number. However, the relative maximum (minimum) value for a continuous function on a closed interval may not be the largest (smallest) value of the function on the interval. The largest and smallest values of a function, if they exist, are called the *absolute maximum* and the *absolute minimum* of the function.

Absolute Maximum; Absolute Minimum Let $y = f(x)$ be a continuous function with domain D. The largest value of $f(x)$ for x in D, if it exists, is called the *absolute maximum* of the function. The smallest value of $f(x)$ for x in D, if it exists, is the *absolute minimum* of the function.

Figure 12 illustrates some of the many situations that can occur. It will help us devise a way for finding the absolute maximum and the absolute minimum when they exist.

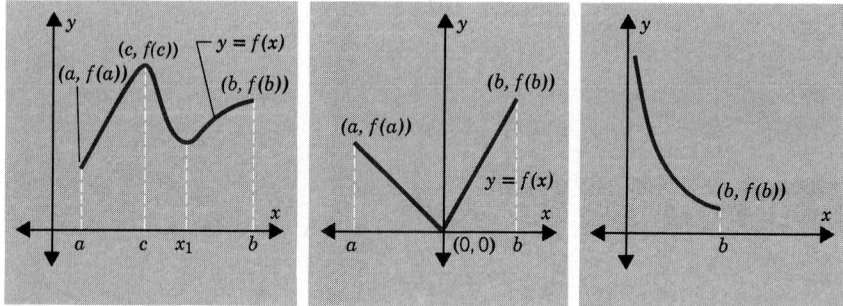

Figure 12

In Figure 12(a) the largest value of $f(x)$ on $[a, b]$ is $f(c)$, and it occurs at the point $(c, f(c))$. This is also a relative maximum of the function. Thus, the absolute maximum may be a relative maximum.

The absolute minimum of the function in Figure 12(a) is $f(a)$, and it occurs at the endpoint $(a, f(a))$. Also, the absolute maximum of the function in Figure 12(b) occurs at the endpoint $(b, f(b))$. Thus, the absolute maximum (minimum) of a function may occur at an endpoint.

In Figure 12(b) we observe that the absolute minimum occurs at $(0, 0)$, a point at which the function has no tangent line. Thus, the absolute maximum (minimum) of a function might occur at numbers where $f'(x)$ does not exist.

In Figure 12(c) the function has no absolute maximum since there is no largest value of $f(x)$ on $(0, b]$.

The following result tells us that a certain type of function always has an absolute maximum and an absolute minimum.

If a continuous function has as its domain a closed interval, the absolute maximum and absolute minimum exist.

2. ABSOLUTE MAXIMUM AND ABSOLUTE MINIMUM

With this in mind, we state a procedure for finding the absolute maximum and absolute minimum of a continuous function.

Test for Absolute Maximum and Absolute Minimum **If a continuous function $y = f(x)$ has a closed interval $[a, b]$ as domain, we can find the absolute maximum (or minimum) by choosing the largest (or smallest) value from among the following:**

1. **Values of $f(x)$ at critical numbers [numbers where $f'(x) = 0$] in (a, b)**
2. $f(a)$
3. $f(b)$
4. **Values of $f(x)$ at numbers where $f'(x)$ does not exist in (a, b)**

If critical numbers of $y = f(x)$ are found that are not in the interval (a, b), these critical numbers should be ignored since we are concerned only with the function in the interval $[a, b]$.

Example 1 Consider the function $f(x) = x^3 - 3x$. If the domain of f is $[0, 2]$, find the absolute maximum and absolute minimum.

Solution The critical numbers obey

$$f'(x) = 3x^2 - 3 = 3(x^2 - 1) = 3(x - 1)(x + 1) = 0$$

So, the critical numbers are

$$x = -1 \quad \text{and} \quad x = 1$$

We ignore the critical number $x = -1$, since it is not in the domain, $0 \le x \le 2$. For the critical number $x = 1$, we have

$$f(1) = -2$$

The values of $f(x)$ at the endpoints 0 and 2 are

$$f(0) = 0 \quad \text{and} \quad f(2) = 2$$

There are no numbers at which $f'(x)$ does not exist. Thus, the absolute maximum of $f(x)$ on $[0, 2]$ is 2 and the absolute minimum is -2.

Example 2 Find the absolute maximum and absolute minimum of the function $f(x) = x^{2/3}$ on the interval $[-1, 8]$.

Solution First, we locate the critical numbers, if any:

$$f'(x) = \frac{2}{3}x^{-1/3} = \frac{2}{3x^{1/3}}$$

Since $f'(x)$ is never zero, there are no critical numbers.

Next, we find the values of $f(x)$ at the endpoints:
$$f(-1) = 1 \quad \text{and} \quad f(8) = 4$$

Since $f'(x)$ does not exist at $x = 0$, we check the value of $f(x)$ at $x = 0$ according to rule 4 of the test for absolute maximum and absolute minimum:
$$f(0) = 0$$

The absolute maximum of $f(x) = x^{2/3}$ on $[-1, 8]$ is 4 and the absolute minimum is 0. Figure 13 illustrates this situation.

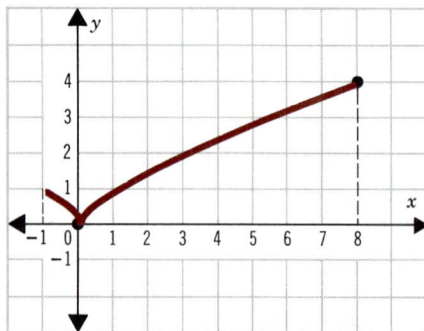

Figure 13

Exercise 2
Solutions to Odd-Numbered Problems begin on page 388.

For the functions in Problems 1–20 find the absolute maximum and absolute minimum of f on the indicated interval.

1. $f(x) = x^2 + 2x$; on $[-3, 3]$
2. $f(x) = x^2 - 8x$; on $[-1, 10]$
3. $f(x) = 1 - 6x - x^2$; on $[0, 4]$
4. $f(x) = 4 - 2x - x^2$; on $[-2, 2]$
5. $f(x) = x^3 - 3x^2$; on $[1, 4]$
6. $f(x) = x^3 - 6x$; on $[-2, 2]$
7. $f(x) = x^4 - 2x^2 + 1$; on $[0, 1]$
8. $f(x) = 3x^4 - 4x^3$; on $[-2, 0]$
9. $f(x) = x^{2/3}$; on $[-1, 1]$
10. $f(x) = x^{1/3}$; on $[-1, 1]$
11. $f(x) = 2\sqrt{x}$; on $[1, 4]$
12. $f(x) = 4 - \sqrt{x}$; on $[0, 4]$
13. $f(x) = x\sqrt{1 - x^2}$; on $[-1, 1]$
14. $f(x) = x^2\sqrt{2 - x}$; on $[0, 2]$
15. $f(x) = \dfrac{x^2}{x - 1}$; on $[-1, \tfrac{1}{2}]$
16. $f(x) = \dfrac{x}{x^2 - 1}$; on $[-\tfrac{1}{2}, \tfrac{1}{2}]$

17. $f(x) = (x + 2)^2(x - 1)^{2/3}$; on $[-4, 5]$
18. $f(x) = (x - 1)^2(x + 1)^3$; on $[-2, 7]$
19. $f(x) = \dfrac{(x - 4)^{1/3}}{x - 1}$; on $[2, 12]$
20. $f(x) = \dfrac{(x + 3)^{2/3}}{x + 1}$; on $[-4, -2]$
21. *Most Economical Speed.* A truck has a top speed of 75 miles per hour and, when traveling at the rate of x miles per hour, consumes gasoline at the rate of $\frac{1}{200}[(1600/x) + x]$ gallon per mile. If the length of the trip is 200 miles and the price of gasoline is $1.60 per gallon, the cost is

$$C(x) = 1.60\left(\dfrac{1600}{x} + x\right)$$

where $C(x)$ is measured in dollars. What is the most economical speed for the truck? Use the interval $[10, 75]$.
22. If the driver of the truck in Problem 21 is paid $8 per hour, what is the most economical speed for the truck?
23. *Maximum Profit.* The relationship between profit P of a firm and the selling price x of its goods is

$$P = 1000x - 25x^2 \qquad 0 \le x \le 40$$

For what range of selling prices is profit increasing? For what selling price is profit maximized?

3. Applied Problems

We begin with an example.

Example 1 A farmer with 4000 meters of fencing wants to enclose a rectangular plot that borders on a straight river. If the farmer does not fence the side along the river, what is the largest area that can be enclosed? See Figure 14.

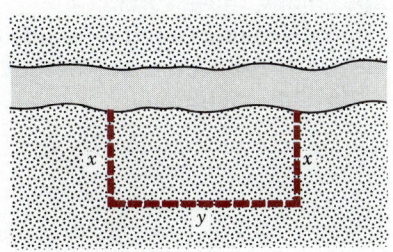

Solution The quantity to be maximized is the area. We denote it by A and denote the dimensions of the rectangle by x and y. The area A is

$$A = xy$$

But x and y are related since the length of fence used is 4000 meters. That is,
$$x + y + x = 4000$$
$$y = 4000 - 2x$$
Thus, the area A is
$$A = A(x) = x(4000 - 2x) = 4000x - 2x^2$$
The restrictions on x are $x \geq 0$ and $x \leq 2000$ (since if $x > 2000$, then the length of fence used, $x + y + x$, would be greater than 4000 meters).

The problem is to maximize $A(x) = 4000x - 2x^2$ on $[0, 2000]$. The critical numbers obey
$$A'(x) = 4000 - 4x = 0$$
$$x = 1000$$
Since this is the only critical number, we use the test for absolute maximum from Section 2, and calculate the values of $A(x)$ at this critical number and at the endpoints of the interval:
$$A(1000) = 2{,}000{,}000 \qquad A(0) = 0 \qquad A(2000) = 0$$
So, the maximum area that can be enclosed is 2,000,000 square meters. ∎

In general, each type of problem we will discuss requires some quantity to be minimized or maximized. We assume that the quantity we want to optimize can be represented by a function. Once this function is determined, the problem can be reduced to the question of determining at what number the function assumes its absolute maximum or absolute minimum.

Even though each applied problem has its unique features, it is possible to outline in a rough way a procedure for obtaining a solution. This five-step procedure is:

1. Identify the quantity for which a maximum or a minimum value is to be found.
2. Assign symbols to represent other variables in the problem. If possible, use an illustration to assist you.
3. Determine the relationships among these variables.
4. Express the quantity to be optimized as a function of one of these variables.
5. Apply the techniques of the previous sections to this function to determine the absolute maximum or absolute minimum.

The following examples illustrate this procedure.

Example 2 Best Dimensions for a Box From each corner of a square piece of sheet metal 18 centimeters on a side, remove a small square of side x centimeters and turn up the edges to form an open box. What should be the dimensions of the box so as to maximize the volume?

Solution The quantity to be maximized is the volume. Therefore, let's denote it by V and denote the dimensions of the side of the small square by x, as shown in Figure

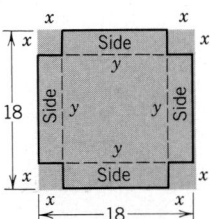

Figure 15

15. Although the area of the sheet metal is fixed, the sides of the square can be changed and thus are treated as variables. Let y denote the portion left after cutting the x's to make the square; we have

$$y = 18 - 2x$$

The height of the box is x, while the area of the base of the box is y^2. The volume V is therefore

$$V = xy^2$$

Since we have a function of two variables, we need to reduce it to a function in one variable. We can do this by substituting for y in the formula for the volume. This gives

$$V = V(x) = x(18 - 2x)^2$$

This is the function to be maximized. Its domain is the set of real numbers. However, physically, the only numbers x that make sense are those between 0 and 9. Thus, we want to find the absolute maximum of

$$V(x) = x(18 - 2x)^2 \qquad 0 \leq x \leq 9$$

To find the number x that maximizes V, we differentiate and find the critical numbers, if any:

$$V'(x) = (18 - 2x)^2 + 2x(18 - 2x)(-2)$$
$$= (18 - 2x)(18 - 6x)$$

Set $V'(x) = 0$ and solve for x:

$$(18 - 2x)(18 - 6x) = 0$$
$$18 - 2x = 0 \quad \text{or} \quad 18 - 6x = 0$$
$$x = 9 \quad \text{or} \quad x = 3$$

The only critical number in $(0, 9)$ is $x = 3$. Thus, we calculate the values of $V(x)$ at this critical number and at the endpoints of the interval:

$$V(0) = 0 \qquad V(3) = 3(18 - 6)^2 = 432 \qquad V(9) = 0$$

The maximum volume is 432 cubic centimeters and the dimensions of the box that yield the maximum volume are $x = 3$ centimeters deep by $y = 12$ centimeters on each side.

■

Example 3 A homeowner wishes to insulate his 1200 square foot attic. If the insulation costs 5 cents per square foot for each inch of thickness, the cost for x inches of thickness is $1200(0.05x)$ dollars. Over an 8-year period, the cost of heating this home with x inches of insulation is $\dfrac{8(292.5)}{x}$ dollars.* How many inches of insulation should be placed in the attic if the total cost is to be a minimum?

Solution The total cost over the 8-year period is

$$C(x) = 1200(0.05x) + \frac{(8)(292.5)}{x} = 60x + \frac{2340}{x}$$

The critical numbers obey

$$C'(x) = 60 - \frac{2340}{x^2} = 0$$

$$x^2 = \frac{2340}{60} = 39, \; x \approx 6.2 \text{ inches}$$

Since $C''(x) = \dfrac{4680}{x^3} > 0$ for all $x > 0$, it follows that this critical number minimizes the cost. Thus, the homeowner should add about 6.2 inches (about R-25) of insulation. ∎

Example 4
Playpen Problem† A certain manufacturer makes a playpen with flexible construction that permits the linkage of its sides (each side is of unit length and the playpen is normally square) to be attached at right angles to a wall (the side of a house, for example). See Figure 16. When the playpen is placed as in Figure 16, the area enclosed is 2 square units, which doubles the child's play area. Is there a configuration that will do better than double the child's play area?

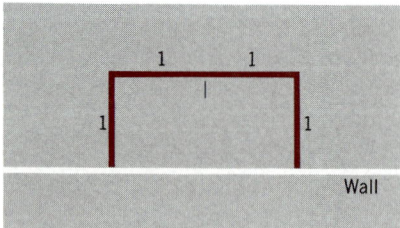

Figure 16

Solution Since the playpen must be attached at right angles to the wall, the possible configurations depend on the amount of wall that is used as a fifth side for the playpen. See Figure 17. Let x represent half the length of wall used as a fifth side.

*We assume a cost of $0.50 per therm for natural gas, 5000 degree days per year, and 1 in. of insulation has an R-4 resistance rating.

†Adapted from *Proceedings, Summer Conference for College Teachers on Applied Mathematics*, University of Missouri—Rolla, 1971.

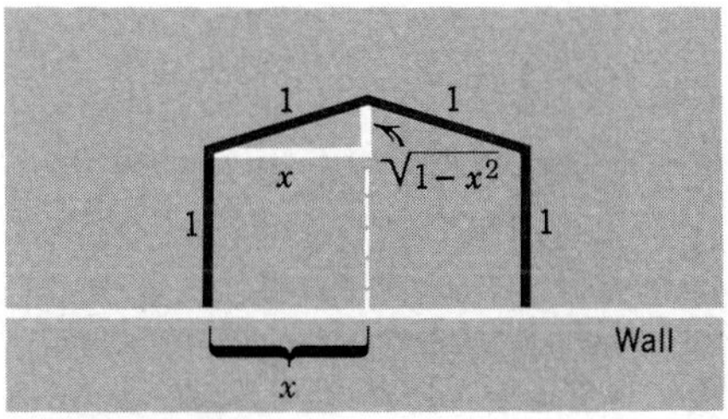

Figure 17

The area A is a function of x and is the sum of two rectangles (with sides 1 and x) and two right triangles (hypotenuse 1 and base x). Thus, the quantity to be maximized is

$$A(x) = 2x + x\sqrt{1 - x^2} \qquad 0 \le x \le 1$$

Now,

$$\begin{aligned} A'(x) &= 2 + \sqrt{1 - x^2} + x(\tfrac{1}{2})(1 - x^2)^{-1/2}(-2x) \\ &= 2 + \sqrt{1 - x^2} - \frac{x^2}{\sqrt{1 - x^2}} \\ &= \frac{2\sqrt{1 - x^2} + 1 - 2x^2}{\sqrt{1 - x^2}} \end{aligned}$$

The critical numbers obey $A'(x) = 0$; that is,

$$\frac{2\sqrt{1 - x^2} + 1 - 2x^2}{\sqrt{1 - x^2}} = 0$$

Ratios equal 0 when the numerator equals 0; thus,

$$\begin{aligned} 2\sqrt{1 - x^2} + 1 - 2x^2 &= 0 \\ 2\sqrt{1 - x^2} &= 2x^2 - 1 \\ 4(1 - x^2) &= 4x^4 - 4x^2 + 1 \\ 4x^4 &= 3 \\ x &= \sqrt[4]{\tfrac{3}{4}} \approx 0.931 \end{aligned}$$

Thus, the only critical number in $(0, 1)$ is 0.931.

Compute $A(x)$ at the endpoints $x = 0$ and $x = 1$, and at the critical number $x = 0.931$. The results are

$$A(0) = 0 \qquad A(1) = 2 \qquad A(0.931) = 2.20$$

Thus, a wall of length $2x = 2(0.931) = 1.862$ will maximize the area, and a configuration like the one in Figure 17 increases the play area by about 10% (from 2 to 2.20). ■

Example 5 Best Dimensions for a Can

A can company wants to produce a cylindrical container with a capacity of 1000 cubic centimeters. The top and bottom of the container must be made of material that costs \$0.05 per square centimeter, while the sides of the container can

160 CH. 4 APPLICATIONS OF THE DERIVATIVE

be made of material costing $0.03 per square centimeter. Find the dimensions that will minimize the total cost of the container.

Solution Figure 18 shows a cylindrical container and the area of its top, bottom, and lateral surfaces.

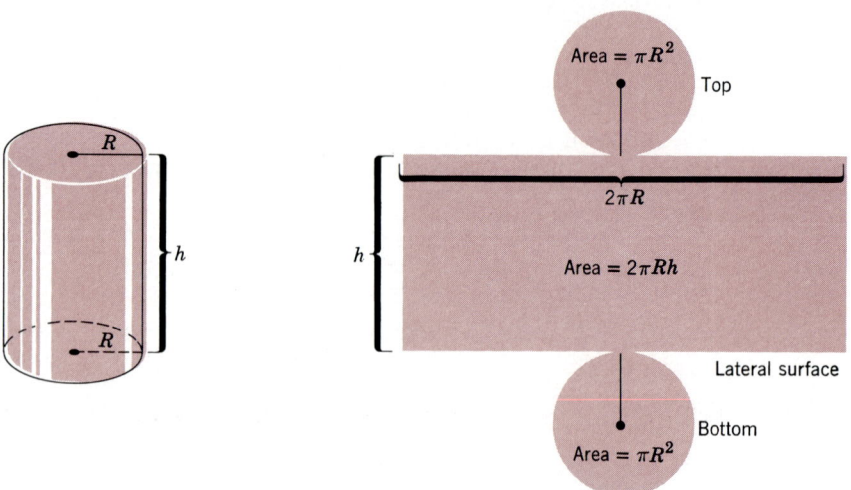

Figure 18

As indicated in the figure, if we let h stand for the height of the can and let R stand for the radius, then the total area of the bottom and top is $2\pi R^2$ and the area of the lateral surface of the can is $2\pi Rh$. The total cost C of manufacturing the can is

$$C = (\$0.05)(2\pi R^2) + (\$0.03)(2\pi Rh) = 0.1\pi R^2 + 0.06\pi Rh$$

This is the function we want to minimize.

The cost function is a function of two variables, h and R. But there is a relationship between h and R, since the volume of the cylinder is fixed at 1000 cubic centimeters. That is,

$$V = 1000 = \pi R^2 h$$

$$h = \frac{1000}{\pi R^2}$$

Substituting this expression for h into the cost function C, we obtain

$$C = C(r) = 0.1\pi R^2 + 0.06\pi R\left(\frac{1000}{\pi R^2}\right) = 0.1\pi R^2 + \frac{60}{R}$$

To find the number R that gives minimum cost, we differentiate C with respect to R:

$$\frac{dC}{dR} = C'(R) = 0.2\pi R - \frac{60}{R^2} = \frac{0.2\pi R^3 - 60}{R^2}$$

3. APPLIED PROBLEMS 161

The critical numbers obey $C'(R) = 0$, that is,

$$0.2\pi R^3 - 60 = 0$$

$$R^3 = \frac{300}{\pi}$$

$$R = \sqrt[3]{300/\pi} \approx 4.57$$

Using the second-derivative test, we obtain

$$C''(R) = 0.2\pi + \frac{120}{R^3}$$

and

$$C''(\sqrt[3]{300/\pi}) = 0.2\pi + \frac{120\pi}{300} > 0$$

Thus, for $R = \sqrt[3]{300/\pi} = 4.57$ centimeters, the cost is a relative minimum.
Since the only physical constraint is that R be positive, this relative minimum value is the absolute minimum. The corresponding height of this can is

$$h = \frac{1000}{\pi R^2} = \frac{1000}{20.89\pi} \approx 15.24 \text{ centimeters}$$

These are the dimensions that will minimize the cost of the material. ∎

If the cost of the material is the same for the top, bottom, and lateral surfaces of a cylindrical container, then the minimum cost occurs when the surface area is minimum. It can be shown (see Problem 19) that for any fixed volume, the minimum surface area is obtained when the height equals twice the radius.

Example 6
Setting Ticket Prices
A company charges $200 for each box of tools on orders of 150 or fewer boxes. The cost to the buyer on every box is reduced by $1 for each order in excess of 150. For what size order is revenue maximum?

Solution For an order of exactly 150 boxes, the company's revenue is

$$\$200(150) = \$30,000$$

For an order of 160 boxes (which is 10 in excess of 150), the per box charge is $200 - 10(1) = 190$ and the revenue is

$$\$190(160) = \$30,400$$

To solve the problem, let x denote the number of boxes sold. The revenue R is

$$R = (\text{Number of boxes})(\text{Cost per box})$$
$$= x(\text{Cost per box})$$

If $x \geq 150$, the charge per box is

$$200 - 1 \begin{pmatrix} \text{Number of boxes} \\ \text{in excess of 150} \end{pmatrix} = 200 - 1(x - 150) = 350 - x$$

$x - 60$

Hence, the revenue R is

$$R = x(350 - x) = 350x - x^2$$

To find the number of boxes leading to maximum revenue, we find the critical numbers of $R(x)$:

$$R'(x) = 350 - 2x$$
$$R'(x) = 0 \quad \text{when} \quad x = 175$$

Since $R''(x) = -2 < 0$ for all x, there is an absolute maximum at $x = 175$. Thus, a purchase of 175 boxes maximizes the company's revenue. The maximum revenue is

$$R = (350)(175) - (175)^2 = \$30{,}625$$

Of course, the company would set this figure as the most it would allow anyone to purchase on *this* plan, since revenue to the company starts to decrease for orders in excess of 175.

■

Exercise 3
Solutions to Odd-Numbered Problems begin on page 389.

1. A farmer with 3000 meters of fencing wants to enclose a rectangular plot that borders on a straight highway. If the farmer does not fence the side along the highway, what is the largest area that can be enclosed?
2. If the farmer in Problem 1 also decides to fence the side along the highway, what is the largest area that can be enclosed?
3. Find the dimensions of the rectangle of the largest area that can be enclosed by 200 meters of fencing.
4. *Cost of Fencing.* A builder wants to fence in 135,000 square meters of land in a rectangular shape. Because of security reasons, the fence in the front will cost $2 per meter, while the fence for the other three sides will cost $1 per meter. How much of each type of fence will be needed in order to minimize the cost? What is the minimum cost?
5. A farmer with 30,000 meters of fencing wants to enclose a rectangular field and then divide it into two plots with a fence parallel to one of the sides. See the illustration. What is the largest area that can be enclosed?

6. A farmer wants to enclose 6000 square meters of land in a rectangular plot and then divide it into two plots with a fence parallel to one

of the sides. What are the dimensions of the rectangular plot that require the least amount of fence?

7. A rectangle has a perimeter of length L. What should the dimensions of the rectangle be if its area is to be a maximum?

8. *Best Dimensions for a Box.* An open box with a square base is to be made from a square piece of cardboard 12 centimeters on a side by cutting out a square from each corner and turning up the sides. Find the dimensions of the box that yield the maximum volume.

9. *Best Dimensions for a Box.* An open box with a square base is to be made from a square piece of cardboard 24 centimeters on a side by cutting out a square from each corner and turning up the sides. Find the dimensions of the box that yield maximum volume.

10. *Best Dimensions for a Box.* A box, open at the top with a square base, is to have a volume of 8000 cubic centimeters. What should the dimensions of the box be if the amount of material used is to be a minimum?

11. If the box in Problem 10 is to be closed on top, what should the dimensions of the box be if the amount of material used is to be a minimum?

12. *Best Dimensions for a Can.* A cylindrical container is to be produced that will have a capacity of 4000 cubic centimeters. The top and bottom of the container are to be made of material that costs $0.50 per square centimeter, while the side of the container is to be made of material costing $0.40 per square centimeter. Find the dimensions that will minimize the total cost of the container.

13. *Best Dimensions for a Can.* A cylindrical container is to be produced that will have a capacity of 10 cubic meters. The top and bottom of the container are to be made of a material that costs $2 per square meter, while the side of the container is made of material costing $1.50 per square meter. Find the dimensions that will minimize the total cost of the container.

14. *Setting Rental Prices.* A car rental agency has 24 identical cars. The owner of the agency finds that at a price of $10 per day, all the cars can be rented. However, for each $1 increase in rental, one of the cars is not rented. What should be charged to maximize income?

15. *Charter Flight Charges.* A charter flight club charges its members $200 per year. But for each new member in excess of 60, the charge for every member is reduced by $2. What number of members leads to a maximum revenue?

16. *Placing Telephone Boxes.* A telephone company is asked to provide telephone service to a customer whose house is located 2 kilometers away from the road along which the telephone lines run. The nearest telephone box is located 5 kilometers down this road. If the cost to connect the telephone line is $50 per kilometer along the road and $60 per kilometer away from the road, where along the road from the box should the company connect the telephone line so as to minimize construction cost? [*Hint:* Let x denote the distance from the box to the connection so that $5 - x$ is

the distance from this point to the point on the road closest to the house.]

17. A small island is 3 kilometers from the nearest point P on the straight shoreline of a large lake. If a woman on the island can row her boat 2.5 kilometers per hour and can walk 4 kilometers per hour, where should she land her boat in order to arrive in the shortest time at a town 12 kilometers down the shore from P?

18. *Most Economical Speed.* A truck has a top speed of 75 miles per hour and, when traveling at the rate of x miles per hour, consumes gasoline at the rate of $\frac{1}{200}[(1600/x) + x]$ gallon per mile. This truck is to be taken on a 200 mile trip by a driver who is to be paid at the rate of b dollars per hour plus a commission of c dollars. Since the time required for this trip at x miles per hour is 200/x, the total cost, if gasoline costs a dollars per gallon, is

$$C(x) = \left(\frac{1600}{x} + x\right)a + \frac{200}{x}b + c$$

Find the most economical possible speed under each of the following sets of conditions:

(a) $b = 0$, $c = 0$ (b) $a = 1.50$, $b = 8.00$, $c = 500$
(c) $a = 1.60$, $b = 10.00$, $c = 0$

*19. Prove that a cylindrical container of fixed volume V requires the least material (minimum surface area) when its height is twice its radius.

20. *Page Layout.* A printer plans on having 50 square inches of printed matter per page and is required to allow for margins of 1 inch on each side and 2 inches on the top and bottom. What are the most economical dimensions for each page if the cost per page depends on the area of the page?

21. *Dimensions for a Window.* A window is to be made in the shape of a rectangle surmounted by a semicircle with diameter equal to the width of the rectangle. See the illustration. If the perimeter of the window is 22 feet, what dimensions will emit the most light?

4. Concavity

In the previous sections, we discussed how to determine where a differentiable function is increasing and decreasing and where it has a relative maximum (minimum). What we are missing is the ability to determine whether the graph of the function opens up or opens down on a given interval; that is, we need to know the *concavity* of the function. For example, the parabola $y = f(x) = x^2 + 1$ obviously opens up, or to be more precise, is *concave up* on its domain. See Figure 3 on page 39. When tangent lines are drawn at every point of this parabola, we notice that they are always below the graph itself. This leads us to formulate the following definition:

Concave Up; Concave Down **A differentiable function is *concave up* on (a, b) if the tangent lines to the graph at every point lie below the graph. A differentiable function is *concave down* on (a, b) if the tangent lines to the graph at every point lie above the graph.**

Figures 19 and 20 illustrate these concepts.

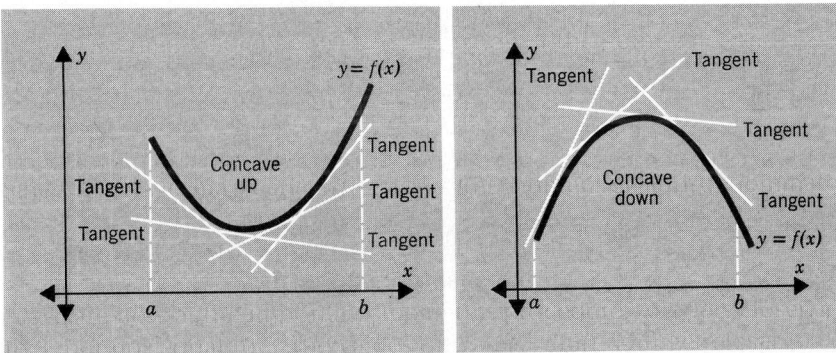

Figure 19   **Figure 20**

Notice that a function may sometimes be concave up and sometimes be concave down. See Figure 21.

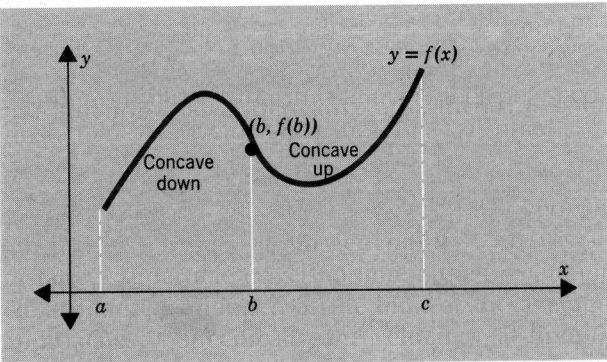

Figure 21

When a function is concave down on (a, b) and concave up on (b, c), the point $(b, f(b))$ is particularly significant, since it is at this point that the graph changes from being concave down to being concave up. This point is called an *inflection point*.

Inflection Point

Observe that the slopes of the tangent lines [measured by $f'(x)$] in Figure 19 are increasing on (a, b). If $f'(x)$ is an increasing function, its derivative $f''(x)$ is positive on (a, b). Similarly, in Figure 20, the slopes of the tangent lines [measured by $f'(x)$] are decreasing on (a, b). If $f'(x)$ is a decreasing function, its derivative $f''(x)$ is negative on (a, b).

The second derivative of a function therefore provides a test for determining where the graph of a function is concave up and where it is concave down.

Test for Concavity **Let $y = f(x)$ be a function and let $f''(x)$ be its second derivative.**

1. **If $f''(x) > 0$ for all x in (a, b), then the graph of f is concave up on (a, b).**
2. **If $f''(x) < 0$ for all x in (a, b), then the graph of f is concave down on (a, b).**

In this test, we are assuming that the second derivative of the function exists and that the function is continuous at every number in the interval (a, b).

Let's return to the notion of an inflection point and find a test for locating such points. At an inflection point $(c, f(c))$, the concavity of the function changes. This means the sign of $f''(x)$ must change as we pass from one side of the inflection point to the other side. It also means that at the inflection point $(c, f(c))$ we must have $f''(c) = 0$.

Test for Inflection Point **To find inflection points of a function:**

1. **Find all points $(c, f(c))$ at which $f''(c) = 0$.**
2. **Determine the concavity on either side of these points. If there is a change in concavity, the point is an inflection point; otherwise, it is not.**

Example 1 Find where the function

$$f(x) = x^3 - 6x^2 + 9x + 3$$

is concave up and concave down. Find all inflection points and sketch the graph.

Solution We compute $f'(x)$ and $f''(x)$:

$$f'(x) = 3x^2 - 12x + 9$$
$$f''(x) = 6x - 12 = 6(x - 2)$$

When $x > 2$, we see that $f''(x) > 0$. Thus, the function is concave up for $x > 2$. Similarly, it is concave down for $x < 2$.

To find the inflection points, set

$$f''(x) = 6(x - 2) = 0$$

The only solution is $x = 2$. Since $f(2) = 5$, the only candidate for an inflection point is $(2, 5)$. Since for $x < 2$, the function is concave down and for $x > 2$, it is concave up, the point $(2, 5)$ is an inflection point.

To graph the function, we use the information above plus whatever we can find out about the function from $f'(x)$. The critical numbers obey

$$f'(x) = 3x^2 - 12x + 9 = 0$$
$$3(x^2 - 4x + 3) = 0$$
$$(x - 1)(x - 3) = 0$$

Thus, $x = 1$ and $x = 3$ are critical numbers and

$$f''(1) = 6(1 - 2) < 0 \quad \text{and} \quad f''(3) = 6(3 - 2) > 0$$

That is, $(1, 7)$ is a relative maximum and $(3, 3)$ is a relative minimum. Figure 22 gives the graph.

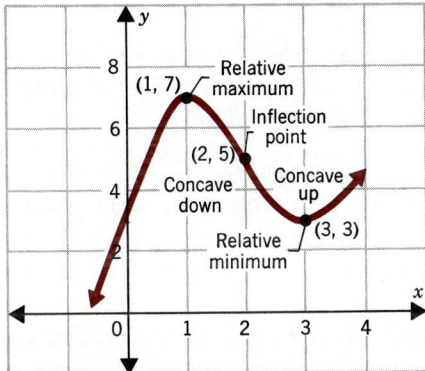

Figure 22

The concavity of a function can change without an inflection point occurring. Consider the function $f(x) = 1/x$. The first and second derivatives are

$$f'(x) = \frac{-1}{x^2} \quad \text{and} \quad f''(x) = \frac{2}{x^3}$$

Clearly,
$$f''(x) < 0 \quad \text{for} \quad x < 0 \qquad f''(x) > 0 \quad \text{for} \quad x > 0$$

That is, the function is concave down for x negative and is concave up for x positive.

Since $f''(x)$ cannot equal zero at any point in the domain, there are no inflection points. See Figure 23 for the graph.

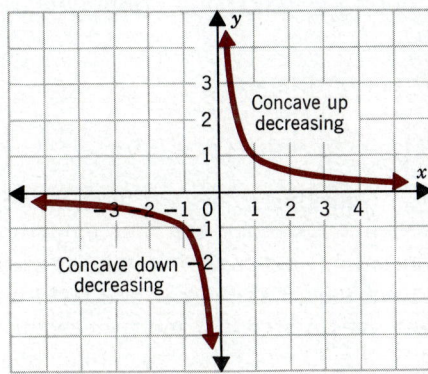

Figure 23

168 CH. 4 APPLICATIONS OF THE DERIVATIVE

Caution
The preceding example shows a function that changes concavity without an inflection point occurring. It is also possible for a function to have an inflection point that cannot be detected by the test for an inflection point. The reason for this is that the test requires determining where the second derivative is zero.

Thus, points at which the second derivative does not exist may be inflection points.

To find inflection points for such a function, we must test the concavity on both sides of points for which the second derivative does not exist. The next example illustrates such a function.

Example 2 Find all inflection points of the function: $f(x) = x^{5/3}$

Solution
$$f'(x) = \frac{5}{3}x^{2/3} \qquad f''(x) = \frac{10}{9}x^{-1/3} = \frac{10}{9x^{1/3}}$$

Here, $f''(x)$ is never zero. But we observe that $f''(x)$ does not exist at $x = 0$, which is in the domain. Since the function is concave down for $x < 0$ and concave up for $x > 0$, the point $(0, 0)$ is an inflection point. See Figure 24. ■

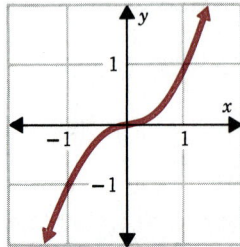

Figure 24

Example 3 Sketch the graph of a function $y = f(x)$ that is continuous for all x and has the following properties:

1. $(1, 3)$, $(3, 5)$, and $(5, 7)$ are on the graph
2. $f'(1) = 0$ and $f'(5) = 0$
3. $f''(x) > 0$ for $x < 3$, $f''(3) = 0$, and $f''(x) < 0$ for $x > 3$

Solution We obtain Figure 25(a) by using the points and first derivatives given above. Since $f''(x) > 0$ for $x < 3$, it follows that $f''(1) > 0$. Similarly, $f''(5) < 0$. So, the point $(1, 3)$ is a relative minimum and $(5, 7)$ is a relative maximum. See Figure 25(b). We also conclude from item 3 above and the test for concavity that the graph of the function is concave up for $x < 3$ and concave down for $x > 3$. Since the concavity changes at 3, the function has an inflection point at $(3, 5)$. See Figure 25(c).

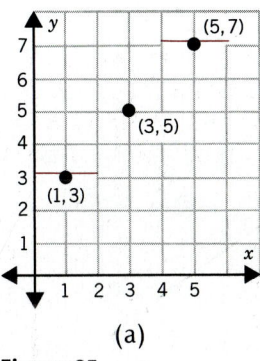
(a)

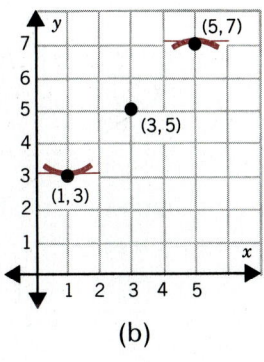
(b)

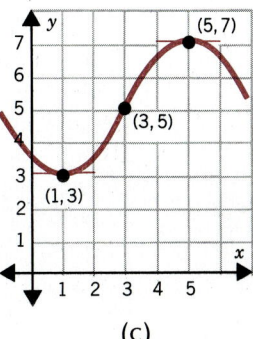
(c)

Figure 25

Exercise 4
Solutions to Odd-Numbered Problems begin on page 392.

In Problems 1–10 find where the function f is concave up or concave down. Find all points of inflection.

1. $f(x) = x^2 - 2x + 5$
2. $f(x) = x^2 + 4x - 2$
3. $f(x) = x^3 - 9x^2 + 2$
4. $f(x) = x^3 - 6x^2 + 9x + 1$
5. $f(x) = x^4 - 4x^3 + 10$
6. $f(x) = 3x^4 - 8x^3 + 6x + 1$
7. $f(x) = \dfrac{x+1}{x}$
8. $f(x) = 2x^2 - \dfrac{1}{x}$
9. $f(x) = 3x^{1/3} + 2x$
10. $f(x) = (x-1)^{3/2}$

In Problems 11–26 graph each function.

11. $f(x) = 2x^3 - 6x^2 + 6x - 3$
12. $f(x) = 2x^3 + 9x^2 + 12x - 4$
13. $f(x) = -2x^3 + 15x^2 - 36x + 7$
14. $f(x) = x^3 + 10x^2 + 25x - 25$
15. $f(x) = x^4 - 4x$
16. $f(x) = x^4 + 4x$
17. $f(x) = 4x^6 - 6x^4$
18. $f(x) = 4x^6 + 6x^4$
19. $f(x) = 3x^5 - 5x^3$
20. $f(x) = 3x^5 + 5x^3$
21. $f(x) = x^{2/3} + x^{1/3}$
22. $f(x) = x^{2/3} - x^{1/3}$
23. $f(x) = x^{2/3}(x - 10)$
24. $f(x) = x^{1/3}(x - 3)$
25. $f(x) = x^{2/3}(x^2 - 8)$
26. $f(x) = x^{1/3}(x^2 - 2)$

27. Sketch the graph of a function $y = f(x)$ that is continuous for all x and has the following properties:

    1. $(0, 10)$, $(6, 15)$, and $(10, 0)$ are on the graph
    2. $f'(6) = 0$ and $f'(10) = 0$
    3. $f''(x) < 0$ for $x < 9$, $f''(9) = 0$, and $f''(x) > 0$ for $x > 9$

28. Sketch the graph of a function $y = f(x)$ that is continuous for all x and has the following properties:

    1. $(-1, 3)$, $(1, 5)$, and $(3, 7)$ are on the graph
    2. $f'(3) = 0$ and $f'(-1) = 0$
    3. $f''(x) > 0$ for $x < 1$, $f''(1) = 0$, and $f''(x) > 0$ for $x > 1$

29. Sketch the graph of a function $y = f(x)$ that is continuous for all x and has the following properties:

   1. $(1, 5)$, $(2, 3)$, and $(3, 1)$ are on the graph
   2. $f'(1) = 0$ and $f'(3) = 0$
   3. $f''(x) > 0$ for $x < 2$, $f''(2) = 0$, and $f''(x) > 0$ for $x > 2$

30. Sketch the graph of a function $y = f(x)$ that is continuous and has the following properties:

   1. Domain of $f(x)$ is $x \geq 0$
   2. $(0, 0)$ and $(6, 7)$ are on the graph
   3. $f'(x) > 0$ for $x > 0$
   4. $f''(x) < 0$ for $x < 6$, $f''(6) = 0$, and $f''(x) > 0$ for $x > 6$

31. The cost (in dollars) of producing x units is given by the function

$$C(x) = 0.001x^3 - 0.3x^2 + 30x + 42$$

   Find where this cost function is concave up and where it is concave down, and find the inflection point.

*32. For the function $f(x) = ax^3 + bx^2$, determine a and b so that the point $(1, 6)$ is a point of inflection of $f(x)$.

*33. Let $f(x) = ax^2 + bx + c$, where $a \neq 0$, and b, c are real numbers. Is it possible for $f(x)$ to have an inflection point? Explain your answer.

*34. Let $f(x) = ax^3 + bx^2 + cx + d$, where $a \neq 0$, and b, c, d are real numbers. Is it possible for $f(x)$ to have an inflection point? Explain your answer.

5. Asymptotes

In Chapter 2 we described $\lim_{x \to c} f(x) = L$ by saying that the value of $f(x)$ can be made as close as we please to L by choosing numbers x sufficiently close to c. It was understood that L and c were numbers. In this section we extend the language of limits to allow c to be $+\infty$ or $-\infty$ (*limits at infinity*) and to allow L to be $+\infty$ or $-\infty$ (*infinite limits*).* These limits, it turns out, are useful for locating *asymptotes* and hence aid in obtaining the graph of a function.

We begin with limits at infinity.

Limits at Infinity

Let's look at the function $f(x) = 1/x$, whose domain is $x \neq 0$. See Figure 26. This function has the property that the value $f(x)$ can be made as close as we please to 0 when the number x is sufficiently positive. The table illustrates this fact for selected numbers x:

x	1	10	100	1000	10,000	100,000
$f(x) = 1/x$	1	0.1	0.01	0.001	0.0001	0.00001

*Remember that the symbols $+\infty$ (plus infinity) and $-\infty$ (minus infinity) are *not* numbers. Plus infinity expresses the idea of unboundedness in the positive direction; minus infinity expresses the idea of unboundedness in the negative direction.

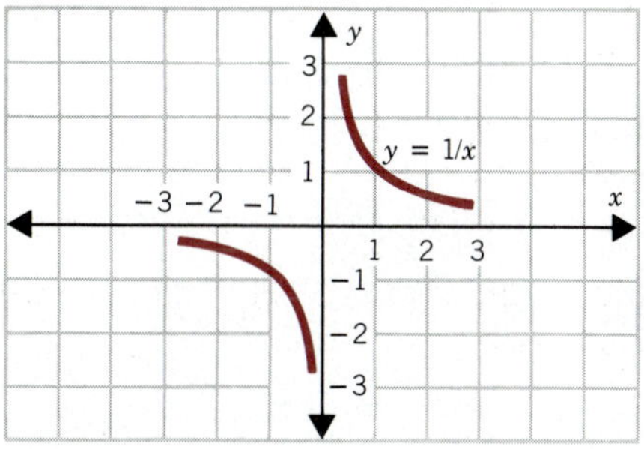

5. ASYMPTOTES

Figure 26

This phenomenon is expressed by saying that $f(x) = 1/x$ has the limit 0 as x approaches $+\infty$ and is symbolized by writing

(1) $$\lim_{x \to +\infty} \frac{1}{x} = 0$$

In the same way, we can write

(2) $$\lim_{x \to -\infty} \frac{1}{x} = 0$$

to indicate that $1/x$ can be made as close as we please to 0 by selecting numbers x sufficiently negative. We summarize statements (1) and (2) by saying that $f(x) = 1/x$ has *limits at infinity*.

Limits at Infinity

Example 1 Find: (a) $\lim_{x \to +\infty} \dfrac{3x - 2}{4x - 1}$ (b) $\lim_{x \to +\infty} \dfrac{5x^4 - 3x^2 + 2}{x^3 + 5}$

Solution (a) We evaluate this limit by first dividing each term of both the numerator and the denominator by the highest power of x that appears in the denominator (in this case x). Then

$$\lim_{x \to +\infty} \frac{3x - 2}{4x - 1} = \lim_{x \to +\infty} \frac{3 - (2/x)}{4 - (1/x)} = \frac{\lim_{x \to +\infty}[3 - (2/x)]}{\lim_{x \to +\infty}[4 - (1/x)]}$$

$$= \frac{\lim_{x \to +\infty} 3 - \lim_{x \to +\infty}(2/x)}{\lim_{x \to +\infty} 4 - \lim_{x \to +\infty}(1/x)} = \frac{3 - 0}{4 - 0} = \frac{3}{4}$$

(b) We follow the same procedure as in Part (a):

$$\lim_{x \to +\infty} \frac{5x^4 - 3x^2 + 2}{x^3 + 5} \underset{\text{Divide by } x^3}{=} \lim_{x \to +\infty} \frac{5x - (3/x) + (2/x^3)}{1 + (5/x^3)} = +\infty$$

Infinite Limits

We again use the function $f(x) = 1/x$, whose graph was given in Figure 26, to introduce the idea of *infinite limits*. We construct a table that gives values of $f(x)$ for selected numbers x that are close to 0:

x	1	0.1	0.01	0.001	0.0001	0.00001
$f(x) = 1/x$	1	10	100	1000	10,000	100,000

Here, we see that as x gets closer to 0 from the right, the value of $f(x) = 1/x$ can be made as positive as we please—that is, $1/x$ becomes unbounded in the positive direction. We express this fact by writing

(3)
$$\lim_{x \to 0^+} \frac{1}{x} = +\infty$$

Similarly, we use the notation

(4)
$$\lim_{x \to 0^-} \frac{1}{x} = -\infty$$

to indicate that $1/x$ can be made as negative as we please by selecting numbers x sufficiently close to 0, but less than 0. We summarize (3) and (4) by saying that $f(x) = 1/x$ has *one-sided infinite limits* at 0.

• **One-Sided Infinite Limits**

We now apply the ideas of limits at infinity and infinite limits to the problem of finding *horizontal* and *vertical asymptotes*.

Horizontal Asymptotes

Limits at infinity have an interesting geometric interpretation. When $\lim\limits_{x \to +\infty} f(x) = L$, it means that as x becomes sufficiently positive, the value of $f(x)$ can be made as close as we please to L; that is, the graph of $y = f(x)$ for x sufficiently positive is as close as we please to the horizontal line $y = L$. Similarly, $\lim\limits_{x \to -\infty} f(x) = M$ means that the values of $f(x)$ can be made as close as we please to M for x sufficiently negative. Thus, the graph of $y = f(x)$ for x sufficiently negative is as close as we please to the horizontal line $y = M$. These lines are called *horizontal asymptotes* of the graph of f. See Figure 27.

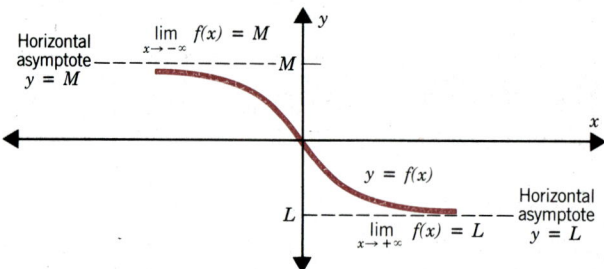

Figure 27

Vertical Asymptotes

Infinite limits are used to find vertical asymptotes. Figure 28 illustrates the possibilities that can occur when a function has an infinite limit. Whenever

$$\lim_{x \to c^-} f(x) = \pm\infty \quad \text{or} \quad \lim_{x \to c^+} f(x) = \pm\infty$$

we call the line $x = c$ a *vertical asymptote* of the graph of f.

5. ASYMPTOTES 173

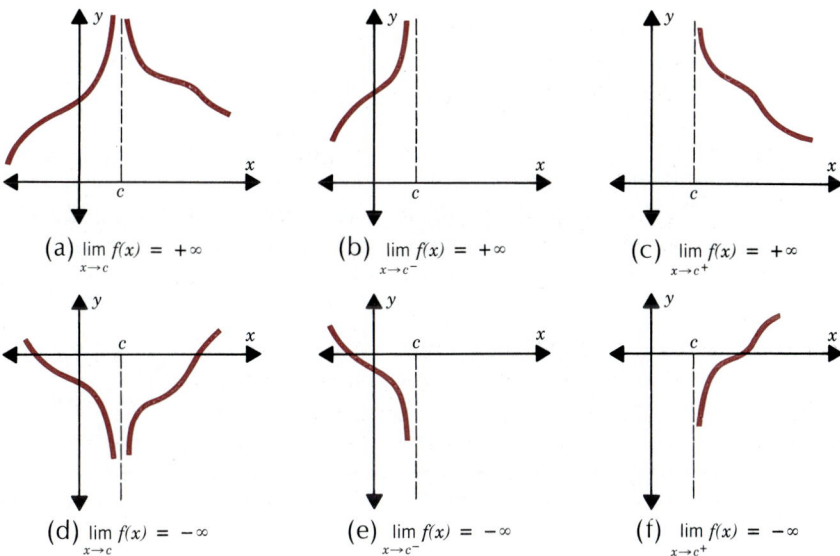

Figure 28

Example 2 Locate all vertical and horizontal asymptotes of the function below, and graph f.

$$f(x) = \frac{x}{x - 3}$$

Solution Since $x = 3$ is the only number for which the denominator of f equals 0, we evaluate the limit of f as $x \to 3$ to determine whether $x = 3$ is a vertical asymptote:

$$\lim_{x \to 3^-} \frac{x}{x - 3} = -\infty \quad \text{and} \quad \lim_{x \to 3^+} \frac{x}{x - 3} = +\infty$$

Hence, the line $x = 3$ is a vertical asymptote for the graph.

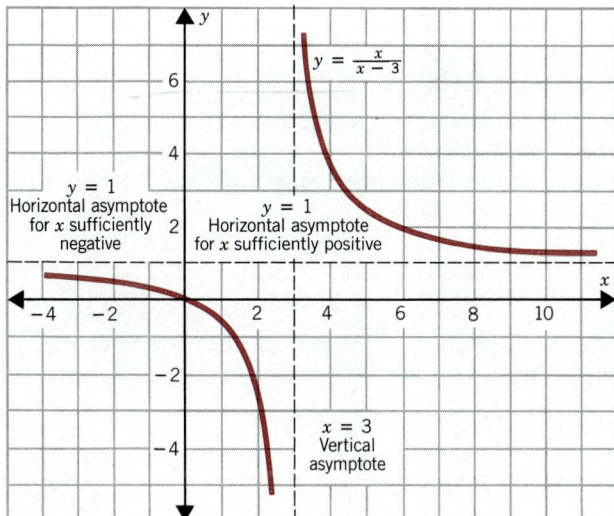

Figure 29

To locate the horizontal asymptotes, if any, we look at the limits at infinity of f:

$$\lim_{x \to +\infty} \frac{x}{x-3} = \lim_{x \to +\infty} \frac{1}{1 - 3/x} = 1 \quad \text{and} \quad \lim_{x \to -\infty} \frac{x}{x-3} = \lim_{x \to -\infty} \frac{1}{1 - 3/x} = 1$$

Thus, the line $y = 1$ is a horizontal asymptote for x sufficiently positive and for x sufficiently negative.

At $x = 0$, we have $f(0) = 0$, and this is the only x-intercept. Putting all this information together, we obtain the graph of f shown in Figure 29. ∎

Sketching Graphs

The next example summarizes how some of the concepts we have discussed in this and earlier chapters can be used to obtain a quite accurate sketch of the graph of a function, while computing only a few points. For the purpose of this example and for Problems 15–22 in Exercise 5 we define "investigate fully the graph of the function" to mean:

Find the following:

1. Intercepts of f
2. Symmetries of f
3. Asymptotes of f
4. Critical numbers of f
5. Relative maxima and relative minima of f
6. Concavity of f
7. Inflection points of f

Example 3   Investigate fully the graph of the function: $f(x) = \dfrac{x^2}{x^2 - 1}$

Solution
1. The only intercept is $(0, 0)$.
2. The graph is symmetric with respect to the y-axis.
3. Since

$$\lim_{x \to +\infty} \frac{x^2}{x^2 - 1} = 1 \quad \text{and} \quad \lim_{x \to -\infty} \frac{x^2}{x^2 - 1} = 1$$

the line $y = 1$ is a horizontal asymptote for x sufficiently positive and for x sufficiently negative. The lines $x = -1$ and $x = 1$ are vertical asymptotes, since

$$\lim_{x \to -1^-} \frac{x^2}{x^2 - 1} = +\infty \quad \text{and} \quad \lim_{x \to -1^+} \frac{x^2}{x^2 - 1} = -\infty$$

$$\lim_{x \to 1^-} \frac{x^2}{x^2 - 1} = -\infty \quad \text{and} \quad \lim_{x \to 1^+} \frac{x^2}{x^2 - 1} = +\infty$$

4. $$f'(x) = \frac{(x^2 - 1)(2x) - x^2(2x)}{(x^2 - 1)^2} = \frac{-2x}{(x^2 - 1)^2}$$

There is a critical number at $x = 0$ ($x = -1$ and $x = 1$ are not in the domain of f).

Since we will need f'' below, we attempt to use the second-derivative test on this critical number:

$$f''(x) = (-2)\left[\frac{(x^2 - 1)^2(1) - (x)(2)(x^2 - 1)(2x)}{(x^2 - 1)^4}\right] = \frac{2(3x^2 + 1)}{(x^2 - 1)^3}$$

5. Since $f''(0) < 0$, it follows that f has a relative maximum at 0.
6. The table provides an analysis of the concavity of f:

Interval	Sign of f''	Concavity
$(-\infty, -1)$	+	Up
$(-1, 1)$	−	Down
$(1, +\infty)$	+	Up

7. Since $f''(x)$ exists for all $x \neq \pm 1$ and is never 0, we conclude that f has no inflection points.

Figure 30 illustrates the graph of f.

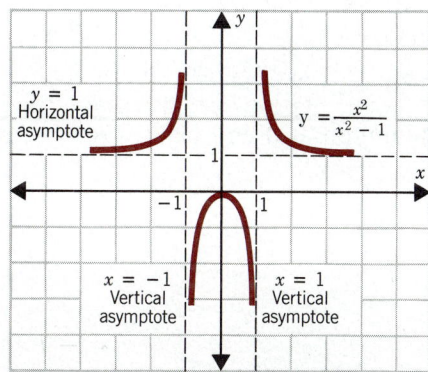

Figure 30

Exercise 5

Solutions to Odd-Numbered Problems begin on page 396.

In Problems 1–8 find the indicated limit.

1. $\lim\limits_{x \to +\infty} \dfrac{x^3 + x^2 + 2x - 1}{x^3 + x + 1}$

2. $\lim\limits_{x \to +\infty} \dfrac{2x^2 - 5x + 2}{5x^2 + 7x - 1}$

3. $\lim\limits_{x \to +\infty} \dfrac{2x + 4}{x - 1}$

4. $\lim\limits_{x \to +\infty} \dfrac{x + 1}{x}$

5. $\lim\limits_{x \to +\infty} \dfrac{3x^2 - 1}{x^2 + 4}$
6. $\lim\limits_{x \to -\infty} \dfrac{x^2 - 2x + 1}{x^3 + 5x + 4}$
7. $\lim\limits_{x \to -\infty} \dfrac{5x^3 - 1}{x^2 + 1}$
8. $\lim\limits_{x \to -\infty} \dfrac{x^2 + 1}{x^3 - 1}$

In Problems 9-14 locate all horizontal and vertical asymptotes, if any, of the function f.

9. $f(x) = 3 + \dfrac{1}{x}$
10. $f(x) = 2 - \dfrac{1}{x^2}$
11. $f(x) = \dfrac{2}{(x-1)^2}$
12. $f(x) = \dfrac{3x - 1}{x + 1}$
13. $f(x) = \dfrac{x^2}{x^2 - 4}$
14. $f(x) = \dfrac{x}{x^2 - 1}$

In Problems 15-22 investigate fully the graph of the function f (see page 174 for directions).

15. $f(x) = \dfrac{2}{x^2 - 4}$
16. $f(x) = \dfrac{1}{x^2 - 1}$
17. $f(x) = \dfrac{2x - 1}{x + 1}$
18. $f(x) = \dfrac{x - 2}{x}$
19. $f(x) = \dfrac{x}{x^2 + 1}$
20. $f(x) = \dfrac{2x}{x^2 - 4}$
21. $f(x) = \dfrac{8}{x^2 - 16}$
22. $f(x) = \dfrac{x^2}{4 - x^2}$

23. Sketch the graph of a function f that is defined and continuous for $-1 \le x \le 2$ and that satisfies the following conditions:

$f(-1) = 1 \qquad f(1) = 2 \qquad f(2) = 3 \qquad f(0) = 0 \qquad f(\tfrac{1}{2}) = 3$

$\lim\limits_{x \to -1^+} f'(x) = -\infty \qquad \lim\limits_{x \to 1^-} f'(x) = -1 \qquad \lim\limits_{x \to 1^+} f'(x) = +\infty$

f has a relative minimum at 0 and f has a relative maximum at $\tfrac{1}{2}$

6. Marginal Analysis

The profit of most firms depends on the revenue derived from the sale of a product and the cost of manufacturing, distributing, and selling this product. If P is the profit, R is the revenue, and C is the cost, then

$$\text{Profit} = \text{Revenue} - \text{Cost}$$
$$P = R - C$$

Revenue is the amount of money derived from the sale of a product and equals the price of the product times the quantity of the product that is actually sold. But the price and the quantity bought and sold are not independent. As the price falls, the demand for the product increases and when the price rises, the demand decreases.

For example, the demand function for corn is usually a decreasing function so that as the quantity x of corn to be sold increases, the price of the corn will decline. See Figure 31.

6. MARGINAL ANALYSIS   177

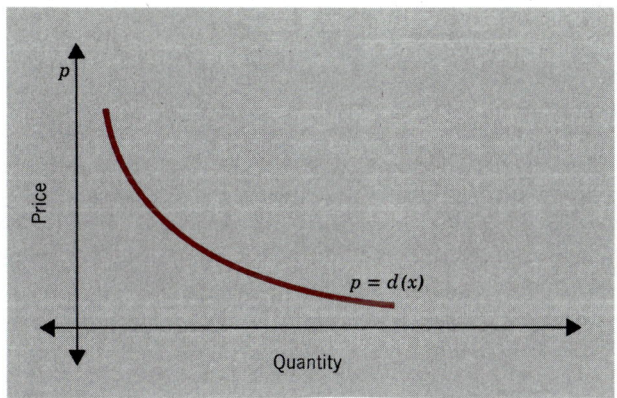

Figure 31

Demand Equation The relationship between the price p of a quantity bought and the amount x of a quantity demanded is called the *demand equation*. If in this equation we solve for p, we have

$$p = d(x)$$

Price Function The function d is called the *price function* and $d(x)$ is the price per unit when x units are demanded.

Even though x can assume only positive integer values in normal economic transactions, we will assume that x is any real number so that the price function $d(x)$ may be treated as a continuous function.

In practice, a price function is found through surveys, analysis of data, history, and other sources available to the economist. The next example illustrates how a linear price function can be constructed. Observe that the fundamental assumption made here is the linear nature of the price function.

Example 1 The manager of Dan's Toy Store has observed that each week 1000 electric trucks
Constructing a are sold at a price of $5 per truck. When there is a special sale, the trucks sell for
Demand Function $4 each and 1200 per week are sold. Assuming a linear price function, construct the price function.

Solution Let p be the price of each truck and let x be the number sold. If the price function $p = d(x)$ is linear, then we know that $(1000, 5)$ and $(1200, 4)$ are two points on the line $p = d(x)$. Thus, the price function is

$$p - 4 = \frac{-1}{200}(x - 1200)$$

$$p = \frac{-1}{200}x + 10$$

The price function obtained in Example 1 is not meant to reflect extreme situations. For example, we do not expect to sell $x = 0$ trucks nor do we expect to sell too many trucks in excess of 1500, since even during a special only 1200 are sold. The price function does represent the relationship between price and quantity in a certain range, in this case, perhaps $500 < x < 1500$.

Our next example illustrates a constant price function.

Example 2 No matter how much wheat a farmer can grow, it can be sold at $2 per bushel. Find the price function.

Solution Since the price per bushel is fixed at $2 per bushel, the price function is

$$p = \$2$$

■

Revenue Function

If x is the number of units sold and $d(x)$ is the price for each unit, the *revenue function* $R(x)$ is defined as

$$R(x) = x\,d(x)$$

Marginal Revenue

The rate of change of revenue as output changes, called the *marginal revenue*, MR, is frequently described by economists as the additional revenue from selling an additional unit of output. If we know the revenue function $R(x)$, the marginal revenue function is the derivative of the revenue function. That is,

$$MR = \lim_{\Delta x \to 0} \frac{\Delta R}{\Delta x} = R'(x)$$

Now we consider the cost to a firm of producing a product. The cost C is composed of two parts, fixed costs and variable costs. Fixed costs remain the same no matter how much is produced or manufactured. Examples of fixed costs are mortgage payments, interest, insurance, real estate taxes, and salaries of people who will be employed even through a time of zero production. Variable costs change as production changes. Examples are costs of raw materials, salaries of people directly related to production, and so on. If C is the cost, VC is the variable cost, and FC is the fixed cost, then

$$C = VC + FC$$

Also, we assume that FC is a constant, that VC is a function of the quantity produced, and that $VC = 0$ when zero quantity is produced.

To increase profits, a company may decide to increase its production. Associated with this increase are added costs, such as new capital investments, additional labor costs, increased storage facilities, added sales and advertising expenditures, and the like. The question that concerns management is how will the cost be affected by an increase in production?

Marginal Cost
Cost Function

What we need to measure is the rate of change of cost as output changes. This quantity, called the *marginal cost*, MC, is frequently described by economists as the additional cost of producing an additional unit of output. If we know the cost function C, the marginal cost function is the derivative of the cost function. That is,

$$MC = \lim_{\Delta x \to 0} \frac{\Delta C}{\Delta x} = C'(x)$$

Profit Function

In a free competition economy the price p is fixed. In a monopoly the price p is determined by a nonconstant price function $d(x)$. In either situation, the revenue derived from selling x units is $R(x) = x\,d(x)$. If $C(x)$ is the cost of producing x units, the *profit function P*, assuming whatever is produced can be sold, is

$$P(x) = R(x) - C(x)$$

What quantity x will maximize profits?

To maximize $P(x)$, we find the critical numbers that obey

$$\frac{d}{dx}P = \frac{d}{dx}(R - C) = 0$$

or

$$\frac{d}{dx}R - \frac{d}{dx}C = 0$$

or

$$MR - MC = 0$$

We apply the second-derivative test to the function P:

$$P''(x) = R''(x) - C''(x)$$

It follows that P has a relative maximum at a number x if $P''(x) < 0$. This will occur at a number x for which the marginal revenue function equals the marginal cost and $R''(x) < C''(x)$. The numbers x are restricted to a closed interval in which the endpoints should be tested separately.

The equality

$$MR = MC$$

is the basis for the classical economic criterion for maximum profit in a monopoly—that marginal revenue and marginal cost be equal.

Example 3
Maximizing Profit

Suppose that the wheat farmer in Example 2, who can sell wheat at a fixed price of $2 per bushel, can produce anywhere from 0 to 15,000 bushels. Suppose the cost function (in dollars) is

$$C = \frac{x^2}{10{,}000} + 500$$

where x represents the number of bushels produced. We interpret this cost function as consisting of total fixed costs of $500 and total variable costs of $x^2/10{,}000$ dollars. The total fixed cost of $500 is due to costs of land, equipment, and so on. The total variable cost represents the cost of planting, fertilizing, and harvesting the crop. For example, the cost of producing 15,000 bushels is

$$C = \frac{(15{,}000)(15{,}000)}{10{,}000} + 500 = 22{,}500 + 500 = \$23{,}000$$

Of particular interest is the marginal cost, which is

$$MC = \frac{dC}{dx} = \frac{x}{5000}$$

For example, for $x = 5000$ bushels, the marginal cost is

$$MC = 1$$

and for $x = 6000$ bushels, the marginal cost is

$$MC = 1.2$$

The difference in marginal cost is an indication that the cost is increasing from $1 per bushel to $1.20 per bushel. This increase is all right provided it is not detrimental to the total profit picture. That is, the combination of cost and revenue is what is critical—not cost alone. Thus, we need to ask how the revenue function changes relative to the quantity produced. Comparing this to the marginal cost will provide valuable information to the farmer.

Assuming the farmer can sell all the wheat that is grown, how much wheat should be produced to maximize profit?

Solution This is a classic example of free competition. The maximum profit occurs when marginal cost equals marginal revenue. Thus,

$$MC = \frac{x}{5000} \qquad MR = 2$$

These are equal when

$$\frac{x}{5000} = 2$$

$$x = 10,000$$

Since

$$P(x) = R(x) - C(x) = 2x - \frac{x^2}{10,000} - 500$$

it follows that

$$P'(x) = 2 - \frac{2x}{10,000} \quad \text{and} \quad P''(x) = -\frac{1}{5000} < 0$$

Thus, at $x = 10,000$ there is a relative maximum. Since

$$P(0) = -500$$

$$P(10,000) = 2(10,000) - \left[\frac{(10,000)^2}{10,000} + 500\right]$$

$$= 20,000 - 10,500 = \$9500$$

$$P(15,000) = 2(15,000) - 23,000 = \$7000$$

the maximum profit occurs for production that is 5000 bushels under the maximum output of 15,000 bushels—that is, at 10,000 bushels.

Exercise 6
Solutions to Odd-Numbered Problems begin on page 397.

1. If the cost function is $C(x) = 2x + 5$ and the revenue function is $R(x) = 8x - x^2$, where x is the number of units produced (in thousands) and R and C are measured in millions of dollars, find the following:
   (a) Marginal revenue
   (b) Marginal revenue at $x = 3$, $x = 4$
   (c) Marginal cost
   (d) Fixed cost
   (e) Variable cost at $x = 4$
   (f) Break-even point, that is, $R(x) = C(x)$
   (g) Profit function
   (h) Most profitable output
   (i) Maximum profit
   (j) Marginal revenue at the most profitable output
   (k) Revenue at the most profitable output
   (l) Variable cost at the most profitable output

2. Suppose the cost function is given by $C(x) = x^2 + 5$ and the price function is $p = 12 - 2x$, where p is the price in dollars and x is the number of units produced (in thousands). Find the answers to parts (a)–(l) in Problem 1 for these functions.

3. During a fixed period, a retail store can sell x units of a product at a price of p cents per unit, where
$$p = 20 - 0.03x$$
The cost of making x units is C cents, where
$$C = 3 + 0.02x$$
What number of units will lead to maximum profits?

4. A certain item can be produced at a unit cost of \$10. The demand for the product is $p = 90 - 0.02x$, where p is the price in dollars and x is the number of units.
   (a) How many units should be produced to maximize profits?
   (b) What is the price that gives maximum profit?
   (c) What is the maximum profit?

5. *Maximizing Profit.* A coal company can produce x tons of coal at a daily cost of C dollars, where
$$C = 200 + 35x + 0.02x^2$$
The coal can be sold at a price of \$39 per ton. How many tons should be produced each day so as to maximize profits?

6. *Maximizing Profit.* A tractor company can manufacture at most 1000 heavy-duty tractors per year. Furthermore, from past demand data, the company knows that the number of heavy-duty tractors it can sell depends only on the price p of each unit. The company also knows that the cost to produce the units is a function of the number x of units sold. Assume that the price function is

$p = 19,000 - 2x$ and the cost function is $C = 1,000,000 + 10,000x + 3x^2$. How many units should be produced to maximize profits?

7. A manufacturer estimates that 500 articles per week can be sold if the unit price is $20, and that weekly sales will rise by 50 units with each $0.50 reduction in price. The cost of producing and selling x articles a week is $C(x) = 4200 + 5.10x + 0.0001x^2$. Find the following:
   (a) Price function
   (b) Level of weekly production for maximum profit
   (c) Price per article at the maximum profit

8. Let x be the total number of items produced per year and let p be the fixed price at which each item is sold. If the cost for producing x items is $C = ax^2 + bx + c$ and $p = r - sx$, what value of x produces maximum profit? Here, $a, b, c, r,$ and s are constants.

9. Viewing the equations $C = 300 + 5x$ and $R = 4x$ in the light of the discussion in this section:
   (a) What do C, R, and x represent?
   (b) What does the 300 represent? Does it have any effect on marginal cost? Why or why not?
   (c) What are marginal cost and marginal revenue?
   (d) In marginal terminology state why the equations represent a losing proposition.
   (e) What type of graph do the equations have?
   (f) Graph the equations. In geometrical terminology state why the graphs depict a losing proposition.

Average Revenue

*10. The quantity $R(x)/x$ is called the *average revenue* from selling x units. If $R''(x) < 0$, show that the average revenue is a maximum when it equals the marginal revenue.

Average Cost

*11. The quantity $C(x)/x$ is called the *average cost* of producing x units. If $C''(x) > 0$, show that the average cost is a minimum when it equals the marginal cost.

7. Models

Model 1: Maximizing Tax Revenue*

In determining the tax rate on cars, telephones, etc., the government is always faced with the following problem: How large should the tax be so that the tax revenue will be as large as possible? Let's examine this situation. When the government places a tax on a product, the price of this product for the consumer may increase and the quantity demanded may decrease accordingly. A very large tax may cause the quantity demanded to diminish to zero, and the result is that no tax revenue is collected. On the other hand, if no tax is levied, there will be no tax revenue at all. Thus, the problem is to find the tax rate that optimizes tax revenue. (Tax revenue is the product of the tax per unit times the actual market quantity consumed.)

*Adapted from P. H. Davis and W. M. Whyburn, *Introduction to Mathematical Analysis with Applications to Problems of Economics,* Addison-Wesley, Reading, Mass., 1958.

Let's assume that because of long-time experience in levying taxes, the government is able to determine that the relationship between the market quantity consumed of a certain product and the related tax is

$$t = \sqrt{27 - 3x^2}$$

where t denotes the amount of tax per unit of a product and x is the market quantity consumed (measured in appropriate units). (It must be pointed out that the relationship between tax rate and consumption is derived by government economists and is subject to both change and criticism.)

Notice that the relationship between tax rate and quantity consumed conforms to the restrictions discussed earlier. For example, when the tax rate $t = 0$, the quantity consumed is $x = 3$; when the tax is at a maximum ($t = 5.2\%$), the quantity consumed is zero. Figure 32 illustrates the graph of this relationship.

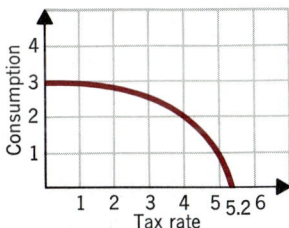

Figure 32

The revenue R due to the tax rate t is the product of tax rate per unit and the market quantity consumed:

$$R = xt = x(27 - 3x^2)^{1/2}$$

where R is measured in dollars. Since both x and t are assumed to be nonnegative, the domain of R is $0 \leq x \leq 3$. Also, $R = 0$ at both $x = 0$ and $x = 3$, so that for some positive number x between 0 and 3, R attains its absolute maximum.

To find the absolute maximum, we take the derivative of R with respect to x using the formula for the derivative of a product and the power rule:

$$R' = (27 - 3x^2)^{1/2} + \frac{1}{2}x(27 - 3x^2)^{-1/2}(-6x)$$

$$= \frac{27 - 3x^2 - 3x^2}{(27 - 3x^2)^{1/2}} = \frac{27 - 6x^2}{(27 - 3x^2)^{1/2}}$$

The critical numbers obey

$$27 - 6x^2 = 0$$
$$x^2 = 4.5$$

Since $0 \leq x \leq 3$, the only critical number is $x = \sqrt{4.5} \approx 2.12$.

To find the absolute maximum, we compare the value of R at the endpoints $x = 0$ and $x = 3$ with its value at $x = 2.12$:

$$R(0) = 0 \qquad R(\sqrt{4.5}) = \sqrt{4.5}\sqrt{13.5} = (2.12)(3.67) = 7.79 \qquad R(3) = 0$$

Thus, the revenue is maximized at $x = 2.12$. The tax rate corresponding to maximum revenue is

$$t = \sqrt{27 - 3x^2} = \sqrt{13.5} \approx 3.67$$

This means that, for a tax rate of 3.67%, a maximum revenue $R = 7.79$ is generated.

Model 2: Optimal Trade-In Time*

Car owners are often faced with the following problem: When is the best time for the car to be traded in?

Two major considerations form the framework of determining optimal trade-in time. The first is the estimated cost of repairs needed to maintain the car; the second is the replacement cost of the car. We want to express these costs as functions of time t and then determine the value of t that minimizes the total cost.

Let's first determine the relationship between time t (measured in months) of ownership and the number of repairs of a car. We can set k, the total number of repairs, equal to

$$k = \frac{1}{10} t^{3/2}$$

Notice that as the time t increases, the total number k of repairs also increases and that a new car ($t = 0$) is expected to have zero repairs. Thus, this relationship between k and t appears reasonable. The average number of repairs per month is k/t, and, if the average cost per repair is $50, the average repair cost per month is $50(k/t)$.

At the end of time t, suppose the car is replaced. Then the average number of replacements per month is $1/t$. If the cost of replacement is $2800, the average replacement cost per month is $2800(1/t)$.

If the original cost of the car is ignored, the function

$$C(t) = 50 \frac{k}{t} + 2800 \frac{1}{t} = 5t^{1/2} + 2800 \frac{1}{t}$$

represents the average repair and replacement costs per month.

Our problem is to determine t ($t > 0$) for which the average cost $C(t)$ of repair and replacement is a minimum. Then

$$C'(t) = \frac{5}{2} t^{-1/2} - \frac{2800}{t^2} = \frac{5t^{3/2} - 5600}{2t^2}$$

The critical numbers obey

$$5t^{3/2} - 5600 = 0$$
$$t^{3/2} = 1120$$
$$t \approx 107.85 \text{ months}$$

The minimum average cost of repair and replacement per month is

$$C(107.85) = 5(10.385) + \frac{2800}{107.85} = 51.93 + 25.96 = \$77.89$$

*Adapted from H. Brems, *Quantitative Economic Theory*, Wiley, New York, 1968.

Model 3: Minimizing Inventory Cost

Tami's Famous Hamburger Palace requires 45,000 hamburger patties every quarter (3 months) based on estimated daily sales of 475 hamburgers plus an additional 25 that are either burned or eaten by employees each day. These hamburgers are packed in boxes of 10 patties each. The manager of the restaurant incurs order charges of 45¢ due to handling and delivery for each order. Also, if a larger order is placed, the patties must be stored in a freezer, and the cost of storing the patties is estimated to be 2¢ per box. How many boxes of patties should be ordered and how many orders should be placed to minimize the cost of storage and of ordering?

To begin, let x denote the number of boxes of patties ordered each time an order is placed. Then, the storage cost for x boxes is

$$\text{Storage cost} = \$0.02x$$

If x boxes are ordered each time an order is placed, and if a total of 4500 boxes (45,000 ÷ 10) are required, then the number of orders necessary is

$$\text{Number of orders} = \frac{4500}{x}$$

(If $4500/x$ is not an integer, we approximate by using the closest integer to $4500/x$.) Each order carries an order charge of $0.45 so that the order cost is

$$\text{Order cost} = \$0.45\left(\frac{4500}{x}\right)$$

The cost function $C(x)$ to be minimized is

$$C(x) = \text{Storage cost} + \text{Order cost} = \$0.02x + \$0.45\left(\frac{4500}{x}\right)$$

Of course, the domain of $C(x)$ is $1 \leq x \leq 4500$. (At most 4500 boxes are required.)

To find the absolute minimum of $C(x)$, we first find the critical numbers. They obey

$$C'(x) = 0.02 - \frac{2025}{x^2} = 0$$

$$0.02x^2 = 2025$$

$$x^2 = \frac{2025}{0.02} = \frac{202{,}500}{2} = 101{,}250$$

$$x \approx 318.2$$

Since partial boxes cannot be ordered, we use $x = 318$, and find the cost:

$$C(318) = 6.36 + 6.37 = \$12.73$$

Testing the endpoints $x = 4500$ and $x = 1$, we find

$$C(4500) = 0.02(4500) + 0.45 = \$90.45$$
$$C(1) = \$2025.02$$

Thus, the minimum cost is achieved when $x = 318$ boxes are ordered and $4500/x = 14.15 \approx 14$ orders are placed.

Model 4: The Response of the Body to a Drug*

The reaction of the body to a dose of a drug can be represented by the following function:

$$R(D) = D^2\left(\frac{C}{2} - \frac{D}{3}\right)$$

where

C = a positive constant
R = the strength of the reaction (for example—if R is the change in blood pressure, it is measured in mm mercury, if R is the change in temperature, it is measured in degrees, and so on.)
D = the amount of the drug

We will assume that whenever the drug is administered, the concentration of the drug already in the body is insignificant; for, if there is already a certain concentration of the drug in the body, the reaction will depend on this initial concentration (Weber-Fechner Law).

D is defined in the range of 0 to C. (In other words, C is the maximum amount that may be given.)

Find the range of dose for which the medicine has maximum sensitivity, in other words, where there is the greatest change in R for a small change in D, or, equivalently, where $R'(D)$ is at a maximum.

To solve the problem, we proceed as follows:

$$R'(D) = DC - D^2$$
$$R''(D) = C - 2D$$
$$R'''(D) = -2$$

$R'(D)$ has a maximum when $D = C/2$. At this value of D the second derivative vanishes and the third derivative is negative. It is easy to see that R at this point ($R = C^3/12$) is one-half of the R obtained when maximum dose is given ($R = C^3/6$).

This phenomenon—that when the strength of the reaction is 50 percent of the maximum strength, the change in the strength of the reaction with respect to a change in dose is greatest—is used to ascertain the exact dose of a drug to be given, since at this dose, small changes in dose will give the greatest changes in the reaction.

Exercise 7
Solutions to *all* these problems begin on page 357.

1. On a particular product, government economists determine that the relationship between tax rate t and the quantity x consumed is

$$t + 3x^2 = 18$$

Graph this relationship and explain how it could be justified. Find the optimal tax rate and the revenue generated by this tax rate.

---

*"The Response of a Body to Drugs," *Some Mathematical Models in Biology*, Robert Thrall, Ed. University of Michigan Press, 1967.

2. The average cost per repair of a television is found to be $30 and the number of repairs k is given by $k = t^{4/3}$, where t is measured in years. The replacement cost is estimated to be $450. Ignoring the original cost of the television, find the optimal time for trading in the television.

3. A hardware store sells 50 lawnmowers per year. It costs the owner $10 to store each lawnmower. For each order placed for lawnmowers, a charge of $20 for handling and delivery is incurred. How many orders should be placed to minimize the cost of storing and ordering?

4. An ice cream store owner wishes to establish an optimal inventory policy for ordering ice cream. It is estimated that a total of 2500 cases will be sold during the next quarter. To minimize storage costs (the cost of keeping the ice cream frozen is significant), the owner does not wish to order all the ice cream at once. On the other hand, ordering small quantities each time incurs additional handling costs. He decides to order equal size quantities consisting of x cases at equally spaced time throughout the quarter. Suppose the cost of each delivery is $90.00 and the storage cost for each case over a quarter is $20.00. Since there is an average of x/2 cases in storage each quarter, the storage cost is $(20)(x/2) = 10x$ dollars. Determine the optimal reorder quantity x that minimizes total cost (delivery cost + storage cost).

5. In many manufacturing operations, the total cost of operation is the sum of the cost of setting up machines plus the cost of operating these machines. If the set up cost is proportional to the number of machines used and the operating cost is inversely proportional to the number of machines used, show that the total cost is a minimum when set up cost equals operating cost.

Chapter Review

Important Terms

increasing function
decreasing function
relative maximum
relative minimum
critical number
critical point
first-derivative test
second-derivative test
*Rolle's theorem
*mean value theorem
absolute maximum
absolute minimum
concave up
concave down
inflection point

limits at infinity
infinite limits
one-sided infinite limits
horizontal asymptote
vertical asymptote
demand equation
price function
revenue function
marginal revenue
marginal cost
cost function
profit function
*average cost
*average revenue

*Discussed in optional problems.

True-False Questions (Answers on page 434.)

T F 1. A differentiable function is increasing on (a, b) if $f'(x) > 0$ throughout (a, b).
T F 2. The absolute maximum of a function equals the value of the function at a critical number.
T F 3. If $f''(x) > 0$ for all x in (a, b), then the graph of f is concave down on (a, b).
T F 4. If $f''(c) = 0$, then there is an inflection point at $(c, f(c))$.
T F 5. If $\lim\limits_{x \to +\infty} f(x) = L$, then the line $y = L$ is a horizontal asymptote of the graph of f.

Fill in the Blanks (Answers on page 435.)

1. The function $f(x)$ is _____ on (a, b) if for any choice of $x_1 < x_2$ in (a, b), $f(x_1) > f(x_2)$.
2. A critical point $(c, f(c))$ is a relative minimum if the function is _____ to the left of c and _____ to the right of c.
3. A differentiable function is _____ _____ on (a, b) if the tangent lines to the graph at every point lie below its graph.
4. At a critical point the graph of f has a _____ tangent line.
5. At an inflection point, the second derivative equals _____.

Review Exercises
Solutions to Odd-Numbered Problems begin on page 384.

1. Sketch the graph of a continuous function having all the given properties:

    1. $(1, 34)$, $(2, 32)$, and $(3, 30)$ are on the graph
    2. $f'(1) = 0$ and $f'(3) = 0$
    3. $f''(x) < 0$ if $x < 2$ and $f''(x) > 0$ if $x > 2$

2. Refer to the graph of $y = f(x)$ below to identify the points or intervals on the x-axis that exhibit the indicated behavior.
    (a) f is increasing (b) $f'(x) < 0$
    (c) Graph of f is concave down (d) Relative minima
    (e) Absolute maximum (f) $f'(x) = 0$
    (g) Inflection points (h) $f'(x)$ does not exist

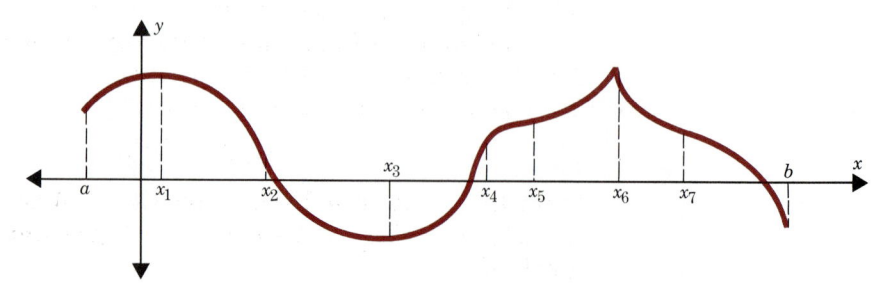

In Problems 3–14 find where each function is increasing and decreasing. Locate all relative maxima and all relative minima. Determine where the function is concave up and concave down. Find all inflection points. Sketch the graph.

3. $f(x) = x^3 - 3x^2 + 3x - 1$
4. $f(x) = 2x^3 - x^2 + 2$
5. $f(x) = x^4 - 2x^2$
6. $f(x) = x^4 + 2x^2$
7. $f(x) = x^5 - 5x$
8. $f(x) = x^5 + 5x^4$
9. $f(x) = x^4 - 4x^3 + 4x^2$
10. $f(x) = x^4 + \frac{4}{3}x^3 - 4x^2$
11. $f(x) = x^{4/3} + 4x^{1/3}$
12. $f(x) = x^{4/3} - 4x^{1/3}$
13. $f(x) = \dfrac{2x}{x^2 + 1}$
14. $f(x) = \dfrac{4x}{x^2 + 4}$

In Problems 15–20 find the absolute maximum and absolute minimum of each function on the domain indicated.

15. $f(x) = x^3 - 3x^2 + 3x - 1$; on $[0, 1]$
16. $f(x) = 2x^3 - x^2 + 2$; on $[0, 1]$
17. $f(x) = x^4 - 2x^2$; on $[-1, 1]$
18. $f(x) = x^4 + 2x^2$; on $[-1, 1]$
19. $f(x) = x^{4/3} + 4x^{1/3}$; on $[-1, 1]$
20. $f(x) = x^{4/3} - 4x^{1/3}$; on $[-1, 1]$

21. **Maximizing Profit.** A company's history shows that profits increase, as a result of advertising, according to
$$P(x) = 150 + 120x - 3x^2$$
where x is the number of dollars spent on advertising (measured in thousands). How much should be spent on advertising to maximize profits?

22. **Best Dimensions of a Can.** A beer can is cylindrical and holds 500 cubic centimeters of beer. If the cost of the material used to make the sides, top, and bottom is the same, what dimensions should the can have to minimize cost?

23. If a function f is continuous for all x and if f has a relative maximum at $(-1, 4)$ and a relative minimum at $(3, -2)$, which of the following statements must be true?
    (a) The graph of f has a point of inflection somewhere between $x = -1$ and $x = 3$.
    (b) $f'(-1) = 0$
    (c) The graph of f has a horizontal asymptote.
    (d) The graph of f has a horizontal tangent line at $x = 3$.
    (e) The graph of f intersects both axes.

24. Let f be a function whose *derivative* is
$$f'(x) = \sqrt{3x^2 + 4}$$
Show that f has an inflection point at $x = 0$. Show that f has no critical points. If $f(0) = 5$ and $f(-3) = 0$, give a rough sketch of the graph of f.

25. Let f be a function whose *derivative* is

$$f'(x) = \frac{5}{x^2 + 1}$$

Show that f has an inflection point at $x = 0$. Show that f has no critical points. If $f(0) = 4$ and $f(-3) = 0$, give a rough sketch of the graph f.

26. If the total cost, C, of producing x gallons of a liquid is

$$C = 0.01x^3 - .1x^2 + 5x$$

(a) What is the expression for marginal cost?
(b) What is marginal cost at 10 gallons output?
(c) Using the second derivative, find how marginal cost is changing at 10 gallons output.
(d) Find the levels of output at which marginal cost is 5 by setting the expression in (a) equal to 5 and solving.
(e) The graph of C has an inflection point. Find x at the inflection point and state the significance of the point in marginal terminology.

27. Study the illustration and answer the following questions:

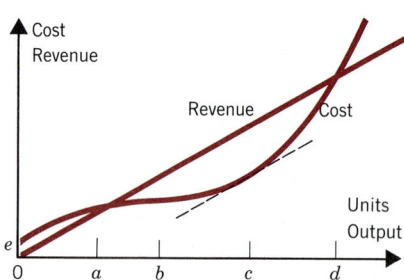

(a) Over what interval is a loss incurred?
(b) Over what interval is a profit incurred?
(c) Over what interval is marginal cost increasing? decreasing?

28. *Maximizing Profit.* The price function of a certain mobile home producer is

$$p = 2402.50 - 0.5x^2$$

where p is the price (in dollars) and x is the number of units sold. The cost of production for x units is

$$C = 1802.5x + 1500$$

How many units need to be sold to maximum profit?

29. *Setting Refrigerator Prices.* A distributor of refrigerators has average monthly sales of 1500 refrigerators, each selling for $300. From past experience, the distributor knows that a special month-long

promotion will enable them to sell 200 additional refrigerators for each $15 decrease in price. What should be charged for each refrigerator during the month of promotion in order to maximize revenue?

Mathematical Questions From CPA Exams (Answers on page 433.)

1. *CPA Exam—November 1975*
   The mathematical notation for the total cost for a business is $2X^3 + 4X^2 + 3X + 5$ where X equals production volume. Which of the following is the mathematical notation for the marginal cost function for this business?
   (a) $2(X^3 + 2X^2 + 1.5X + 2.5)$
   (b) $6X^2 + 8X + 3$
   (c) $2X^3 + 4X^2 + 3X$
   (d) $3X + 5$

2. *CPA Exam—November 1974*
   The mathematical notation for the total cost function for a business is $4X^3 + 6X^2 + 2X + 10$ where X equals production volume. Which of the following is the mathematical notation for the average cost function for that business?
   (a) $2(2X^2 + 3X + 2)$
   (b) $2X^3 + 3X^2 + X + 5$
   (c) $0.4X^3 + 0.6X^2 + 0.2X + 1$
   (d) $4X^2 + 6X + 2 + \dfrac{10}{X}$

3. *CPA Exam—November 1976*
   The mathematical notation for the average cost function for a business is $6X^3 + 4X^2 + 2X + 8 + 2/X$ where X equals production volume. What would be the mathematical notation for the total cost function for the business?
   (a) The average cost function multiplied by X.
   (b) The average cost function divided by X.
   (c) The average cost function divided by $X/2$.
   (d) The first derivative of the average cost function.

4. *CPA—Review*
   To find a minimum cost point given a total cost equation, the initial steps are to find the first derivative, set this derivative equal to zero, and solve the equation. Using the solution(s) so derived, what additional steps must be taken, and what result indicates a minimum?
   (a) Substitute the solution(s) in the first derivative equation and a positive solution indicates a minimum.
   (b) Substitute the solution(s) in the first derivative equation and a negative solution indicates a minimum.
   (c) Substitute the solution(s) in the second derivative equation and a positive solution indicates a minimum.
   (d) Substitute the solution(s) in the second derivative equation and a negative solution indicates a minimum.

Other Books

Batschelet, E., *Introduction to Mathematics for Life Sciences,* Springer-Verlag, New York, 1972.

Chiang, Alpha, *Fundamental Methods of Mathematical Economics,* 2nd ed., McGraw-Hill, New York, 1974.

Ferguson, C. E., *Microeconomic Theory,* 3rd ed., Richard D. Irwin Press, Homewood, Ill., 1972.

Grossman, Stanley, and James Turner, *Mathematics for the Biological Sciences,* Macmillan, New York, 1974.

Springer, C. H., R. E. Herlihy, and R. I. Beggs, *Advanced Methods and Models,* Mathematics for Management Series, Vol. 2, Richard D. Irwin Press, Homewood, Ill., 1965.

Thrall, Robert M., Ed., *Some Mathematical Models In Biology,* University of Michigan, Ann Arbor, 1967.

5 The Exponential and Logarithm Functions

1. The Exponential Function
2. The Derivative of the Exponential Function
3. The Logarithm Function and Its Derivative
4. Logistic Curves

Chapter Review
   Mathematical Questions from Actuary Exams

194 CH. 5 THE EXPONENTIAL AND LOGARITHM FUNCTIONS

Until now, our study of calculus has been concerned, for the most part, with polynomial functions and ratios of polynomial functions (the rational functions). Although these functions occur in many practical situations, there are other functions that are equally important and useful. Two of these functions are the *exponential function* and the *logarithm function*.

1. The Exponential Function

We begin with a brief review of exponents.* Recall that if a is a real number and $n \geq 1$ is a positive integer, then a^n means a times itself n times; thus, for example,

$$2^3 = 2 \cdot 2 \cdot 2 \qquad 5^2 = 5 \cdot 5 \qquad \pi^4 = \pi \cdot \pi \cdot \pi \cdot \pi$$

For a^n, a is the *base* and n is the *exponent*.

Laws of Exponents Two useful properties of exponents are

$$a^m \cdot a^n = a^{m+n} \qquad \text{and} \qquad (a^m)^n = a^{mn} \qquad m, n \text{ positive integers}$$

We define raising a to the 0 power and to a negative power in such a way that the above laws of exponents are preserved. Thus, if $a \neq 0$,

$$a^0 = 1 \qquad \text{and} \qquad a^{-n} = \frac{1}{a^n} \qquad n \geq 1 \text{ an integer}$$

For example,

$$5^0 = 1 \qquad \text{and} \qquad 2^{-3} = \frac{1}{2^3} = \frac{1}{8}$$

With this meaning of a^n, where n is any integer, we turn to the question of raising a to a rational number exponent. Again we define the meaning of $a^{p/q}$ so that the laws of exponents are obeyed:

$$a^{p/q} = (\sqrt[q]{a})^p \qquad q \geq 2 \text{ is an integer and } p \text{ is any integer}$$

where $\sqrt[q]{a}$ denotes the qth root of a. Here, we need to restrict the base a somewhat, since even roots of negative numbers do not exist in the real number system. Thus, $a^{p/q}$ has meaning except when $a < 0$ and q is even. (We assume that p/q is in lowest terms.)

For example,

$$8^{2/3} = (\sqrt[3]{8})^2 = 2^2 = 4$$
$$(-27)^{2/3} = (\sqrt[3]{-27})^2 = (-3)^2 = 9$$

Expressions such as $(-4)^{1/2}$ and $(-25)^{3/4}$ are undefined.

The next logical step is to raise a to any real number exponent. When we do this, we obtain a^x, where x is a real number. The restrictions on the base a are

---

*You should review these concepts in the Appendix before studying this chapter.

1. THE EXPONENTIAL FUNCTION

that $a > 0$ and $a \neq 1$. As a result, for our purposes, expressions such as $(-3)^{\sqrt{2}}$ are not defined.

Exponential Function An *exponential function* is a function of the form

$$y = f(x) = a^x$$

where $a > 0$, $a \neq 1$ is a real constant.

The reason for the name given to this particular function is that the independent variable *x* appears as an exponent.

We assume that the laws of exponents are obeyed by a^x for any real number *x*. That is, if x_1 and x_2 are two real numbers and a is a positive real number, then

$$a^{x_1} \cdot a^{x_2} = a^{x_1+x_2} \quad \text{and} \quad (a^{x_1})^{x_2} = a^{x_1 x_2}$$

To gain insight into the nature of the graph of an exponential function, we study a few examples.

Example 1 Graph the exponential function: $f(x) = 2^x$

Solution We begin by looking at some of the points on the graph of $f(x) = 2^x$:

x	−10	−5	−4	−3	−2	−1	−0.5
$f(x) = 2^x$	$2^{-10} = 0.00098$	$2^{-5} = 0.031$	$2^{-4} = 0.0625$	$2^{-3} = 0.125$	$2^{-2} = 0.25$	$2^{-1} = 0.5$	$2^{-0.5} = 0.707$
x	0	0.5	1	2	$\sqrt{7}$	3	5	10
$f(x) = 2^x$	$2^0 = 1.0$	$2^{0.5} = 1.414$	$2^1 = 2.0$	$2^2 = 4.0$	$2^{\sqrt{7}} = 6.258$	$2^3 = 8$	$2^5 = 32$	$2^{10} = 1024$

Notice in the table that for increasingly negative numbers *x*, the value of 2^x is very close to zero, but positive. For large positive numbers *x*, the value of 2^x is very large and positive. From the table, we conclude that the domain of $f(x) = 2^x$ is the set of real numbers and the range is the set of positive real numbers. Figure 1 illustrates the graph.

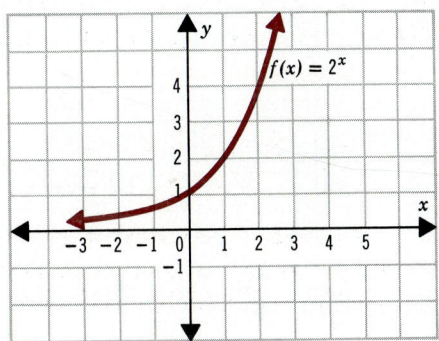

Figure 1

The graph of $f(x) = 2^x$ in Figure 1 is typical of all exponential functions that have a base larger than 1. Such functions are positive everywhere and are increasing functions on their domain. Their graphs pass through the point $(0, 1)$ and thereafter rise rapidly as x increases. In addition, $f(x) = a^x$, $a > 1$, is a continuous function. See the graphs in Figure 2.

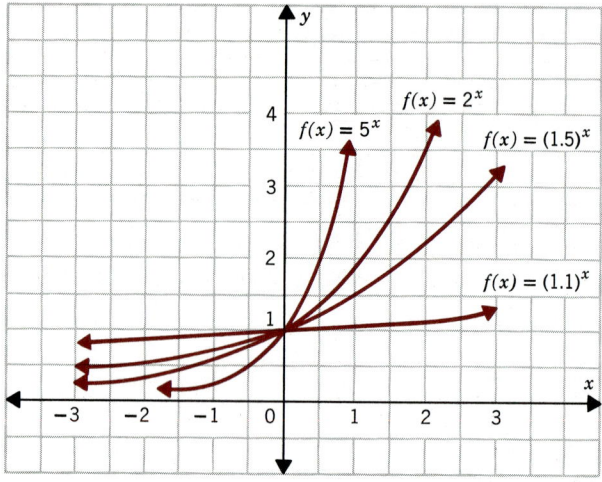

Figure 2

Growth Curves

Exponential functions such as those shown in Figure 2 are often used as models for describing such phenomena as the growth of bacteria in a culture, population growth, and compound interest. We refer to these curves as *growth curves*.

For example, $1000 invested in 1980 at a rate of 10% per annum compounded annually will increase in value over the years according to the graph given in Figure 3. The graph is that of an exponential function whose base is 1.1. Some

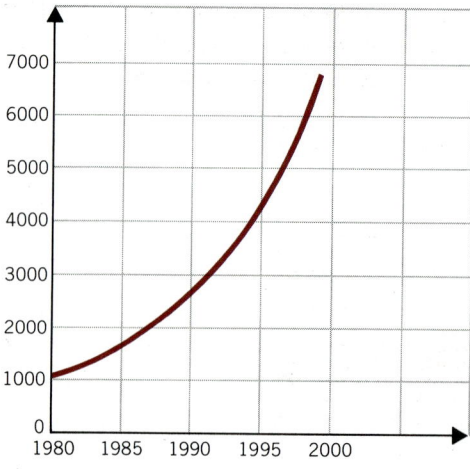

Figure 3

points on this graph are (1985, 1611), (1990, 2594), and (2000, 6727), indicating the value of the original $1000 is $1611 in 1985, $2594 in 1990, and $6727 in 2000.

Example 2 Graph the exponential function: $f(x) = \left(\dfrac{1}{2}\right)^x$

Solution Again, we look at some points on the graph:

x	−10	−5	−2	−1	0
$f(x) = (\tfrac{1}{2})^x$	$(\tfrac{1}{2})^{-10} = 1024$	$(\tfrac{1}{2})^{-5} = 32$	$(\tfrac{1}{2})^{-2} = 4$	$(\tfrac{1}{2})^{-1} = 2$	$(\tfrac{1}{2})^0 = 1$

x	1	2	5	10
$f(x) = (\tfrac{1}{2})^x$	$(\tfrac{1}{2})^1 = 0.5$	$(\tfrac{1}{2})^2 = 0.25$	$(\tfrac{1}{2})^5 = 0.031$	$(\tfrac{1}{2})^{10} = 0.00098$

Thus, for increasingly negative numbers x, the value of $(\tfrac{1}{2})^x$ becomes large and positive; for large positive numbers x, $(\tfrac{1}{2})^x$ is close to zero and positive. Figure 4 illustrates the graph.

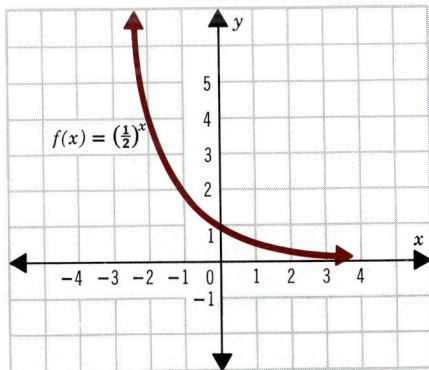

Figure 4

The graph of $f(x) = (\tfrac{1}{2})^x$ in Figure 4 is typical of all exponential functions that have a base smaller than 1. Such functions are positive everywhere and are decreasing on their domain. Their graphs pass through the point (0, 1) and thereafter decrease as x increases. In addition, $f(x) = a^x$, $0 < a < 1$, is a continuous function. See the graphs in Figure 5.

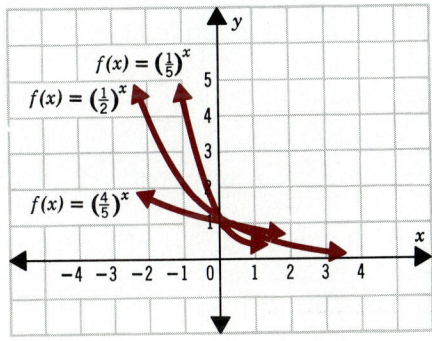

Figure 5

198 CH. 5 THE EXPONENTIAL AND LOGARITHM FUNCTIONS

Decay Curves

Exponential functions such as those shown in Figure 5 are often used as models for describing such phenomena as radioactive decay, depreciation, and price-demand curves. We refer to these curves as *decay curves*.

For example, the price-demand curve for a particular commodity may be described by the graph in Figure 6. Here, we measure price along the vertical axis and demand along the horizontal axis. The curve itself closely resembles an exponential function that has a base less than 1.

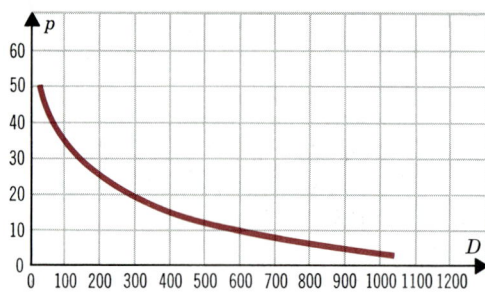

Figure 6

The Number e

The exponential function $f(x) = e^x$, where e is the irrational number given approximately by $e = 2.71828\ldots$, has many applications. We will look at the number e more carefully in the next section. (A partial table of values for e^x is found in Table 1 on page 205.)

One application using the number e is illustrated in the following example.

Example 3
Product Reaction

The proportion of people responding to the advertisement of a new product after it has been on the market t days is

$$1 - e^{-0.06t}$$

The marketing area contains 10,000,000 potential customers, and each response to the advertisement results in profit to the company of $0.70 (on the average). This profit is exclusive of advertising cost. The fixed cost of producing the advertising is $30,000 and the variable cost is $5000 for each day the advertisement runs.

(a) Find the value of $1 - e^{-0.06t}$ when t is very large and interpret the answer.
(b) What percentage of customers respond after 10 days of advertising?
(c) What is the cost function $C(t)$?
(d) After 30 days of advertising, what is the net profit?

Solution
(a) For large t, $e^{-0.06t} = 1/e^{0.06t}$ tends to zero. Thus, for large t, $1 - e^{-0.06t}$ tends to 1. The interpretation is that eventually everyone will respond to the advertisement.
(b) The proportion of customers that respond after 10 days is found by substituting $t = 10$ into the response function $1 - e^{-0.06t}$. Thus, for $t = 10$, we consult Table 1, page 205, and find that

$$1 - e^{-0.6} = 1 - 0.549 = 0.451$$

Thus, 45.1% of the potential customers have responded after 10 days of advertising.

(c) The cost function is
$$C(t) = \$30{,}000 + \$5000t$$

(d) The net profit function is the profit from sales less advertising cost. The profit from sales is
$$R(t) = 10{,}000{,}000(1 - e^{-0.06t})(\$0.70) = \$7{,}000{,}000(1 - e^{-0.06t})$$

Thus, the net profit is
$$P(t) = R(t) - C(t) = \$7{,}000{,}000(1 - e^{-0.06t}) - \$30{,}000 - \$5000t$$

For $t = 30$,
$$P(30) = \$7{,}000{,}000(1 - e^{-1.8}) - \$30{,}000 - \$150{,}000$$
$$= \$7{,}000{,}000(0.835) - \$180{,}000$$
$$= \$5{,}845{,}000 - \$180{,}000$$
$$= \$5{,}665{,}000$$

Exercise 1
Solutions to Odd-Numbered Problems begin on page 403.

In Problems 1–16 calculate each expression.

1. $27^{2/3}$
2. $8^{2/3}$
3. $9^{3/2}$
4. $16^{-3/4}$
5. $16^{-1/2}$
6. $8^{4/3}$
7. $27^{-2/3}$
8. $16^{1/4}$
9. $(\frac{1}{9})^{1/2}$
10. $(\frac{1}{8})^{-2}$
11. $(\frac{1}{8})^{-1/3}$
12. $(\frac{9}{16})^{-1/2}$
13. $(9^{1/3})(3)^{1/3}$
14. $(27^{3/2})(\frac{1}{3})^{3/2}$
15. $[(8^{-1})(8^{1/3})]^3$
16. $27^{-1}(27^{1/3})^3$

In Problems 17–22 use a hand calculator to compute each expression, rounding off your answer to three decimal places, if necessary.

17. 2^x, for $x = 4$
18. 2^x, for $x = -4$
19. 2^x, for $x = \frac{1}{3}$
20. 2^x, for $x = 3.1$
21. 2^x, for $x = 0.1$
22. 2^x, for $x = 0.9$

In Problems 23–28 graph each function.

23. $f(x) = 3^x$
24. $f(x) = 5^x$
25. $f(x) = (\frac{1}{3})^x$
26. $f(x) = (\frac{1}{5})^x$
27. $f(x) = 4^{0.5x}$
28. $f(x) = (\frac{1}{4})^{0.5x}$

29. *Product Reaction.* If, in Example 3, the response function is
$$1 - e^{-0.1t}$$
and the other information remains the same, answer questions (b) and (d) of the example.

30. *Product Reaction.* If, in Example 3, the response function is
$$1 - e^{-0.05t}$$
and the profit per response is $0.10, while all other data are the same, answer questions (b) and (d) of the example.

31. The earnings of 3M Co. have been increasing at an annual rate of 10%. Find a function that expresses the relationship between time and earnings.

32. *Reliability of a Product.* The proportion of batteries that still maintain a charge after x years of use is

$$f(x) = (\tfrac{3}{4})^x$$

   (a) What proportion of the batteries still maintain a charge after 2 years?
   (b) What proportion of the batteries will fail to hold a charge between the second and third year of use?
   (c) What proportion of the batteries will fail to hold a charge within the first year of use?

33. The annual profit of a company due to a particular item is found to be

$$P(x) = \$10{,}000 + \$25{,}000(\tfrac{1}{4})^{0.5x}$$

   where x is the number of years the item has been on the market. Graph this function and estimate the profit for $x = 1, 3,$ and 5 years.

34. The demand for a new product increases rapidly at first and then levels off. Suppose the percentage of actual buyers obeys

$$f(x) = 100 - 90(\tfrac{1}{3})^x$$

   where x is the number of months the product is on the market. Graph the function and determine to what proportion of the market the new product sells for $x = 2, 3,$ and 5 months.

*35. If $f(x) = a^x$, show that:
   (a) $f(x + 1) = af(x)$
   (b) $f(x + 1) - f(x) = (a - 1)f(x)$
   (c) $f(x + h) = a^h f(x)$

2. The Derivative of the Exponential Function

At this point, you may find it helpful to refer to Chapter 3, where the derivative is first introduced. This review will, of course, serve to reinforce the important ideas of calculus, and it will enable you to follow the discussion in this section more easily.

We begin the discussion of the derivative of the exponential function by considering the function itself:

$$f(x) = a^x \qquad a > 0, \qquad a \neq 1$$

To find the derivative of $f(x) = a^x$, we use the alternate formula for finding derivatives [see (6) on page 103], where, for convenience, we have replaced Δx by h:

$$f'(x) = \lim_{h \to 0} \frac{f(x + h) - f(x)}{h}$$

2. THE DERIVATIVE OF THE EXPONENTIAL FUNCTION

Thus, for $f(x) = a^x$, we have

$$f'(x) = \frac{d}{dx}a^x = \lim_{h \to 0} \frac{a^{x+h} - a^x}{h} = \lim_{h \to 0} a^x\left(\frac{a^h - 1}{h}\right)$$

$$= a^x \lim_{h \to 0} \frac{a^h - 1}{h}$$

Assuming the limit on the right exists and equals some number, it follows (since $a^0 = 1$) that the derivative of a^x at 0 is

$$f'(0) = \lim_{h \to 0} \frac{a^h - 1}{h}$$

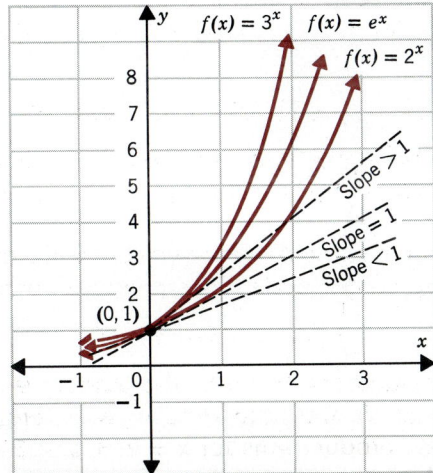

Figure 7

The limit above is the slope of the tangent line of $f(x) = a^x$ at the point $(0, 1)$, and the value of this limit depends upon the choice of a. Observe in Figure 7 that the slope of the tangent line to $f(x) = 2^x$ at $(0, 1)$ is less than 1, and that the slope of the tangent line to $f(x) = 3^x$ at $(0, 1)$ is greater than 1. From this, we conjecture that there should be a number a, $2 < a < 3$, for which the slope of the tangent line to $f(x) = a^x$ at $(0, 1)$ is exactly 1. This particular number a is the number $e = 2.71828\ldots$, referred to earlier. Thus,

$$\lim_{h \to 0} \frac{e^h - 1}{h} = 1$$

Using this result, we find that

$$\frac{d}{dx}e^x = \lim_{h \to 0} \frac{e^{x+h} - e^x}{h} = \lim_{h \to 0} \frac{e^x(e^h - 1)}{h} = e^x \lim_{h \to 0} \frac{e^h - 1}{h} = e^x(1) = e^x$$

Hence,

$$\frac{d}{dx}e^x = e^x$$

The derivative of the exponential function e^x is itself.

202   CH. 5   THE EXPONENTIAL AND LOGARITHM FUNCTIONS

The function e^x behaves in the same way as 2^x and 3^x except that computing the derivative of e^x is much easier. This is why the function e^x is very commonly used in all applications that require an exponential type of function to describe natural phenomena. As a result, e^x is usually called *the* exponential function. Scientific calculators enable us to compute e^x for a wide range of numbers x.

Example 1 Find the derivative of:

(a) $f(x) = x^2 + e^x$ (b) $f(x) = xe^x$ (c) $f(x) = \dfrac{e^x}{x}$

Solution (a) $f'(x) = \dfrac{d}{dx}(x^2 + e^x) = 2x + e^x$

(b) We use the formula for the derivative of a product:

$$f'(x) = \frac{d}{dx}xe^x = x\frac{d}{dx}e^x + e^x\frac{d}{dx}x = xe^x + e^x(1) = e^x(x+1)$$

(c) We write $f(x) = e^x/x = e^x x^{-1}$ so that

$$f'(x) = \frac{d}{dx}(e^x x^{-1})$$

$$= e^x x^{-1} - x^{-2}e^x = \frac{e^x}{x} - \frac{e^x}{x^2} = e^x\left(\frac{1}{x} - \frac{1}{x^2}\right)$$

■

The Derivative of $y = e^{g(x)}$

In Section 6 of Chapter 3 the power rule was used to find the derivative of certain kinds of functions.

For example, the function $y = e^{x^2+1}$ is composed of the two functions $y = e^{g(x)}$ and $g(x) = x^2 + 1$. As another example, $y = e^{x^3}$ is composed of $y = e^{g(x)}$ and $g(x) = x^3$.

The formula* for finding the derivative of the function $y = e^{g(x)}$, where $g(x)$ is a differentiable function, is

(1) $$\frac{d}{dx}e^{g(x)} = e^{g(x)}g'(x)$$

Example 2 Find the derivative of:
(a) $f(x) = 4e^{5x}$ (b) $f(x) = e^{x^2+1}$ (c) $f(x) = e^{2x^2-(1/x)}$

Solution (a) Using (1), where $g(x) = 5x$, we find

$$f'(x) = \frac{d}{dx}(4e^{5x}) = 4\frac{d}{dx}e^{5x} = 4e^{5x}(5) = 20e^{5x}$$

---

*For those of you who covered the optional section on the Chain Rule in Chapter 3, formula (1) is a result of applying the Chain Rule. For example, if $y = e^u$ and $u = g(x)$, then $y = e^{g(x)}$ and

$$\frac{dy}{dx} = \frac{dy}{du}\frac{du}{dx} = (e^u)(g'(x)) = e^{g(x)}g'(x)$$

(b) Using (1), where $g(x) = x^2 + 1$, we find

$$f'(x) = \frac{d}{dx} e^{x^2+1} = e^{x^2+1}(2x)$$

(c) Here, $g(x) = 2x^2 - (1/x)$, so that $g'(x) = 4x + (1/x^2)$. By (1),

$$f'(x) = \frac{d}{dx} e^{2x^2-(1/x)} = e^{2x^2-(1/x)}\left(4x + \frac{1}{x^2}\right)$$

■

Example 3
Product Reaction

For the situation described in Example 3 of Section 1, how many days should the advertisement run to maximize profits?

Solution Profit is a maximum when marginal revenue equals marginal cost. From parts (c) and (d) of Example 3 in Section 1,

$$MR = \$7{,}000{,}000(0.2)e^{-0.06t} \quad \text{and} \quad MC = \$5000$$

Setting $MC = MR$, we have

$$\$5000 = \$1{,}400{,}000 e^{-0.06t}$$
$$e^{0.06t} = 280$$

From Table 1, page 205, we find that $e^{5.63} \approx 280$. Thus,

$$0.06t \approx 5.63$$

$$t \approx \frac{563}{6} = 94$$

Thus, the company should advertise the product for approximately 94 days to maximize profit.

■

Example 4 Graph the function: $f(x) = xe^{-x}$

Solution The domain of this function is all real numbers and it is continuous on its domain.

Since e^{-x} is always positive, the function is negative when $x < 0$ and is positive when $x > 0$. Also, $f(0) = 0$.

The first derivative is

$$f'(x) = x(-e^{-x}) + (1)e^{-x} = (1-x)e^{-x}$$

Again, since e^{-x} is always positive, $f'(x)$ has the same sign as $1 - x$. We conclude that the function is decreasing for $x > 1$ and increasing for $x < 1$. Also, $x = 1$ is a critical number.

The second derivative is

$$f''(x) = \frac{d}{dx} f'(x) = (1-x)(-e^{-x}) + (-1)(e^{-x})$$
$$= (x-2)e^{-x}$$

This is clearly positive for $x > 2$ and negative for $x < 2$. We conclude that the graph is concave up for $x > 2$ and concave down for $x < 2$. Since $f''(2) = 0$, the point $(2, 2e^{-2}) \approx (2, 0.27)$ is an inflection point.

To locate the relative maxima and relative minima, we use the second-derivative test. Thus, at the critical number $x = 1$, the sign of the second derivative is negative so that the point $(1, f(1))$ is a relative maximum. The value of the function at $x = 1$ is $f(1) = e^{-1} = 0.368$.

Put all this information together and draw the graph. See Figure 8.

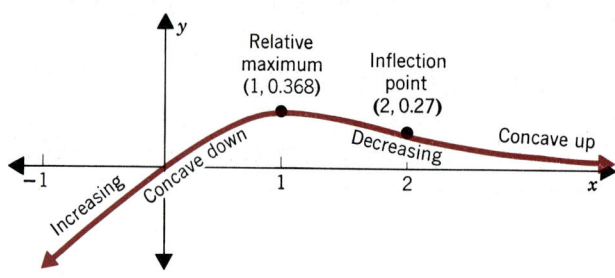

Figure 8

The Number e as a Limit

Since the derivative of $f(x) = e^x$ is $f'(x) = e^x$ and therefore $f'(0) = 1$, it follows from the definition of the derivative [see (6) on page 103] that

$$f'(0) = \lim_{h \to 0} \frac{f(0 + h) - f(0)}{h} = \lim_{h \to 0} \frac{e^h - 1}{h} = 1$$

Equivalently, when h is very close to 0, we may write

$$\frac{e^h - 1}{h} \approx 1$$

$$e^h - 1 \approx h$$

$$e^h \approx 1 + h$$

Raising both sides to the power $1/h$, we have

$$e \approx (1 + h)^{1/h}$$

In other words, for h sufficiently close to 0, e is close to $(1 + h)^{1/h}$. In terms of limits this can be expressed in the following way*:

(2)
$$e = \lim_{h \to 0} (1 + h)^{1/h}$$

Now set $n = 1/h$. If h is close to 0, the number n is very large. That is, as $h \to 0^+$, then $n \to +\infty$. Thus,

(3)
$$e = \lim_{h \to 0^+} (1 + h)^{1/h} = \lim_{n \to +\infty} \left(1 + \frac{1}{n}\right)^n$$

---

*In many calculus books, the number e is defined by the limit in (2).

2. THE DERIVATIVE OF THE EXPONENTIAL FUNCTION

Table 1 lists some approximate values for e that were calculated using (3). Correct to nine decimal places, it turns out that

$$e = 2.718281828\ldots$$

We hasten to point out that this pattern for e does *not* repeat. In fact, it can be shown that e is an irrational number.

Table 1

$n = 100$	$\left(1 + \frac{1}{n}\right)^n = \left(1 + \frac{1}{100}\right)^{100}$	≈ 2.704814
$n = 10{,}000$	$\left(1 + \frac{1}{n}\right)^n = \left(1 + \frac{1}{10{,}000}\right)^{10{,}000}$	≈ 2.7181459
$n = 100{,}000$	$\left(1 + \frac{1}{n}\right)^n = \left(1 + \frac{1}{100{,}000}\right)^{100{,}000}$	≈ 2.7182818

Continuously Compounded Interest

One use of the fact that $e = \lim_{h \to 0^+}(1 + h)^{1/h}$ is found in finance. Suppose a principal P is to be invested at an annual rate of interest r, which is compounded n times per year. The interest earned on a principal P at each compounding period is then $P \cdot (r/n)$. The amount A after one compounding period is

$$A = P + P \cdot \left(\frac{r}{n}\right) = P \cdot \left(1 + \frac{r}{n}\right)$$

After 2 compoundings

$$A = P\left(1 + \frac{r}{n}\right) + P\left(1 + \frac{r}{n}\right)\left(\frac{r}{n}\right) = P\left(1 + \frac{r}{n}\right)^2$$

After 3 compoundings

$$A = P\left(1 + \frac{r}{n}\right)^2 + P\left(1 + \frac{r}{n}\right)^2\left(\frac{r}{n}\right) = P\left(1 + \frac{r}{n}\right)^3$$

$\vdots$

After n compoundings (1 year) $A = P\left(1 + \frac{r}{n}\right)^n$

The amount A after 1 year accrued on a principal P when it is invested at an annual rate of interest r and is compounded n times per year is

(4) $$A = P\left(1 + \frac{r}{n}\right)^n$$

Compound Interest Formula

Formula (4) is usually referred to as the *compound interest formula*.
For example, the result of investing $1000 at an annual rate of 10% yields the amounts listed in Table 2 for various compounding periods.

206 CH. 5 THE EXPONENTIAL AND LOGARITHM FUNCTIONS

Table 2

	P = Principal = \$1000	r = Annual rate of interest = 10% = 0.10
	n = Number of times compounded per year	A = Amount after 1 year
1	Annual compounding	$A = P(1 + r) = 1000(1 + 0.1) = \$1100.00$
2	Semiannual compounding	$A = P\left(1 + \dfrac{r}{2}\right)^2 = 1000(1 + 0.05)^2 = \$1102.50$
4	Quarterly compounding	$A = P\left(1 + \dfrac{r}{4}\right)^4 = 1000(1 + 0.025)^4 = \$1103.81$
12	Monthly compounding	$A = P\left(1 + \dfrac{r}{12}\right)^{12} = 1000(1 + 0.00833)^{12} = \$1104.71$
365	Daily compounding	$A = P\left(1 + \dfrac{r}{365}\right)^{365} = 1000(1 + 0.000274)^{365} = \$1105.16$

Now we ask what happens to the amount after 1 year as the number of times n that the interest is compounded per year gets larger and larger. In other words, we want to calculate $\lim_{n \to +\infty} [P(1 + r/n)^n]$:

$$\lim_{n \to +\infty}\left[P\left(1 + \frac{r}{n}\right)^n\right] = P \lim_{n \to +\infty}\left[\left(1 + \frac{r}{n}\right)^n\right] = P \lim_{n \to +\infty}\left[\left(1 + \frac{r}{n}\right)^{n/r}\right]^r$$

$$= P\left[\lim_{n \to +\infty}\left(1 + \frac{r}{n}\right)^{n/r}\right]^r = Pe^r$$

Thus, no matter how often the interest is compounded during the year, the amount after 1 year has a definite ceiling, Pe^r. When interest is compounded so that the amount after 1 year is Pe^r, we say that the interest is *compounded continuously*.

Continuous Compounding

For example, the amount A due to investing \$1000 for 1 year at an annual rate of 10% compounded continuously is

$$A = 1000e^{0.1} = \$1105.17$$

The formula $A = Pe^r$ gives the amount A after 1 year resulting from investing a principal P at the annual rate of interest r compounded continuously.

The amount A due to investing a principal P for a period of t years at the annual rate of interest r compounded continuously is

(5) $$A = Pe^{rt}$$

Example 5 If \$1000 is invested at 10% compounded continuously, the amount A after 3 years is

$$A = Pe^{rt} = 1000e^{(0.1)(3)} = 1000e^{0.3} \approx \$1350$$

After 5 years, the amount A is

$$A = 1000e^{(0.1)(5)} \approx \$1649$$

Effective Rate of Interest

The term *effective rate of interest* is often used. This is the equivalent annual rate of interest due to compounding. When interest is compounded annually, there is no difference between the annual rate and the effective rate; however, when interest is compounded more than once a year, the effective rate always exceeds the annual rate. For example, using the results in Table 2 and equation (5), we find that when the annual rate is 10%, the effective rates are those listed in Table 3. It is worth noting that although the difference between a bank's paying interest yearly (almost none do now) versus compounding quarterly or monthly is fairly substantial, the difference between daily and continuous compounding is practically negligible.

Table 3

	Annual Rate (%)	Effective Rate (%)
Annual compounding	10	10
Semiannual compounding	10	10.25
Quarterly compounding	10	10.381
Monthly compounding	10	10.471
Daily compounding	10	10.516
Continuous compounding	10	10.517

Exercise 2

Solutions to Odd-Numbered Problems begin on page 404.

In Problems 1–30 find the derivative of each function.

1. $f(x) = 5e^x$
2. $f(x) = 2e^x$
3. $f(x) = e^{5x}$
4. $f(x) = e^{-2x}$
5. $f(x) = 8e^{-x/2}$
6. $f(x) = \dfrac{1}{e^{-4x}}$
7. $f(x) = xe^x$
8. $f(x) = x^2 e^x$
9. $f(x) = e^{x^2}$
10. $f(x) = e^{x^3}$
11. $f(x) = e^{\sqrt{x}}$
12. $f(x) = e^{-\sqrt{x}}$
13. $f(x) = \sqrt{e^x}$
14. $f(x) = (e^x)^{1/2}$
15. $f(x) = 1 - e^x$
16. $f(x) = 1 - e^{-2x}$
17. $f(x) = \dfrac{e^x + e^{-x}}{2}$
18. $f(x) = \dfrac{e^x - e^{-x}}{2}$
19. $f(x) = e^{-3x} - 3x$
20. $f(x) = e^{-2x} - 2x^2 + x$
21. $f(x) = e^x(e^{3x} - e^{-x})$
22. $f(x) = \dfrac{e^x + e^{-x}}{e^x}$
23. $f(x) = e^{2x^2 + x + 1}$
24. $f(x) = e^{x^3 + x - (1/x)}$
25. $f(x) = \dfrac{e^x}{x}$
26. $f(x) = \dfrac{e^{-x}}{x}$
27. $f(x) = e^{x - (1/x)}$
28. $f(x) = e^{x + (1/x)}$
29. $f(x) = \sqrt{1 + e^x}$
30. $f(x) = \sqrt{1 - e^{-x}}$

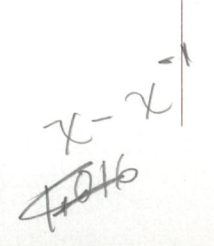

In Problems 31 and 32 find y', y'', and y''', where a is a constant.

31. $y = e^{ax}$
32. $y = e^{-ax}$
33. If $y = e^{2x}$, show that $y'' - 4y = 0$.
34. If $y = e^{-2x}$, show that $y'' - 4y = 0$.

In Problems 35 and 36 graph each function.

35. $f(x) = xe^x$
36. $f(x) = x^2 e^x$

37. Find the absolute maximum and the absolute minimum of each function on the given interval.
    (a) $f(x) = e^x$; on $[-10, 10]$ (b) $f(x) = e^{-x}$; on $[-10, 10]$
38. Find the amount after 1 year if $500 is invested at 6% compounded continuously for 1 year. Use a calculator to obtain the answer if the rate is $6\frac{1}{4}\%$ compounded quarterly. Which is better?
39. Find the amount after 1 year if $1000 is invested at 8% compounded continuously for 2 years. Use a calculator to obtain the amount if the rate is $8\frac{1}{2}\%$ compounded quarterly. Which is better?
40. What principal P should be invested at 6% compounded continuously in order to have $1000 after 1 year? What principal is required if the interest is compounded quarterly?
41. What principal P should be invested at 10% compounded continuously in order to have $2000 after 3 years? What principal is required if the interest is compounded quarterly?
42. How long (in months) does it take for a principal P to double if it is invested at 10% compounded continuously? How long does it take if the compounding is quarterly?
43. Rework Problem 42 if the rate of interest is 8%.
*44. *Normal Density.* The function

$$f(x) = \frac{1}{\sqrt{2\pi}} e^{-x^2/2}$$

is often encountered in probability theory and is called the *normal density function.* Determine where this function is increasing and decreasing, find all relative maxima and relative minima, find all inflection points, and determine intervals of concavity. Graph the function.

*45. Find the number x that maximizes the function

$$E(x) = 75{,}000(1 - 0.15^x) - 500x$$

3. The Logarithm Function and Its Derivative

We turn now to the logarithm function. Consider the relationship

(1) $$x = a^y \qquad a > 0, \qquad a \neq 1$$

between x and y. Of course, this is recognizable as an exponential relationship between x and y.

3. THE LOGARITHM FUNCTION AND ITS DERIVATIVE

Logarithm Function If, in (1), we solve for y in terms of x, we obtain the *logarithm function*, denoted by $y = \log_a x$ and read as "y is the log to the base a of x."

Thus,

$$y = \log_a x \quad \text{means} \quad a^y = x$$

For example,

$y = \log_2 1$	means	$2^y = 1,$	that is,	$y = 0$
$y = \log_2 4$	means	$2^y = 4,$	that is,	$y = 2$
$y = \log_3 \frac{1}{3}$	means	$3^y = \frac{1}{3},$	that is,	$y = -1$
$y = \log_a 1$	means	$a^y = 1,$	that is,	$y = 0$
$y = \log_a a$	means	$a^y = a,$	that is,	$y = 1$

The last two results may be rewritten as

(2) $$\log_a 1 = 0 \qquad \log_a a = 1 \qquad \text{for } a > 0, a \neq 1$$

The logarithm function, $y = f(x) = \log_a x$, is a special name for $a^y = x$, so that the domain of the logarithm function is the set of positive real numbers x, while the range is the set of all real numbers y. The graphs of some logarithm functions are given in Figure 9.

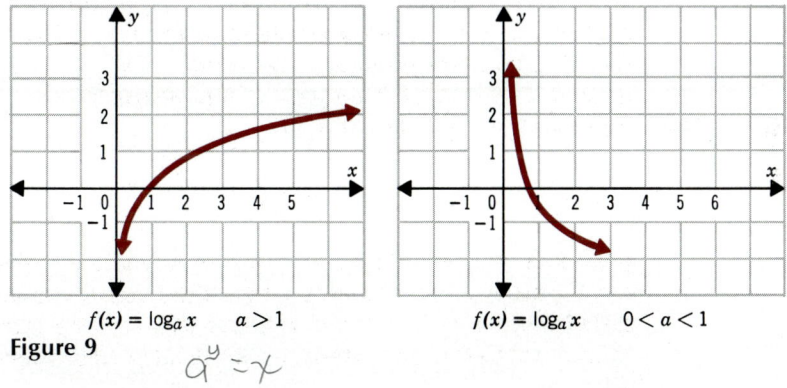

$f(x) = \log_a x \quad a > 1 \qquad\qquad f(x) = \log_a x \quad 0 < a < 1$

Figure 9

$a^y = x$

Note that $f(x) = \log_a x$, $a > 1$, is increasing, while $f(x) = \log_a x$, $0 < a < 1$, is decreasing. Each function is continuous on its domain, namely, $x > 0$, and passes through the point $(1, 0)$.

Logarithms have certain properties that can be derived from their definition. These properties are often used to facilitate computations.

(3) If M and N represent positive real numbers and r is any real number, then

(a) $\log_a(MN) = \log_a M + \log_a N$

(b) $\log_a \dfrac{M}{N} = \log_a M - \log_a N$

(c) $\log_a M^r = r \log_a M$

Here are some examples of these properties:
$$\log_a[x(x+1)] = \log_a x + \log_a(x+1)$$
$$\log_a \frac{x^2+4}{x+1} = \log_a(x^2+4) - \log_a(x+1)$$
$$\log_a x^{3/2} = \tfrac{3}{2}\log_a x$$

Example 1 Using the properties listed in (3), simplify the following expressions:
(a) $\log_a 7 + 3\log_a 2$
(b) $\tfrac{1}{3}\log_a 75 - \log_a(5^3 - 1)$
(c) $\log_a x + \log_a 4 + \log_a(x^2 - 1) - \log_a 7$

Solution (a) $\log_a 7 + 3\log_a 2 \underset{\text{By (3c)}}{=} \log_a 7 + \log_a 2^3 \underset{\text{By (3a)}}{=} \log_a(7)(8) = \log_a 56$

(b) $\tfrac{1}{3}\log_a 75 - \log_a(5^3 - 1) = \log_a 75^{1/3} - \log_a 124$
$$= \log_a \sqrt[3]{75} - \log_a 124$$
$$\underset{\text{By (3b)}}{=} \log_a\left(\frac{\sqrt[3]{75}}{124}\right)$$

(c) $\log_a x + \log_a 4 + \log_a(x^2 - 1) - \log_a 7 \underset{\text{By (3a)}}{=} \log_a(4x) + \log_a(x^2 - 1) - \log_a 7$
$$\underset{\text{By (3a)}}{=} \log_a[4x(x^2 - 1)] - \log_a 7$$
$$\underset{\text{By (3b)}}{=} \log_a \frac{4x(x^2 - 1)}{7}$$

∎

Common Logarithms Until recently, *common logarithms,* or *logarithms to the base 10,* were used to facilitate arithmetical computations. For example, from Table 2 (see page 206), we find that
$$\log_{10} 2 = 0.3010 \quad \text{and} \quad \log_{10} 3 = 0.4771$$

These values and the properties of logarithms enable us to compute expressions such as those in Example 2.

Example 2 Compute:
(a) $\log_{10} 6$ (b) $\log_{10} \tfrac{2}{3}$ (c) $\log_{10} 32$ (d) $\log_{10} 200$

Solution (a) $\log_{10} 6 = \log_{10}(2 \cdot 3) = \log_{10} 2 + \log_{10} 3 = 0.3010 + 0.4771 = 0.7781$
(b) $\log_{10} \tfrac{2}{3} = \log_{10} 2 - \log_{10} 3 = 0.3010 - 0.4771 = -0.1761$

(c) $\log_{10} 32 = \log_{10} 2^5 = 5 \log_{10} 2 = 5(0.3010) = 1.5050$

(d) $\log_{10} 200 = \log_{10}(2 \cdot 100) = \log_{10} 2 + \log_{10} 100$
$= \log_{10} 2 + \log_{10} 10^2 = \log_{10} 2 + 2 \log_{10} 10$
$= 0.3010 + 2 = 2.3010$

Natural Logarithm

The development of hand calculators has made this particular use of common logarithms less important than it once was. On the other hand, *natural logarithms* are encountered as the result of natural phenomena. As a result, we find it appropriate to study natural logarithms, which are *logarithms to the base e*, where e is the irrational number approximately equal to 2.71828.

For convenience when using natural logarithms, we omit the base e and instead write "ln x." That is,

$$\log_e x = \ln x$$

As a result,

$$\ln e = 1 \quad \text{and} \quad \ln 1 = 0$$

Other values of ln x, taken from Table 3, page 355, are

$$\ln 2 = 0.6931 \qquad \ln 3 = 1.0986 \qquad \ln 10 = 2.3026$$

The Derivative of $y = \ln x$

To find the derivative of ln x, we observe that if $y = \ln x$, then $e^y = x$. That is,

$$e^{\ln x} = x$$

If we differentiate both sides with respect to x, using formula (1) from page 202, we obtain

$$e^{\ln x} \frac{d}{dx} \ln x = 1$$

$$\frac{d}{dx} \ln x = \frac{1}{e^{\ln x}}$$

But $e^{\ln x} = x$. Thus,

(4) $$\frac{d}{dx} \ln x = \frac{1}{x}$$

Example 3 Find the derivative of:
(a) $f(x) = x^2 + \ln x$ (b) $f(x) = x \ln x$

Solution (a) $f'(x) = \frac{d}{dx}(x^2 + \ln x) = 2x + \frac{1}{x}$

(b) $f'(x) = \frac{d}{dx} x \ln x = x \frac{d}{dx} \ln x + \ln x \frac{d}{dx} x$

$= (x)\left(\frac{1}{x}\right) + (\ln x)(1) = 1 + \ln x$

Sometimes it is necessary to differentiate the natural logarithm of a function, such as $\ln g(x)$.

The rule* for finding the derivative of functions such as $f(x) = \ln g(x)$, where $g(x)$ is a differentiable function, is

(5)
$$\frac{d}{dx} \ln g(x) = \frac{g'(x)}{g(x)}$$

Example 4 Find the derivative of:
(a) $f(x) = \ln(x^2 + 1)$ (b) $f(x) = \ln \sqrt{x^2 + 1}$ (c) $f(x) = (\ln x)^2$

Solution (a) Here, $g(x) = x^2 + 1$ and $g'(x) = 2x$, so that

$$f'(x) = \frac{d}{dx} \ln(x^2 + 1) = \frac{2x}{x^2 + 1}$$

(b) Always simplify first if possible! Here we may use part (c) of (3) to write $\ln \sqrt{x^2 + 1}$ as $\frac{1}{2} \ln(x^2 + 1)$. Then we may use the solution to part (a) above to get $f'(x) = x/(x^2 + 1)$.

(c) We use the power rule with

$$f(x) = [g(x)]^2 \quad \text{and} \quad g(x) = \ln x$$

$$f'(x) = 2 \ln x \left(\frac{d}{dx} \ln x \right) = \frac{2 \ln x}{x}$$

The properties of logarithms (3) can sometimes be used to simplify the work needed to find the derivative of certain algebraic functions.

Example 5 Find the derivative of: $f(x) = \ln[(2x - 1)^3 (2x + 1)^5]$

Solution Rather than attempt to use (5), we first use the fact that a logarithm transforms products into sums. That is, we may write

$$f(x) \underset{\text{By (3a)}}{=} \ln(2x - 1)^3 + \ln(2x + 1)^5 \underset{\text{By (3c)}}{=} 3 \ln(2x - 1) + 5 \ln(2x + 1)$$

Now we differentiate using (5):

$$f'(x) = (3)\left(\frac{2}{2x - 1}\right) + (5)\left(\frac{2}{2x + 1}\right) = \frac{6}{2x - 1} + \frac{10}{2x + 1} = \frac{4(8x - 1)}{4x^2 - 1}$$

As Example 5 illustrates, some thought should be given to the possibility of simplification before differentiating. The next example illustrates a somewhat more subtle procedure.

---

*For those of you who covered the optional section on the Chain Rule in Chapter 3, formula (5) is a result of applying the Chain Rule. For example, if $y = \ln u$ and $u = g(x)$, then $y = \ln g(x)$ and

$$\frac{dy}{dx} = \frac{dy}{du} \frac{du}{dx} = \left(\frac{1}{u}\right)(g'(x)) = \frac{g'(x)}{g(x)}$$

3. THE LOGARITHM FUNCTION AND ITS DERIVATIVE

Example 6 Find the derivative of: $f(x) = \dfrac{x^2\sqrt{5x + 1}}{(3x - 2)^3}$

Solution As you will see, it is easier to take the natural logarithm of both sides before differentiating. That is, look instead at

$$\ln f(x) = \ln \dfrac{x^2\sqrt{5x + 1}}{(3x - 2)^3}$$

Using the properties of logarithms, we may write the above expression as

$$\ln f(x) = \ln x^2 + \ln \sqrt{5x + 1} - \ln(3x - 2)^3$$
$$= 2 \ln x + \tfrac{1}{2}\ln(5x + 1) - 3 \ln(3x - 2)$$

We now use (5) to find $f'(x)$:

$$\dfrac{f'(x)}{f(x)} = \dfrac{2}{x} + \dfrac{5}{2(5x + 1)} - \dfrac{(3)(3)}{3x - 2}$$

$$f'(x) = \dfrac{x^2\sqrt{5x + 1}}{(3x - 2)^3}\left[\dfrac{2}{x} + \dfrac{5}{2(5x + 1)} - \dfrac{9}{3x - 2}\right]$$

∎

Logarithmic Differentiation We refer to the procedure used in Example 6 as *logarithmic differentiation*. Let's do another example to illustrate the procedure.

Example 7 Find the derivative of: $f(x) = \dfrac{\sqrt{4x + 3}}{(2x - 5)^3}$

Solution
$$\ln f(x) = \ln \dfrac{(4x + 3)^{1/2}}{(2x - 5)^3} = \ln(4x + 3)^{1/2} - \ln(2x - 5)^3$$
$$= \tfrac{1}{2}\ln(4x + 3) - 3 \ln(2x - 5)$$

$$\dfrac{f'(x)}{f(x)} = \dfrac{1}{2}\left(\dfrac{4}{4x + 3}\right) - 3\left(\dfrac{2}{2x - 5}\right)$$

$$f'(x) = \dfrac{\sqrt{4x + 3}}{(2x - 5)^3}\left(\dfrac{2}{4x + 3} - \dfrac{6}{2x - 5}\right)$$

∎

Change-of-Base Formula The problem of determining the derivative of the logarithm function $\log_a x$ for any base a is solved by using the change-of-base formula:

$$\log_a M = \dfrac{\log_b M}{\log_b a}$$

To develop the formula for the derivative of $f(x) = \log_a x$, we change over to the natural base e. Then,

$$f(x) = \log_a x = \dfrac{\log_e x}{\log_e a} = \dfrac{\ln x}{\ln a}$$

214 CH. 5 THE EXPONENTIAL AND LOGARITHM FUNCTIONS

Since ln a is a constant, we have

$$f'(x) = \frac{d}{dx}\log_a x = \frac{1}{\ln a}\frac{d}{dx}\ln x = \frac{1}{\ln a}\frac{1}{(x)}$$

Thus,

(6) $$\frac{d}{dx}\log_a x = \frac{1}{x \ln a}$$

Example 8 Find the derivative of: $f(x) = \log_2 x$

Solution Using (6), we have

$$f(x) = \frac{d}{dx}\log_2 x = \frac{1}{x \ln 2}$$

■

Finally, we want to find the derivative of $f(x) = a^x$, where $a > 0, a \neq 1$, is any real constant.

To solve this problem, we use the definition of a logarithm. Then,

$$y = a^x$$

$$x = \log_a y = \frac{\ln y}{\ln a} = \frac{\ln a^x}{\ln a}$$

Now, we differentiate with respect to x and use (5), where $g(x) = a^x$:

$$1 = \frac{1}{\ln a}\frac{d}{dx}\ln a^x = \left(\frac{1}{\ln a}\right)\left(\frac{1}{a^x}\right)\frac{d}{dx}a^x$$

Thus,

(7) $$\frac{d}{dx}a^x = (\ln a)a^x$$

Example 9 Find the derivative of: $f(x) = 2^x$

Solution Using (7), we have

$$f'(x) = \frac{d}{dx}2^x = (\ln 2)2^x$$

■

Exercise 3
Solutions to Odd-Numbered Problems begin on page 405.

In Problems 1–4 graph each function.

1. $f(x) = \log_5 x$
2. $f(x) = \log_3 x$
3. $f(x) = \log_{1/3} x$
4. $f(x) = \log_{1/5} x$

3. THE LOGARITHM FUNCTION AND ITS DERIVATIVE   215

In Problems 5-10 write each logarithm expression using exponential notation.

5. $\log_3 9 = 2$
6. $\log_2 16 = 4$
7. $\log_3 \frac{1}{81} = -4$
8. $\log_{1/2} \frac{1}{16} = 4$
9. $\log_a P = Q$
10. $\log_b Y = C$

In Problems 11-14 write each exponential expression using logarithmic notation.

11. $10^3 = 1000$
12. $10^{-3} = 0.001$
13. $a^{1/2} = 3$
14. $x^3 = 6$

In Problems 15-20 calculate the indicated quantity.

15. $\log_2 32$
16. $\log_{1/2} 32$
17. $\log_{10} 10^{-3}$
18. $\log_{10} 10^4$
19. $\log_2 24 - \log_2 12$
20. $\log_3 15 - \log_3 5$

In Problems 21-28 use $\log_{10} 2 = 0.3010$, $\log_{10} 3 = 0.4771$, and $\log_{10} 5 = 0.6990$ to compute each quantity.

21. $\log_{10} 12$
22. $\log_{10} 250$
23. $\log_{10} 7.5$
24. $\log_{10} 12.5$
25. $\log_{10} 36$
26. $\log_{10} 30$
27. $\log_{10} \frac{1}{8}$
28. $\log_{10} \frac{1}{2}$

In Problems 29-34 simplify each expression using the properties of logarithms.

29. $\ln 3 + \ln x$
30. $\ln x^6 - \ln x^3$
31. $\frac{1}{2} \ln 16$
32. $\ln 2 - \ln x + \ln 4$
33. $\ln(x + 1) + \ln(x + 2) + \ln(x + 3)$
34. $\ln(x + 1) - 3 \ln(x + 3)$

In Problems 35-56 find the derivative of each function.

35. $f(x) = 6 \ln x$
36. $f(x) = -2 \ln x$
37. $f(x) = \ln 3x$
38. $f(x) = \ln 5x$
39. $f(x) = 8 \ln \frac{x}{2}$
40. $f(x) = \frac{1}{2} \ln 4x$
41. $f(x) = x \ln x$
42. $f(x) = x^2 \ln x$
43. $f(x) = \ln x^2$
44. $f(x) = \ln x^3$
45. $f(x) = \ln \sqrt{x}$
46. $f(x) = \ln \sqrt[3]{x}$
47. $f(x) = \sqrt{\ln x}$
48. $f(x) = \sqrt[3]{\ln x}$
49. $f(x) = \frac{1}{x} \ln x$
50. $f(x) = e^x + x \ln x$
51. $f(x) = e^{\ln x}$
52. $f(x) = \ln(\ln x)$
53. $f(x) = x \ln(x^2 + 4)$
54. $f(x) = x \ln(x^2 + 5x + 1)$
55. $f(x) = x \ln \sqrt{x^2 + 1}$
56. $f(x) = x \ln \sqrt[3]{3x + 1}$

In Problems 57–62 use logarithmic differentiation to find the derivative of each function.

57. $f(x) = (x^2 + 1)^2(2x^3 - 1)^4$
58. $f(x) = (3x^2 + 4)^3(x^2 + 1)^4$
59. $f(x) = (x^3 + 1)(x - 1)(x^4 + 5)$
60. $f(x) = \sqrt{x^2 + 1}(x^3 - 5)(3x + 4)$
61. $f(x) = \dfrac{x^2(x^3 + 1)}{\sqrt{x^2 + 1}}$
62. $f(x) = \dfrac{\sqrt{x}(x^3 + 2)^2}{\sqrt[3]{3x + 4}}$

In Problems 63–66 find the derivative of each function.

63. $f(x) = 2^x$
64. $f(x) = 3^x$
65. $f(x) = \log_2 x$
66. $f(x) = \log_3 x$

In Problems 67 and 68 graph each function.

67. $f(x) = x \ln x$
68. $f(x) = \dfrac{1}{x} \ln x$

In Problems 69 and 70 find y', y'', and y''', where a is a constant.

69. $y = e^{ax}$
70. $y = \ln(ax)$

71. *Marginal Cost.* The cost (in dollars) of producing x units (measured in thousands) of a certain product is found to be

$$C(x) = \$20 + \ln(x + 1)$$

Find the marginal cost.

72. *Weber–Fechner Law.* When a certain drug is administered, the reaction R to the dose x is given by the *Weber–Fechner law*:

$$R = 5.5 \ln x + 10$$

Find the reaction rate for a dose of 5 units.

73. *Maximum Profit.* The revenue (in dollars) derived from the sale of x units (measured in thousands) of the product in Problem 71 is

$$R(x) = 0.1x$$

Find the number of units that will maximize profit.

74. Find the absolute maximum and the absolute minimum of the following functions on the given interval:

(a) $f(x) = \dfrac{\ln x}{x}$; on $[1, 3]$
(b) $f(x) = xe^x$; on $[0, 4]$

75. Find the derivative of order n of:

(a) $f(x) = e^{ax}$, where a is a constant
(b) $f(x) = \ln x$

76. If $f(x) = \log_a x$, show that

$$f(x + h) - f(x) = \log_a\left(1 + \dfrac{h}{x}\right)$$

77. *Concentration of a drug.* A drug administered intravenously is eliminated from the blood by organs, such as the liver, that consume it. If the initial concentration of the drug is A (in moles/liter),

then the concentration C (in moles/liter) of the drug at some future time t (in minutes) is given by $C = A2^{-kt}$, where k is a positive constant that depends on the drug administered. Find:

(a) the time in minutes needed to eliminate $\frac{1}{2}$ the drug;
(b) the rate of decrease of the concentration of the drug at any time t.

4. Logistic Curves

We have seen that many models in business and economics require the use of exponential functions. The graphs of these functions are called *growth curves* or *decay curves,* depending on whether they are increasing or decreasing functions.

For example, we used a growth curve to measure the increase in gross national product per unit time, and we saw that the relationship between price and demand is a decay curve.

Such curves can be used to reflect accurately situations in which the values of the function increase without bound (or decrease to zero). However, curves of this type cannot be used if the growth (or decay), after a fast start, begins to level off, approaching a maximum (or minimum) value. For example, the value of a car depreciates very fast initially, but then levels off and, of course, can never depreciate more than its original cost. The demand for a new product, such as video games, hula-hoops, big wheels, and so on, starts high, increases rapidly, and then levels off. Curves of this kind are called *modified growth curves,* and their general equation is

Modified Growth Curve

$$f(x) = C(1 - b^{-ax})$$

where a and C are positive real numbers and $b > 1$ is a real number. Figure 10 illustrates a modified growth curve. Notice that the line $y = C$ is an asymptote and that it is approached but never reached.

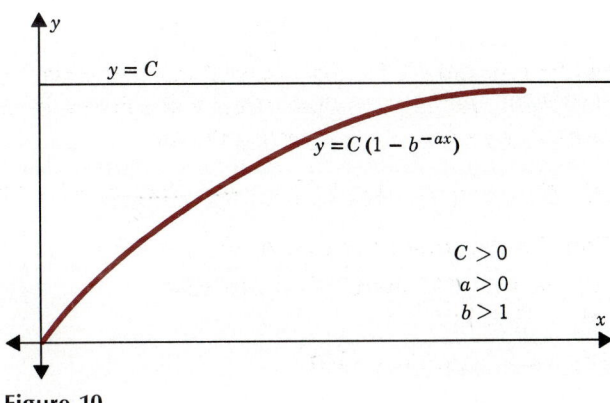

Figure 10

If the modified growth curve is used as a model for the growth in sales of a new product, it may not accurately reflect the real situation. Sometimes, the initial rate of growth is slow, and, as time progresses, the rate increases to a

218 CH. 5 THE EXPONENTIAL AND LOGARITHM FUNCTIONS

Logistic Curve

maximum value and then begins to decline. In other words, the modified growth curve may only be a good model for what happens at the end of the sales cycle. Curves that describe a situation in which the rate of growth is slow at first, increases to a maximum rate, and then decreases are called *logistic curves,* or *saturation curves*. These curves are best characterized by their "S" shape. Figure 11 illustrates a typical logistic curve.

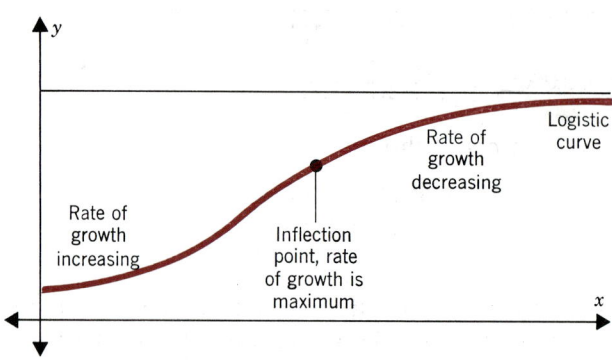

Figure 11

Example 1
Maximum Sales Rate

The sales of a new line of color televisions over a period of time is expected to follow the relationship

$$f(x) = \frac{10{,}000}{1 + 100e^{-x}}$$

where x is measured in years. Analyze the graph of this function, determine the year in which a maximum sales rate is achieved, and find the upper limit to sales in any year.

Solution

The domain of this function is $x \geq 0$ and the function is continuous on its domain. The y-intercept is

$$y = f(0) = \frac{10{,}000}{1 + 100} = 99.01$$

This represents the predicted number of television sets sold when production begins.

As x increases without bound, the value of e^{-x} gets closer to zero. As a result, the behavior of $f(x)$ as x increases without bound is

$$\frac{10{,}000}{1 + 100e^{-x}} \approx \frac{10{,}000}{1 + 100(0)} = 10{,}000$$

This number represents an upper estimate for sales.

The derivative of the function is

$$f'(x) = \frac{10{,}000}{(1 + 100e^{-x})^2} \cdot 100e^{-x} = \frac{1{,}000{,}000e^{-x}}{(1 + 100e^{-x})^2}$$

Clearly, $f'(x) > 0$ for $x \geq 0$. Thus, the function is increasing, which means that sales are increasing each year.

The second derivative is

$$f''(x) = 1{,}000{,}000 \frac{d}{dx}[e^{-x}(1+100e^{-x})^{-2}]$$

$$= 1{,}000{,}000[-e^{-x}(1+100e^{-x})^{-2} + 2e^{-x}100e^{-x}(1+100e^{-x})^{-3}]$$

$$= 1{,}000{,}000\left[\frac{-e^{-x}(1+100e^{-x}) + 200e^{-2x}}{(1+100e^{-x})^3}\right]$$

$$= 1{,}000{,}000 e^{-x}\left[\frac{100e^{-x} - 1}{(1+100e^{-x})^3}\right]$$

The inflection points, if there are any, obey

$$100e^{-x} - 1 = 0$$
$$e^x = 100$$
$$x \approx 4.6$$

There is an inflection point at this number x. Moreover, at $x = 4.6$, the first derivative $f'(x)$ achieves its maximum value. Thus, at 4.6 years, the rate of sales achieves its maximum. The graph is given in Figure 12. Observe that 10,000 sales are the most that can be achieved.

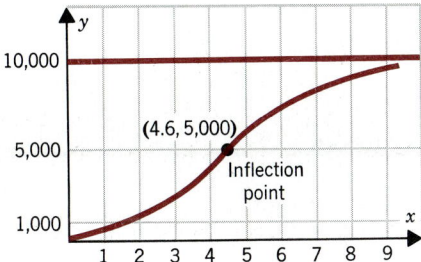

Figure 12

Exercise 4

Solutions to Odd-Numbered Problems begin on page 407.

1. The rate at which a grocery store clerk can stock cans on a shelf follows a learning curve (modified growth curve) of the form

   $$S(t) = 2000 - 1000e^{-0.4t}$$

   where t is the number of weeks the clerk has been employed and S is the number of cans stacked per hour.

   (a) How many cans can a new employee stack per hour?
   (b) At what rate is the employee learning after 5 weeks on the job?
   (c) What is the upper limit of the number of cans an employee can stack per hour?

2. Follow the directions of Problem 1 if the learning curve is of the form

   $$S(t) = 2000 - 1200e^{-0.6t}$$

3. *Maximum Sales Rate.* The sales of a new car model over a period of time is expected to follow the relationship

$$f(x) = \frac{20{,}000}{1 + 50e^{-x}}$$

where x is measured in months. Determine the month in which the sales rate is a maximum. Graph the function.

4. *Maximum Sales Rate.* The sales of a new stereo system over a period of time is expected to follow the relationship

$$f(x) = \frac{5000}{1 + 5e^{-x}}$$

where x is measured in years. Determine the year in which the sales rate is a maximum. Graph the function.

5. *Spread of Disease.* In a town of 50,000 people, the number of people at time t who have influenza is

$$N(t) = \frac{10{,}000}{1 + 9999e^{-t}}$$

where t is measured in days. Note that the flu is spread by the one person who has it at $t = 0$. At what time t is the rate of spreading of flu the greatest? Graph the function.

*6. In general, a logistic curve is of the form

$$f(x) = \frac{M}{1 + ae^{-bx}}$$

where $a > 0, M > 0, b > 0$ are real numbers. At what number x is the derivative $f'(x)$ a maximum?

Chapter Review

Important Terms

exponential function
laws of exponents
growth curves
decay curves
the number e
continuously compounded interest
effective rate of interest
*normal density

logarithm function
properties of logarithms
common logarithm
natural logarithm
logarithmic differentiation
change-of-base formula
modified growth curve
logistic curve

True-False Questions (Answers on page 434.)

T F 1. The derivative of e^x is e^x.
T F 2. One of the properties of a logarithm is that $\log_a (M + N) = \log_a M + \log_a N$.
T F 3. If $x = e^y$, then $y = \ln x$.
T F 4. The graph of $f(x) = e^x$ is increasing.
T F 5. The graph of $f(x) = \ln x$ is decreasing.

*Discussed in optional problem.

Fill in the Blanks
(Answers on page 435.)

1. A function of the form $f(x) = a^x$, $a > 0$, $a \neq 1$, is called an _____ function.
2. $\log_2 4 + \log_2 8 = \log_2$ _____.
3. If $y = e^{x^2}$, then $y' =$ _____.
4. The formula for continuously compounded interest involves the number _____.
5. The function $f(x) = \ln x$ is called the _____ _____ function.

Review Exercises
Solutions to Odd-Numbered Problems begin on page 408.

In Problems 1–10 find the derivative of each function.

1. $f(x) = 4e^{5x}$
2. $f(x) = 2e^{-3x}$
3. $f(x) = 15 \ln \dfrac{x}{3}$
4. $f(x) = \tfrac{1}{2} \ln(6x)$
5. $f(x) = e^{2x^2+5}$
6. $f(x) = e^{x^3+1}$
7. $f(x) = \ln(2x^2 + 5)$
8. $f(x) = \ln(x^3 + 1)$
9. $f(x) = (x^2 + 1)^2(x^2 - 1)^3$
10. $f(x) = (2x + 1)^2(x^2 + 4)^3$

11. How long will it take $10,000 to double if it can be invested at 12% compounded continuously?

12. *Demand for Oil.* The demand for oil in the United States increases at a rate of 4% per year. Assuming this rate remains constant after 1983, when will the demand double? Assume compounding takes place continuously.

13. Analyze and graph the function:
$$f(x) = \dfrac{2000}{1 + 4e^{-x}}$$

14. *Spread of Rumor.* In a city of 50,000 people, the number of people at time t who have heard a certain rumor obeys
$$N(t) = \dfrac{50{,}000}{1 + 49{,}999e^{-t}}$$
At what time t is the rate of spreading of the rumor greatest? Graph the function $N(t)$.

15. *Maximum Sales Rate.* An advertising company conducts a special campaign to promote sales of a certain product. They estimate that the benefits of the campaign will result in extra sales and, when the campaign is over, the extra sales will obey a decay curve of the form
$$S = 4000e^{-0.3t}$$
where S is the amount of extra sales and t is the time in days after the advertising campaign is over. How many extra sales are obtained 10 days after the close of the advertising campaign? What is the rate of extra sales at $t = 10$?

Mathematical Questions From Actuary Exams (Answers on page 433.)

1. Actuary Exam—Part I

   $\frac{d}{dx}(x^2 e^{x^2}) =$

   (a) $2xe^{x^2}$ (b) $\frac{x^3}{3} e^{x^2}$ (c) $4x^2 e^{x^2}$ (d) $2xe^{x^2} + x^4 e^{x^2-1}$
   (e) $2x^3 e^{x^2} + 2xe^{x^2}$

2. Actuary Exam—Part I
   If $\log_6 2 = b$, which of the following is equal to $\log_6(4 \cdot 27)$?

   (a) $3 - b$ (b) $2 + b$ (c) $5b$ (d) $3b$ (e) $2b + \frac{27}{6}$

3. Actuary Exam—Part I
   Figure A below could represent the graph of which of the following?
   (a) $y = x^3 e^{x^2}$ (b) $y = xe^x$ (c) $y = xe^{-x}$
   (d) $y = x^2 e^x$ (e) $y = x^2 e^{x^2}$

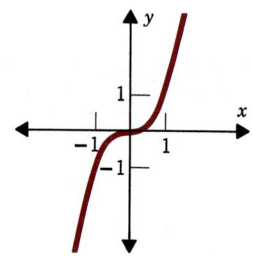

Figure A

4. Actuary Exam—Part I
   If b and c are real numbers and there are two different real numbers x such that $(e^x)^2 + be^x + c = 0$, which of the following must be true?

   I. $b^2 - 4c > 0$ II. $b < 0$ III. $c > 0$

   (a) None   (b) I only   (c) I and II only
   (d) I and III only   (e) I, II, and III

Other Books

Batschelet, E., *Introduction to Mathematics for Life Sciences,* Springer-Verlag, New York, 1972.

Chiang, Alpha, *Fundamental Methods of Mathematical Economics,* 2nd ed., McGraw-Hill, New York, 1974.

Ferguson, C. E., *Microeconomic Theory,* 3rd ed., Richard D. Irwin Press, Homewood, Ill., 1972.

Grossman, Stanley, and James Turner, *Mathematics for the Biological Sciences,* Macmillan, New York, 1974.

Springer, C. H., R. E. Herlihy, and R. I. Beggs, *Advanced Methods and Models,* Mathematics for Management Series, Vol. 2, Richard D. Irwin Press, Homewood, Ill., 1965.

Thrall, Robert M., Ed., *Some Mathematical Models in Biology,* University of Michigan, Ann Arbor, 1967.

6 Functions of Two Variables

1. Introduction
2. Partial Derivatives
3. Maxima and Minima
4. Lagrange Multipliers
5. Least Squares

Chapter Review

1. Introduction

In our discussion so far we have considered only functions of one independent variable, expressed by $y = f(x)$. Quite often, models in business give rise to important problems for which a single dependent variable is related to more than one independent variable.

For example, the cost of producing a certain item may depend on variables such as labor and material. In economic theory, supply and demand of a commodity often depend not only on its own price, but also on the prices of related commodities and some other factors (such as income level, time of year, etc.).

Another example of a function of more than one variable arises in physiology and concerns the relationship between body surface area and the weight and height of a person. If w denotes a person's weight (in kilograms), and h is a person's height (in centimeters), the surface area A (in square meters) has been found to be $A = (.007184)\, w^{0.425}\, h^{0.725}$.

Function of Two Variables   A functional relationship for two independent variables is denoted by $z = f(x, y)$. Here, $z = f(x, y)$ is termed a *function of two variables*, x and y, where x and y are the independent variables and z is a single dependent variable. For each pair of numbers x and y, the dependent variable assumes a single number as specified by the relationship. For example, for the function

$$z = f(x, y) = x^2 + y^3$$

if the independent variables are $x = 2$ and $y = 4$, the dependent variable z is

$$z = f(2, 4) = 2^2 + 4^3 = 4 + 64 = 68$$

In the same way, we can define functions of three independent variables, $w = f(x, y, z)$; four independent variables, $u = f(w, x, y, z)$; and so on. In this chapter, we will mainly concern ourselves with functions of two independent variables.

In Chapter 1 we established that each point on a line can be associated with a real number and that each point in the plane can be associated with an ordered pair of real numbers. We now continue the development of these concepts one step further—to the representation of points in three-dimensional space. We set up a coordinate system by selecting a fixed point, called the *origin*. Through the origin, we draw three mutually perpendicular lines, called the *coordinate axes*, and label them as the x-axis, y-axis, and z-axis. On each of the three axes, we choose one direction as positive and select an appropriate scale. See Figure 1.

Ordered Triple   Just as we did in one and two dimensions, we assign coordinates to each point P in three dimensions. Specifically, we identify a point P with an *ordered triple* of real numbers (x, y, z), and we refer to it as *the point* (x, y, z). Thus, the point $(3, 5, 7)$ is the point for which $x = 3$, $y = 5$, and $z = 7$; that is, starting from the origin, we reach P by moving 3 units along the positive x-axis, then 5 units in the direction of the positive y-axis, and finally, 7 units in the direction of the positive z-axis. Figure 2 illustrates the location of the point $(3, 5, 7)$, as well as the points $(3, 5, 0)$ and $(0, 5, 0)$. Observe that any point on the x-axis will have the form $(x, 0, 0)$. Similarly, $(0, y, 0)$ and $(0, 0, z)$ represent points on the y-axis and z-axis, respectively.

In addition, all points of the form $(x, y, 0)$ are contained in a plane, called the *xy-plane*. Similarly, the points $(0, y, z)$ are in the *yz-plane* and the points $(x, 0, z)$ are in the *xz-plane*. See Figure 3.

1. INTRODUCTION

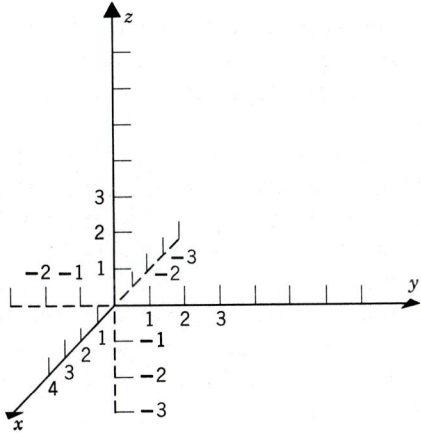

Figure 1

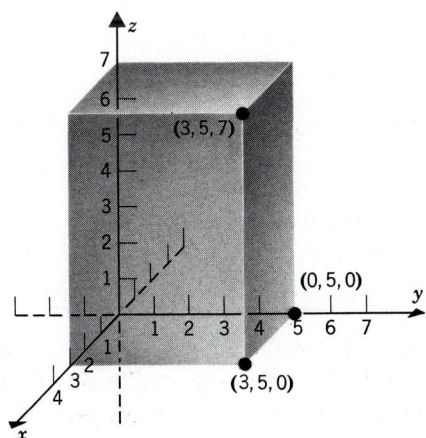

Figure 2

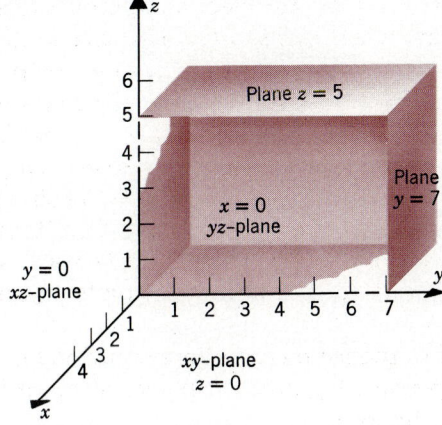

Figure 3

226　CH. 6　FUNCTIONS OF TWO VARIABLES

Figure 3 also illustrates that points of the form (x, y, z) where $z = 5$ lie in a plane parallel to the xy-plane. Also illustrated are points (x, y, z) where $y = 7$. This is a plane parallel to the xz-plane.

The graph of a function of a single variable is a curve consisting of the set of points (x, y) in the xy-plane for which $y = f(x)$. Similarly, the graph of a function f of two variables, called a *surface*, consists of all points (x, y, z) in three-dimensional space for which $z = f(x, y)$.

• **Surface**

Obtaining the shape, or even a rough sketch, of the graph of most functions of two variables is a difficult task and is not taken up in this book. Computers may be used to generate *computer graphics* of mathematical surfaces in three dimensions.

• **Computer Graphics**

For example, a portion of a surface $z = f(x, y)$ can be viewed in perspective and delineated by a rectangular mesh over a rectangular portion of the xy-plane. By changing the computer program, the point of view and other characteristics of the computer graphic may be altered. See Figure 4.

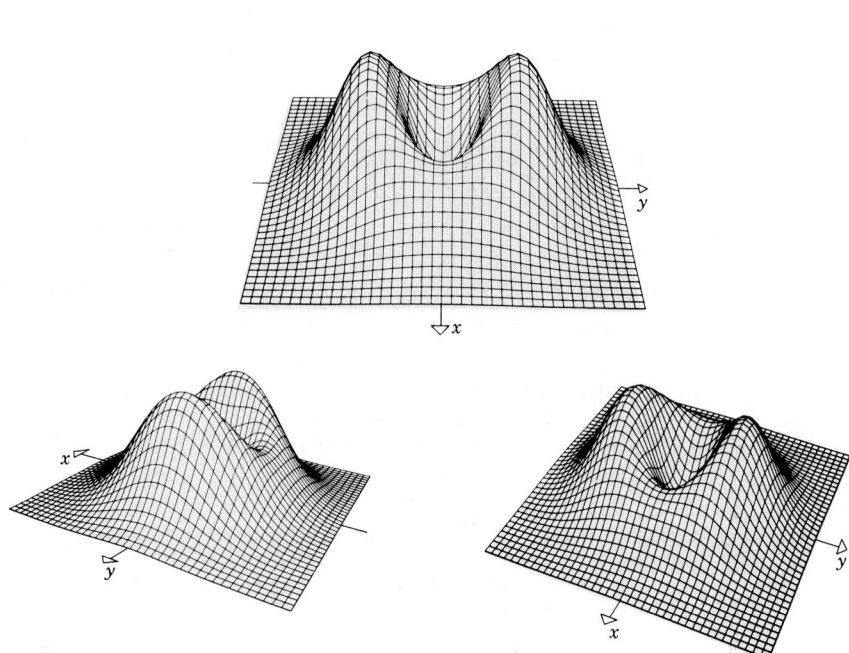

Figure 4

Three views of the surface
$z = (x^2 + 2y^2)e^{1-x^2-y^2}$

Exercise 1

Solutions to Odd-Numbered Problems begin on page 408.

In Problems 1–10, evaluate $f(2, 1)$.

1. $f(x, y) = x^2 + y$
2. $f(x, y) = x - y^2$
3. $f(x, y) = \sqrt{xy}$
4. $f(x, y) = x\sqrt{y}$

5. $f(x, y) = \dfrac{1}{2x + y}$ 6. $f(x, y) = \dfrac{x}{x - 3y}$

7. $f(x, y) = \dfrac{x^2 - y}{x - y}$ 8. $f(x, y) = \dfrac{x + y^2}{x^2 - y^2}$

9. $f(x, y) = \sqrt{4 - x^2 y^2}$ 10. $f(x, y) = \sqrt{9 - x^2 y^2}$

11. Let $f(x, y) = 3x + 2y + xy$. Find:
    (a) $f(1, 0)$ (b) $f(0, 1)$ (c) $f(2, 1)$
    (d) $f(x + \Delta x, y)$ (e) $f(x, y + \Delta y)$

12. Let $f(x, y) = x^2 y + x + 1$. Find:
    (a) $f(0, 0)$ (b) $f(0, 1)$ (c) $f(2, 1)$
    (d) $f(x + \Delta x, y)$ (e) $f(x, y + \Delta y)$

13. Let $f(x, y) = \sqrt{xy} + x$. Find:
    (a) $f(0, 0)$ (b) $f(0, 1)$ (c) $f(a^2, t^2)$; $a > 0, t > 0$
    (d) $f(x + \Delta x, y)$ (e) $f(x, y + \Delta y)$

14. Let $f(x, y) = e^{x+y}$. Find:
    (a) $f(0, 0)$ (b) $f(1, -1)$ (c) $f(x + \Delta x, y)$
    (d) $f(x, y + \Delta y)$

In Problems 15–20 describe the set of all points (x, y, z) that satisfy the given conditions.

15. $y = 3$ 16. $z = -3$
17. $x = 0$ 18. $z = 5$
19. $x = 1$ and $y = 0$ 20. $x = y$ and $z = 0$

21. For the function $z = f(x, y) = 3x + 4y$, find:
    (a) $f(x + \Delta x, y)$
    (b) $f(x + \Delta x, y) - f(x, y)$
    (c) $\dfrac{f(x + \Delta x, y) - f(x, y)}{\Delta x}$, $\Delta x \neq 0$
    (d) $\lim\limits_{\Delta x \to 0} \dfrac{f(x + \Delta x, y) - f(x, y)}{\Delta x}$, $\Delta x \neq 0$

22. For the function $z = f(x, y) = 4x + 5y$, find:
    (a) $f(x, y + \Delta y)$
    (b) $f(x, y + \Delta y) - f(x, y)$
    (c) $\dfrac{f(x, y + \Delta y) - f(x, y)}{\Delta y}$, $\Delta y \neq 0$
    (d) $\lim\limits_{\Delta y \to 0} \dfrac{f(x, y + \Delta y) - f(x, y)}{\Delta y}$, $\Delta y \neq 0$

23. *Cost of Construction.* The cost of the bottom and top of a cylindrical tank is $300 per square meter and the cost of the sides is $500 per square meter. Write the total cost of constructing such a tank as a function of the radius R and height h (both in meters).

24. *Distance Formula.* In a three-dimensional coordinate system, the formula for computing the distance between the point (x_1, y_1, z_1) and the point (x_2, y_2, z_2) is

$$d = \sqrt{(x_2 - x_1)^2 + (y_2 - y_1)^2 + (z_2 - z_1)^2}$$

> Use this formula to find the distance between the following pairs of points:
> (a) $(0, 0, 0)$ and $(-4, 0, 3)$
> (b) $(1, 2, -1)$ and $(3, -1, 5)$
> (c) $(0, 2, -1)$ and $(1, 5, -3)$

2. Partial Derivatives

Consider a function $z = f(x, y)$ of two variables. If we think of y as a constant, then $z = f(x, y)$ is a function of the single variable x. If we proceed to differentiate with respect to x, we obtain the *partial derivative of z with respect to x*, denoted by the symbol $\partial z / \partial x$ (or by f_x).

For example, if $z = f(x, y) = x^2 + xy^2$, then

$$\frac{\partial z}{\partial x} = 2x + y^2$$

Note that y is treated as if it were a constant.

Similarly, for the same function, $z = f(x, y) = x^2 + xy^2$, the *partial derivative of z with respect to y* (x is held constant) is

$$\frac{\partial z}{\partial y} = 2xy$$

Compare the following definition of partial derivatives with the definition of the derivative of a function of a single variable found on page 103.

Partial Derivatives **If** $z = f(x, y)$**, then the** *partial derivative of z with respect to x at* (x, y) **is**

$$\frac{\partial z}{\partial x} = \lim_{\Delta x \to 0} \frac{f(x + \Delta x, y) - f(x, y)}{\Delta x}$$

provided the limit exists.

The *partial derivative of z with respect to y at* (x, y) **is**

$$\frac{\partial z}{\partial y} = \lim_{\Delta y \to 0} \frac{f(x, y + \Delta y) - f(x, y)}{\Delta y}$$

provided the limit exists.

In the one-variable case, the derivative measures the instantaneous rate of change of the function. In the two-variable case, $\partial z / \partial x$ measures the instantaneous rate of change of f with respect to x (y is held constant) and $\partial z / \partial y$ measures the instantaneous rate of change of f with respect to y (x is held constant).

Example 1 Find $\partial z / \partial x$ and $\partial z / \partial y$ if:
(a) $z = f(x, y) = x^3 + 2x^2y + y^2$
(b) $z = f(x, y) = x \ln y + y e^x$

Solution (a) $\dfrac{\partial z}{\partial x} = f_x = 3x^2 + 4xy$ $\qquad \dfrac{\partial z}{\partial y} = f_y = 2x^2 + 2y$

(b) $\dfrac{\partial z}{\partial x} = f_x = \ln y + ye^x$ $\qquad \dfrac{\partial z}{\partial y} = f_y = \dfrac{x}{y} + e^x$

■

Geometrical Interpretation

Let's look at a geometrical interpretation of the partial derivatives $\partial z/\partial x$ and $\partial z/\partial y$. Suppose z is a function of x and y, and we are considering the partial derivative of z with respect to x at the point (x_1, y_1, z_1). When we compute $\partial z/\partial x$, y is taken to be a constant, and in this case $y = y_1$. This is geometrically equivalent to considering the intersection of the plane $y = y_1$ with the surface $z = f(x, y)$. See Figure 5(a).

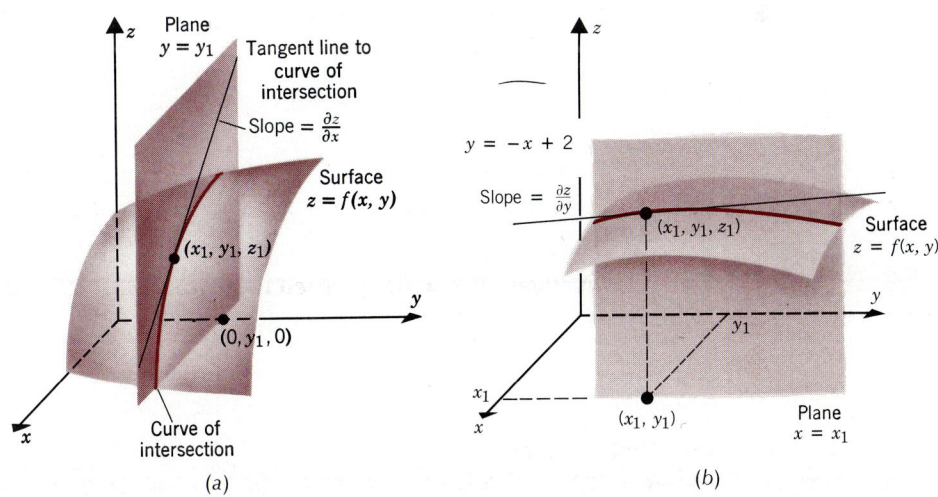

Figure 5

The partial derivative $\partial z/\partial x$ at (x_1, y_1, z_1) equals the slope of the tangent line of the curve that is the intersection of the plane $y = y_1$, and the original surface.

Similarly, the partial derivative $\partial z/\partial y$ at (x_1, y_1, z_1) equals the slope of the tangent line of the curve that is the intersection of the plane $x = x_1$ and the original surface. See Figure 5(b).

Example 2 Find the slope of the tangent line to the curve of intersection of the surface

$$z = 16 - x^2 - y^2$$

and the plane $y = 2$ at the point $(1, 2, 11)$.

Solution The slope of the tangent line to the curve of intersection of $z = 16 - x^2 - y^2$ and the plane $y = 2$ is

$$\frac{\partial z}{\partial x} = -2x$$

At the point $(1, 2, 11)$, the slope is -2. See Figure 6.

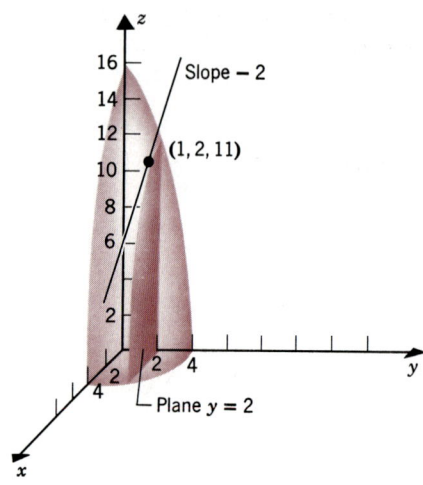

Figure 6

Example 3 In many production processes, the cost of manufacturing consists of a fixed cost and two variable costs: the cost of raw materials and the cost of labor. If the total cost is given by

$$C(x, y) = \$180 + 18x + 40y$$

where x is the cost (in dollars) of raw materials and y is the cost (in dollars) of labor, find $\partial C/\partial x$ and $\partial C/\partial y$ and give an interpretation of your answer.

Solution

$$\frac{\partial C}{\partial x} = 18 \quad \text{and} \quad \frac{\partial C}{\partial y} = 40$$

We interpret $\partial C/\partial x = 18$ to mean that when the cost of labor is held fixed, an increase of \$1 in the cost of raw materials causes an increase of \$18 in the total cost of the product. The partial derivative $\partial C/\partial x$ thus measures the incremental cost due to an increase of \$1 in the cost of raw material, while labor costs are held fixed. The partial derivative $\partial C/\partial y$ is similarly interpreted. The partial derivative $\partial C/\partial x$ is called the *marginal cost of raw material* and $\partial C/\partial y$ is the *marginal cost of labor*.

Marginal Cost

There are four second-order partial derivatives for a function of two variables. These four partial derivatives are

$$\frac{\partial^2 z}{\partial x^2} = \frac{\partial}{\partial x}\left(\frac{\partial z}{\partial x}\right) = f_{xx}$$

$$\frac{\partial^2 z}{\partial y^2} = \frac{\partial}{\partial y}\left(\frac{\partial z}{\partial y}\right) = f_{yy}$$

$$\frac{\partial^2 z}{\partial x\, \partial y} = \frac{\partial}{\partial x}\left(\frac{\partial z}{\partial y}\right) = f_{yx}$$

$$\frac{\partial^2 z}{\partial y\, \partial x} = \frac{\partial}{\partial y}\left(\frac{\partial z}{\partial x}\right) = f_{xy}$$

The notation $\partial^2 z/\partial x\, \partial y = f_{yx}$ means that starting with $z = f(x, y)$, the partial derivatives are taken first with respect to y and then with respect to x, in that order. A similar interpretation is given to $\partial^2 z/\partial y\, \partial x = f_{xy}$.

Example 4 For the functions given in Example 1, find f_{xx}, f_{yy}, f_{xy}, and f_{yx}.

Solution (a) $f_{xx} = 6x + 4y$
$f_{yy} = 2$
$f_{yx} = \frac{\partial}{\partial x}\left(\frac{\partial z}{\partial y}\right) = \frac{\partial}{\partial x}(2x^2 + 2y) = 4x$
$f_{xy} = \frac{\partial}{\partial y}\left(\frac{\partial z}{\partial x}\right) = \frac{\partial}{\partial y}(3x^2 + 4xy) = 4x$

(b) $f_{xx} = ye^x$
$f_{yy} = -\frac{x}{y^2}$
$f_{yx} = \frac{\partial}{\partial x}\left(\frac{\partial z}{\partial y}\right) = \frac{\partial}{\partial x}\left(\frac{x}{y} + e^x\right) = \frac{1}{y} + e^x$
$f_{xy} = \frac{\partial}{\partial y}\left(\frac{\partial z}{\partial x}\right) = \frac{\partial}{\partial y}(\ln y + ye^x) = \frac{1}{y} + e^x$

Note that for both functions in Example 4, we have

$$\frac{\partial^2 z}{\partial x\, \partial y} = \frac{\partial^2 z}{\partial y\, \partial x}$$

Although there are functions $z = f(x, y)$ for which this equality does not hold true, such functions are rare and are not encountered in this book.

Exercise 2
Solutions to Odd-Numbered Problems begin on page 409.

In Problems 1–10 find f_x, f_y, f_{xx}, f_{yy}, f_{yx}, and f_{xy}.

1. $f(x, y) = y^3 - 2xy + y^2 - 12x^2$
2. $f(x, y) = x^3 - xy + 10y^2 x$
3. $f(x, y) = xe^y + ye^x + x$
4. $f(x, y) = xe^x + xe^y + y$
5. $f(x, y) = \dfrac{x}{y}$
6. $f(x, y) = \dfrac{y}{x}$
7. $f(x, y) = \ln(x^2 + y^2)$
8. $f(x, y) = \ln(x^2 - y^2)$

9. $f(x, y) = \dfrac{10 - x + 2y}{xy}$ 10. $f(x, y) = \dfrac{5 + 3x - 2y}{xy}$

In Problems 11–16 verify that $f_{xy} = f_{yx}$.

11. $f(x, y) = x^3 + y^2$ 12. $f(x, y) = x^2 - y^3$
13. $f(x, y) = 3x^4y^2 + 7x^2y$ 14. $f(x, y) = 5x^3y - 8xy^2$
15. $f(x, y) = \dfrac{y}{x^2}$ 16. $f(x, y) = \dfrac{x}{y^2}$

17. Find the slope of the tangent line to the curve of intersection of the cylinder $4z = 5\sqrt{16 - x^2}$ and the plane $y = 3$ at the point $(2, 3, 5\sqrt{3}/2)$.
18. Find the slope of the tangent line to the curve of intersection of the surface $3z = \sqrt{36x^2 + 16y^2 - 144}$ and the plane $y = 0$ at the point $(3, 0, 2\sqrt{5})$.
19. *Demand for Butter.* In a large town, the demand for butter (measured in pounds) is given by the formula

$$z = 2400 - 50x + 90y$$

where x is the average price per pound (in cents) of butter and y is the average price per pound (in cents) of margarine. Find the two first-order partial derivatives of z and interpret your answer.
20. *Marginal Productivity.* The production function of a certain commodity is given by $P = 8l - l^2 + 3lk + 50k - k^2$, where l and k are the labor and capital inputs, respectively. Find the *marginal productivities* of l and k at $l = 2$ and $k = 5$.

3. Maxima and Minima

In Section 1 of Chapter 4 we discussed a technique for determining the conditions under which a function of one variable $y = f(x)$ has a relative maximum or a relative minimum. Recall that if $y = f(x)$ has a relative maximum or a relative minimum at c, then $f'(c) = 0$. A corresponding result holds true for a function of two variables.

Relative Maximum

Relative Minimum

A function $z = f(x, y)$ of two variables is said to have a *relative maximum* at (c, d) if $f(c, d) > f(x, y)$ for all numbers x and y close to (c, d). A function $z = f(x, y)$ has a *relative minimum* at (c, d) if $f(c, d) < f(x, y)$ for all numbers x and y close to (c, d).

Figure 7 illustrates these definitions.

If $z = f(x, y)$ has a relative maximum or a relative minimum at (c, d), the two first-order partial derivatives are zero at (c, d); that is,

(1) $$f_x(c, d) = 0 \quad \text{and} \quad f_y(c, d) = 0$$

The converse of (1) is not necessarily true. That is, if $f_x(c, d) = 0$ and $f_y(c, d) = 0$, then f may or may not have a relative maximum or a relative mini-

3. MAXIMA AND MINIMA

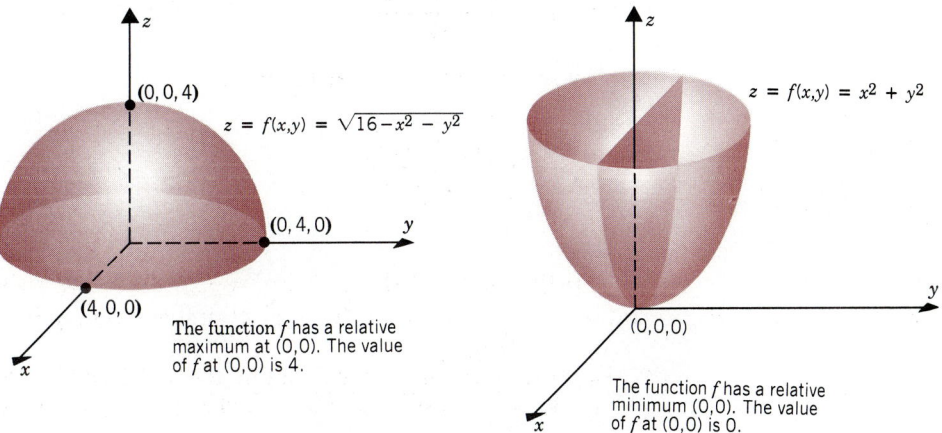

Figure 7

mum at (c, d). Statement (1) gives us *necessary, but not sufficient, conditions* for f to have a relative maximum or a relative minimum at (c, d).

Critical Point A point (c, d) at which $f_x(c, d) = f_y(c, d) = 0$ is called a *critical point* of the function $z = f(x, y)$. A critical point (c, d) that is neither a relative maximum nor

Saddle Point a relative minimum is called a *saddle point*.

The next test, using second-derivative tests, gives us *sufficient conditions* for a function to have a relative maximum, a relative minimum, or a saddle point at a critical point.

Test for a Relative Maximum
If (c, d) is a critical point of $z = f(x, y)$ and if

1. $f_{xx}(c, d) < 0$
2. $f_{xx}(c, d) \cdot f_{yy}(c, d) > [f_{xy}(c, d)]^2$

then the function has a *relative maximum* at (c, d).

Test for a Relative Minimum
If (c, d) is a critical point of $z = f(x, y)$ and if

1. $f_{xx}(c, d) > 0$
2. $f_{xx}(c, d) \cdot f_{yy}(c, d) > [f_{xy}(c, d)]^2$

then the function has a *relative minimum* at (c, d).

Test for Saddle Point
Suppose (c, d) is a critical point of $z = f(x, y)$ and suppose

$$f_{xx}(c, d) \cdot f_{yy}(c, d) < [f_{xy}(c, d)]^2$$

In this case, the point $(c, d, f(c, d))$ is a *saddle point*.

Figure 8 illustrates the graph of a function that has c saddle points. Such points will not be discussed any further in this book. Interested students should refer to texts in advanced calculus.

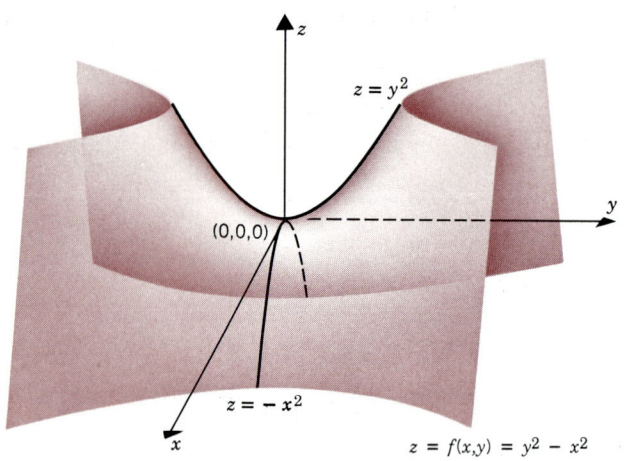

The origin (0,0,0) is a saddle part of f. The reason for this name should now be obvious.

Figure 8

Example 1 For the function

$$z = f(x, y) = x^2 + xy + y^2 - 6x + 6$$

find all relative maxima and all relative minima.

Solution First, we compute the first-order partial derivatives of $z = f(x, y)$.

$$f_x = 2x + y - 6 \qquad f_y = x + 2y$$

The critical points, if there are any, obey the conditions

$$2x + y - 6 = 0 \quad \text{and} \quad x + 2y = 0$$

Solving these equations simultaneously, we find that $x = 4$, $y = -2$ is a critical point.

To determine whether this critical point yields a relative minimum or a relative maximum, or neither, we find the values of f_{xx}, f_{yy}, f_{xy}, and f_{yx} at $(4, -2)$:

$$f_{xx}(4, -2) = 2 \qquad f_{xy}(4, -2) = f_{yx}(4, -2) = 1 \qquad f_{yy}(4, -2) = 2$$

The condition

$$f_{xx}(4, -2) \cdot f_{yy}(4, -2) > [f_{xy}(4, -2)]^2$$

is satisfied and $f_{xx}(4, -2) = 2 > 0$. So, the function has a relative minimum at $(4, -2)$.

Example 2 The demand functions for two products are

$$p = 12 - 2x \quad \text{and} \quad q = 20 - y$$

where p and q are the respective prices (in thousands of dollars) for each product, and x and y are the respective amounts (in thousands of units) of each sold. Suppose the joint cost function is

$$C(x, y) = x^2 + 2xy + 2y^2$$

Find the revenue function and the profit function. Determine the prices and amounts that will maximize profit. What is the maximum profit?

Solution The revenue function R is the sum of the revenues due to each product. Thus,

$$R = xp + yq = x(12 - 2x) + y(20 - y)$$

The profit function P is

$$\begin{aligned} P = P(x, y) &= R - C \\ &= x(12 - 2x) + y(20 - y) - (x^2 + 2xy + 2y^2) \\ &= -3x^2 - 3y^2 - 2xy + 12x + 20y \end{aligned}$$

The first-order partial derivatives of P are

$$\frac{\partial P}{\partial x} = -6x - 2y + 12 \qquad \frac{\partial P}{\partial y} = -6y - 2x + 20$$

The critical points obey

$$-6x - 2y + 12 = 0 \quad \text{and} \quad -6y - 2x + 20 = 0$$

Solving these equations, we find that the only critical point is $(1, 3)$.

The second-order partial derivatives of P are

$$P_{xx}(x, y) = -6 \qquad P_{xy}(x, y) = P_{yx}(x, y) = -2 \qquad P_{yy}(x, y) = -6$$

for any value (x, y). At the critical point $(1, 3)$, we see that

$$[P_{xx}(1, 3)][P_{yy}(1, 3)] > [P_{xy}(1, 3)]^2 \quad \text{and} \quad P_{xx}(1, 3) < 0$$

Thus, P has a relative maximum at $(1, 3)$. For these quantities sold, the corresponding prices p and q are $p = \$10{,}000$ and $q = \$17{,}000$. The maximum profit is $P(1, 3) = \$36{,}000$.

Exercise 3
Solutions to Odd-Numbered Problems begin on page 410.

In Problems 1–18 find all relative maxima and all relative minima of each function.

1. $f(x, y) = 3x^2 - 2xy + y^2$
2. $f(x, y) = x^2 - 2xy + 3y^2$
3. $f(x, y) = x^2 + y^2 - 3x + 12$
4. $f(x, y) = x^2 + y^2 - 6y + 10$
5. $f(x, y) = x^2 - y^2 + 4x + 8y$
6. $f(x, y) = x^2 - y^2 - 2x + 4y$

7. $f(x, y) = x^2 + 4y^2 - 4x + 8y - 1$
8. $f(x, y) = x^2 + y^2 - 4x + 2y - 4$
9. $f(x, y) = x^2 + y^2 + xy - 6x + 6$
10. $f(x, y) = x^2 + y^2 + xy - 8y$
11. $f(x, y) = 2 + x^2 - y^2 + xy$
12. $f(x, y) = x^2 - y^2 + 2xy$
13. $f(x, y) = x^3 - 6xy + y^3$
14. $f(x, y) = x^3 - 3xy - y^3$
15. $f(x, y) = x^3 + x^2y + y^2$
16. $f(x, y) = 3y^3 - x^2y + x$
17. $f(x, y) = \dfrac{y}{x + y}$
18. $f(x, y) = \dfrac{x}{x + y}$

19. **Economics.** The demand functions for two products are $p = 12 - x$ and $q = 8 - y$, where p and q are the respective prices (in thousands of dollars), and x and y are the respective amounts (in thousands of units) of each product sold. If the joint cost function is $C(x, y) = x^2 + 2xy + 3y^2$, determine the quantities x, y and prices p, q that maximize profit. What is the maximum profit?

20. **Economics.** The labor cost of a firm is given by the function

$$Q(x, y) = x^2 + y^3 - 6xy + 3x + 6y - 5$$

where x is the number of days required by a skilled worker and y is the number of days required by a semiskilled worker. Find the values of x and y for which the labor cost is a minimum.

21. A certain mountain is in the shape of the surface

$$z = 2xy - 2x^2 - y^2 - 8x + 6y + 4$$

(The unit of distance is 1000 feet.) If sea level is the xy-plane, how high is the mountain?

22. **Reaction to Drugs.** Two drugs are used simultaneously as a treatment for a certain disease. The reaction R (measured in appropriate units) to x units of the first drug and y units of the second drug is

$$R(x, y) = x^2y^2(a - x)(b - y) \qquad 0 \leq x \leq a,\ 0 \leq y \leq b$$

For a fixed amount x of the first drug, what amount y of the second drug produces the maximum reaction? For a fixed amount y of the second drug, what amount x of the first drug produces the maximum reaction? If x and y are both variable, what amount of each maximizes the reaction?

23. **Reaction to Drugs.** The reaction $R(x, t)$ to x units of a drug t hours after the drug has been administered is given by

$$R(x, t) = x^2(a - x)t^2 e^{-t} \qquad 0 \leq x \leq a$$

For what amount x is the reaction as large as possible? When does the maximum reaction occur?

*24. **Expansion of Gas.** The volume of a fixed amount of gas varies proportionally with the temperature and inversely with the pressure. Therefore, $V = k(T/P)$, where $k > 0$ is a constant and V, T, and P are the volume, temperature, and pressure, respectively. Calculate $\partial V/\partial T$ and $\partial V/\partial P$. Prove that $P \cdot (\partial V/\partial P) + T \cdot (\partial V/\partial T) = 0$.

4. Lagrange Multipliers

In Section 3 we introduced a method to find the relative maximum and relative minimum of a function of two variables without any constraints or extra conditions on the function or the variables. However, in many practical problems we are faced with maximizing or minimizing a function subject to existing conditions or constraints on the variables involved.

For example, a consumer may want to maximize utility derived from the consumption of commodities, subject to budget constraints; or a manufacturer may want to produce a box with a fixed volume so that the least amount of material is used.

Constraint
Lagrange Multipliers

When we want to maximize or minimize $z = f(x, y)$ subject to the *constraint* $g(x, y) = 0$, we use the method of *Lagrange multipliers*. The purpose of this section is to demonstrate its use.

Consider a function $z = f(x, y)$ of two variables x and y, subject to a single constraint $g(x, y) = 0$. We introduce a new variable λ, called the *Lagrange multiplier*, and construct the function

$$F(x, y, \lambda) = f(x, y) + \lambda g(x, y)$$

This new function F is a function of three variables, x, y, and λ. It can be shown that a relative maximum (minimum) of F is also a relative maximum (minimum) of $z = f(x, y)$, subject to the constraint $g(x, y) = 0$. Thus, the critical points (x, y) at which the function $z = f(x, y)$ has a relative maximum (minimum) obey

(1)
$$\frac{\partial F}{\partial x} = \frac{\partial f}{\partial x} + \lambda \frac{\partial g}{\partial x} = 0$$

$$\frac{\partial F}{\partial y} = \frac{\partial f}{\partial y} + \lambda \frac{\partial g}{\partial y} = 0$$

$$\frac{\partial F}{\partial \lambda} = g(x, y) = 0$$

Any solution of the above system of equations is a critical point of the function $z = f(x, y)$ to be maximized or minimized.

Example 1 Find the maximum value of

$$f(x, y) = xy$$

subject to the constraint

$$g(x, y) = x + y - 16 = 0$$

Solution First, we construct the function F:

$$F(x, y, \lambda) = f(x, y) + \lambda g(x, y) = xy + \lambda(x + y - 16)$$

Then, the system of equations (1) is

$$\frac{\partial f}{\partial x} + \lambda \frac{\partial g}{\partial x} = 0 \qquad \frac{\partial f}{\partial y} + \lambda \frac{\partial g}{\partial y} = 0 \qquad x + y - 16 = 0$$

238 CH. 6 FUNCTIONS OF TWO VARIABLES

But
$$\frac{\partial f}{\partial x} = y \qquad \frac{\partial f}{\partial y} = x \qquad \frac{\partial g}{\partial x} = 1 \qquad \frac{\partial g}{\partial y} = 1$$

Hence, the system (1) may be rewritten as
$$y + \lambda = 0 \qquad x + \lambda = 0 \qquad x + y - 16 = 0$$

Since $x = -\lambda$ and $y = -\lambda$, the third equation may be written as
$$-\lambda - \lambda - 16 = 0$$
$$\lambda = -8$$

The only critical point is $x = 8$, $y = 8$ (since $x = -\lambda$, $y = -\lambda$). Therefore, $f(x, y) = xy$ has a relative maximum at $(8, 8)$, and the maximum is $f(8, 8) = 64$. ∎

Example 2 Find the minimum value of
$$f(x, y) = xy$$
subject to the constraint
$$g(x, y) = x^2 + y^2 - 4 = 0$$

Solution First, we construct the function F:
$$F(x, y, \lambda) = f(x, y) + \lambda g(x, y) = xy + \lambda(x^2 + y^2 - 4)$$

The system of equations (1) is

(2)
$$\frac{\partial F}{\partial x} = y + \lambda \cdot 2x = 0$$
$$\frac{\partial F}{\partial y} = x + \lambda \cdot 2y = 0$$
$$\frac{\partial F}{\partial \lambda} = x^2 + y^2 - 4 = 0$$

From the first of these, we find that

(3)
$$y = -2x\lambda$$

Substituting into the second equation of (2), we obtain
$$x - 4x\lambda^2 = 0$$
$$x(1 - 4\lambda^2) = 0$$

Thus, either
$$x = 0 \qquad \text{or} \qquad 1 - 4\lambda^2 = 0$$

But $x = 0$ implies that $y = 0$, and these two numbers do not satisfy the constraint $g(x, y) = 0$. Hence, we must have
$$1 - 4\lambda^2 = 0$$
$$\lambda = \pm\tfrac{1}{2}$$

Substituting these values for λ into equation (3), we find
$$y = x \quad \text{or} \quad y = -x$$
Since x and y are subject to the constraint $g(x, y) = 0$, we must have
$$x^2 + y^2 - 4 = x^2 + x^2 - 4 = 0$$
$$2x^2 = 4$$
$$x = \pm\sqrt{2}$$
Similarly,
$$y = \pm\sqrt{2}$$
Thus, the critical points are
$$(\sqrt{2}, \sqrt{2}), \quad (\sqrt{2}, -\sqrt{2}), \quad (-\sqrt{2}, \sqrt{2}), \quad (-\sqrt{2}, -\sqrt{2})$$
Checking the value of $f(x, y)$ at each of these points, we get
$$f(\sqrt{2}, \sqrt{2}) = \sqrt{2}\sqrt{2} = 2$$
$$f(\sqrt{2}, -\sqrt{2}) = \sqrt{2}(-\sqrt{2}) = -2$$
$$f(-\sqrt{2}, \sqrt{2}) = -\sqrt{2}\sqrt{2} = -2$$
$$f(-\sqrt{2}, -\sqrt{2}) = (-\sqrt{2})(-\sqrt{2}) = 2$$
Hence, $f(x, y)$ attains its minimum value at the two critical points $(-\sqrt{2}, \sqrt{2})$ and $(\sqrt{2}, -\sqrt{2})$. The minimum value is -2. ■

Example 3 A manufacturer produces two types of engines, x and y, and the joint cost function is given by
$$C(x, y) = x^2 + 3xy - 6y$$
To minimize cost, how many engines of each type should be produced if there must be a total of 42 engines?

Solution The condition of a total of 42 engines constitutes the constraint of the problem. Thus, the constraint is
$$g(x, y) = x + y - 42 = 0$$
The function F is
$$F(x, y, \lambda) = C(x, y) + \lambda g(x, y)$$
$$= x^2 + 3xy - 6y + \lambda(x + y - 42)$$
The system (1) is
$$\frac{\partial F}{\partial x} = 2x + 3y + \lambda = 0$$
$$\frac{\partial F}{\partial y} = 3x - 6 + \lambda = 0$$
$$\frac{\partial F}{\partial \lambda} = x + y - 42 = 0$$

The solution of this system is

$$x = 33 \quad y = 9 \quad \lambda = -93$$

Thus, a minimum cost is achieved for $x = 33$ and $y = 9$.

Exercise 4
Solutions to Odd-Numbered Problems begin on page 411.

In Problems 1–4 use the method of Lagrange multipliers to find the maximum value of the function $z = f(x, y)$ subject to the constraint $g(x, y) = 0$.

1. $f(x, y) = 3x + 4y$ $g(x, y) = x^2 + y^2 - 9 = 0$
2. $f(x, y) = 3xy$ $g(x, y) = x^2 + y^2 - 4 = 0$
3. $f(x, y) = 12xy - 3y^2 - x^2$ $g(x, y) = x + y - 16 = 0$
4. $f(x, y) = xy$ $g(x, y) = x + y - 8 = 0$

In Problems 5–8 use the method of Lagrange multipliers to find the minimum value of the function $z = f(x, y)$ subject to the constraint $g(x, y) = 0$.

5. $f(x, y) = x^2 + y^2$ $g(x, y) = x + y - 1 = 0$
6. $f(x, y) = 3x + 4y$ $g(x, y) = x^2 + y^2 - 9 = 0$
7. $f(x, y) = 5x^2 + 6y^2 - xy$ $g(x, y) = x + 2y - 24 = 0$
8. $f(x, y) = x^2 + y^2$ $g(x, y) = 2x + 3y - 4 = 0$

9. *Joint Cost Function.* Let x and y be two types of items produced by a factory, and let

$$C = 18x^2 + 9y^2$$

be the joint cost of production of x and y. If $x + y = 54$, find x, y that minimizes cost.

10. Find two real numbers x and y such that their product is maximum, subject to the constraint that their sum is 40.

11. *Production Function.* The production function of ABC Manufacturing is $P(x, y) = x^2 + 3xy - 6x$, where x and y represent two different types of input. Find the amounts of x and y that maximize production if $x + y = 40$.

*12. *Minimizing Materials.* A container producer wants to build a closed rectangular box with a volume of 175 cubic feet. Determine what the dimensions of the container should be so as to use the least amount of material in construction.

5. Least Squares

In this section we study how to derive certain functional relationships between two variables. The technique used to determine such relationships is called *regression analysis*, a subject that is an integral part of most statistics courses.

When the relationship between the two variables is nearly linear, we use *linear regression* techniques to find the relationships.

Consider the following table that describes annual crop yield y in terms of annual rainfall x.

Rainfall x (in.)	Crop Yield y (tons)
10	4
14	9.3
19	13.0
24	13.6
33	13.9
39	14.0

If we use rainfall (x) as the horizontal axis and crop yield (y) as the vertical axis and plot each pair of points in rectangular coordinates, we obtain a so-called *scatter diagram*. See Figure 9.

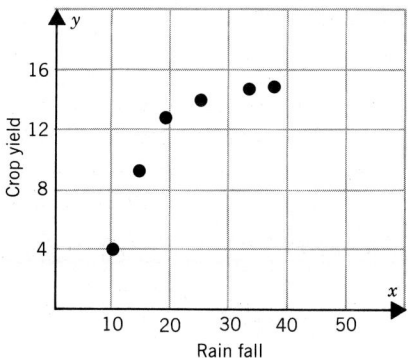

Figure 9

We try now to find a suitable function whose graph closely fits the geometrical configuration of the scatter diagram. It is clear that a straight line cannot be used to describe the relationship.

When a scatter diagram does result in a nearly linear pattern, we use a straight line to fit the data. The question becomes "What is the best line to use?"

One method for fitting a straight line to the data is to simply draw a straight line through the pattern of points that looks good. This technique is called the freehand method. It has the obvious good features of being quick and easy to apply. However, there are many situations when it is rather difficult to decide what is a good line to draw through the points. The individual judgment that is required to apply this method can be a disadvantage.

Another method for determining a good straight line through the data that does not possess the disadvantage of being subjective, as was the freehand method, is called the *method of least squares*.

Least Squares

The method of least squares is based on the principle that the amount of variation of a line from a set of plotted points can be measured by the sum of the

242 CH. 6 FUNCTIONS OF TWO VARIABLES

Goodness of fit squares of the vertical distances of the line from the points in the scatter diagram. See Figure 10. Based on this, we define the *goodness of fit* for a line to be the quantity

$$d_1^2 + d_2^2 + \cdots + d_n^2$$

where $d_1, d_2, \ldots, d_n$ are the vertical distances shown in Figure 10.

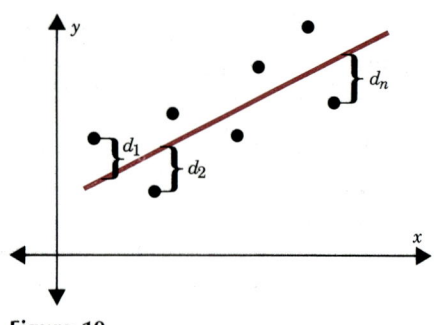

Figure 10

Line of best fit The principle of least squares states that the *line of best fit* for the data is the one that makes the sum of the squares of the deviations from the line a minimum.

Example 1 Compare the goodness of fit of the lines

$$L: \quad y = 2x + 1 \qquad M: \quad y = 3x + 2$$

for the data $(0, 1), (1, 4), (2, 6), (3, 10)$.

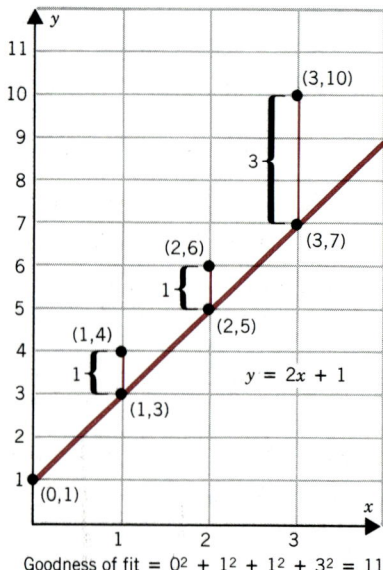
Goodness of fit = $0^2 + 1^2 + 1^2 + 3^2 = 11$

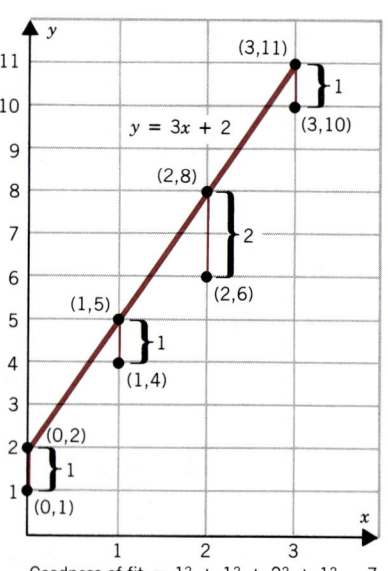
Goodness of fit = $1^2 + 1^2 + 2^2 + 1^2 = 7$

Figure 11

Solution Figure 11 illustrates the situation. For the line L the goodness of fit is
$$0^2 + 1^2 + 1^2 + 3^2 = 11$$
For the second line M, the goodness of fit is
$$1^2 + 1^2 + 2^2 + 1^2 = 7$$
Thus, the line $y = 3x + 2$ provides a better fit for the data. ∎

But how do we get the line of best fit? Suppose the data to be fitted consist of the points
$$(x_1, y_1), (x_2, y_2), \ldots, (x_n, y_n)$$
Using the criterion that the sum of the squares of the vertical distances should be a minimum, suppose $y = mx + b$ is the line of best fit for the data. (We assume this line is nonvertical.) The vertical distance from the line $y = mx + b$ to the point (x_1, y_1) is $d_1 = |mx_1 + b - y_1|$; the vertical distance from $y = mx + b$ to (x_2, y_2) is $d_2 = |mx_2 + b - y_2|$; and so on. The sum of the squares of these distances is

$$d_1^2 + d_2^2 + \cdots + d_n^2$$
$$= (mx_1 + b - y_1)^2 + (mx_2 + b - y_2)^2 + \cdots + (mx_n + b - y_n)^2$$

This expression is a function F of the two variables m and b. If we can find numbers m and b that make this expression minimum, we will have the slope m and y-intercept b of the line of best fit.

Set

(1) $\quad F(m, b) = (mx_1 + b - y_1)^2 + (mx_2 + b - y_2)^2 + \cdots + (mx_n + b - y_n)^2$

The critical numbers obey

$$\frac{\partial F}{\partial m} = 2x_1(mx_1 + b - y_1) + 2x_2(mx_2 + b - y_2) + \cdots + 2x_n(mx_n + b - y_n) = 0$$

$$\frac{\partial F}{\partial b} = 2(mx_1 + b - y_1) + 2(mx_2 + b - y_2) + \cdots + 2(mx_n + b - y_n) = 0$$

Rearranging the terms, these expressions may be written as

(2) $\quad (x_1^2 + x_2^2 + \cdots + x_n^2)m + (x_1 + x_2 + \cdots + x_n)b = x_1 y_1 + x_2 y_2 + \cdots + x_n y_n$
$\quad\quad\ (x_1 + x_2 + \cdots + x_n)m + nb = y_1 + y_2 + \cdots + y_n$

Since $x_1, x_2, \ldots, x_n, y_1, y_2, \ldots, y_n$ and n are all known, this is a system of two linear equations in two unknowns, m and b, which, it turns out, has a unique solution.

It can be shown that the solution of (2) for m and b actually minimizes the function F in (1). Thus,

the solution of (2) for m and b give the slope and y-intercept of the line of best fit for the data $(x_1, y_1), (x_2, y_2), \ldots, (x_n, y_n)$

Example 2 Find the line of best fit for the data
$$(0, 1), (1, 4), (2, 6), (3, 10)$$

Solution We use this data in (2) to obtain

$$14m + 6b = 46$$
$$6m + 4b = 21$$

The solution is found to be $m = 2.9$, $b = .9$. Hence, the line of best fit is $y = 2.9x + .9$. See Figure 12. The goodness of fit of $y = 2.9x + .9$ is $(.1)^2 + (.2)^2 + (.7)^2 + (.4)^2 = .7$.

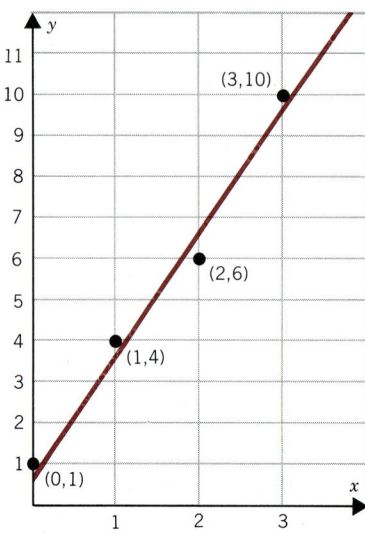

Figure 12

Example 3 The table below lists the height and weight of ten male high school juniors. Plot a scatter diagram for the data and find the equation of the line of best fit for the data.

Student	Height x	Weight y	x^2	xy
1	63	145	3969	9135
2	69	165	4761	11385
3	72	190	5184	13680
4	60	125	3600	7500
5	68	150	4624	10200
6	70	170	4900	11900
7	61	130	3721	7930
8	69	140	4761	9660
9	73	180	5329	13140
10	65	155	4225	10075
Total	670	1550	45074	104605

Solution For convenience, we have also listed in the table the quantities we will need to use in equation (2) to get the line of best fit. Thus, for insertion in equation (2) we have

$$n = 10; \quad x_1 + x_2 + \cdots + x_{10} = 670; \quad y_1 + y_2 + \cdots + y_{10} = 1550;$$
$$x_1^2 + x_2^2 + \cdots + x_{10}^2 = 45074; \quad x_1y_1 + x_2y_2 + \cdots + x_{10}y_{10} = 104605$$

The equations we need to solve are

$$45074m + 670b = 104605$$
$$670m + 10b = 1550$$

A little effort yields the solution

$$m = 4.1 \qquad b = -119.7$$

The equation of the line of best fit is

$$y = 4.1x - 119.7$$

Figure 13 illustrates the data and the line of best fit.

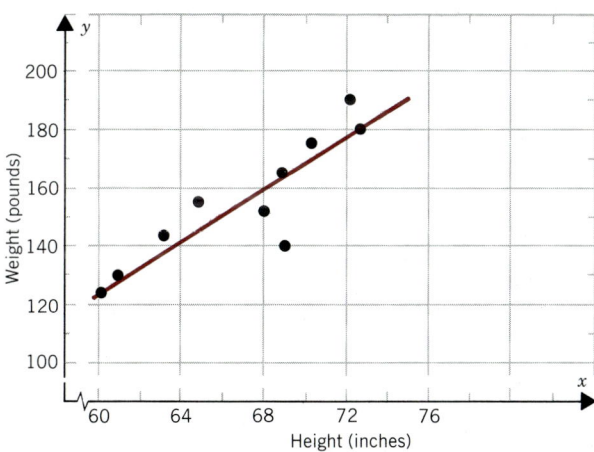

Figure 13

Care should be exercised when a scatter diagram is fitted by a straight line. The line should be drawn within the region of the scatter diagram and should not be extended or extrapolated outside this region. Also, when the scales are "broken," as they are in Figure 13, it is not obvious from the graph of the line what the y-intercept is. The y-intercept is -119.7 in this example; however, one might make the mistake of thinking that it is 110 by not carefully looking at the graph. If the x-axis were not broken, the y-axis would be drawn much further to the left of where it is located.

Exercise 5
Solutions to Odd-Numbered Problems begin on page 412.

In Problems 1–4 plot a scatter diagram for the data. Find the equation of the line of best fit and plot it on the same coordinate system.

1. $(0, 7), (1, 4), (2, 2), (3, -1)$
2. $(1, 1), (2, 5), (3, 9), (4, 11)$
3. $(0, 1), (1, 2), (3, 6), (6, 8)$
4. $(1, 1), (2, 3), (5, 7), (6, 9)$
5. $(1, 0), (2, 1), (3, 3), (5, 4)$
6. $(0, 2), (1, 2), (2, 2), (4, 4)$
7. $(1, 0), (2, 1), (3, 4), (6, 6)$

8. (0, 2), (1, 2), (2, 2), (5, 5)
9. (1, 1), (3, 4), (5, 3), (6, 5), (8, 8)
10. (1, 1), (3, 3), (5, 4), (6, 8), (7, 8)
11. An automobile dealer has kept a monthly record of hours of advertising time x on a local television station and the number of cars sold per month y.

Month	x	y
January	3	21
February	4	27
March	6	36
April	3	23

(a) Find the line of best fit for the data.
(b) If the dealer plans to use 4 hours of television time in May, estimate the number of cars that will be sold.

12. The Red Ruby Nut Co. sells three oz. cans of cashews. The number of cans y sold during the first week of each month is tabulated below versus the sales price x (in cents).

Month	x	y
April	90	200
May	80	240
June	70	300
July	60	350
August	50	420

(a) Find the line of best fit for the data.
(b) Estimate the sales price that will permit the company to sell exactly 400 cans per week.

13. The following data pertain to the advertising expenses (expressed as a percentage of total expenses) and the net operating profits (expressed as a percentage of total sales) in a sample of twelve sporting goods stores:

Advertising Expenses x	Profit y
1.5	3.1
0.8	1.9
2.6	4.2
1.0	2.3
0.6	1.2
2.8	4.9
1.2	2.8
0.9	2.1
0.4	1.4
1.3	2.4
1.2	2.4
2.0	3.8

Find the line of best fit and use it to predict the profit of such a store (expressed as a percentage of total sales) when its advertising expenses are 1.2 percent of its total expenses.

14. The following data pertain to the growth of a cactus graft under controlled environmental conditions:

Months after Grafting x	Height (Inches) y
1	0.8
2	2.4
3	4.0
4	5.1
5	7.3
6	9.4

Find the line of best fit and use it to predict the cactus graft's height after 9 months.

Chapter Review

Important Terms

function of two variables
ordered triple
surface
computer graphics
partial derivative
marginal cost
relative maximum
relative minimum

critical point
saddle point
constraint
Lagrange multipliers
goodness of fit
line of best fit
least squares

Review Exercises

Solutions to Odd-Numbered Problems begin on page 414.

1. For the function $z = f(x, y) = x^2 + xy$, find:
   (a) $f(x + \Delta x, y)$
   (b) $f(x + \Delta x, y) - f(x, y)$
   (c) $\dfrac{f(x + \Delta x, y) - f(x, y)}{\Delta x}$, $\Delta x \neq 0$
   (d) $\lim\limits_{\Delta x \to 0} \dfrac{f(x + \Delta x, y) - f(x, y)}{\Delta x}$, $\Delta x \neq 0$

2. Find f_x, f_y, f_{xx}, f_{xy}, and f_{yy} for each function:
   (a) $z = f(x, y) = x^2 y + 4x$
   (b) $z = f(x, y) = x^2 + y^2 + 2xy$
   (c) $z = f(x, y) = y^2 e^x + x \ln y$

3. For each of the following surfaces, find all relative maxima, relative minima, and saddle points:
   (a) $z = f(x, y) = xy - 6x - x^2 - y^2$
   (b) $z = f(x, y) = x^2 + 2x + y^2 + 4y + 10$
   (c) $z = f(x, y) = 2x - x^2 + 4y - y^2 + 10$
   (d) $z = f(x, y) = xy$
   (e) $z = f(x, y) = x^2 - 9y + y^3$
4. Show that the plane $z = 3x + 4y - 2$ has no critical points.
5. Use the method of Lagrange multipliers to find the maximum value of each of the following functions $z = f(x, y)$ subject to the constraint $g(x, y) = 0$:
   (a) $f(x, y) = 5x^2 - 3y^2 + xy$ $g(x, y) = 2x - y - 20 = 0$
   (b) $f(x, y) = x\sqrt{y}$ $g(x, y) = 2x + y - 3000 = 0$
6. Use the method of Lagrange multipliers to find the minimum value of each of the following functions $z = f(x, y)$ subject to the constraint $g(x, y) = 0$:
   (a) $f(x, y) = x^2 + y^2$ $g(x, y) = 2x + y - 4 = 0$
   (b) $f(x, y) = xy^2$ $g(x, y) = x^2 + y^2 - 1 = 0$
7. *Joint Profit Function.* A company produces two products at a total cost

$$C(x, y) = x^2 + 200x + y^2 + 100y - xy$$

where x and y represent the units produced of each product. The revenue function is

$$R(x, y) = 2000x - 2x^2 + 100y - y^2 + xy$$

Find the number of units of each product that will maximize profit.

Other Books

Mizrahi, A., and M. Sullivan, *Calculus and Analytic Geometry,* Wadsworth, Belmont, Ca., 1982.

7
Antiderivatives

1. Antiderivatives; the Indefinite Integral
2. Integration by Substitution
3. Integration by Parts
4. Differential Equations
*5. Application to Marginal Analysis
6. Table of Integrals
Chapter Review

*This section may be omitted without loss of continuity

With this chapter we begin the study of *integral calculus,* which may be subdivided into two parts: The *indefinite integral* is concerned with the inverse process of differentiation, usually referred to as finding the *antiderivative of a function.* The *definite integral* plays a major role in applications to geometry (finding the area under a curve), to business and economics (marginal analysis, consumer's surplus, maximizing profit over time), and to probability.

In this chapter we concentrate on the techniques for finding antiderivatives of functions; defining the indefinite integral; and citing applications to business (marginal analysis), to biology (bacterial growth), and to anthropology (determining ages). In Chapter 8 we will define the definite integral and discuss various applications in business, geometry, economics, and probability.

1. Antiderivatives; the Indefinite Integral

We have already learned that to each differentiable function $f(x)$ there corresponds a derivative function $f'(x)$. It is also possible to ask the following question: If a function $f(x)$ is given, can we find a function $F(x)$ whose derivative is $f(x)$? That is, is it possible to find a function $F(x)$ so that $F'(x) = f(x)$? If such a function $F(x)$ can be found, it is called *an antiderivative of $f(x)$.*

Antiderivative **A function $F(x)$ is called an *antiderivative* of the function $f(x)$ if**

$$F'(x) = f(x).$$

For example, an antiderivative of $f(x) = 2x$ is x^2, since

$$\frac{d}{dx} x^2 = 2x$$

Another function whose derivative is $2x$ is $x^2 + 3$, since

$$\frac{d}{dx}(x^2 + 3) = 2x$$

This leads us to suspect that the function $f(x) = 2x$ has an unlimited number of antiderivatives. Indeed, any of the functions $x^2, x^2 + \frac{1}{2}, x^2 + 2, x^2 + \sqrt{5}, x^2 - \pi, x^2 - 1, \ldots, x^2 + K$, where K is any constant, has the property that its derivative is $2x$. We conjecture that all the antiderivatives of $2x$ are given by $x^2 + K$, where K is any constant. The following result assures us that there cannot be antiderivatives of the function $f(x) = 2x$ other than those of the form $x^2 + K$, where K is a constant.

Once one antiderivative $F(x)$ of $f(x)$ is found, all the antiderivatives of $f(x)$ are of the form

$$F(x) + K$$

where K is a constant.

1. ANTIDERIVATIVES; THE INDEFINITE INTEGRAL

See Problem 26 in Exercise 1 for an outline of the proof of this result. For example, all the antiderivatives of $f(x) = x^5$ are of the form

$$F(x) = \frac{x^6}{6} + K$$

where K is a constant. By differentiating, we can check the answer:

$$F'(x) = \frac{d}{dx}\left(\frac{x^6}{6} + K\right) = x^5$$

Example 1 Find all the antiderivatives of: $f(x) = x^{1/2}$

Solution Recall that the derivative of the function $\frac{2}{3}x^{3/2}$ is

$$\frac{2}{3}(\frac{3}{2}x^{1/2}) = x^{1/2}$$

So, all the antiderivatives of $f(x) = x^{1/2}$ are of the form

$$F(x) = \frac{2}{3}x^{3/2} + K$$

where K is a constant.

In Example 1, you may ask how we knew that we should choose the function $\frac{2}{3}x^{3/2}$. First, we know that

$$\frac{d}{dx}x^r = rx^{r-1}$$

That is, differentiation reduces the exponent by 1. Antidifferentiation is the inverse process, so it should increase the exponent by 1. This is how we obtained the $x^{3/2}$ part of $\frac{2}{3}x^{3/2}$. Second, the $\frac{2}{3}$ factor is needed so that, when we differentiate, we get $x^{1/2}$ and not $\frac{3}{2}x^{1/2}$.

Thus, because

$$\frac{d}{dx}x^{r+1} = (r+1)x^r$$

for any real number r, it follows that

All the antiderivatives of x^r are $\dfrac{x^{r+1}}{r+1} + K$

for $r \neq -1$ and K any constant. The case for which $r = -1$ requires special attention and is considered a little later.

We use a special symbol to represent all the antiderivatives of a function—

Integral Sign   the *integral sign*, $\int$.

Indefinite Integral   Let $F(x)$ be an antiderivative of the function $f(x)$. The *indefinite integral of $f(x)$*, denoted by $\int f(x)\,dx$, is

$$\int f(x)\,dx = F(x) + K$$

where K is a constant.

252 CH. 7 ANTIDERIVATIVES

Thus, the indefinite integral of $f(x)$ is a symbol for all the antiderivatives of $f(x)$.

In the expression $\int f(x)\, dx$, the integral sign $\int$ indicates that the operation of antidifferentiation is to be performed on the function $f(x)$; and the dx reinforces the fact that the operation is to be performed with respect to the variable x. The function $f(x)$ is called the *integrand*.

• Integrand

Example 2 (a) $\int x^2\, dx = \dfrac{x^3}{3} + K$ (b) $\int x^5\, dx = \dfrac{x^6}{6} + K$ (c) $\int x^{1/2}\, dx = \tfrac{2}{3} x^{3/2} + K$ ■

Based on the relationship between the process of differentiation and that of indefinite integration, or antidifferentiation, we can construct a list of formulas. These formulas may be verified by differentiating the right-hand side.

If c is a real number and K is a constant,

(1) $$\int c\, dx = cx + K$$

Example 3 (a) $\int 5\, dx = 5x + K$ (b) $\int dx = x + K$ ■

For any real number $r \neq -1$ and K a constant,

(2) $$\int x^r\, dx = \dfrac{x^{r+1}}{r+1} + K$$

Example 4 (a) $\int \dfrac{1}{\sqrt{x}}\, dx = \int x^{-1/2}\, dx = \dfrac{x^{(-1/2)+1}}{-\tfrac{1}{2}+1} + K = 2x^{1/2} + K$

(b) $\int x^\pi\, dx = \dfrac{x^{\pi+1}}{\pi+1} + K$ ■

Because of the derivative formulas

$$\dfrac{d}{dx} e^x = e^x \quad \text{and} \quad \dfrac{d}{dx} \ln x = \dfrac{1}{x}$$

we are led to the following two indefinite integrals:

(3) $$\int e^x\, dx = e^x + K$$

K is a constant

(4) $$\int \dfrac{1}{x}\, dx = \ln x + K$$

Formula (4) takes care of finding $\int x^{-1}\, dx$. Now the formula for $\int x^r\, dx$ is complete.

Rules (5) and (6), on page 253 enable us to evaluate additional indefinite

1. ANTIDERIVATIVES; THE INDEFINITE INTEGRAL

integrals when used in conjunction with formulas (1)–(4). Each of these rules is a consequence of a rule for differentiation.

(5) $$\int cf(x)\,dx = c\int f(x)\,dx \qquad c \text{ is a real number}$$

In words, the indefinite integral of a real number c times $f(x)$ equals c times the indefinite integral of $f(x)$.

Example 5
$$\int 3x^4\,dx = 3\int x^4\,dx = 3\left(\frac{x^5}{5}\right) + K = \frac{3x^5}{5} + K$$

(6) $$\int [f(x) \pm g(x)]\,dx = \int f(x)\,dx \pm \int g(x)\,dx$$

In words, the indefinite integral of a sum (or difference) equals the sum (or difference) of the indefinite integrals.

Example 6
$$\int (x^2 + x^3)\,dx = \int x^2\,dx + \int x^3\,dx = \frac{x^3}{3} + \frac{x^4}{4} + K$$

Properties (5) and (6) can be combined and used for sums and differences of three or more functions.

Example 7
$$\int \left(7x^5 + \frac{1}{2x} - x^{1/2}\right) dx = \int 7x^5\,dx + \int \frac{1}{2x}\,dx - \int x^{1/2}\,dx$$
$$= 7\int x^5\,dx + \frac{1}{2}\int \frac{1}{x}\,dx - \int x^{1/2}\,dx$$
$$= \frac{7x^6}{6} + \frac{1}{2}\ln x - \frac{x^{3/2}}{\frac{3}{2}} + K$$
$$= \tfrac{7}{6}x^6 + \tfrac{1}{2}\ln x - \tfrac{2}{3}x^{3/2} + K$$

Sometimes, it is helpful to use some algebra to put the expression in a convenient form before applying integration formulas.

Example 8 (a) $\displaystyle\int \frac{15\,dx}{x^5} \underset{\text{By (5)}}{=} 15\int \frac{1}{x^5}\,dx \underset{\text{Algebra}}{=} 15\int x^{-5}\,dx \underset{\text{By (2)}}{=} \frac{15x^{-4}}{-4} + K = \frac{-15}{4x^4} + K$

(b) $\displaystyle\int \frac{x^2 - 2x + 5}{x}\,dx = \int \left(x - 2 + \frac{5}{x}\right) dx = \int x\,dx - \int 2\,dx + \int \frac{5}{x}\,dx$
$$= \tfrac{1}{2}x^2 - 2x + 5\ln x + K$$

254   CH. 7   ANTIDERIVATIVES

At this point, some remarks are in order. First, there are no simple integration formulas for the integral of a product and the integral of a quotient, as you will see in Problem 25 (below). Second, in general, integration is more difficult than differentiation because of the lack of straightforward formulas. More thought and more experience are the keys to doing integration!

Exercise 1
Solutions to Odd-Numbered Problems begin on page 415.

In Problems 1–24 evaluate each indefinite integral.

1. $\int 3\,dx$
2. $\int -4\,dx$
3. $\int x\,dx$
4. $\int x^2\,dx$
5. $\int x^{1/3}\,dx$
6. $\int x^{4/3}\,dx$
7. $\int x^{-2}\,dx$
8. $\int x^{-3}\,dx$
9. $\int x^{-1/2}\,dx$
10. $\int x^{-2/3}\,dx$
11. $\int (x^2 + 2e^x)\,dx$
12. $\int (3x + 5e^x)\,dx$
13. $\int \left(\dfrac{x-1}{x}\right)dx$
14. $\int \left(\dfrac{x+1}{x}\right)dx$
15. $\int \left(\dfrac{3\sqrt{x}+1}{\sqrt{x}}\right)dx$
16. $\int \left(\dfrac{2\sqrt{x}-4}{\sqrt{x}}\right)dx$
17. $\int \dfrac{x^2-4}{x+2}\,dx$
18. $\int \dfrac{x^2-1}{x-1}\,dx$
19. $\int x(x-1)\,dx$
20. $\int x(x+2)\,dx$
21. $\int \dfrac{3x^5+2}{x}\,dx$
22. $\int \dfrac{x^6+x^2+1}{x^3}\,dx$
23. $\int \dfrac{4e^x+e^{2x}}{e^x}\,dx$
24. $\int \dfrac{3e^x+xe^{2x}}{xe^x}\,dx$

*25. Verify the following:

(a) $\int (x \cdot \sqrt{x})\,dx \neq \int x\,dx \cdot \int \sqrt{x}\,dx$

(b) $\int x(x^2+1)\,dx \neq x \int (x^2+1)\,dx$

(c) $\int \dfrac{x^2-1}{x-1}\,dx \neq \dfrac{\int (x^2-1)\,dx}{\int (x-1)\,dx}$

*26. Prove that if $F(x)$ is an antiderivative of $f(x)$, then any other antiderivative $G(x)$ of $f(x)$ is given by

$$G(x) = F(x) + K \qquad K \text{ is a constant}$$

An outline of a proof follows: Since $F(x)$ and $G(x)$ are each an antiderivative of $f(x)$, then

$$F'(x) = f(x) \quad \text{and} \quad G'(x) = f(x)$$

Thus,

$$G'(x) - F'(x) = 0 \quad \text{or} \quad \dfrac{d}{dx}[G(x) - F(x)] = 0$$

2. Integration by Substitution

Indefinite integrals that cannot be evaluated by using formulas (1)–(4) of Section 1 may sometimes be evaluated by the *substitution technique*. This method involves the introduction of a function that changes the integrand into a form to which the formulas apply.

An example will help illustrate the procedure.

Example 1 Evaluate $\int (x^2+5)^3\, 2x\, dx$ by using the substitution $u = x^2 + 5$.

Solution First, calculate the derivative of u:

$$\dfrac{du}{dx} = 2x$$

Treat the derivative as a quotient* (multiply both sides by dx) and write

$$du = 2x\, dx$$

Now, we substitute by replacing all the x's and dx by u's and du:

$$\int (x^2+5)^3\, 2x\, dx = \int u^3\, du = \tfrac{1}{4}u^4 + K = \tfrac{1}{4}(x^2+5)^4 + K$$
$\uparrow$ $\qquad\qquad\qquad\qquad\qquad\uparrow$
$u = x^2 + 5 \qquad\qquad\qquad\qquad u = x^2 + 5$
$du = 2x\, dx$

■

*When the derivative du/dx is treated as a quotient, the quantity du is called the *differential of u* and dx is called the *differential of x*.

Although we could have done the above integration by multiplying out the $(x^2 + 5)^3$, the substitution technique is easier. Sometimes, substitution is the only method available.

Example 2 Evaluate: $\int e^{2x+1} dx$

Solution We use the substitution $u = 2x + 1$. Then

$$\frac{du}{dx} = 2 \quad \text{and} \quad du = 2\, dx$$

$$\int e^{2x+1} dx \underset{\substack{\uparrow \\ u = 2x+1 \\ du = 2\, dx}}{=} \int \frac{e^u\, du}{2} = \tfrac{1}{2} e^u + K \underset{\substack{\uparrow \\ u = 2x+1}}{=} \tfrac{1}{2} e^{2x+1} + K$$

Example 3 Evaluate: $\int e^{-x} dx$

Solution Set $u = -x$. Then $du = -dx$ and

$$\int e^{-x} dx \underset{\substack{\uparrow \\ u = -x \\ du = -dx}}{=} \int e^u (-du) = -e^u + K \underset{\substack{\uparrow \\ u = -x}}{=} -e^{-x} + K$$

Example 4 Evaluate: $\int x \sqrt{x^2 + 1}\, dx$

Solution We use the substitution $u = x^2 + 1$. Then

$$\frac{du}{dx} = 2x \quad \text{and} \quad du = 2x\, dx \quad \text{and} \quad \frac{du}{2} = x\, dx$$

Then

$$\int x \sqrt{x^2 + 1}\, dx = \int \sqrt{x^2 + 1}\, x\, dx = \int \sqrt{u}\, \frac{du}{2} = \frac{1}{2} \int u^{1/2}\, du$$

$$= \frac{1}{2} \frac{u^{3/2}}{\frac{3}{2}} + K = \frac{(x^2 + 1)^{3/2}}{3} + K$$

Note that in using the substitution $u = x^2 + 1$ in Example 4, we must substitute not only for the integrand $x \sqrt{x^2 + 1}$, but also for dx. In fact, it is the existence of x as part of the integrand that makes the substitution $u = x^2 + 1$ work. For example, if we use this same substitution to try to evaluate $\int \sqrt{x^2 + 1}\, dx$, we obtain

$$\int \sqrt{x^2 + 1}\, dx = \int \sqrt{u}\, \frac{du}{2\sqrt{u-1}} = \int \frac{\sqrt{u}}{2\sqrt{u-1}}\, du$$

2. INTEGRATION BY SUBSTITUTION

In this case, the substitution results in an integrand that is *more complicated* than the original one.

Thus, the idea behind the substitution method is to obtain an integral $\int h(u)\, du$ that is simpler than the original integral $\int f(x)\, dx$. When a substitution does not simplify the integral, other substitutions should be tried. If these do not work, other integration methods must be applied.* Since integration, unlike differentiation, has no prescribed method, a lot of practice in integration is required.

To illustrate that the integral has no prescribed method for evaluation, we use two different substitutions in the next example.

Example 5 Evaluate: $\int x\sqrt{4+x}\, dx$

Solution SUBSTITUTION I: Let $u = 4 + x$. Then $du = dx$ and $x = u - 4$, so that

$$\int x\sqrt{4+x}\, dx = \int (u - 4)\sqrt{u}\, du$$

$$= \int (u^{3/2} - 4u^{1/2})\, du$$

$$= \frac{u^{5/2}}{\frac{5}{2}} - \frac{4u^{3/2}}{\frac{3}{2}} + K$$

$$= \frac{2(4+x)^{5/2}}{5} - \frac{8(4+x)^{3/2}}{3} + K$$

SUBSTITUTION II: Let $u^2 = 4 + x$. Then $2u\, du = dx$ and $x = u^2 - 4$, so that

$$\int x\sqrt{4+x}\, dx = \int (u^2 - 4)u\, 2u\, du$$

$$= 2\int (u^4 - 4u^2)\, du$$

$$= \frac{2u^5}{5} - \frac{8u^3}{3} + K$$

$$= \frac{2(4+x)^{5/2}}{5} - \frac{8(4+x)^{3/2}}{3} + K$$

Example 6 Evaluate: $\int \dfrac{dx}{2\sqrt{x}(1+\sqrt{x})^3}$

Solution We use the substitution

$$u = 1 + \sqrt{x} \quad \text{and} \quad du = \frac{dx}{2\sqrt{x}}$$

---

*One such method, called *integration by parts,* is the subject of the next section.

Then
$$\int \frac{dx}{2\sqrt{x}(1+\sqrt{x})^3} = \int \frac{du}{u^3} = \int u^{-3}\, du = \frac{u^{-2}}{-2} + K = \frac{-1}{2u^2} + K$$
$$= \frac{-1}{2(1+\sqrt{x})^2} + K$$

Example 7 Evaluate: $\int \dfrac{dx}{x \ln x}$

Solution Set $u = \ln x$. Then $du = (1/x)\, dx$ and
$$\int \frac{dx}{x \ln x} = \int \frac{du}{u} = \ln u + K = \ln(\ln x) + K$$

A summary of the method of integration by substitution is given in the flow-chart in Figure 1.

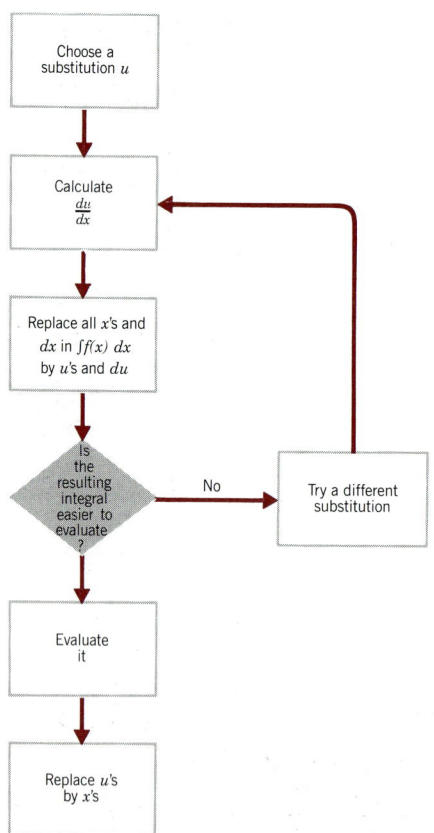

Figure 1

Exercise 2
Solutions to Odd-Numbered Problems begin on page 416.

In Problems 1-34 evaluate each indefinite integral.

1. $\int (2x+1)^5\, dx$
2. $\int (3x-5)^4\, dx$
3. $\int e^{2x-3}\, dx$
4. $\int e^{3x+4}\, dx$
5. $\int (-2x+3)^{-2}\, dx$
6. $\int (5-2x)^{-3}\, dx$
7. $\int (x^2+4)^2 x\, dx$
8. $\int (x^2-2)^3 x\, dx$
9. $\int e^{x^3+1} x^2\, dx$
10. $\int e^{2x^2+1} x\, dx$
11. $\int (e^x + e^{-x})\, dx$
12. $\int (e^x - e^{-x})\, dx$
13. $\int (x^3+2)^6 x^2\, dx$
14. $\int (x^3-1)^4 x^2\, dx$
15. $\int \dfrac{x}{\sqrt[3]{1+x^2}}\, dx$
16. $\int \dfrac{x}{\sqrt[5]{1-x^2}}\, dx$
17. $\int x\sqrt{x+3}\, dx$
18. $\int x\sqrt{x-3}\, dx$
19. $\int \dfrac{e^x}{e^x+1}\, dx$
20. $\int \dfrac{e^{-x}}{e^{-x}+4}\, dx$
21. $\int \dfrac{e^{\sqrt{x}}\, dx}{\sqrt{x}}$
22. $\int \dfrac{e^{\sqrt[3]{x}}\, dx}{x^{2/3}}$
23. $\int \dfrac{(x^{1/3}-1)^6\, dx}{x^{2/3}}$
24. $\int \dfrac{(x^{1/3}+2)^3}{x^{2/3}}\, dx$
25. $\int \dfrac{(x+1)\, dx}{(x^2+2x+3)^2}$
26. $\int \dfrac{(x+4)\, dx}{(x^2+8x+2)^3}$
27. $\int \dfrac{dx}{\sqrt{x}(1+\sqrt{x})^4}$
28. $\int \dfrac{(3-2\sqrt{x})^2}{\sqrt{x}}\, dx$
29. $\int \dfrac{dx}{2x+3}$
30. $\int \dfrac{dx}{3x-5}$
31. $\int \dfrac{x\, dx}{4x^2+1}$
32. $\int \dfrac{x\, dx}{5x^2-2}$
33. $\int \dfrac{x+1}{x^2+2x+2}\, dx$
34. $\int \dfrac{2x-1}{x^2-x+4}\, dx$

*35. Use a substitution to verify the formula
$$\int (ax+b)^n\, dx = \dfrac{(ax+b)^{n+1}}{a(n+1)} + K \qquad a \neq 0, \quad n \neq -1$$

3. Integration by Parts

In this section we discuss another technique for evaluating indefinite integrals so that we can handle problems such as the following:

$$\int xe^x \, dx \qquad \int \ln x \, dx \qquad \int x \ln x \, dx$$

This method, called *integration by parts,* is based on the product rule for differentiation and is an effective and versatile technique for integration.

Recall that if u and v are differentiable functions of x, then

$$\frac{d}{dx}(uv) = \frac{du}{dx}v + u\frac{dv}{dx}$$

Integrating both sides, we get

$$uv = \int v\frac{du}{dx}\,dx + \int u\frac{dv}{dx}\,dx$$

$$\int u\frac{dv}{dx}\,dx = uv - \int v\frac{du}{dx}\,dx$$

In abbreviated form, this may be written as

(1) $$\int u\,dv = uv - \int v\,du$$

By-Parts Formula

This formula is usually referred to as the *by-parts formula.*

To apply this formula, we separate the integrand into two parts. We call one u and the other dv. We differentiate u to obtain du and integrate dv to obtain v. If we can then integrate $\int v\,du$, the problem is solved. The goal of this procedure, then, is to choose u and dv so that the term $\int v\,du$ is easier to solve than the original problem. As the examples will illustrate, this usually happens when u is simplified by differentiation.

Example 1 Evaluate: $\int xe^x\,dx$

Solution To use the by-parts formula, we choose u and dv so that

$$\int u\,dv = \int xe^x\,dx$$

and $\int v\,du$ is easier to evaluate than $\int u\,dv$. In this example, we decide to choose

$$u = x \qquad \text{and} \qquad dv = e^x\,dx$$

As a result of this choice,

$$du = dx \qquad \text{and} \qquad v = \int dv = \int e^x\,dx = e^x$$

Note that we only require a particular antiderivative of *dv* at this stage; we will add the constant of integration later. Substitution in (1) results in

$$\int \overset{u}{x} \overset{dv}{e^x \, dx} = \overset{u}{x} \overset{v}{e^x} - \int \overset{v}{e^x} \overset{du}{dx} = xe^x - e^x + K = e^x(x - 1) + K$$

Let's look once more at Example 1. Suppose we had chosen *u* and *dv* differently:

$$u = e^x \quad \text{and} \quad dv = x \, dx$$

This choice would have resulted in

$$du = e^x \, dx \quad \text{and} \quad v = \frac{x^2}{2}$$

and equation (1) would have yielded

$$\int xe^x \, dx = \frac{x^2}{2} e^x - \int \frac{x^2 e^x}{2} \, dx$$

As you can see, instead of obtaining an integral that is easier to evaluate, we obtain one that is more complicated than the original. This means that an unwise choice of *u* and *dv* has been made.

Unfortunately, there are no general directions for choosing *u* and *dv* except that:

1. **dx is always a part of *dv***
2. **it must be possible to integrate *dv***
3. ***u* and *dv* are chosen so that $\int v \, du$ is easier to evaluate than the original integral $\int u \, dv$; this often happens when *u* is simplified by differentiation**

In making an initial choice for *u* and *dv*, a certain amount of trial and error is used. If a selection appears to hold little promise, abandon it and try some other choice. If no choices work, it may be that some other technique of integration should be tried.

Let's look at some more examples.

Example 2 Evaluate: $\int x \ln x \, dx$

Solution We choose

$$u = \ln x \qquad dv = x \, dx$$
$$du = \frac{1}{x} dx \qquad v = \frac{x^2}{2}$$

Then

$$\int x \ln x \, dx = \frac{x^2}{2} \ln x - \int \frac{x^2}{2} \frac{1}{x} dx = \frac{1}{2} x^2 \ln x - \frac{1}{2} \int x \, dx = \frac{1}{2} x^2 \ln x - \frac{1}{4} x^2 + K$$
$$= \frac{x^2}{2} \left(\ln x - \frac{1}{2} \right) + K$$

■

Sometimes it is necessary to integrate by parts more than once to solve a particular problem, as illustrated by the next example.

Example 3 Evaluate: $\int x^2 e^x \, dx$

Solution Let

$$u = x^2 \quad \text{and} \quad dv = e^x \, dx$$

so that

$$du = 2x \, dx \quad \text{and} \quad v = e^x$$

Then

$$\int x^2 e^x \, dx = x^2 e^x - \int 2x e^x \, dx$$

We must still evaluate $\int x e^x \, dx$. In Example 1 we found (by using the integration by parts technique) that

$$\int x e^x \, dx = e^x(x - 1) + K$$

Thus,

$$\int x^2 e^x \, dx = x^2 e^x - 2[e^x(x - 1)] + K = x^2 e^x - 2x e^x + 2e^x + K$$
$$= e^x(x^2 - 2x + 2) + K$$

■

Example 4 Find all the antiderivatives of ln x.

Solution To find all the antiderivatives of ln x, we use the by-parts formula. Taking $u = \ln x$ and $dv = dx$, we find

$$\int \ln x \, dx = x \ln x - \int x \frac{1}{x} dx = x \ln x - x + K = x(\ln x - 1) + K$$

■

The integration by parts formula is useful for the evaluation of indefinite integrals that have integrands composed of e^x times a polynomial function of x or

In x times a polynomial function of x. It can also be used for other types of indefinite integrals that will not be discussed in this book.

Exercise 3

Solutions to Odd-Numbered Problems begin on page 417.

In Problems 1-13 evaluate each indefinite integral.

1. $\int xe^{2x}\, dx$
2. $\int xe^{-3x}\, dx$
3. $\int x^2 e^{-x}\, dx$
4. $\int x^2 e^{2x}\, dx$
5. $\int \sqrt{x}\ln x\, dx$
6. $\int x(\ln x)^2\, dx$
7. $\int (\ln x)^2\, dx$
8. $\int \frac{\ln x}{x^2}\, dx$
9. $\int x^2 \ln 3x\, dx$
10. $\int x^2 \ln 5x\, dx$
11. $\int x^2(\ln x)^2\, dx$
12. $\int x^3(\ln x)^2\, dx$
13. $\int \frac{\ln x}{x}\, dx$

---

4. Differential Equations

In studies of physical, chemical, biological, and other natural phenomena, scientists attempt, on the basis of long observation, to deduce mathematical laws that will describe and predict nature's behavior. Such laws often involve the derivatives of some unknown function F, and it is required to find this unknown function F. For example, it may be required to find all functions $y = F(x)$ so that

(1) $$\frac{dy}{dx} = f(x)$$

This equation is an example of what is called a *differential equation*, and a function $y = F(x)$ for which $dy/dx = f(x)$ is a *solution* of the differential equation. The *general solution* of $dy/dx = f(x)$ consists of all the antiderivatives of f. For example, the general solution of the differential equation

(2) $$\frac{dy}{dx} = 5x^2 + 2$$

is

$$y = F(x) = \frac{5x^3}{3} + 2x + K$$

Particular Solution

A *particular solution* of $dy/dx = f(x)$ occurs when K is assigned a particular value. When a particular solution is required, we use a *boundary condition on F*. For example, in the differential equation (2) we might require the general solution to obey the condition that $y = 5$ when $x = 3$. Then

$$y = \frac{5x^3}{3} + 2x + K$$
$$5 = (\tfrac{5}{3})(27) + (2)(3) + K$$
$$K = -46$$

The particular solution of (2) with boundary condition that $y = 5$ when $x = 3$ is therefore

$$y = \tfrac{5}{3}x^3 + 2x - 46$$

Example 1 Solve the differential equation below with the boundary condition that $y = -1$ when $x = 3$.

$$\frac{dy}{dx} = x^2 + 2x + 1$$

Solution The general solution of the differential equation is

$$y = \frac{x^3}{3} + x^2 + x + K$$

To determine the number K, we use the boundary condition. Then

$$-1 = \frac{3^3}{3} + 3^2 + 3 + K$$
$$K = -22$$

The particular solution of the differential equation with boundary condition that $y = -1$ when $x = 3$ is

$$y = \frac{x^3}{3} + x^2 + x - 22$$

∎

Applications

The statement below has many applications in physics, chemistry, and biology.

The amount A of a substance varies with time t in such a way that the time rate of change of A is proportional to A itself.

We may state this in the form of the *differential equation*

(3) $$\frac{dA}{dt} = kA$$

where $k \neq 0$ is a real number. If $k > 0$, then (3) asserts that the time rate of change of A is positive so that the amount A of the substance is increasing; if

$k < 0$, then (3) asserts that the time rate of change of A is negative so that the amount A of the substance is decreasing.

We seek a solution to (3). We begin by rewriting the equation in terms of differentials as

$$\frac{dA}{A} = k\, dt$$

Then we integrate each side to get

$$\int \frac{dA}{A} = \int k\, dt$$

from which we obtain

$$\ln A = kt + K$$

To determine the constant K, we use the boundary condition that when $t = 0$, the initial amount present is A_0. As a result, $\ln A_0 = K$, so that

$$\ln A = kt + \ln A_0$$
$$\ln A - \ln A_0 = kt$$
$$\ln \frac{A}{A_0} = kt$$
$$e^{\ln(A/A_0)} = e^{kt}$$
$$\frac{A}{A_0} = e^{kt}$$

The solution of the differential equation (3) is therefore

(4) $$A = A_0 e^{kt}$$

When a function $A = A(t)$ varies according to the law given by (3), or its equivalent (4), it is said to follow the *exponential law*, or the *law of uninhibited growth or decay*, or the *law of continuously compounded interest*.*

As previously noted, the sign of k determines whether $A(t)$ is increasing (if $k > 0$) or decreasing (if $k < 0$). Figure 2 illustrates the graphs of (4) for both $k > 0$ and $k < 0$.

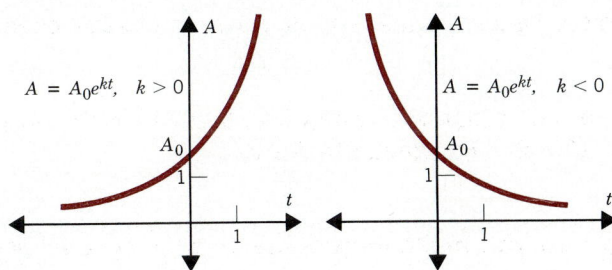

Figure 2

*Refer to Chapter 5, Section 2.

Bacterial Growth

Our first application is to bacterial growth.

Example 2 Assume that a colony of bacteria *increases at a rate proportional to the number present.** If the number of bacteria doubles in 5 hours, how long will it take for the bacteria to triple?

Solution Let $N(t)$ be the number of bacteria present at time t. Then the assumption that this colony of bacteria increases at a rate proportional to the number present can be mathematically written as

(5)
$$\frac{dN}{dt} = kN$$

where k is a positive constant of proportionality. The solution of (5) is

$$N(t) = N_0 e^{kt}$$

where N_0 is the initial number of bacteria in this colony. Since the number of bacteria doubles in 5 hours, we have

$$N(5) = 2N_0$$
$$N_0 e^{5k} = 2N_0$$
$$e^{5k} = 2$$
$$k = (\tfrac{1}{5})\ln 2$$

The time t that is required for this colony to triple must satisfy the equation

$$N(t) = 3N_0$$
$$N_0 e^{kt} = 3N_0$$
$$e^{kt} = 3$$
$$kt = \ln 3$$
$$t = \left(\frac{1}{k}\right)\ln 3 = 5\,\frac{\ln 3}{\ln 2} \underset{\text{Table 9}}{=} 5\,\frac{1.0986}{0.6931} \approx 7.925 \text{ hours}$$

Radioactive Decay

Our second application is to radioactive decay, and in particular its use in carbon dating. For a radioactive substance, *the rate of decay is proportional to the amount present at a given time t.* That is, if A represents the amount of a radioactive substance at time t, we have

$$\frac{dA}{dt} = kA$$

Half-Life

where the constant k is negative and depends on the radioactive substance. The *half-life* of a radioactive substance is the time required for half of the substance to decay.

---

*This is a model of uninhibited growth. However, after enough time has passed, growth will not continue at a rate proportional to the number present. Other factors, such as lack of living space, dwindling food supply, and so on, will start to affect the rate of growth. The model presented accurately reflects the way growth occurs in the early stages.

In carbon dating, we use the fact that all living organisms contain two kinds of carbon, carbon-12 (a stable carbon) and carbon-14 (a radioactive carbon). As a result, when an organism dies, the amount of carbon-12 present remains unchanged, while the amount of carbon-14 begins to decrease. This change in the amount of carbon-14 present relative to the amount of carbon-12 present makes it possible to calculate the time at which the organism lived.

Example 3 In the skull of an animal found in an archaeological dig, it was determined that about 20% of the original amount of carbon-14 was still present. If the half-life of carbon-14 is 5600 years, find the appproximate age of the animal.

Solution Let A be the amount of carbon-14 present in the skull at time t. Then A satisfies the differential equation $dA/dt = kA$, whose solution is

$$A = A_0 e^{kt}$$

where A_0 is the amount of carbon-14 present at time $t = 0$. To determine the constant k, we use the fact that when $t = 5600$, half of the original amount A_0 will remain. Thus,

$$\tfrac{1}{2} A_0 = A_0 e^{5600k}$$
$$\tfrac{1}{2} = e^{5600k}$$
$$5600k = \ln \tfrac{1}{2}$$
$$k = -0.000124$$

The relationship between the amount A of carbon-14 and time t is therefore

$$A = A_0 e^{-(0.000124)t}$$

If the amount A of carbon-14 is 20% of the original amount A_0, we have

$$0.2 A_0 = A_0 e^{-(0.000124)t}$$
$$0.2 = e^{-(0.000124)t}$$

By taking the natural logarithm of both sides, we have

$$-(0.000124)t = \ln 0.2$$
$$-(0.000124)t = -1.6094$$
$$t = \frac{1.6094}{0.000124} \approx 12{,}979 \text{ years}$$

Thus, the animal lived approximately 13,000 years ago.

Exercise 4
Solutions to Odd-Numbered Problems begin on page 418.

In Problems 1–10 solve each differential equation using the indicated boundary condition.

1. $\dfrac{dy}{dx} = x^2 - 1,$

   $y = 0$ when $x = 0$

2. $\dfrac{dy}{dx} = x^2 + 4,$

   $y = 1$ when $x = 0$

3. $\dfrac{dy}{dx} = x^2 - x$,

   $y = 3$ when $x = 3$

4. $\dfrac{dy}{dx} = x^2 + x$,

   $y = 5$ when $x = 3$

5. $\dfrac{dy}{dx} = x^3 - x + 2$,

   $y = 1$ when $x = -2$

6. $\dfrac{dy}{dx} = x^3 + x - 5$,

   $y = 1$ when $x = -2$

7. $\dfrac{dy}{dx} = e^x$,

   $y = 4$ when $x = 0$

8. $\dfrac{dy}{dx} = \dfrac{1}{x}$,

   $y = 0$ when $x = 1$

9. $\dfrac{dy}{dx} = \dfrac{x^2 + x + 1}{x}$,

   $y = 0$ when $x = 1$

10. $\dfrac{dy}{dx} = x + e^x$,

    $y = 4$ when $x = 0$

11. *Bacterial Growth.* The rate of growth of bacteria is proportional to the amount present. If initially there are 100 bacteria and 5 minutes later there are 150 bacteria, how many bacteria will be present after 1 hour? How many are present after 90 minutes? How long will it take for the number of bacteria to reach 1,000,000?

12. Answer the questions posed in Problem 11 if after 8 minutes the number of bacteria present grows from 100 to 150.

13. *Radioactive Decay.* The half-life of radium is 1690 years. If 8 grams of radium are present now, how many grams will be present in 100 years?

14. *Radioactive Decay.* If 25% of a radioactive substance disappears in 10 years, what is the half-life of the substance?

15. *Age of a Tree.* A piece of charcoal is found to contain 30% of the carbon-14 it originally had. When did the tree from which the charcoal came die? Use 5600 years as the half-life of carbon-14.

16. *Age of a Fossil.* A fossilized leaf contains 70% of a normal amount of carbon-14. How old is the fossil?

17. *Population Growth.* The population growth of a colony of mosquitoes obeys the uninhibited growth equation. If there are 1500 mosquitoes initially, and there are 2500 mosquitoes after 24 hours, what is the size of the mosquito population after 3 days?

18. *Population Growth.* The population of a suburb doubled in size in an 18 month period. If this growth continues and the current population is 8000, what will the population be in 4 years?

19. *Bacterial Growth.* The number of bacteria in a culture is growing at a rate of $3000e^{2t/5}$ per unit of time t. At $t = 0$, the number of bacteria present was 7500. Find the number present at $t = 5$.

20. *Bacterial Growth.* At any time t, the rate of increase in the area of a culture of bacteria is twice the area of the culture. If the initial area of the culture is 10, then what is the area at time t?

*21. *Bacterial Growth.* The rate of change in the number of bacteria in a culture is proportional to the number present. In a certain labora-

tory experiment, a culture had 10,000 bacteria initially, 20,000 bacteria at time t_1 minutes, and 100,000 bacteria at $(t_1 + 10)$ minutes.
(a) In terms of t only, find the number of bacteria in the culture at any time t minutes ($t \geq 0$).
(b) How many bacteria were there after 20 minutes?
(c) At what time were 20,000 bacteria observed? That is, find the value of t_1.

*22. *Chemistry.* Salt (NaCl) decomposes in water into sodium (Na$^+$) and chloride (Cl$^-$) ions at a rate proportional to its mass. If the initial amount of salt is 25 kilograms, and after 10 hours, 15 kilograms are left:
(a) How much salt would be left after 1 day?
(b) After how many hours would there be less than $\frac{1}{2}$ kilogram of salt left?

*23. *Age of a Fossil.* Radioactive beryllium is sometimes used to date fossils found in deep-sea sediment. The decay of radioactive beryllium satisfies the equation $dA/dt = -\alpha A$, where $\alpha = 1.5 \times 10^{-7}$, and t is measured in years. What is the half-life of radioactive beryllium?

*24. *Pressure.* Atmospheric pressure is a function of altitude above sea level and is given by the equation $dP/da = \beta P$, where β is a constant. The pressure is measured in millibars (mb). At sea level ($a = 0$), $P(0)$ is 1013.25 mb, which means that the atmosphere at sea level will support a column of mercury 1013.25 millimeters high at a standard temperature of 15°C. At an altitude of $a = 1500$ meters, the pressure is 845.6 mb.
(a) What is the pressure at $a = 4000$ meters?
(b) What is the pressure at 10 kilometers?
(c) In California, the highest and lowest points are Mount Whitney (4418 meters) and Death Valley (86 meters below sea level). What is the difference in their atmospheric pressures?
(d) What is the atmospheric pressure at Mount Everest (elevation 8848 meters)?
(e) At what elevation is the atmospheric pressure equal to 1 mb?

*5. Application to Marginal Analysis

As we discussed in Chapter 4, marginal revenue and marginal cost are defined as the first derivative of the revenue function and cost function, respectively. As a result, if the marginal revenue (or marginal cost) is a known function, the revenue function $R(x)$ (or cost function $C(x)$) may be found by using the process of antidifferentiation. That is,

$$\int MR\, dx = R(x) + K \quad \text{and} \quad \int MC\, dx = C(x) + K$$

The presence of a constant K in the antiderivative of a function offers us a great deal of flexibility, since the undetermined constant K can assume any value that a particular practical situation demands.

*This section may be omitted without loss of continuity.

Example 1 By experimenting with various production techniques, a manufacturer finds that the marginal cost of production is given by the function

$$MC = 2x + 6$$

where x is the number of units produced and MC is the marginal cost in dollars. The fixed cost of production is known to be \$9. Find the cost of production.

Solution The antiderivative of the marginal cost of production MC is the cost $C(x)$ of production. That is,

$$\int MC\, dx = \int (2x + 6)\, dx$$

so that

$$C(x) = x^2 + 6x + K$$

where K is a constant. We can find the value of the constant K by observing that of all the cost functions with derivative $2x + 6$, only *one* has a fixed cost of production of \$9. See Figure 3. This means that when $x = 0$ items are produced, the fixed cost will be \$9. That is,

$$C(0) = (0)^2 + 6(0) + K = 9$$
$$K = 9$$

Thus, the cost function is

$$C(x) = x^2 + 6x + 9$$

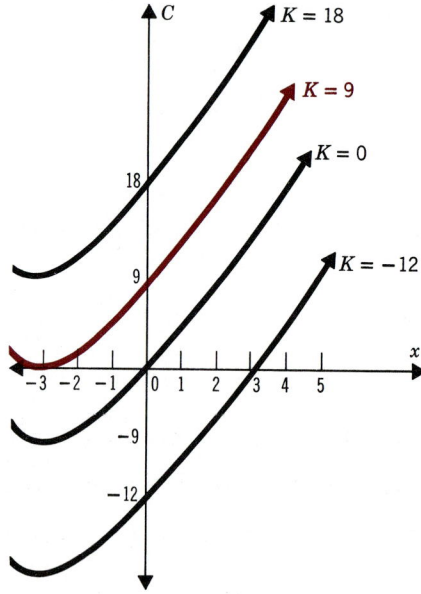

Figure 3

Example 2 Suppose the manufacturer in Example 1 receives a per unit price of \$60. This means the instantaneous rate of change of revenue with respect to quantity is the constant 60, that is,

$$MR = 60$$

(a) Find the revenue function $R(x)$.
(b) Find the profit function $P(x)$.
(c) Find the sales volume that yields maximum profit.
(d) What is the profit at this sales volume?

Solution (a) The revenue function $R(x)$ is the antiderivative of the marginal revenue function MR. That is,

$$R(x) = \int MR\, dx = \int 60\, dx$$

So,

$$R(x) = 60x + K$$

Now, of all these revenue functions, there is only one for which sales equal zero for $x = 0$ units sold. See Figure 4. Thus,

$$0 = 60(0) + K$$
$$K = 0$$

This means the revenue function is

$$R(x) = 60x$$

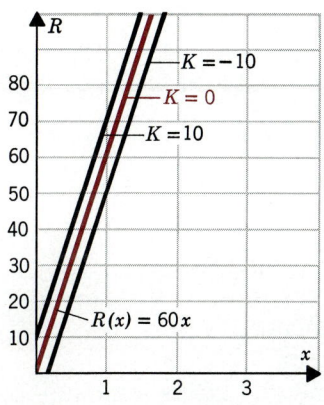

Figure 4

(b) The profit function $P(x)$ is the difference between revenue and cost. That is,

$$P(x) = R(x) - C(x) = 60x - (x^2 + 6x + 9) = -x^2 + 54x - 9$$

(c) The maximum profit is obtained when marginal revenue equals marginal cost:

$$MR = MC$$
$$60 = 2x + 6$$
$$2x = 54$$
$$x = 27$$

Thus, when sales total 27 units, a maximum profit is obtained.

(d) The profit for sales of 27 units is

$$P(27) = -(27)(27) + 54(27) - 9 = 720$$

The profit of $720 for sales totaling 27 units can also be found by differentiating the profit function $P(x)$ and finding those x for which $P'(x) = 0$. To verify this,

$$P'(x) = -2x + 54 = 0$$
$$x = 27$$

■

Exercise 5
Solutions to Odd-Numbered Problems begin on page 419.

In Problems 1–4 find the revenue function. Assume that revenue is zero when zero units are sold.

1. $MR = 600$
2. $MR = 350$
3. $MR = 20x + 5$
4. $MR = 50x - x^2$

In Problems 5–8 find the cost function and determine where the cost is a minimum.

5. $MC = 14x - 2800$
   Fixed cost $= \$4300$
6. $MC = 6x - 2400$
   Fixed cost $= \$800$
7. $MC = 20x - 8000$
   Fixed cost $= \$500$
8. $MC = 15x - 3000$
   Fixed cost $= \$1000$

9. The marginal cost of production is found to be

$$MC = 1000 - 20x + x^2$$

where x is the number of units produced. The fixed cost of production is $9000. Find the cost function.

10. In Problem 9 the manufacturer fixes the price per unit at $3400.
    (a) Find the revenue function.
    (b) Find the profit function.
    (c) Find the sales volume that yields maximum profit.
    (d) What is the profit at this sales volume?

11. A company determines that the marginal cost of producing x units of a particular commodity during 1 day of operation is $MC = 16x - 1591$, where the production cost is in dollars. The selling price of the commodity is fixed at $9 per unit and the fixed cost is $1800 per day.
    (a) Find the cost function.
    (b) Find the revenue function.
    (c) Find the profit function.
    (d) What is the maximum profit that can be obtained in 1 day of operation?

6. Table of Integrals

In this section, we present formulas for finding the indefinite integral of functions that occur frequently. Many of these formulas you will be able to derive by using techniques discussed previously in the book. Some of them (19 and 20) are *recursive*, meaning that they must be repeated to find the antiderivative.

Table of Integrals

1. $\int x(ax+b)^n \, dx = \dfrac{(ax+b)^{n+1}}{a^2}\left(\dfrac{ax+b}{n+2} - \dfrac{b}{n+1}\right) + K$

   for $n \neq -1, -2$

2. $\int x(ax+b)^{-1} \, dx = \dfrac{x}{a} - \dfrac{b}{a^2}\ln|ax+b| + K$

3. $\int x(ax+b)^{-2} \, dx = \dfrac{1}{a^2}\left(\ln|ax+b| + \dfrac{b}{ax+b}\right) + K$

4. $\int \dfrac{1}{x(ax+b)} \, dx = \dfrac{1}{b}\ln\left|\dfrac{x}{ax+b}\right| + K$

5. $\int \dfrac{dx}{x\sqrt{ax+b}} = \dfrac{1}{\sqrt{b}}\ln\left|\dfrac{\sqrt{ax+b} - \sqrt{b}}{\sqrt{ax+b} + \sqrt{b}}\right| + K$ for $b > 0$

6. $\int \dfrac{\sqrt{ax+b}}{x} \, dx = 2\sqrt{ax+b} + b\int \dfrac{1}{x\sqrt{ax+b}} \, dx$ for $b > 0$

7. $\int \dfrac{\sqrt{ax+b}}{x^2} \, dx = -\dfrac{\sqrt{ax+b}}{x} + \dfrac{a}{2}\int \dfrac{1}{x\sqrt{ax+b}} \, dx$ for $b > 0$

8. $\int \dfrac{1}{x^2\sqrt{ax+b}} \, dx = -\dfrac{\sqrt{ax+b}}{bx} - \dfrac{a}{2b}\int \dfrac{1}{x\sqrt{ax+b}} \, dx$ for $b > 0$

9. $\int \dfrac{1}{a^2 - x^2} \, dx = \dfrac{1}{2a}\ln\left|\dfrac{x+a}{x-a}\right| + K$

10. $\int \dfrac{1}{(a^2 - x^2)^2} \, dx = \dfrac{x}{2a^2(a^2 - x^2)} + \dfrac{1}{2a^2}\int \dfrac{1}{a^2 - x^2} \, dx$

11. $\int \dfrac{1}{x\sqrt{a^2 + x^2}} \, dx = -\dfrac{1}{a}\ln\left|\dfrac{a + \sqrt{a^2 + x^2}}{x}\right| + K$

12. $\int \dfrac{1}{x^2\sqrt{a^2 + x^2}} \, dx = -\dfrac{\sqrt{a^2 + x^2}}{a^2 x} + K$

13. $\int \dfrac{\sqrt{a^2 - x^2}}{x} \, dx = \sqrt{a^2 - x^2} - a\ln\left|\dfrac{a + \sqrt{a^2 - x^2}}{x}\right| + K$

14. $\int \dfrac{1}{x\sqrt{a^2 - x^2}} \, dx = -\dfrac{1}{a}\ln\left|\dfrac{a + \sqrt{a^2 - x^2}}{x}\right| + K$

15. $\int \dfrac{1}{x^2\sqrt{a^2-x^2}}\,dx = -\dfrac{\sqrt{a^2-x^2}}{a^2 x} + K$

16. $\int \dfrac{1}{\sqrt{x^2-a^2}}\,dx = \ln\left|x + \sqrt{x^2-a^2}\right| + K$

17. $\int \dfrac{1}{x^2\sqrt{x^2-a^2}}\,dx = \dfrac{\sqrt{x^2-a^2}}{a^2 x} + K$

18. $\int \dfrac{1}{x\sqrt{2ax-x^2}}\,dx = -\dfrac{1}{a}\sqrt{\dfrac{2a-x}{x}} + K$

19. $\int x^n\, e^{ax}\,dx = \dfrac{1}{a}x^n\, e^{ax} - \dfrac{n}{a}\int x^{n-1}\, e^{ax}\,dx$

20. $\int x^n b^{ax}\,dx = \dfrac{x^n b^{ax}}{a\ln b} - \dfrac{n}{a\ln b}\int x^{n-1} b^{ax}\,dx$ for $b > 0$ and $b \neq 1$

Example 1 Find

$$\int x(5x+6)^4\,dx$$

Solution We examine the integrand and find that it matches the integrand of formula (1) in the Table of Integrals, when $n = 4$, $a = 5$, $b = 6$. Thus

$$\int x(5x+6)^4\,dx = \dfrac{(5x+6)^5}{25}\left[\dfrac{5x+6}{6} - \dfrac{6}{5}\right] + K$$

∎

Exercise 6
Solutions to Odd-Numbered Problems begin on page 419.

Use the Table of Integrals to evaluate each of the following integrals.

1. $\int x(3x+5)^2\,dx$

2. $\int \dfrac{1}{x\sqrt{2x+3}}\,dx$

3. $\int \dfrac{\sqrt{x+2}}{x^2}\,dx$

4. $\int \dfrac{1}{x^2\sqrt{9-x^2}}\,dx$

5. $\int \dfrac{1}{x\sqrt{2x-x^2}}\,dx$

6. $\int \dfrac{1}{(4-x^2)^2}\,dx$

7. $\int x^3\, e^{-x}\,dx$

8. $\int \dfrac{\sqrt{1-x^2}}{x}\,dx$

Chapter Review

Important Terms
antiderivative
integral sign
indefinite integral
integrand
integration by substitution
integration by parts
differential equation
general solution
particular solution
boundary condition
bacterial growth
radioactive decay
half-life
table of integrals

Review Exercises
Solutions to Odd-Numbered Problems begin on page 420.

In Problems 1–18 evaluate each indefinite integral.

1. $\int (x^3 - 3x + 1)\, dx$
2. $\int (x^3 + 4x - 2)\, dx$
3. $\int (x^{1/3} - 4x^{1/2})\, dx$
4. $\int (x^{3/2} + 5x^{1/2})\, dx$
5. $\int (1 + e^{-x})\, dx$
6. $\int (1 - e^x)\, dx$
7. $\int x\sqrt{x^2 - 1}\, dx$
8. $\int x\sqrt{3x^2 - 1}\, dx$
9. $\int \sqrt{3x - 2}\, dx$
10. $\int \sqrt{2 - 3x}\, dx$
11. $\int \dfrac{x^3\, dx}{(x^4 + 1)^{3/2}}$
12. $\int \dfrac{x^2\, dx}{(x^3 - 1)^{1/2}}$
13. $\int \dfrac{5x\, dx}{x^2 + 1}$
14. $\int \dfrac{5x^2}{x^3 - 1}\, dx$
15. $\int xe^{x/2}\, dx$
16. $\int x \ln 3x\, dx$
17. $\int x^2 e^{x^3}\, dx$
18. $\int \dfrac{x^{3/2}\, dx}{x^{5/2} + 2}$

In Problems 19–22 solve each differential equation using the indicated boundary condition.

19. $\dfrac{dy}{dx} = \dfrac{x + 1}{x}$, $y = 1$ when $x = 1$
20. $\dfrac{dy}{dx} = e^x + x$, $y = 5$ when $x = 0$
21. $\dfrac{dy}{dx} = x\sqrt{x^2 + 1}$, $y = 6$ when $x = 0$
22. $\dfrac{dy}{dx} = x\sqrt{x^2 - 16}$, $y = 0$ when $x = 5$

23. Find a function $f(x)$ satisfying the conditions $f'(x) = 1/\sqrt{x}$ with $f(1) = 1$.
24. The marginal revenue of a company is found to be

$$MR = 64x - x^2$$

where x is the number (in thousands) of units sold. Find the revenue function and determine the number of sales that maximizes revenue.
25. *Bacterial Growth.* Bacteria grown in a certain culture increase at a rate proportional to the amount present. If there are 2000 bacteria present initially and the amount triples in 2 hours, how many bacteria will there be in $4\frac{1}{2}$ hours?
26. *Population Growth.* In 1970 the population of Glenwood was 3000. By 1980 the population doubled. Assuming that this growth continues and assuming that the rate of increase is proportional to the population, what will the population be in 1985?
27. *Radioactive Decay.* The half-life of a certain radioactive material is 1000 years. How long will it take for the amount of radioactive material present to decay to 20% of its original amount?
28. *Age of an Animal.* The skeleton of an animal is found to contain 35% of the original amount of carbon-14. What is the approximate age of this animal? (The half-life of carbon-14 is 5600 years.)
29. Crater Lake in Oregon was formed by the eruption of Mt. Mozama. Wood from a tree burned in that eruption contains 2.8×10^{10} atoms of carbon 14 per gram of carbon. A living tree contains approximately 6×10^{10} atoms of carbon 14 per gram of carbon. How long ago did the eruption occur? (Use 5600 years as the half-life of carbon 14.)

8
The Definite Integral and Its Applications

1. The Definite Integral
2. Application in Geometry: Area under a Graph
3. Applications in Business
4. The Definite Integral as the Limit of a Sum
5. Average Value of a Function
6. Application to Probability
*7. Models
   Model 1: The Learning Curve
   Model 2: Consumer's Surplus
   Model 3: Maximizing Profit over Time
   Model 4: Average Rate Measure of Synchrony.
Chapter Review
   Mathematical Questions from Actuary Exams

*This section may be omitted without loss of continuity.

1. The Definite Integral

We begin with an example illustrating the general idea of a *definite integral*.

Example 1 The marginal cost of a certain firm is given by the equation

$$MC = 4 - 0.2x \qquad 0 \le x \le 10$$

where MC is in units of thousands of dollars and the quantity x produced is in hundreds of units per day. If the number of units produced in a given day changes from 2 hundred to 5 hundred units, what is the change in cost?

Solution If $C(x)$ is the cost function, the change in cost from $x = 2$ to $x = 5$ is

$$C(5) - C(2)$$

This is the number we seek. Since we know that the derivative of C is the marginal cost, we have

$$MC = C'(x) = 4 - 0.2x$$

The cost C is an antiderivative of $C'(x) = 4 - 0.2x$. Thus,

$$C(x) = \int C'(x)\, dx = \int (4 - 0.2x)\, dx = 4x - 0.1x^2 + K$$

We use this to compute $C(5) - C(2)$:

(1) $$C(5) - C(2) = [4(5) - (0.1)(25) + K] - [4(2) - (0.1)(4) + K] = 9.9$$

Thus, the change in cost is 9.9 thousand dollars.

In this example, the change in C was computed by using an antiderivative of $C'(x)$, which is symbolized by $\int C'(x)\, dx$. To indicate that the change is from $x = 2$ to $x = 5$, we add to this notation, as follows:

$$\text{change in C from 2 to 5} = \int_2^5 C'(x)\, dx$$

This form is called a *definite integral*.

Definite Integral The *definite integral* **from a to b of a continuous function f equals the change from a to b in an antiderivative of f. Thus, if F is an antiderivative of f, the definite integral of f from a to b is**

$$\int_a^b f(x)\, dx = F(b) - F(a)$$

In $\int_a^b f(x)\, dx$, **the numbers a and b are called the** *lower* **and** *upper* **limits of** *integration,* **respectively.**

Example 2 The definite integral from 2 to 3 of $f(x) = x^2$ is computed by first finding an antiderivative of $f(x)$. One such antiderivative is $F(x) = x^3/3$. Thus, the definite integral from 2 to 3 of x^2 is

$$F(3) - F(2) = \frac{27}{3} - \frac{8}{3} = \frac{19}{3}$$

That is,

$$\int_2^3 x^2 \, dx = \tfrac{19}{3}$$

In computing $\int_a^b f(x) \, dx$, we find that the choice of an antiderivative of $f(x)$ does not matter. Look back at equation (1) in the solution to Example 1. The constant K canceled out. Now look at Example 2. If we had used $F(x) = x^3/3 + K$ as the antiderivative of x^2, we would have found that

$$\int_2^3 x^2 \, dx = F(3) - F(2) = [\tfrac{27}{3} + K] - [\tfrac{8}{3} + K] = \tfrac{19}{3}$$

Again, the constants cancel out. This will always be the case.

Fundamental Formula of Integral Calculus **Any antiderivative of $f(x)$ can be used to evaluate $\int_a^b f(x) \, dx$ and if F is an antiderivative of f, we have the formula**

(2) $$\int_a^b f(x) \, dx = F(b) - F(a)$$

In place of $F(b) - F(a)$, we may use the notation

$$\int_a^b f(x) \, dx = F(x) \Big|_a^b = F(b) - F(a)$$

In terms of this new notation, to calculate $F(x)\Big|_a^b$, first replace x by the upper limit b to obtain $F(b)$, and from this subtract $F(a)$, obtained by setting $x = a$.

Example 3 (a) $\int_{-1}^5 6x \, dx = 3x^2 \Big|_{-1}^5 = 3(5)^2 - 3(-1)^2 = 75 - 3 = 72$

(b) $\int_1^2 x^3 \, dx = \dfrac{x^4}{4}\Big|_1^2 = \dfrac{(2)^4}{4} - \dfrac{(1)^4}{4} = \dfrac{16}{4} - \dfrac{1}{4} = \dfrac{15}{4}$

It is important to distinguish between the indefinite integral and the definite integral. The indefinite integral, a symbol for all the antiderivatives of a function, is a function. On the other hand, the definite integral is a number.

Example 4 Evaluate: $\int_1^4 \sqrt{x}\, dx$

Solution Since $\frac{2}{3}x^{3/2}$ is an antiderivative of $\sqrt{x}$, we get

$$\int_1^4 \sqrt{x}\, dx = \frac{2}{3}x^{3/2}\Big|_1^4 = \frac{2}{3}(4)^{3/2} - \frac{2}{3}(1)^{3/2} = \frac{16}{3} - \frac{2}{3} = \frac{14}{3}$$

We list some properties of definite integrals below.

If $f(x)$ is continuous on the interval $[a, b]$ and has an antiderivative on $[a, b]$, then

(3) $$\int_a^b f(x)\, dx = -\int_b^a f(x)\, dx$$

(4) $$\int_a^a f(x)\, dx = 0$$

Example 5 (a) $\int_4^1 \sqrt{x}\, dx = -\int_1^4 \sqrt{x}\, dx = -\frac{14}{3}$ (b) $\int_1^1 x\, dx = 0$

Formulas (3) and (4) are an immediate consequence of the fundamental formula of integral calculus (2). Specifically, if $F(x)$ is an antiderivative of $f(x)$, then

$$\int_a^b f(x)\, dx = F(b) - F(a) = -[F(a) - F(b)] = -\int_b^a f(x)\, dx$$

and

$$\int_a^a f(x)\, dx = F(a) - F(a) = 0$$

If $f(x)$ is continuous and has an antiderivative on the interval $[a, b]$, and if c is between a and b, then

(5) $$\int_a^b f(x)\, dx = \int_a^c f(x)\, dx + \int_c^b f(x)\, dx$$

If $f(x)$ is continuous and has an antiderivative on the interval $[a, b]$ and c is a constant, then

(6) $$\int_a^b cf(x)\, dx = c\int_a^b f(x)\, dx$$

Example 6 (a) If $\int_1^3 f(x)\,dx = 5$ and $\int_3^6 f(x)\,dx = 7$, then

$$\int_1^6 f(x)\,dx = \int_1^3 f(x)\,dx + \int_3^6 f(x)\,dx = 5 + 7 = 12$$

(b) $\int_1^2 16x^2\,dx = 16\int_1^2 x^2\,dx = 16\left(\dfrac{x^3}{3}\bigg|_1^2\right) = 16\left(\dfrac{8}{3} - \dfrac{1}{3}\right) = (16)\left(\dfrac{7}{3}\right) = \dfrac{112}{3}$ ∎

If $f(x)$ and $g(x)$ are continuous and have antiderivatives on the interval $[a, b]$, then

(7)
$$\int_a^b [f(x) \pm g(x)]\,dx = \int_a^b f(x)\,dx \pm \int_a^b g(x)\,dx$$

Example 7
$$\int_1^2 (x^2 + \sqrt{x})\,dx = \int_1^2 x^2\,dx + \int_1^2 \sqrt{x}\,dx = \dfrac{x^3}{3}\bigg|_1^2 + \dfrac{2}{3}x^{3/2}\bigg|_1^2$$
$$= \dfrac{7}{3} + \dfrac{2}{3}(2\sqrt{2} - 1) = \dfrac{4\sqrt{2}}{3} + \dfrac{5}{3}$$ ∎

Example 8 Evaluate: $\int_1^2 3x(x^2 - 1)\,dx$

Solution
$$\int_1^2 3x(x^2 - 1)\,dx = \int_1^2 (3x^3 - 3x)\,dx = \int_1^2 3x^3\,dx - \int_1^2 3x\,dx$$
$$= 3\int_1^2 x^3\,dx - 3\int_1^2 x\,dx = 3\left(\dfrac{x^4}{4}\bigg|_1^2\right) - 3\left(\dfrac{x^2}{2}\bigg|_1^2\right)$$
$$= 3\left(4 - \dfrac{1}{4}\right) - 3\left(2 - \dfrac{1}{2}\right) = 3\left(\dfrac{15}{4}\right) - 3\left(\dfrac{3}{2}\right) = \dfrac{27}{4}$$ ∎

Example 9 Evaluate: $\int_0^2 x\sqrt{4 - x^2}\,dx$.

Solution First we evaluate the indefinite integral $\int x\sqrt{4 - x^2}\,dx$, using the substitution $u = 4 - x^2$. Then $du = -2x\,dx$, and

$$\int x\sqrt{4 - x^2}\,dx = \int \sqrt{4 - x^2}\,x\,dx = \int \sqrt{u}\left(-\dfrac{du}{2}\right)$$
$$= -\dfrac{1}{2}\left(\dfrac{u^{3/2}}{\frac{3}{2}}\right) + C = \dfrac{-(4 - x^2)^{3/2}}{3} + C$$

By the fundamental formula, we find

$$\int_0^2 x\sqrt{4 - x^2}\,dx = \dfrac{-(4 - x^2)^{3/2}}{3}\bigg|_0^2 = 0 + \dfrac{4^{3/2}}{3} = \dfrac{8}{3}$$ ∎

282 CH. 8 THE DEFINITE INTEGRAL AND ITS APPLICATIONS

When the substitution method is used to evaluate definite integrals, it is sometimes easier to change the limits of integration. For example, to evaluate $\int_0^2 x\sqrt{4-x^2}\,dx$, as we did in Example 9, we let $u = 4 - x^2$ and $du = -2x\,dx$. But now we use the equation $u = 4 - x^2$ to change the limits of integration. Thus, when $x = 0$ (the *old* lower limit of integration), we have $u = 4 - 0^2 = 4$ (the *new* lower limit of integration). Similarly, when $x = 2$ (the old upper limit of integration), we have $u = 0$ (the new upper limit of integration). Using this technique,

$$\int_0^2 x\sqrt{4-x^2}\,dx = \int_4^0 \sqrt{u}\left(-\frac{du}{2}\right) = -\frac{1}{2}\left(\frac{u^{3/2}}{\frac{3}{2}}\right)\Big|_4^0 = -\frac{1}{3}(0 - 4^{3/2}) = \frac{8}{3}$$

where $u = 4 - x^2 = 4 - 2^2 = 0$ and $u = 4 - x^2 = 4 - 0 = 4$.

A word to the wise! This second way of evaluating a definite integral may be faster, but it also requires more care. Make certain not to forget to change the limits of integration when using this method.

Exercise 1
Solutions to Odd-Numbered Problems begin on page 421.

In Problems 1–34 apply the fundamental formula of integral calculus (2) to evaluate each definite integral.

1. $\int_1^2 (3x - 1)\,dx$
2. $\int_1^2 (2x + 1)\,dx$
3. $\int_0^1 (3x^2 + e^x)\,dx$
4. $\int_{-2}^0 (e^x + x^2)\,dx$
5. $\int_0^1 \sqrt{u}\,du$
6. $\int_1^4 \sqrt{u}\,du$
7. $\int_0^1 (t^2 - t^{3/2})\,dt$
8. $\int_1^4 (\sqrt{x} - a^2 x)\,dx$
9. $\int_{-2}^3 (x - 1)(x + 3)\,dx$
10. $\int_0^1 (z^2 + 1)^2\,dz$
11. $\int_1^2 \dfrac{x^2 - 1}{x^4}\,dx$
12. $\int_1^3 \dfrac{2 - x^2}{x^4}\,dx$
13. $\int_1^4 \left(\sqrt[5]{t^2} + \dfrac{1}{t}\right)dt$
14. $\int_1^4 \left(\sqrt{u} + \dfrac{1}{u}\right)du$
15. $\int_1^4 \dfrac{x + 1}{\sqrt{x}}\,dx$
16. $\int_1^9 \dfrac{\sqrt{x} + 1}{x^2}\,dx$
17. $\int_3^3 (5x^4 + 1)^{3/2}\,dx$
18. $\int_{-1}^1 (x + 1)^3\,dx$

19. $\int_{-1}^{1} (x+1)^2 \, dx$

20. $\int_{-1}^{-1} \sqrt[3]{x^2+4} \, dx$

21. $\int_{1}^{e} \left(x - \frac{1}{x}\right) dx$

22. $\int_{1}^{e} \left(x + \frac{1}{x}\right) dx$

23. $\int_{0}^{1} e^{-x} \, dx$

24. $\int_{0}^{1} x^2 e^{x^3} \, dx$

25. $\int_{1}^{3} \frac{dx}{x+1}$

26. $\int_{-2}^{2} e^{-7x/2} \, dx$

27. $\int_{0}^{1} \frac{\sqrt{x}}{x^{3/2}+1} \, dx$

28. $\int_{2}^{3} \frac{dx}{x \ln x}$

29. $\int_{1}^{3} x e^{2x} \, dx$

30. $\int_{0}^{4} (1 + x e^{-x}) \, dx$

31. $\int_{1}^{2} x e^{-3x} \, dx$

32. $\int_{1}^{3} x^2 \ln x \, dx$

33. $\int_{1}^{5} \ln x \, dx$

34. $\int_{1}^{2} x \ln x \, dx$

*35. **Even Function.** A continuous function $f(x)$ is said to be an *even* function if $f(-x) = f(x)$. It can be shown that if $f(x)$ is an even function, then

$$\int_{-a}^{a} f(x) \, dx = 2 \int_{0}^{a} f(x) \, dx \qquad a > 0$$

Verify the above formula by evaluating the following definite integrals:

(a) $\int_{-1}^{1} x^2 \, dx$ (b) $\int_{-1}^{1} (x^4 + x^2) \, dx$

*36. **Odd Function.** A continuous function $f(x)$ is said to be an *odd* function if $f(-x) = -f(x)$. It can be shown that if $f(x)$ is an odd function, then

$$\int_{-a}^{a} f(x) \, dx = 0 \qquad a > 0$$

Verify the above formula by evaluating the following definite integrals:

(a) $\int_{-1}^{1} x \, dx$ (b) $\int_{-1}^{1} x^3 \, dx$

2. Application in Geometry: Area under a Graph

The development of the integral, like that of the derivative, was originally motivated to a large extent by attempts to solve a basic problem in geometry—namely, the *area problem*. The question is: Given a nonnegative function f whose domain is the closed interval $[a, b]$, what is the area enclosed by the graph of f, the x-axis, and the vertical lines $x = a$ and $x = b$? Figure 1 illustrates the area to be found.

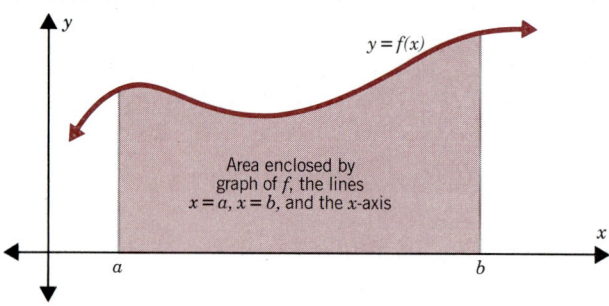

Figure 1

In this section we illustrate how the antiderivative, definite integral, and area under a graph are related to each other. What is area? In plane geometry, we learn how to find the area of certain geometric figures, such as squares, rectangles, and circles. For example, the area of a square with a side of length 3 feet is 9 square feet. The reason is that the square can be subdivided into 9 smaller squares, each having sides of length 1 foot.

We also know that the area of a rectangle with length a units and width b units, is ab square units.

All area problems have certain features in common. For example, whenever the area of an object is computed, it is expressed as a number of square units; this number is never negative. *Thus, one property of area is that it is nonnegative.*

Consider the trapezoid shown in Figure 2. This trapezoid has been decomposed into two nonoverlapping geometric figures, a triangle (with area A_1) and a rectangle (with area A_2). Clearly, the area of the trapezoid is the sum $A_1 + A_2$ of the two component areas. Thus, as long as two regions do not overlap (except perhaps for a common boundary), the total area can be found by adding the component areas. We sometimes call this the *additive property of area*.

Additive Property of Area

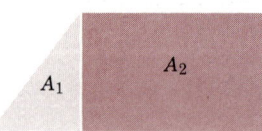

Figure 2

Properties of Area
Two properties of area are:

(I) Area ≥ 0
(II) If A and B are two nonoverlapping regions with areas that are known, then

$$\text{Total area of } A \text{ and } B = \text{Area of } A + \text{Area of } B$$

The above two properties enable us to compute the areas of polygons and, in fact, any region enclosed by straight lines. However, we still are not able to calculate the area of a region enclosed by an arbitrary graph. For example, the problem of determining the area "under the graph of $f(x) = x^2$ from $x = 0$ to $x = 1$," that is, the area of the region enclosed by $f(x) = x^2$, the x-axis, and the vertical line $x = 1$, cannot be solved using the methods of plane geometry. See Figure 3.

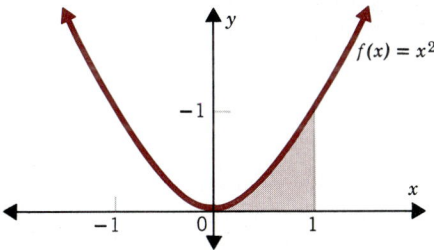

Figure 3

The next result gives a technique for evaluating areas such as the shaded region shown in Figure 3.

Suppose $y = f(x)$ is a continuous function defined on a closed interval I and $f(x) \geq 0$ for all points x in I. Then, for a and b in I, the definite integral

$$\int_a^b f(x)\, dx$$

is the area under the graph of $y = f(x)$ and above the x-axis between the lines $x = a$ and $x = b$.

Figure 4 on page 286 illustrates this statement, which is justified at the end of this section.

The significance of this result is that it enables us to find the area under the graph of $y = f(x)$, provided two conditions are met: f must be continuous and f must be nonnegative. If both conditions hold, then:

(1) **Area enclosed by $y = f(x)$, the x-axis, $x = a$, and $x = b$ is $\int_a^b f(x)\, dx$**

We return now to solve the area problem illustrated in Figure 3.

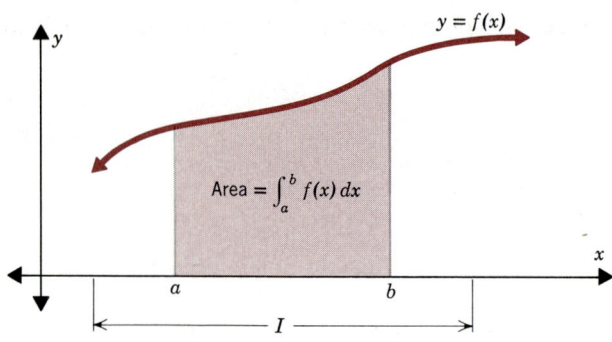

Figure 4

Example 1 Find the area enclosed by $f(x) = x^2$, the x-axis, $x = 0$, and $x = 1$.

Solution The area we seek is given by the definite integral

$$\int_0^1 x^2 \, dx = \frac{x^3}{3} \Big|_0^1 = \frac{1}{3}$$

Thus, the area illustrated in Figure 3 is $\frac{1}{3}$ square unit.

A drawback to this result is the condition that the function must be nonnegative on I. Suppose a function is continuous on the interval I, $a \leq x \leq b$, and has an antiderivative on I, but is negative for $a \leq x \leq c$ and is positive for $c \leq x \leq b$. How, in this situation, do we compute the area enclosed by $y = f(x)$, the x-axis, $x = a$, and $x = b$? See Figure 5.

Notice in Figure 5 that the area A in question is composed of two nonoverlapping areas, A_1 and A_2, so that, by the additive property of area,

$$A = A_1 + A_2$$

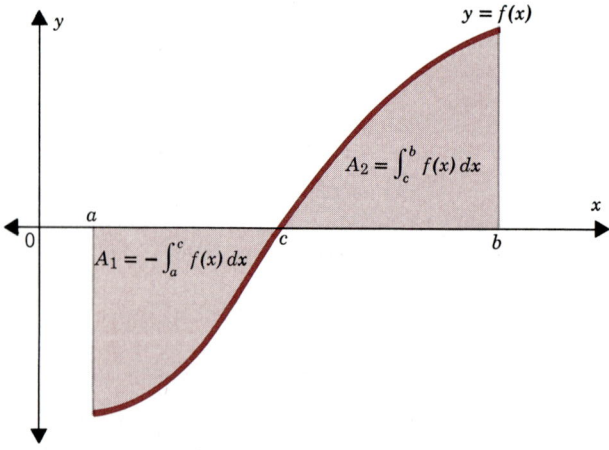

Figure 5

Also, we know that on the interval $[c, b]$, the function is nonnegative so that

$$A_2 = \int_c^b f(x)\, dx$$

To find the area A_1, we note that, since $f(x) \leq 0$ on $a \leq x \leq c$, then $-f(x) \geq 0$, and, by symmetry, the area A_1 equals

$$A_1 = \int_a^c [-f(x)]\, dx = -\int_a^c f(x)\, dx$$

The total area A we seek is therefore

$$A = A_1 + A_2 = -\int_a^c f(x)\, dx + \int_c^b f(x)\, dx$$

The next example illustrates this procedure for calculating area.

Example 2 Find the area enclosed by $f(x) = x^3$, the x-axis, $x = -1$, and $x = \frac{1}{2}$.

Solution The desired area is indicated by the shaded region in Figure 6. Notice that it is composed of two regions: A_1, in which $f(x) < 0$ over the interval $[-1, 0)$; and A_2, in which $f(x) > 0$ over the interval $(0, \frac{1}{2}]$. To solve the problem, we use the additive property of area. Since $f(x) < 0$ for $-1 \leq x < 0$,

$$A_1 = -\int_{-1}^0 x^3\, dx = -\left.\frac{x^4}{4}\right|_{-1}^0 = \frac{1}{4}$$

For the area A_2, we have

$$A_2 = \int_0^{1/2} x^3\, dx = \left.\frac{x^4}{4}\right|_0^{1/2} = \frac{1}{64}$$

The total area A (since the regions do not overlap) is

$$A = A_1 + A_2 = \tfrac{1}{4} + \tfrac{1}{64} = \tfrac{17}{64}$$

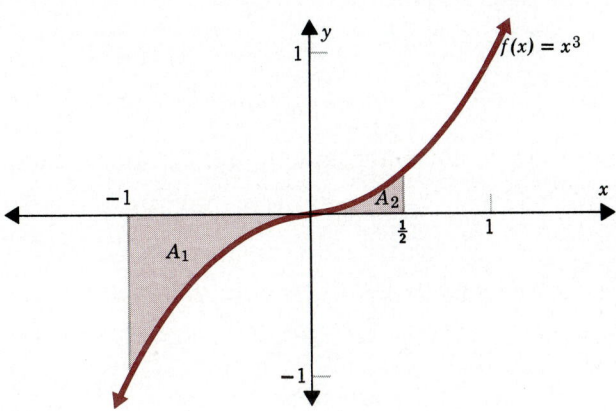

Figure 6

Example 2 illustrates the necessity of graphing the function before any attempt is made to compute area. In the subsequent examples we shall always graph the function before doing anything else.

Example 3 Find the area enclosed by $f(x) = x^2 - 4$ and the x-axis from $x = 0$ to $x = 4$.

Solution On the interval $[0, 4]$, the graph crosses the x-axis at $x = 2$ since $f(2) = 0$. Also, $f(x) < 0$ from $x = 0$ to $x = 2$, and $f(x) > 0$ from $x = 2$ to $x = 4$. The areas A_1 and A_2 as depicted in Figure 7 are

$$A_1 = -\int_0^2 (x^2 - 4)\, dx = -\left(\frac{x^3}{3} - 4x\right)\Big|_0^2 = -\left(\frac{8}{3} - 8\right) = \frac{16}{3}$$

$$A_2 = \int_2^4 (x^2 - 4)\, dx = \left(\frac{x^3}{3} - 4x\right)\Big|_2^4 = \left(\frac{64}{3} - 16\right) - \left(\frac{8}{3} - 8\right)$$

$$= \frac{56}{3} - 8 = \frac{32}{3}$$

The total area is

$$A = A_1 + A_2 = \tfrac{16}{3} + \tfrac{32}{3} = \tfrac{48}{3} = 16$$

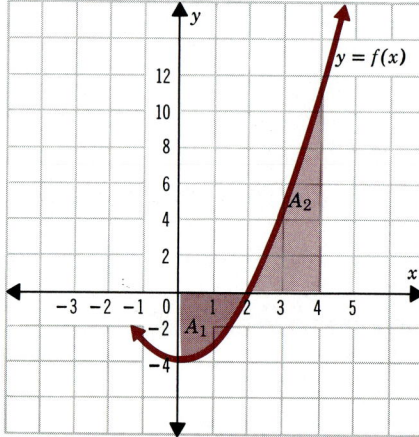

Figure 7

The next example illustrates how to find the area enclosed by the graphs of two functions.

Example 4 Find the area enclosed by the graphs of the functions

$$f(x) = 2x^2 \quad \text{and} \quad g(x) = 2x + 4$$

Solution First, we graph each of the functions, as shown in Figure 8.

2. APPLICATION IN GEOMETRY: AREA UNDER A GRAPH

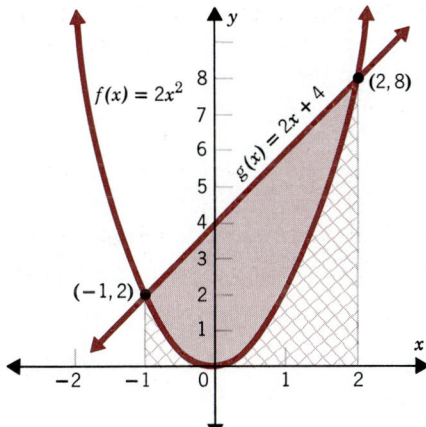

Figure 8

The area to be calculated (the shaded portion of Figure 8) lies under the graph of the line $g(x) = 2x + 4$ and above the graph of $f(x) = 2x^2$. To find this area, we first need to find the numbers x at which the graphs intersect, that is, all numbers x for which $2x^2 = 2x + 4$. The solutions of this equation are obtained as follows:

$$2x^2 - 2x - 4 = 0$$
$$x^2 - x - 2 = 0$$
$$(x + 1)(x - 2) = 0$$
$$x + 1 = 0 \qquad x - 2 = 0$$
$$x = -1 \qquad x = 2$$

Thus, the points of intersection of the two graphs are $(-1, 2)$ and $(2, 8)$, as shown in Figure 8.

From Figure 8 we can see that if we subtract the area under $f(x) = 2x^2$, between $x = -1$ and $x = 2$, from the area under $g(x) = 2x + 4$, between $x = -1$ and $x = 2$, we will have the area A we seek. Thus,

$$A = \int_{-1}^{2} g(x)\, dx - \int_{-1}^{2} f(x)\, dx = \int_{-1}^{2} [g(x) - f(x)]\, dx$$

$$= \int_{-1}^{2} [(2x + 4) - 2x^2]\, dx = \left(x^2 + 4x - \frac{2x^3}{3} \right) \bigg|_{-1}^{2}$$

$$= \left(4 + 8 - \frac{16}{3} \right) - \left(1 - 4 + \frac{2}{3} \right) = 9$$

The technique used in Example 4 can be used whenever we are asked to determine the area enclosed by the graphs of two continuous nonnegative functions $f(x)$ and $g(x)$ from $x = a$ to $x = b$.

Suppose, as depicted in Figure 9 on page 290, $f(x) \geq g(x)$ for x in $[a, b]$, and we wish to determine the area enclosed by $f(x)$, $g(x)$, $x = a$, and $x = b$. If we denote

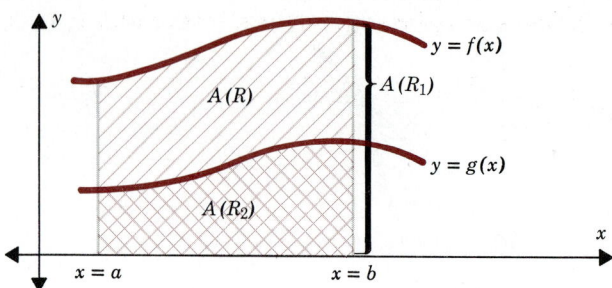

Figure 9

this area by $A(R)$, the area under $f(x)$ by $A(R_1)$, and the area under $g(x)$ by $A(R_2)$, then

$$A(R) = A(R_1) - A(R_2)$$
$$= \int_a^b f(x)\,dx - \int_a^b g(x)\,dx = \int_a^b [f(x) - g(x)]\,dx$$

The next example illustrates this formula.

Example 5 Find the area enclosed by the graphs of the functions

$$f(x) = 10x - x^2 \quad \text{and} \quad g(x) = 30 - 3x$$

Solution First, we graph the two functions. See Figure 10, where the points of intersection of the two graphs were obtained by finding all numbers x for which

$$g(x) = f(x)$$
$$30 - 3x = 10x - x^2$$
$$x^2 - 13x + 30 = 0$$
$$(x - 3)(x - 10) = 0$$

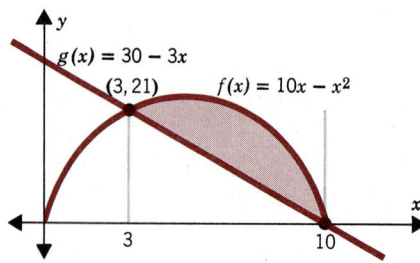

Figure 10

Thus, the points where the two curves meet are $(3, 21)$ and $(10, 0)$. We also see that for $3 \leq x \leq 10$,

$$f(x) \geq g(x)$$

2. APPLICATION IN GEOMETRY: AREA UNDER A GRAPH   291

Thus, the required area, indicated by the shaded portion in Figure 10, is

$$\int_3^{10} [(10x - x^2) - (30 - 3x)]\, dx = \int_3^{10} [-x^2 + 13x - 30]\, dx$$
$$= \left(\frac{-x^3}{3} + \frac{13x^2}{2} - 30x \right) \Big|_3^{10} = \frac{343}{6}$$

Remember, when computing area using the formula $\int_a^b [f(x) - g(x)]\, dx$, it must be true that $f(x) \geq g(x)$ on $[a, b]$. If this condition is not met, break the area up into pieces on which the inequality does hold and compute each one separately.

Justification of Formula (1)

Look at Figure 11. Choose a number c in I so that $c < a$. Suppose x in I is an arbitrary number for which $x > c$. Let $A(x)$ denote the area enclosed by $y = f(x)$ and the x-axis from c to x. See Figure 11. We want to show that $A'(x) = f(x)$ for all x in I, $x > c$.

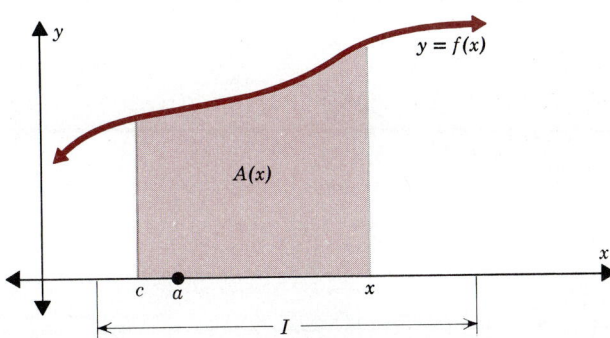

Figure 11

Now, choose $h > 0$ so that $x + h$ is in I. Then $A(x + h)$ is the area enclosed by $y = f(x)$ and the x-axis from c to $x + h$. See Figure 12 on page 292. The difference $A(x + h) - A(x)$ is just the area enclosed by $y = f(x)$ and the x-axis from x to $x + h$. See Figure 13 on page 292.

If we construct a rectangle with base equal to h and area equal to $A(x + h) - A(x)$, the height of the rectangle is then

$$\frac{A(x + h) - A(x)}{h}$$

since the above represents the area of the rectangle divided by its base h.

Next, we superimpose this rectangle on Figure 13 to obtain Figure 14.

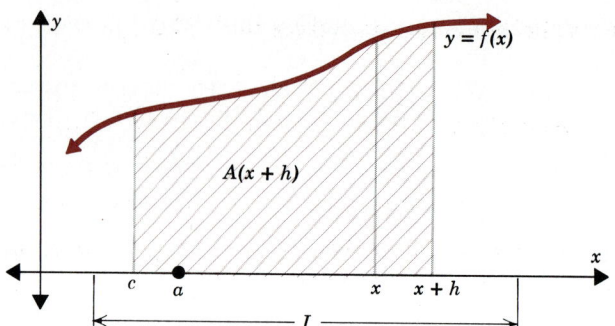

Figure 12

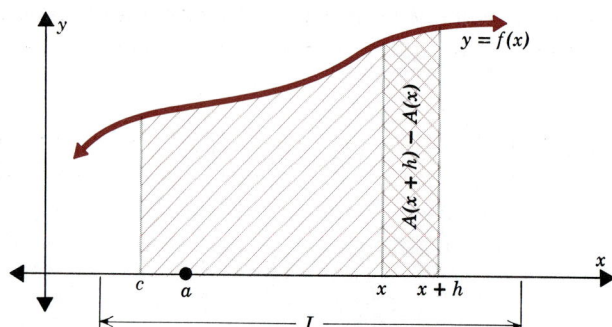

Figure 13

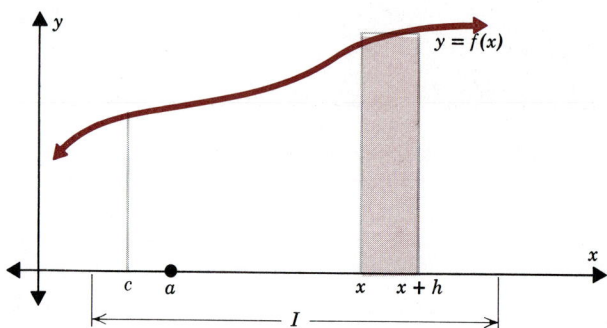

Figure 14

Since $y = f(x)$ is assumed to be a continuous function, and since both the rectangle and the shaded area have the same base and the same area, the upper edge of the rectangle must cross the graph of $y = f(x)$.

As we let $h \to 0^+$, the height of the rectangle tends to $f(x)$; that is,

$$\frac{A(x+h) - A(x)}{h} \to f(x) \quad \text{as} \quad h \to 0^+$$

A similar argument applies if we choose $h < 0$ and let $h \to 0^-$. Thus,

$$\lim_{h \to 0} \frac{A(x+h) - A(x)}{h} = f(x)$$

2. APPLICATION IN GEOMETRY: AREA UNDER A GRAPH

The limit on the left is the derivative of A. Thus,

$$A'(x) = f(x)$$

Since the choice of x is arbitrary (except for the condition that $x > c$), it follows that

$$A'(x) = f(x) \qquad \text{for all } x \text{ in } I, x > c$$

In other words, we have shown that the area $A(x)$ is an antiderivative of $f(x)$ on I. Hence,

$$\int_a^b f(x)\, dx = A(x)\Big|_a^b = A(b) - A(a)$$

But the area we want to find is the area enclosed by $y = f(x)$ and the x-axis from a to b. Since a and b are in I, this is the quantity $A(b) - A(a)$. See Figure 15. Hence,

Area enclosed by $y = f(x)$ and the x-axis from a to b is $\int_a^b f(x)\, dx$

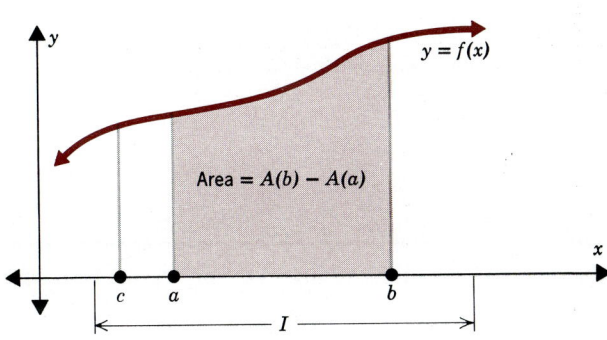

Figure 15

Exercise 2
Solutions to Odd-Numbered Problems begin on page 422.

In Problems 1–10 find the area described. Be sure to sketch the graph first.

1. Enclosed by $f(x) = 3x + 2$, the x-axis, and the lines $x = 2$ and $x = 6$
2. Enclosed by $f(x) = 3 - x$, the x-axis, and the lines $x = 0$ and $x = 3$
3. Enclosed by $f(x) = x^2$, the x-axis, and the lines $x = 0$ and $x = 2$
4. Enclosed by $f(x) = x^2$, the x-axis, and the lines $x = -2$ and $x = 1$
5. Enclosed by $f(x) = x^2 + 2$, the x-axis, and the lines $x = -2$ and $x = 1$
6. Enclosed by $f(x) = x^2 - 4$, the x-axis, and the lines $x = 2$ and $x = 4$
7. Enclosed by $f(x) = x$, the x-axis, and the lines $x = 1$ and $x = 2$
8. Enclosed by $f(x) = 1/x$, the x-axis, and the lines $x = 1$ and $x = 2$

9. Enclosed by $f(x) = e^x$, the x-axis, and the lines $x = 0$ and $x = 1$
10. Enclosed by $f(x) = x^3$, the x-axis, and the lines $x = 0$ and $x = 1$

In Problems 11-28 find the area enclosed by the graphs of the given functions and lines. Draw a sketch first.

11. $f(x) = x$, $g(x) = 2x$, $x = 0$, $x = 1$
12. $f(x) = x$, $g(x) = 3x$, $x = 0$, $x = 3$
13. $f(x) = x^2$, $g(x) = x$
14. $f(x) = x^2$, $g(x) = 4x$
15. $f(x) = x^2 + 1$, $g(x) = x + 1$
16. $f(x) = x^2 + 1$, $g(x) = 4x + 1$
17. $f(x) = \sqrt{x}$, $g(x) = x^3$
18. $f(x) = x^2$, $g(x) = x^3$
19. $f(x) = x^2$, $g(x) = x^4$
20. $f(x) = \sqrt{x}$, $g(x) = x^2$
21. $f(x) = x^2 - 4x$, $g(x) = -x^2$
22. $f(x) = x^2 - 8x$, $g(x) = -x^2$
23. $f(x) = 4 - x^2$, $g(x) = x + 2$
24. $f(x) = 2 + x - x^2$, $g(x) = -x - 1$
25. $f(x) = x^3$, $g(x) = 4x$
26. $f(x) = x^3$, $g(x) = 16x$
27. $y = x^2$, $y = x$, $y = -x$
28. $y = x^2 - 1$, $y = x - 1$, $y = -x - 1$

*29. Show that the shaded area in the figure is $\frac{2}{3}$ of the area of the parallelogram ABCD. (This illustrates a result due to Archimedes concerning sectors of parabolas.)

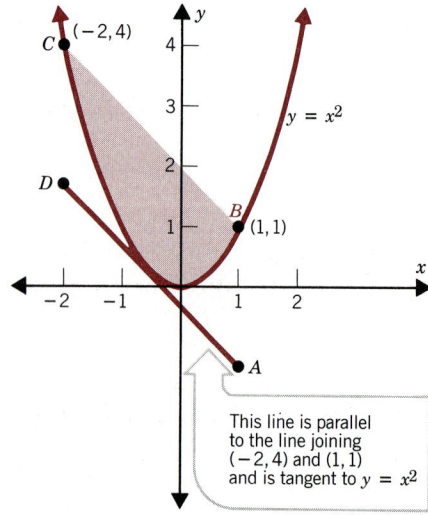

*30. **Mean Value Theorem for Integrals.** If $y = f(x)$ is continuous on an interval I, $a \leq x \leq b$, then there is a number c, $a < c < b$, so that

$$\int_a^b f(x)\,dx = f(c)(b-a)$$

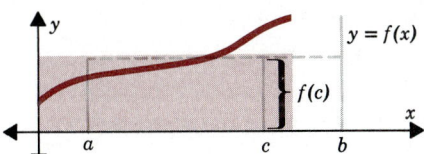

The interpretation of this result is that there is a rectangle with base $(b-a)$ and height $f(c)$, whose area is numerically equal to the area $\int_a^b f(x)\,dx$. See the illustration. Verify this result by finding c for the functions below. Graph each function.

(a) $f(x) = x^2$, $a = 0$, $b = 1$
(b) $f(x) = 1/x^2$, $a = 1$, $b = 4$

*31. If $y = f(x)$ is continuous on the interval I and if it has an antiderivative on I, then for some a in I,

$$\frac{d}{dx}\int_a^x f(t)\,dt = f(x) \qquad \text{for } x > a \text{ in } I$$

This result gives us a technique for finding the derivative of a definite integral in which the lower limit is fixed and the upper limit is variable. Use this result to find:

(a) $\dfrac{d}{dx}\displaystyle\int_1^x t^2\,dt$ (b) $\dfrac{d}{dx}\displaystyle\int_2^x \sqrt{t^2 - 2}\,dt$

(c) $\dfrac{d}{dx}\displaystyle\int_5^x \sqrt{t^2 + 2t}\,dt$

3. Applications in Business
Variable Cost

The determination of the variable cost of producing a consecutive number of units is important to manufacturers. The cost can be computed once the marginal cost MC is known as a function of the quantity x of units produced.

The *variable cost of producing a through b units* **is given by the formula**

(1) $$VC = \int_{a-1}^b MC(x)\,dx$$

Example 1 The marginal cost for a certain commodity is

$$MC(x) = 2x^2 - 3x + 2$$

Determine the variable cost of producing 12 through 16 units.

Solution Here, $a - 1 = 12 - 1 = 11$, $b = 16$, and

$$VC = \int_{11}^{16} (2x^2 - 3x + 2) \, dx = \left(2\frac{x^3}{3} - \frac{3x^2}{2} + 2x\right)\bigg|_{11}^{16}$$
$$= (2730.67 - 384 + 32) - (887.33 - 181.5 + 22) = 1650.84$$

Thus, the variable cost of producing 12 through 16 units is $1650.84.

Depletion

Natural resources such as oil, gas, coal, and so on, are in limited quantity, and their total depletion depends on the rate at which each resource is being consumed. Suppose A_0 represents the amount of a natural resource used at time $t = 0$ and suppose the annual rate of depletion for this natural resource is k percent.

If we assume that the rate of depletion compounds continuously,* the amount A used after time T (in years) is given by

(2)
$$A = \int_0^T A_0 e^{kt} \, dt$$

Example 2
Depletion of Natural Resources

In 1976 ($t = 0$) the world use of oil was 21 billion barrels.† Suppose the annual rate of depletion of oil is 8%. Assuming this rate stays the same in the future:
(a) How many barrels of oil will the world use from 1976 to 1984?
(b) How long will it take to use all the available oil, if it is known that in 1976 there were 550 billion barrels† of proven reserves? (Assume that no new oil is discovered.)

Solution (a) Using equation (2) with $T = 8$, $k = 0.08$, and $A_0 = 21$ (billion barrels), we find the amount A used between 1976 and 1984:

$$A = \int_0^8 21 e^{0.08t} \, dt = \frac{21}{0.08} e^{0.08t} \bigg|_0^8 = \frac{21}{0.08} (e^{0.08(8)} - 1)$$
$$= 262.5(1.8965 - 1) = 262.5(0.8965) = 235.33$$

Thus, from 1976 to 1984, the world will use 235.33 billion barrels.
(b) Here, we need to know how long it will take to use 550 billion barrels of oil.

---
*Consult Section 2 of Chapter 5 for a discussion of continuous compounding.
†*Source:* Bureau of Mines, United States Department of the Interior.

Thus, we use equation (2) to determine T with $A = 550$, $A_0 = 21$, and $k = 0.08$:

$$550 = \int_0^T 21e^{0.08t}\,dt = \frac{21}{0.08}e^{0.08t}\Big|_0^T = \frac{21}{0.08}(e^{0.08T} - 1)$$

To solve for T, we proceed as follows:

$$550 = \frac{21}{0.08}(e^{0.08T} - 1)$$

$$\frac{(550)(0.08)}{21} + 1 = e^{0.08T}$$

$$3.095 = e^{0.08T}$$

so that

$$\ln 3.095 = 0.08T$$
$$T = 14.12$$

Thus, 14.12 years from 1976, or by 1990, the world reserve of oil will be depleted (assuming that no new oil is discovered). ■

Rates

When the rate of sales of a product is a known function of x, say $f(x)$, where x is a time measure, the total sales of this product over a time period T is

(3) $$\text{Total sales over time } T = \int_0^T f(x)\,dx$$

For example, suppose the rate of sales of a new product is given by

$$f(x) = 100 - 90e^{-x}$$

where x is the number of days the product is on the market. The total sales during the first 4 days is

$$\int_0^4 f(x)\,dx = \int_0^4 (100 - 90e^{-x})\,dx = (100x + 90e^{-x})\Big|_0^4$$
$$= 400 + 90e^{-4} - 90 = 310 + 90e^{-4}$$
$$= 310 + 90(0.018) = 311.62 \text{ units}$$

Example 3
Rate of Sales A company has current sales of $1,000,000 per month, and profit to the company averages 10% of sales. The company's past experience with a certain advertising strategy is that sales will increase by 2% per month over the length of the advertising campaign (12 months). The company now needs to decide whether to embark on a similar campaign that will have a total cost of $130,000. The decision will be Yes, provided the increase in sales due to the campaign results in profits that exceed $13,000. (This is a 10% return on the advertising investment.)

Solution The monthly rate of sales during the advertising campaign obeys a growth curve of the form

$$\$1{,}000{,}000 e^{0.02t}$$

where t is measured in months. The total sales after 12 months (the length of the campaign) is

$$\text{Total sales} = \int_0^{12} 1{,}000{,}000 e^{0.02t}\, dt = \left.\frac{1{,}000{,}000 e^{0.02t}}{0.02}\right|_0^{12}$$
$$= 50{,}000{,}000(e^{0.24} - 1) = 50{,}000{,}000(0.271)$$
$$= \$13{,}550{,}000 \quad \text{(approximately)}$$

The profit to the company is 10% of sales so that the profit due to the increase in sales is

$$0.10(13{,}550{,}000 - 12{,}000{,}000) = \$155{,}000$$

This $155,000 profit was achieved through the expenditure of $130,000 in advertising. Thus, the advertising yielded a true profit of

$$\$155{,}000 - \$130{,}000 = \$25{,}000$$

Since this represents more than a 10% return on the cost of the advertising, the company should proceed with the advertising campaign. ■

Annuity

An *annuity* is a sequence of equal periodic payments. For example, a 20 payment life insurance policy is an annuity in which 20 equal premiums, or payments, are paid on an annual basis earning a fixed interest rate. In this section we discuss the situation in which the interest earned is compounded continuously.

Amount of an Annuity The *amount of an annuity* is the sum of all payments made, plus all interest accumulated.

If an annuity consists of equal annual payments P in which an interest rate of r% per annum is compounded continuously, the *amount A of the annuity* after N payments is

(4) $$A = \int_0^N P e^{rt}\, dt$$

Example 4 A savings and loan association pays 6% per annum compounded continuously. If a person places $1000 in a savings account each year, how much will be in the account after 3 years?

Solution Here, $P = 1000$, $N = 3$, and $r = 0.06$. The amount A after 3 years is

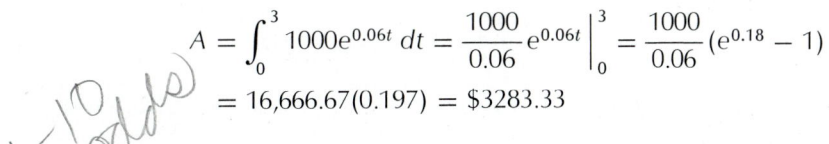

$$= 16{,}666.67(0.197) = \$3283.33$$

Exercise 3
Solutions to Odd-Numbered Problems begin on page 425.

1. **Variable Cost.** The marginal cost of producing a certain kind of light bulb is found to be $MC = e^{-0.01x}$. What is the variable cost of producing 81 to 100 units? What is the variable cost of producing 801 to 1000 units?

2. **Rate of Sales.** The rate of sales of a new product is given by
$$f(x) = 1200 - 950e^{-x}$$
where x is the number of months the product is on the market. Find the total sales during the first year.

3. **Rate of Sales.** A company whose annual sales are currently $300,000 has been experiencing sales increases of 10% per year. Assuming this rate of growth continues, what will annual sales be in 4 years?

4. **Rate of Sales.** In Example 3, what decision should the company make if sales due to advertising increase by only 1.5% per month?

5. **Depletion of Natural Gas.** In 1976 ($t = 0$) the annual world use of natural gas was 50 trillion cubic feet.* The annual rate of depletion of gas is 3%. Assuming this rate stays the same in the future:
   (a) How many cubic feet will the world use from 1976 to 1986?
   (b) How long will it take to use all the available gas, if it is known that in 1976 there were 2200 trillion cubic feet* of proven reserves? (Assume that no new discoveries are made.)

6. **Depletion of Coal.** In 1976 ($t = 0$) the annual world use of coal was 4.2 billion short tons.* (Each short ton is 2000 pounds.) The annual rate of depletion of coal is 10%. Assuming this rate stays the same in the future:
   (a) How many short tons will the world use from 1976–1986?
   (b) How long will it take to use all the available coal, if it is known that in 1976 there were 660 billion short tons* of proven reserves?

7. **Annuity.** If $500 is deposited each year in a savings account paying 5.5% per annum compounded continuously, how much is in the account after 4 years?

8. **Annuity.** If $1200 is deposited each year in a savings account paying 5% per annum compounded continuously, how much is in the account after 3 years?

*Source: Bureau of Mines, United States Department of the Interior.

9. *Annuity.* How much needs to be saved each year in a savings account paying 6% per annum compounded continuously in order to accumulate $6000 in 3 years? [*Hint:* Use (4) to find the payment P required.]

10. *Annuity.* Answer Problem 9 if the rate of interest is 8%.

4. The Definite Integral as the Limit of a Sum

In this section we use the fact that the definite integral $\int_a^b f(x)\,dx$ equals the area under the graph of a continuous nonnegative function f from a to b to show that the definite integral can be written as the limit of a certain sum.

We begin by considering Figure 16. The area A enclosed by $y = f(x)$, the x-axis, $x = a$, and $x = b$, is between the area of the rectangle r_1 (whose height is the minimum of f on $[a, b]$) shown on the left and the area of the rectangle R_1 (whose height is the maximum of f on $[a, b]$) shown on the right. That is,

$$\text{Area } r_1 < A < \text{Area } R_1$$

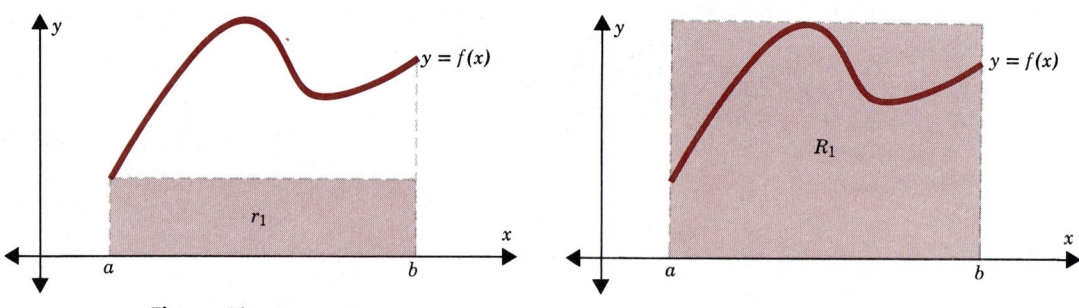

Figure 16

But these are not very good approximations to the area A. A better approximation can be obtained by selecting a number between a and b, say x_1, and forming additional rectangular strips. See Figure 17. Under these conditions, we see that

$$\text{Area } r_1 + \text{Area } r_2 < A < \text{Area } R_1 + \text{Area } R_2$$

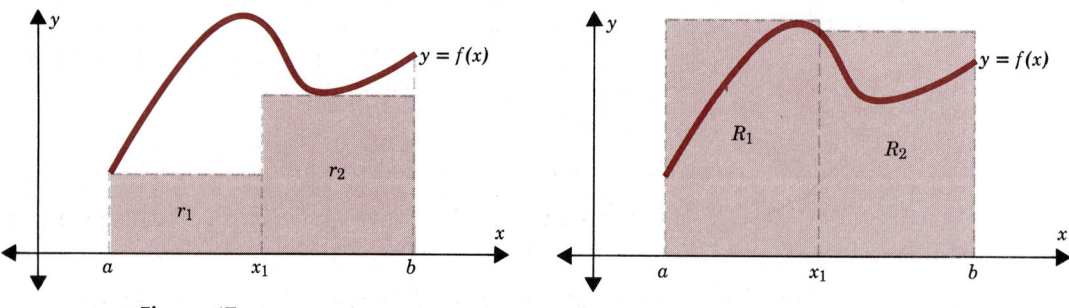

Figure 17

4. THE DEFINITE INTEGRAL AS THE LIMIT OF A SUM 301

Suppose we select two numbers between a and b, say x_1 and x_2, and form the corresponding rectangles. See Figure 18. Under these conditions, we have

$$\text{Area } r_1 + \text{Area } r_2 + \text{Area } r_3 < A < \text{Area } R_1 + \text{Area } R_2 + \text{Area } R_3$$

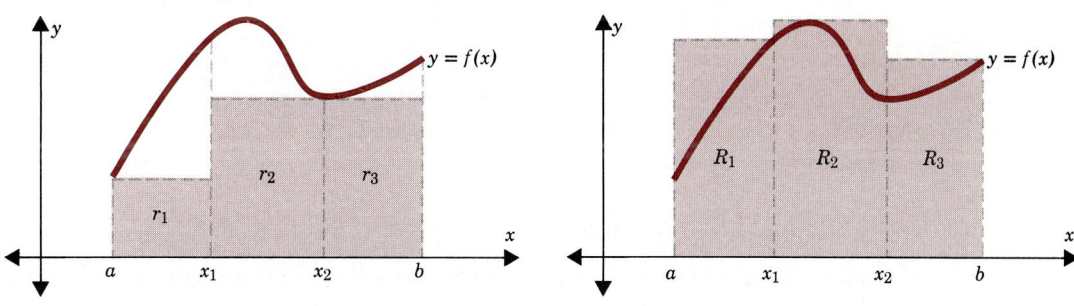

Figure 18

As we move from Figure 16 to Figure 18, we see that the more numbers we select between a and b, the better is our approximation for A.

Suppose we choose $n - 1$ numbers between a and b, say $x_1, x_2, \ldots, x_{n-1}$, and form the corresponding rectangles. See Figure 19. Then

$$\text{Area } r_1 + \text{Area } r_2 + \cdots + \text{Area } r_n < A < \text{Area } R_1 + \text{Area } R_2 + \cdots + \text{Area } R_n$$

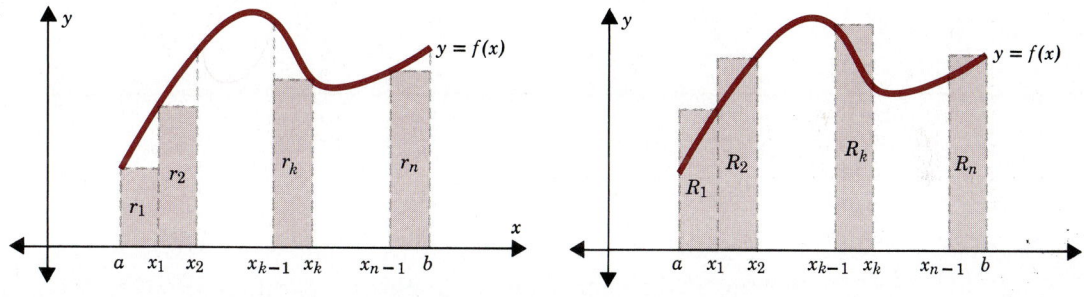

Figure 19

The area of each rectangle is the product of its height by its base. Compare the areas r_1 and R_1. Their bases are the same $(x_1 - a)$ and the height of R_1 is larger than the height of r_1. The area enclosed by $y = f(x)$, the x-axis, $x = a$, and $x = x_1$ may be approximated by a rectangle with base $(x_1 - a)$ and height $f(X_1^*)$, where X_1^* is some number between a and x_1. Repeat this argument for each pair of rectangles r_1, R_1; r_2, R_2; r_3, R_3; etc. See Figure 20 on page 302.

The area A, then, is approximated by

$$A \approx f(X_1^*)(x_1 - a) + f(X_2^*)(x_2 - x_1) + \cdots + f(X_n^*)(b - x_{n-1})$$

To obtain the actual area A, we let the numbers chosen between a and b get arbitrarily large; that is, we let $n \to +\infty$. Then

$$\text{Area } A = \lim_{n \to +\infty} [f(X_1^*)(x_1 - a) + f(X_2^*)(x_2 - x_1) + \cdots + f(X_n^*)(b - x_{n-1})]$$

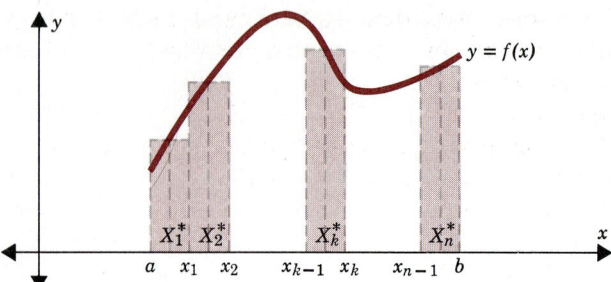

Figure 20

This limit, if it exists, is the area A enclosed by $y = f(x)$, the x-axis, $x = a$, and $x = b$. It is also the definite integral from a to b of $f(x)$. Thus,

$$\int_a^b f(x)\, dx = \lim_{n \to +\infty} [f(X_1^*)(x_1 - a) + f(X_2^*)(x_2 - x_1) + \cdots + f(X_n^*)(b - x_{n-1})]$$

Let's review this procedure. We consider a continuous function $y = f(x)$ with domain $a \leq x \leq b$. See Figure 21.

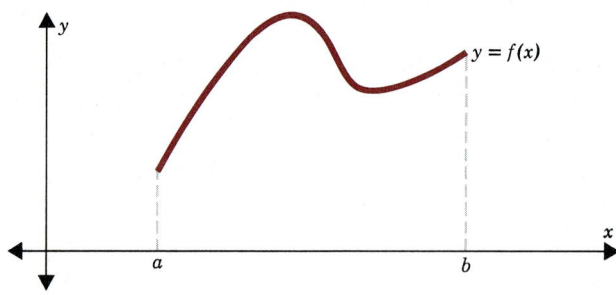

Figure 21

Concentrate on the interval $[a, b]$. Suppose we partition, or divide, this interval into n subintervals. The point $a = x_0$ is the initial point, the first point of the subdivision is x_1, the second is $x_2, \ldots,$ and the nth point is $b = x_n$. See Figure 22.

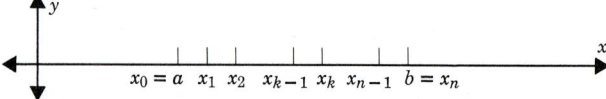

Figure 22

The original interval $[a, b]$ now consists of n subintervals, and the length of each one is

First,   Second,   Third, ...,
$$\Delta x_1 = x_1 - x_0, \qquad \Delta x_2 = x_2 - x_1, \qquad \Delta x_3 = x_3 - x_2, \ldots,$$

kth, ...,   nth
$$\Delta x_k = x_k - x_{k-1}, \ldots, \qquad \Delta x_n = x_n - x_{n-1}$$

4. THE DEFINITE INTEGRAL AS THE LIMIT OF A SUM

Norm

We use the symbol Δ to denote the largest such length. (The value of Δ, of course, depends on how the partition itself has been chosen.) We call Δ the *norm of the partition*.

Next, we concentrate on the function. Pick a number in each subinterval (you may select a number in the interval or either endpoint, if you wish) and evaluate the function at this number. To fix our ideas, let X_k^* denote the chosen number. The corresponding value of the function is $f(X_k^*)$. This represents the height of the function at X_k^*.

Multiply $f(X_1^*)$ times $\Delta x_1 = x_1 - x_0$, $f(X_2^*)$ times $\Delta x_2 = x_2 - x_1, \ldots, f(X_n^*)$ times $\Delta x_n = x_n - x_{n-1}$, and add up these products. The result is the sum

$$f(X_1^*)\Delta x_1 + f(X_2^*)\Delta x_2 + \cdots + f(X_n^*)\Delta x_n$$

Riemann Sum

This is called a *Riemann sum* for the function f on $[a, b]$.

Finally, we take the limit of this sum as the norm $\Delta \to 0$. If this limit exists, it is the definite integral of $f(x)$ from a to b,† that is

(1) $$\int_a^b f(x)\, dx = \lim_{\Delta \to 0}[f(X_1^*)\Delta x_1 + f(X_2^*)\Delta x_2 + \cdots + f(X_n^*)\Delta x_n]$$

Example 1 Compute, by using Riemann sums: $\int_0^1 x\, dx$

Solution Choose a partition that divides the interval $[0, 1]$ into subintervals of equal length. This requires that each subinterval be of length $1/n$. The partition is

$$x_0 = 0, \quad x_1 = \frac{1}{n}, \quad x_2 = \frac{2}{n}, \quad \ldots, \quad x_k = \frac{k}{n}, \quad \ldots, \quad x_n = \frac{n}{n} = 1$$

See Figure 23.

Figure 23

The norm of this partition is $\Delta = 1/n$.

Next, we select a number X_k^* in each subinterval. Suppose we agree to select the right endpoint of each subinterval. Then,

$$X_1^* = \frac{1}{n}, \quad X_2^* = \frac{2}{n}, \quad \ldots, \quad X_n^* = 1$$

The Riemann sum for $f(x) = x$ is

$$f(X_1^*)\Delta x_1 + f(X_2^*)\Delta x_2 + \cdots + f(X_n^*)\Delta x_n = \frac{1}{n}\cdot\frac{1}{n} + \frac{2}{n}\cdot\frac{1}{n} + \cdots + \frac{n}{n}\cdot\frac{1}{n}$$

$$= \frac{(1 + 2 + \cdots + n)}{n}\cdot\frac{1}{n}$$

†It can be shown that this approach to the definite integral will lead to the result $\int_a^b f(x)\, dx = F(b) - F(a)$, if F is an antiderivative of f.

It can be shown that

$$1 + 2 + \cdots + n = \frac{n(n + 1)}{2}$$

Thus, the Riemann sum reduces to

$$\frac{n(n + 1)}{2n^2} = \frac{n^2 + n}{2n^2}$$

If we compute the limit of this quantity as $\Delta \to 0$, observing that since $\Delta = 1/n$, $\Delta \to 0$ and $n \to +\infty$ mean the same thing, we have

$$\int_0^1 x \, dx = \lim_{\Delta \to 0} [f(X_1^*) \Delta x_1 + f(X_2^*) \Delta x_2 + \cdots + f(X_n^*) \Delta x_n]$$

$$= \lim_{n \to +\infty} \frac{n^2 + n}{2n^2} = \lim_{n \to +\infty} \frac{1 + 1/n}{2} = \frac{1}{2}$$

■

Since $f(x) = \frac{1}{2}x^2$ is an antiderivative of x, we can check the answer obtained in Example 1 by applying the fundamental formula of integral calculus, to get

$$\int_0^1 x \, dx = \frac{1}{2}x^2 \Big|_0^1 = \frac{1}{2}.$$

The method of evaluating a definite integral by finding the limit of a sum is certainly long and tedious when compared to using the fundamental formula of integral calculus. However, it does have advantages. First, it provides a way to evaluate definite integrals for which no antiderivative can be found. Second, many applications lead to limits of sums and therefore it is important to know that such expressions are actually definite integrals. (See the next section for an application.)

Since the evaluation of Riemann sums requires knowing some sums, we list common sums below. These will be needed when you work the problems in Exercise 4.

1. $1 + 2 + 3 + \cdots + n = \dfrac{n(n + 1)}{2}$

2. $1^2 + 2^2 + 3^2 + \cdots + n^2 = \dfrac{n(n + 1)(2n + 1)}{6}$

3. $1^3 + 2^3 + 3^3 + \cdots + n^3 = \dfrac{n^2(n + 1)^2}{4}$

Exercise 4
Solutions to Odd-Numbered Problems begin on page 425.

In Problems 1–4 evaluate each integral by the technique of Riemann sums, using a partition that divides the interval into n equal parts and choosing the point X^* as the right endpoint of each subinterval. Check your answer by using the fundamental formula of integral calculus.

1. $\int_0^1 3\,dx$
2. $\int_0^2 x\,dx$
3. $\int_0^1 x^2\,dx$
4. $\int_0^1 x^3\,dx$

5. Find an approximate value of $\int_1^2 dx/x$ by computing the Riemann sums corresponding to a partition of $[1, 2]$ into four subintervals of equal length and evaluating the integrand at the midpoint of each subinterval. Compare your answer with the true value, which is $0.6931\ldots$.

6. If the interval $[1, 5]$ is divided into eight subintervals of equal length, what is the largest Riemann sum of $f(x) = x^2$ that can be computed using this partition? The smallest? Compute the average of these sums. What integral has been approximated and what is its exact value?

5. Average Value of a Function

In this section we examine how the definite integral can be applied to calculate averages. For example, at the United States Weather Bureau, a continuous reading of the temperature over a 24 hour period is taken daily. To obtain the average daily temperature, twelve readings may be taken at 2 hour intervals beginning at midnight: $f(0), f(2), f(4), \ldots, f(20), f(22)$. The average temperature is then calculated as

$$\frac{f(0) + f(2) + f(4) + \cdots + f(20) + f(22)}{12}$$

This number represents a good approximation to the true average as long as there were no drastic temperature changes over short periods of time. To improve the approximation, readings may be taken every hour. The average in this case would be

$$\frac{f(0) + f(1) + \cdots + f(22) + f(23)}{24}$$

An even better approximation would be obtained if readings were recorded every half hour.

In general, if $y = f(x)$ is a continuous function defined on the interval $[a, b]$, we can obtain the *average of f on* $[a, b]$ as follows: Partition the closed interval $[a, b]$ into n subintervals

$$[a, x_1], [x_1, x_2], \ldots, [x_{k-1}, x_k], \ldots, [x_{n-1}, b]$$

each of length $\Delta x = (b - a)/n$. This is the norm Δ of the partition. Pick a number in each subinterval and let these numbers be $X_1^*, X_2^*, \ldots, X_n^*$. An approximation of the average value of f over the interval $[a, b]$ is the sum

(1) $$\frac{f(X_1^*) + f(X_2^*) + \cdots + f(X_n^*)}{n}$$

If we multiply and divide the expression in (1) by $b - a$, we get

$$\frac{f(X_1^*) + f(X_2^*) + \cdots + f(X_n^*)}{n} = \frac{1}{b-a}\left[f(X_1^*)\frac{b-a}{n} + f(X_2^*)\frac{b-a}{n} + \cdots + f(X_n^*)\frac{b-a}{n}\right]$$

$$= \frac{1}{b-a}[f(X_1^*)\Delta x + f(X_2^*)\Delta x + \cdots + f(X_n^*)\Delta x]$$

The sum obtained gives an approximation to the average value. As the norm $\Delta \to 0$, this sum is a better and better approximation to the average value of f on $[a, b]$. However, this sum is a Riemann sum, so that its limit is a definite integral. This suggests the following definition:

Average Value of a Function over an Interval The average value AV of a continuous function f over $[a, b]$ is

(2) $$AV = \frac{1}{b-a}\int_a^b f(x)\,dx$$

Example 1 The average value of $f(x) = x^3$ over the interval $[0, 2]$ is

$$AV = \frac{\int_0^2 x^3\,dx}{2-0} = \frac{4}{2} = 2$$

■

The average value AV of a function f, as defined in (2), has an interesting geometric interpretation. If we rearrange the formula for AV, we obtain

(3) $$(AV)(b - a) = \int_a^b f(x)\,dx$$

If $f(x) \geq 0$ on $[a, b]$, the right side of (3) represents the area enclosed by the graph of $y = f(x)$, the x-axis, the line $x = a$, and the line $x = b$. The left side of the equation can be interpreted as the area of a rectangle of height AV and base $b - a$. Hence (3) asserts that the average value of the function is the height of a rectangle with base $b - a$ and area equal to the area under the graph of f. See Figure 24.

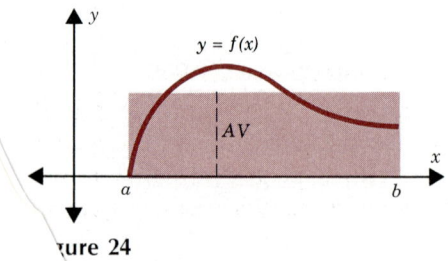

Figure 24

5. AVERAGE VALUE OF A FUNCTION

Example 2 Suppose the current world population is 5×10^9 and the population in t years is assumed to grow exponentially according to the law

$$P(t) = (5 \times 10^9) e^{0.02t}$$

What is the average world population during the next 20 years? (This number is helpful for long-range planning of agricultural and industrial output.)

Solution The average value of the population function $P(t)$ from $t = 0$ to $t = 20$ is

$$AV = \frac{1}{20 - 0} \int_0^{20} P(t) \, dt = \frac{1}{20} \int_0^{20} (5 \times 10^9) e^{0.02t} \, dt$$

$$= \frac{5 \times 10^9}{20} \int_0^{20} e^{0.02t} \, dt = \left(\frac{10^9}{4}\right) \frac{e^{0.02t}}{0.02} \bigg|_0^{20}$$

$$= \left(\frac{10^9}{0.08}\right)(e^{0.4} - 1) \approx (6.15 \times 10^9)$$

Exercise 5

Solutions to Odd-Numbered Problems begin on page 426.

In Problems 1–10 find the average value of each function f over the given interval.

1. $f(x) = x^2$, over $[0, 1]$
2. $f(x) = 2x^2$, over $[4, 2]$
3. $f(x) = 1 - x^2$, over $[-1, 1]$
4. $f(x) = 16 - x^2$, over $[-4, 4]$
5. $f(x) = 3x$, over $[1, 5]$
6. $f(x) = 4x$, over $[-5, 5]$
7. $f(x) = -5x^4 + 4x - 10$, over $[-2, 2]$
8. $f(x) = 10x^4 - 2x + 7$, over $[-1, 2]$
9. $f(x) = e^x$, over $[0, 1]$
10. $f(x) = e^{-x}$, over $[0, 1]$

11. Rework Example 2 if the population function is given by $P(t) = (5 \times 10^9) e^{0.03t}$. (This is a 3% growth rate.)
12. Rework Example 2 if the growth rate is 1%.
13. A rod 3 meters long is heated to $25x$ degrees Celsius, where x is the distance (in meters) from one end of the rod. Calculate the average temperature of the rod.
14. The rainfall per day, measured in centimeters, x days after the beginning of the year, is $0.00002(6511 + 366x - x^2)$. By integration, estimate the average daily rainfall for the first 180 days of the year.
15. A car starting from rest accelerates at the rate of 3 meters per second per second. Find its average speed over the first 8 seconds.
16. What is the average area of all circles with radii between 1 and 3 meters?

6. Application to Probability

Suppose an experiment involves weighing individuals. The weight of each individual is treated as a continuous variable. Likewise, in the experiment of determining the time between arrival of customers at a gas station, the time between arrivals is treated as a continuous variable.

Suppose we want to compute the probability* that a person selected at random in the United States will be between 22 and 24 years old, if the only known information about the distribution of the population by ages is given by data grouped in 10 year intervals. See Figure 25.

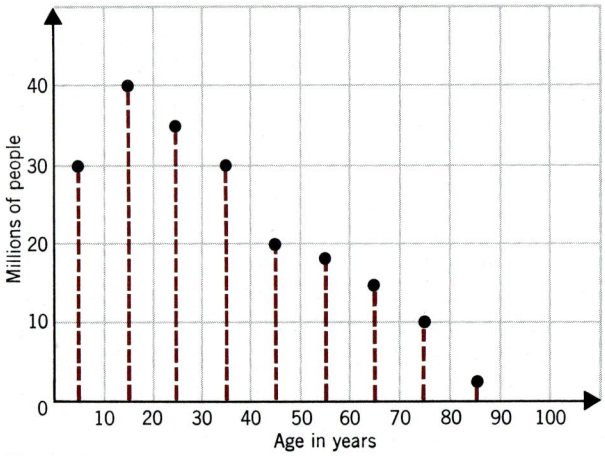

Figure 25

In this illustration there are 30 million people in the age group 0–10, 40 million between the ages of 10 and 20, and so on. We also know that the total population is 200 million.

Thus, since there are 40 million people in the age group 10–20, the probability that a person is in this age group is $\frac{40}{200} = .20$. Figure 26 illustrates the distribution of probabilities for each age group.

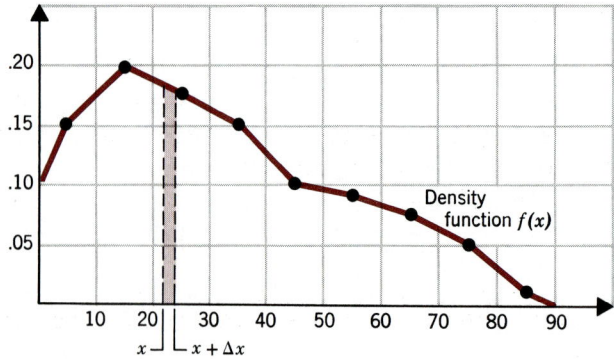

Figure 26

---

*For our purposes, it is sufficient to think of the probability of an event as a number from 0 to 1 that equals its chance of occurring. Thus, a probability represents the likelihood of an event happening.

Density Function

The function constructed by connecting the probability values by a smooth curve is called a *density function*.

When probabilities are associated with intervals, it is reasonable to assume that their values depend not only on the lengths of the intervals, but also on their locations. For instance, there is no reason why the probability that a person is in the age group 10–20 should equal the probability that a person is in the age group 60–70, even though the two intervals have the same length. If we assume that there exists a function with the values $f(x)$, then the probability that a person will be in a certain age group on a small interval from x to $x + \Delta x$ is approximately $f(x)\Delta x$, and the size of Δx determines how good the approximation is. This approximation is given by the area of the shaded rectangle shown in Figure 26, where the height of the rectangle is $f(x)$ and the base is Δx.

In a similar manner, we obtain the probabilities of other age groups by computing the areas corresponding to different subintervals. The desired probability for the whole interval is approximately the sum of the areas

$$f(x_1)\Delta x + f(x_2)\Delta x + \cdots + f(x_n)\Delta x$$

for $x_1, x_2, \ldots, x_n$ on the whole interval.

Using the fact that the exact area under the graph of $f(x)$ is given by a definite integral, the exact probability that a person is between the ages c and d, denoted by $P(c \leq X \leq d)$, is given by

$$P(c \leq X \leq d) = \int_c^d f(x)\,dx$$

Thus, for the density function illustrated in Figure 26, the probability that a person is between 22 and 24 years of age is

$$\int_{22}^{24} f(x)\,dx$$

Probability Density

The density function in Figure 26 is called the *probability density*. In this case, it is the probability per unit age.

The argument we have presented here leads to the following definition of probability:

The probability P that the outcome of an experiment results in a value of a variable X between c and d is given by

(1) $$P(c \leq X \leq d) = \int_c^d f(x)\,dx$$

where $f(x)$ denotes the values of an appropriate function called a *probability density*.

Thus, we have reduced the problem of finding probabilities to that of finding the area enclosed by the graph of a probability density function and the x-axis on the interval from a to b. When probability density functions are integrated between any two values, they yield probabilities, so they possess the properties given in the following restricted definition*:

Density Function A function $y = f(x)$ is termed a *density function* if it has two properties:

(I) $\quad \int_a^b f(x)\, dx = 1$

where [a, b] is an interval containing all values that the variable X can assume,

(II) $\quad f(x) \geq 0$

With (1) in mind, the rationale behind condition (I) becomes apparent. Since the interval [a, b] contains all the values the variable can assume, the probability that a variable lies between a and b must equal 1.

Example 1 Show that the function $f(x) = \frac{3}{56}(5x - x^2)$ is a density function over the interval [0, 4].

Solution If $f(x)$ is indeed a density function, it has to satisfy conditions (I) and (II). To verify condition (I), we evaluate

$$\int_0^4 \frac{3}{56}(5x - x^2)\, dx = \frac{3}{56}\left(\frac{5x^2}{2} - \frac{x^3}{3}\right)\Big|_0^4$$

$$= \frac{3}{56}\left[\frac{(5)16}{2} - \frac{64}{3}\right] = \frac{3}{56}\left(\frac{56}{3}\right) = 1$$

Hence, condition (I) is satisfied.
Condition (II) is also satisfied, since

$$f(x) = \frac{3}{56}(5x - x^2) = \frac{3}{56}x(5 - x) \geq 0$$

for all x in the interval [0, 4].
Thus, $f(x) = \frac{3}{56}(5x - x^2)$, for x in $0 \leq x \leq 4$, is a density function. ∎

Example 2 Compute the probability that the variable X with density function $\frac{3}{56}(5x - x^2)$ assumes values between 1 and 2.

Solution To compute $P(1 \leq X \leq 2)$, we use (1):

$$P(1 \leq X \leq 2) = \int_1^2 \frac{3}{56}(5x - x^2)\, dx = \frac{3}{56}\left(\frac{5x^2}{2} - \frac{x^3}{3}\right)\Big|_1^2$$

$$= \frac{3}{56}\left[\frac{(5)(4)}{2} - \frac{8}{3}\right] - \frac{3}{56}\left(\frac{5}{2} - \frac{1}{3}\right) = \frac{31}{112}$$ ∎

*Our definition is restricted, since we are considering a probability density function defined on a finite interval [a, b], rather than one defined over the interval $(-\infty, +\infty)$. However, the more general definition requires improper integrals, a topic that is not taken up in this book.

6. APPLICATION TO PROBABILITY

How do we obtain the density function $f(x)$? For individual random experiments, it is possible to construct them as we indicated in the example on age probabilities. However, the construction of a density function is usually a tedious and difficult task and often depends on the nature of the problem. Fortunately, several relatively simple density functions are available that can be used to fit most experiments. In every example we discuss, the probability density function is given.

Uniform Density Function One such function is the *uniform density function,* or the *uniform distribution.* This function is the simplest of probability functions and is one in which the variable assumes all its values with equal probability. The density function for this variable is

$$f(x) = \begin{cases} \dfrac{1}{b-a} & \text{if } a \leq x \leq b \\ 0 & \text{if } x < a \text{ or } x > b \end{cases}$$

The graph of this function is given in Figure 27.

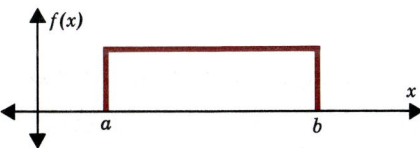

Figure 27

Notice that the function has the value zero outside the interval $a \leq x \leq b$.

To verify that the uniform density function is a density function, we need to show that conditions (I) and (II) are satisfied.

(I) The definite integral of $f(x)$ with limits of integration a and b is equal to 1. That is,

$$\int_a^b \frac{1}{b-a}\,dx = \frac{x}{b-a}\bigg|_a^b = \frac{b-a}{b-a} = 1$$

(II) Since $b > a$, $f(x) = 1/(b-a) > 0$, for $a \leq x \leq b$.

Example 3 Trains leave a terminal every 40 minutes. What is the probability that a passenger arriving at a random time to catch a train will have to wait more than 10 minutes?

Solution Let T (time) be a variable and assume it is uniformly distributed for $0 \leq T \leq 40$. The probability that the passenger must wait at least 10 minutes is

$$P(T \geq 10) = \int_{10}^{40} \frac{1}{40}\,dt = \frac{t}{40}\bigg|_{10}^{40} = \frac{40}{40} - \frac{10}{40} = \frac{3}{4}$$ ∎

The next density function we discuss is the exponential density function. It is a quite useful function and is used often in many applications.

Exponential Density Function Let X be a continuous variable. Then X is said to be *exponentially distributed* if X has the density function

$$f(x) = \begin{cases} \lambda e^{-\lambda x} & \text{if } x \geq 0 \\ 0 & \text{if } x < 0 \end{cases}$$

where λ is a positive constant.

The graph of the exponential density function is given in Figure 28.

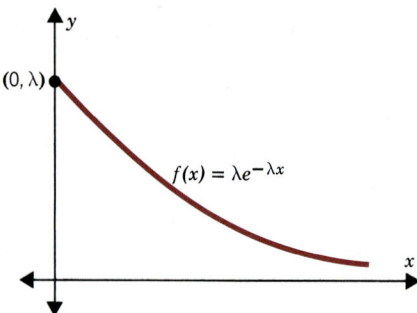

Figure 28

To show that the exponential density function satisfies all the properties of a density function requires a knowledge of improper integrals, a topic treated in most calculus and analytic geometry textbooks.

Two common situations that lead to the use of an exponential density function are listed below:

1. The life of a product, such as an automobile tire, where X is the life expectancy in years
2. The time between the arrival of customers at a service station, where X is this time

In general, any situation that deals with *waiting time* between successive events will lead to an exponential density function. In the exponential density function, the constant λ plays the role of the average number of arrivals per unit time.

Example 4 Airplanes arriving at an airport follow a pattern similar to the exponential density function with an average of $\lambda = 15$ arrivals per hour. Determine the probability of an arrival within 0.1 hour (6 minutes).

Solution The probability is

$$P(t \leq 0.1) = \int_0^{0.1} 15e^{-15t}\, dt = -e^{-15t} \Big|_0^{0.1} = -e^{-1.5} + 1$$
$$= 1 - .223 = .777$$

Thus, the probability of an arrival within 0.1 hour is .777. ■

The next example illustrates how the average λ of an exponential density function can be calculated.

6. APPLICATION TO PROBABILITY

Example 5 From past data it is known that a certain machine normally produces one defective product in every 200. To detect defective products, an inspector tests a continuous stream of products and records the interval in which a defective product appears. Use an exponential density function to find the probability that, after a defective product is found, the next 200 are nondefective.

Solution If we start from a defective item, the probability that the $(x + 1)$st product will be the next defective is

$$f(x) = \lambda e^{-\lambda x}$$

The average defective rate, $\frac{1}{200}$, equals λ, so that the exponential density function is

$$f(x) = \tfrac{1}{200} e^{-x/200}$$

The probability that a defective product will be found within the first 200 items following a defective one is

$$\int_0^{200} \tfrac{1}{200} e^{-x/200}\, dx = -e^{-x/200}\Big|_0^{200} = -e^{-1} + 1 = .632$$

Thus, the probability that the next defective product will not be within the next 200 is $1 - .632 = .368$. ∎

Exercise 6
Solutions to Odd-Numbered Problems begin on page 426.

In Problems 1–8 verify that each function is a probability density function over the indicated interval.

1. $f(x) = \tfrac{1}{2}$, over $[0, 2]$
2. $f(x) = \tfrac{1}{5}$, over $[0, 5]$
3. $f(x) = 2x$, over $[0, 1]$
4. $f(x) = \tfrac{1}{8}x$, over $[0, 4]$
5. $f(x) = \tfrac{3}{250}(10x - x^2)$, over $[0, 5]$
6. $f(x) = \tfrac{6}{27}(3x - x^2)$, over $[0, 3]$
7. $f(x) = \dfrac{1}{x}$, over $[1, e]$
8. $f(x) = \dfrac{4}{3(x + 1)^2}$, over $[0, 3]$

If $f(x) \geq 0$ is not a probability density function, we can find a constant k such that $kf(x)$ satisfies the condition $\int_a^b kf(x)\, dx = 1$.

For the functions in Problems 9–16 determine the constant k that will make each one a probability density function over the interval indicated.

9. $f(x) = 1$, over $[0, 3]$
10. $f(x) = 1$, over $[0, 4]$
11. $f(x) = x$, over $[0, 2]$
12. $f(x) = x$, over $[0, \tfrac{1}{2}]$
13. $f(x) = 10x - x^2$, over $[0, 5]$
14. $f(x) = 10x - x^2$, over $[0, 8]$
15. $f(x) = \dfrac{1}{x}$, over $[1, 2]$
16. $f(x) = \dfrac{1}{(x + 1)^3}$, over $[3, 7]$

17. A number x is selected at random from the interval $[0, 5]$. The probability density function for x is

$$f(x) = \tfrac{1}{5} \qquad \text{for } 0 \leq x \leq 5$$

Find the probability that a number is selected in the subinterval $[1, 3]$.

18. A number x is selected at random from the interval $[0, 10]$. The probability density function for x is

$$f(x) = \tfrac{1}{10} \qquad \text{for } 0 \leq x \leq 10$$

Find the probability that a number is selected in the subinterval $[6, 9]$.

19. The time between incoming telephone calls at a hotel switchboard has an exponential density function with $\lambda = 0.5$ minute. What is the probability that there is an interval of at least 6 minutes between incoming calls?

20. The demand for an inventory item has a density function given by

$$f(x) = 0.2 e^{-0.2x}$$

where $f(x)$ is the probability that x items will be in demand over a 1 week period. What is the probability that less than 5 items will be in demand? Less than 100? More than 10?

21. Let T be the variable that a subject in a psychological testing program will make a certain choice after t seconds. If the probability density function is

$$f(t) = 0.4 e^{-0.4t}$$

what is the probability that the subject will make the choice in less than 5 seconds?

22. Buses on a certain route run every 50 minutes. What is the probability that a person arriving at a random stop along the route will have to wait at least 30 minutes? Assume that the variable T is the time the person will have to wait and assume that T is uniformly distributed.

23. A manufacturer of educational games for children finds through extensive psychological research that the average time it takes for a child in a certain age group to learn the rules of the game is predicted by a *beta probability density function*,

$$f(x) = \begin{cases} \dfrac{1}{4500}(30x - x^2) & \text{if } 0 \leq x \leq 30 \\ 0 & \text{if } x < 0 \text{ or } x > 30 \end{cases}$$

where x is the time in minutes. What is the probability a child will learn how to play the game within 10 minutes? What is the probability a child will learn the game after 20 minutes? What is the probability the game is learned in at least 10 minutes, but no more than 20 minutes?

*7. Models

Model 1: The Learning Curve

Quite often, the managerial planning and control component of a production industry is faced with the problem of predicting labor time requirements and cost per unit of product. The tool used to achieve such predictions is the *learning curve*. The basic assumption made here is that, in certain production industries such as the assembling of televisions and cars, the worker learns from experience. As a result, the more often a worker repeats an operation, the more efficiently the job is performed. Hence, direct labor input per unit of product declines. If the *rate* of improvement is regular enough, the learning curve can be used to predict future reductions in labor requirements.

The general form of the function describing such a situation is

$$f(x) = cx^k$$

where $f(x)$ is the number of hours of direct labor required to produce the xth unit, $-1 \leq k < 0$, and $c > 0$. The choice of x^k, with $-1 \leq k < 0$, guarantees that, as the number x of units produced increases, the direct labor input decreases. See Figure 29.

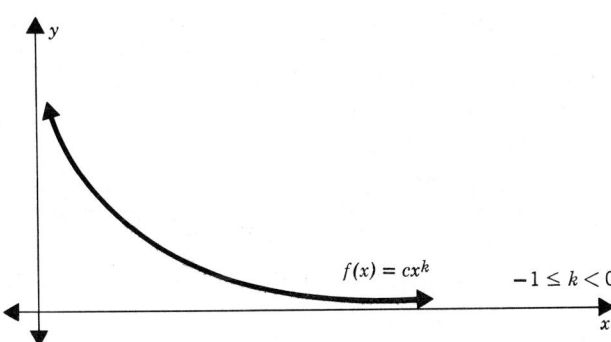

Figure 29

The function $f(x) = cx^k$ describes a rate of learning per unit produced. This rate is measured in terms of labor-hours per unit. As Figure 29 illustrates, the number of direct labor-hours declines as more items are produced.

Once a learning curve has been determined for a gross production process, it can be used as a predictor to determine the number of production hours for future work.

If a learning curve is known, the total number of labor-hours required to produce units numbered a through b is

(1)
$$N = \int_a^b f(x)\,dx = \int_a^b cx^k\,dx$$

---

*This section may be omitted without loss of continuity.

Example 1 Caryl's Air Conditioning Company manufactures air conditioners on an assembly line. From experience, it was determined that the first 100 air conditioners required 1272 labor-hours. For each subsequent 100 air conditioners (1 unit), less labor-hours were required according to the learning curve

$$f(x) = 1272x^{-0.25}$$

where $f(x)$ is the rate of labor-hours required to assemble the xth unit (each unit being 100 air conditioners). This curve was determined after 30 units had been manufactured.

The company is in the process of bidding for a large contract involving 5000 additional air conditioners, or 50 additional units. The company can estimate the labor-hours required to assemble these units by evaluating

$$N = \int_{30}^{80} 1272x^{-0.25}\, dx = \frac{1272x^{0.75}}{0.75}\bigg|_{30}^{80}$$
$$= 1696(80^{0.75} - 30^{0.75}) = 1696(26.75 - 12.82)$$
$$= 1696(13.93) = 23{,}625.28$$

Thus, the company can bid estimating the total labor-hours needed as 23,625.28. ∎

Model 2: Consumer's Surplus

Suppose the price p a consumer is willing to pay for a quantity x of a particular commodity is governed by the demand curve

$$p = D(x)$$

In general, the function $D(x)$ is a decreasing function, indicating that, as the price of the commodity increases, the quantity the consumer is willing to buy declines.

Suppose the price p that a producer is willing to charge for a quantity x of a particular commodity is governed by the supply curve

$$p = S(x)$$

In general, the function $S(x)$ is an increasing function since, as the price p of a commodity increases, the more the producer is willing to supply the commodity.

• Equilibrium Point
• Market Price
• Demand Level

The point of intersection of the demand curve and the supply curve is called the *equilibrium point E*. If the coordinates of E are (x^*, p^*), then p^*, the *market price*, is the price a consumer is willing to pay for and a producer is willing to sell for a quantity x^*, the *demand level*, of the commodity. See Figure 30.

The total revenue of the producer at a market price p^* and a demand level x^* is p^*x^* (the price per unit times the number of units). This revenue can be interpreted geometrically as the area of the rectangle $0p^*Ex^*$ in Figure 30.

In a free market economy, there are times when some consumers would be willing to pay more for a commodity than the market price p^* that they actually do pay. The benefit of this to consumers—that is, the difference between what consumers *actually* paid and what they were *willing* to pay—is called the *consumer's surplus CS*. To obtain a formula for consumer's surplus CS, we use Figure 31 as a guide.

• Consumer's Surplus

7. MODELS   317

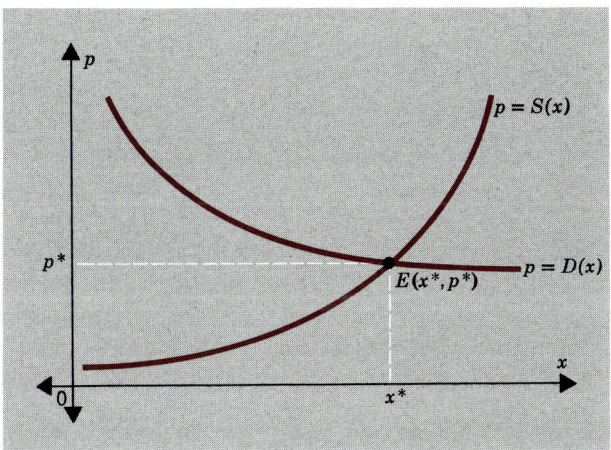

Figure 30

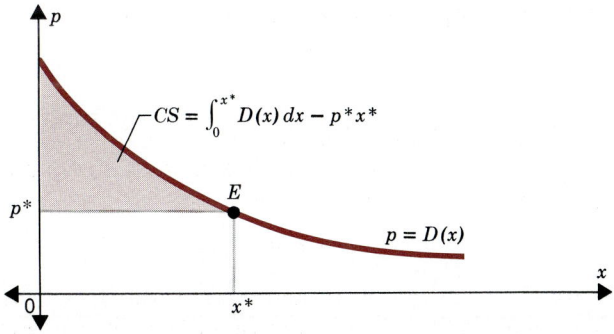

Figure 31

The quantity $\int_0^{x^*} D(x)\,dx$ is the area under the demand curve $D(x)$ from $x = 0$ to $x = x^*$, and represents the total revenue that would have been generated by the willingness of some consumers to pay more. By subtracting p^*x^* (the revenue actually achieved), the result is a surplus CS to the consumer. Thus, we have the formula

(2) $$CS = \int_0^{x^*} D(x)\,dx - p^*x^*$$

Producer's Surplus

In a free market economy, there are also times when some producers would be willing to sell at a price below the market price p^* than the consumer actually pays. The benefit of this to the producer—that is, the difference between the revenue producers *actually* receive and what they would have been willing to receive—is called the *producer's surplus PS*. To obtain a formula for *PS*, we use Figure 32 on page 318 as a guide.

The quantity $\int_0^{x^*} S(x)\,dx$ is the area under the supply curve $S(x)$ from $x = 0$ to

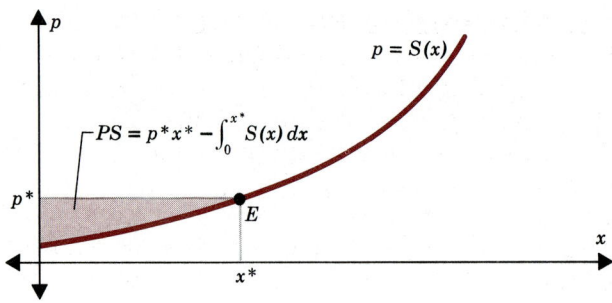

Figure 32

$x = x^*$, and represents the total revenue that would have been generated by some producer's willingness to sell at a lower price. If we subtract this amount from p^*x^* (the revenue actually achieved), the result is a surplus to the producer, PS. Thus, the formula for PS is

(3) $$PS = p^*x^* - \int_0^{x^*} S(x)\, dx$$

Example 2 illustrates a situation in which both the supply and demand curves are linear.

Example 2 Find CS and PS for the demand curve $D(x) = 18 - 3x$ and the supply curve $S(x) = 3x + 6$, where $p^* = D(x^*) = S(x^*)$.

Solution We first determine the equilibrium point E, by solving the equation

$$D(x) = S(x)$$
$$18 - 3x = 3x + 6$$
$$6x = 12$$
$$x^* = 2$$

To find p^*, we compute $D(x^*)$:

$$p^* = D(x^*) = D(2) = 18 - 6 = 12$$

To find CS and PS, we use formulas (2) and (3):

$$CS = \int_0^2 (18 - 3x)\, dx - (2)(12) = \left(18x - \frac{3x^2}{2}\right)\bigg|_0^2 - 24$$
$$= 36 - 6 - 24 = 6$$

$$PS = (2)(12) - \int_0^2 (3x + 6)\, dx = 24 - \left(\frac{3x^2}{2} + 6x\right)\bigg|_0^2$$
$$= 24 - (6 + 12) = 6$$

Thus, in this example, the consumer's surplus and producer's surplus each equal $6.

Model 3: Maximizing Profit over Time

The model introduced here is concerned with business operations of a special character. In oil drilling, mining, and other depletion operations, the initial revenue rate is generally higher than the revenue rate after a period of time has passed. That is, revenue rate, as a function of time, is a decreasing function (this is because depletion is occurring).

The cost rate of such operations generally increases with time because of inflation and other reasons. That is, cost rate, as a function of time, is an increasing function. The problem that management faces is to determine the time t^* that maximizes the profit function $P(t)$.

To construct a model, we denote the cost and the revenue function by $C(t)$ and $R(t)$, respectively, where t denotes time. This representation of cost and revenue deviates from the usual economic definitions of cost per unit times number of units, and price per unit times number of units. The derivatives $C'(t)$ and $R'(t)$, taken with respect to time, represent cost and revenue as time rates. Furthermore, we make the natural assumption that the revenue rate, say dollars per week, is greater than the cost rate at the beginning of the business operation under consideration. Also, as time goes on, we assume the cost rate increases to the revenue rate, and thereafter exceeds it. The optimum time at which the business operation should terminate is that point in time where the rates are equal. That is, the optimum time t^* obeys

$$C'(t^*) = R'(t^*)$$

The profit rate $P'(t)$ is the difference between the revenue rate and the cost rate. That is,

$$P'(t) = R'(t) - C'(t)$$

Hence,

$$P(t) - P(0) = \int_0^t [R'(t) - C'(t)]\, dt$$

The maximum profit is obtained when $t = t^*$, since $P'(t^*) = R'(t^*) - C'(t^*) = 0$. Thus, the maximum profit is $P(t^*)$. Geometrically, the maximum profit $P(t^*)$ is the area enclosed by the graphs of $C'(t)$ and $R'(t)$ from $t = 0$ to $t = t^*$. See Figure 33.

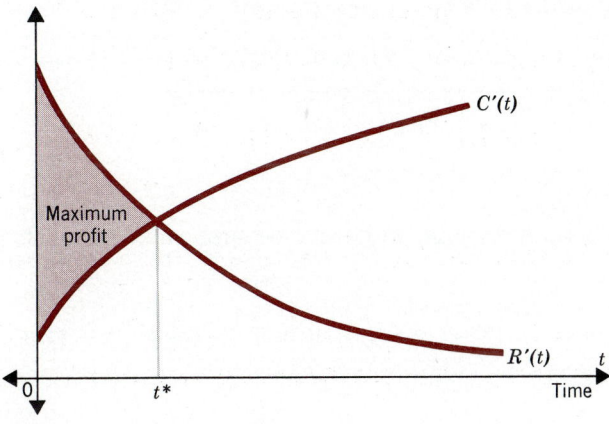

Figure 33

Notice that in Figure 33 the revenue rate function obeys the assumptions made in constructing the model: It is decreasing and it is very high initially. Also, the cost rate function is increasing and is concave down, indicating that the cost rate eventually levels off.

Example 3 The G-B Oil Company's revenue rate (in millions of dollars per year) at time t years is

$$R'(t) = 9 - t^{1/3}$$

and the corresponding cost rate function (also in millions of dollars) is

$$C'(t) = 1 + 3t^{1/3}$$

Determine how long the oil company should continue to operate and what the total profit will be at the end of the operation.

Solution Recall that the time t^* of optimal termination is found when

$$R'(t) = C'(t)$$
$$9 - t^{1/3} = 1 + 3t^{1/3}$$
$$8 = 4t^{1/3}$$
$$2 = t^{1/3}$$
$$t^* = 8 \text{ years}$$

The revenue and cost rate functions are given in Figure 34.

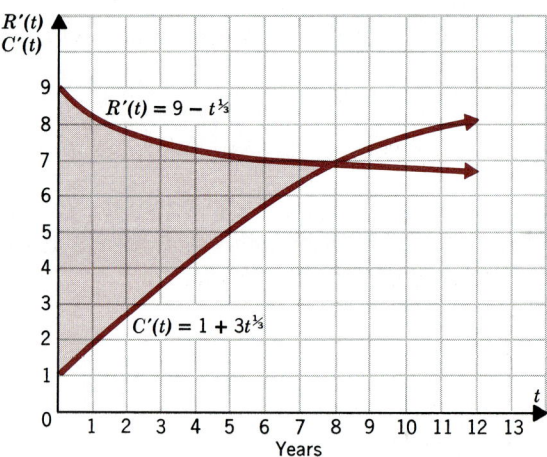

Figure 34

At $t^* = 8$, both revenue and cost rates are 7 million dollars per year. The profit $P(t^*)$ is

$$P(t^*) = \int_0^8 [R'(t) - C'(t)]\, dt = \int_0^8 [(9 - t^{1/3}) - (1 + 3t^{1/3})]\, dt$$

$$= (8t - 3t^{4/3})\Big|_0^8 = 16 \text{ million dollars}$$

In Example 3 we were forced to overlook the *fixed* cost for the cost function at time $t = 0$. This is because, if $C(t)$ contains a constant (the fixed cost), then it becomes zero when we take the derivative $C'(t)$. Thus, in the final analysis of the problem, total profit should be reduced by the amount corresponding to the fixed cost.

Model 4: Average Rate Measure of Synchrony*

If, in most of the cells of a population, simultaneous or nearly simultaneous changes occur, that population is described as "synchronous" with respect to those changes. For example, in a cell population synchronized with respect to cell division, nearly all cells divide within some small time interval. Because of the usefulness of synchronized populations in experimental investigations, it is desirable to have a mathematical measure of the degree of synchrony. Such a measure might also be of value in assessing various techniques used to induce synchrony in population of cells.

The following measure has been developed for evaluating synchrony of cell division in growing populations:

Let E be an event that occurs with fixed periodicity, T, in each cell of a population. Denote by $\epsilon(t)$, the number of times the event E has occurred in time t, and by $N(t)$ the number of cells in the population at time t. Then $f(t) = (d\epsilon/dt)/N(t)$ represents the instantaneous rate of occurrence of the event per individual in the population. If the population is sufficiently large, then f can be assumed to be continuous over the interval of measurement (t_1, t_2).

Let R be the average rate of occurrence of the event per cell over (t_1, t_2). Then

$$R = \frac{1}{t_2 - t_1} \int_{t_1}^{t_2} f(t)\, dt$$

Figure 35 shows the relationship of $f(t)$ to R on (t_1, t_2) for a hypothetical population. From the definition of R it follows that if the population is at all synchronized, then part of the curve for $f(t)$ lies above R.

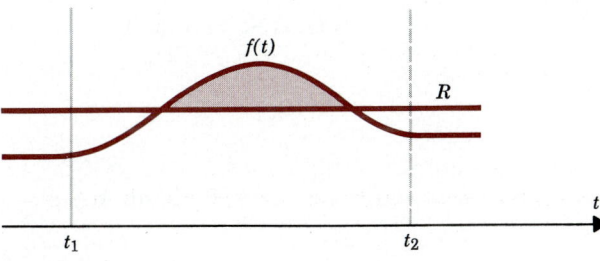

Figure 35

*James Mortimer, "Some Measures of Synchrony in Cell Populations" *Some Mathematics Models in Biology*, University of Michigan Press, Ann Arbor, 1967.

If the population were perfectly synchronized and $t_2 - t_1 = T$, then $f(t)$ would be an impulse of strength RT occurring at some time t, $t_1 \leq t \leq t_2$.

On the other hand, for perfect asynchrony, $f(t) = R$ for all t.

A measure of synchrony should, at least, give appropriate values for these two extremes. Therefore, we will require the measure to give a value of 100 percent for perfect synchrony and 0 percent for perfect asynchrony. The following measure satisfies this condition:

$$\text{degree of synchrony} = M(f) = \frac{\text{area under } f \text{ above } R}{R(t_2 - t_1)} \cdot 100 \text{ percent}$$

It can be shown that this measure also gives appropriate values for intermediate levels of synchrony and, therefore, might be useful for the evaluation of synchrony in cell populations.

Exercise 7
Solutions to Odd-Numbered Problems begin on page 427.

1. After producing 35 units, a company determines that its production facility is following a learning curve of the form

$$f(x) = 1000x^{-0.5}$$

   where $f(x)$ is the rate of labor-hours required to assemble the xth unit. How many total labor-hours should they estimate are required to produce an additional 25 units?

2. Danny's Auto Shop has found that, after tuning up 50 cars, a learning curve of the form

$$f(x) = 1000x^{-1}$$

   is being followed. How many total labor-hours should they estimate are required to tune up an additional 50 cars?

*3. In the construction of a learning curve, how would you interpret the situation in which $k \to 0^-$?

4. Find CS and PS for the demand curve

$$D(x) = -5x + 20$$

   and the supply curve

$$S(x) = 4x + 8$$

   Sketch the graphs.

5. Follow the same directions as in Problem 4 if

$$D(x) = -0.4x + 15 \quad \text{and} \quad S(x) = 0.8x + 0.5$$

6. Find the consumer's surplus for the demand curve

$$D(x) = 50 - 0.025x^2$$

   if it is known that the market quantity x^* is 20 units.

7. The revenue and the cost rate of Gold Star mining operation are, respectively,

$$R'(t) = 19 - t^{1/2} \quad \text{and} \quad C'(t) = 3 + 3t^{1/2}$$

where t is measured in years and R and C are measured in millions of dollars. Determine how long the operation should continue and the profit that can be generated during this period.

Chapter Review

Important Terms

definite integral
lower limit of integration
upper limit of integration
fundamental formula of integral calculus
*even function
*odd function
additive property of area
*mean value theorem for integrals
variable cost
depletion
rate
density function
probability density

annuity
amount of annuity
norm of a partition
Riemann sum
average value of a function
*learning curve
*equilibrium point
*market price
*demand level
*consumer's surplus
*producer's surplus
uniform density function
exponential density function

Review Exercises
Solutions to Odd-Numbered Problems begin on page 427.

In Problems 1–8 evaluate each definite integral.

1. $\displaystyle\int_{2}^{4} x^2 \, dx$

2. $\displaystyle\int_{0}^{3} x^3 \, dx$

3. $\displaystyle\int_{-1}^{1} (x^2 + x - 1) \, dx$

4. $\displaystyle\int_{0}^{2} (2x^2 - x + 4) \, dx$

5. $\displaystyle\int_{0}^{4} e^x \, dx$

6. $\displaystyle\int_{1}^{2} \frac{1}{x} \, dx$

7. $\displaystyle\int_{0}^{2} x^2 \sqrt{x^3 + 1} \, dx$

8. $\displaystyle\int_{0}^{1} x \sqrt{x^2 + 1} \, dx$

9. *Annuity.* If a person saves $2000 per year for 5 years in a savings account that pays 6% per annum compounded continuously, will there be enough in the account to make a 20% down payment on a $100,000 house?

*Discussed in optional problems or section.

10. How much should the person in Problem 9 save each year in order to have the required down payment?
11. *Depletion of Coal.* In 1976 ($t = 0$) the annual world use of coal was 4.2 billion short† tons (each short ton is 2000 pounds). The annual rate of depletion of coal is 10%. Assuming this rate stays the same in the future, how many short tons will the world use by 1996?
12. The rate of sales of a certain product obeys

$$f(x) = 1340 - 850e^{-x}$$

where x is the number of years the product is on the market. Find the total sales during the first 5 years.
13. Find the area enclosed by $f(x) = x^2$ and $g(x) = x^3$.
14. Find the area enclosed by $f(x) = -x^2 + 2x + 2$ and $g(x) = x^2 - 4x + 2$.
15. Find the area enclosed by $f(x) = x^2$ and $g(x) = \sqrt{x}$.
16. Find the area enclosed by $f(x) = x^3 - x$ and the x-axis from $x = -1$ to $x = 2$.
17. Find the area enclosed by $f(x) = 2x/(x^2 + 1)$ and the x-axis from $x = 0$ to $x = 2$.
18. Find the area enclosed by $f(x) = xe^{3x^2}$ and the x-axis from $x = 0$ to $x = 1$.
19. Find CS and PS for the demand curve $D(x) = 12 - (x/50)$ and the supply curve $S(x) = (x/20) + 5$.
20. Show that the function

$$f(x) = \frac{3}{688,000}(-x^2 + 200x - 5000)$$

is a probability density function over the interval $[20, 100]$.
21. A man who is currently 20 years old wants to purchase life insurance. The insurance company is interested in determining at what age X (in years) he is likely to die. If the probability density function given in Problem 20 measures this likelihood, find the probability that the man is likely to die on or before age 40. What is the probability he will die on or before age 60?
22. An experiment has the probability density function $f(x) = 6(x - x^2)$ and outcomes lying between 0 and 1. Determine the probability that an outcome:
    (a) Lies between $\frac{1}{3}$ and $\frac{1}{2}$ (b) Lies between 0 and $\frac{3}{4}$
23. Suppose the outcome X of an experiment lies between 0 and 2, and the probability density function for X is $f(x) = \frac{1}{2}x$. Find:
    (a) $P(X \leq 1)$ (b) $P(1 \leq X \leq 1.5)$ (c) $P(1.5 \leq X)$
24. At a fast-food counter it takes an average of 3 minutes to get serviced. Suppose that the service time X for a customer has an exponential probability density function.
    (a) What fraction of the customers are serviced within 2 minutes?

---

†*Source:* Bureau of Mines, United States Department of the Interior.

(b) What is the probability that a customer will have to wait at least 3 minutes?

25. A toy machine produces a toy every 2 minutes. An inspector arrives at a random time and must wait X minutes for a part.
   (a) Find the probability density function for X.
   (b) Find the probablity that the inspector has to wait at least 1 minute.
   (c) Find the probability that the inspector has to wait no more than 1 minute.

Mathematical Questions From Actuary Exams (Answers on page 433.)

1. Actuary Exam—Part I
$$\int_1^e \frac{1}{x} \ln x \, dx =$$
   (a) $1/e$ (b) $\frac{1}{2}$ (c) 1 (d) e (e) e^2

2. Actuary Exam—Part I
$$\int_0^1 x \ln x \, dx =$$
   (a) $-\infty$ (b) -2 (c) -1 (d) $-\frac{1}{4}$ (e) $-\frac{2}{9}$

3. Actuary Exam—Part I
   If $\int_1^b f(x) \, dx = b^2 e^b - e$ for all $b > 0$, then for all $x > 0$, $f(x) =$
   (a) $x^2 e^x$ (b) $\frac{x^3}{3} e^x$ (c) $x^2 e^x + 2xe^x$ (d) $2xe^x$
   (e) $x^2 e^x - e^{x-1}$

4. Actuary Exam—Part I
   If the area of the region bounded by $y = f(x)$, the x-axis, and the lines $x = a$ and $x = b$ is given by $\int_a^b f(x) \, dx$, which of the following must be true?
   (a) $a < b$ and $f(x) > 0$
   (b) $a < b$ and $f(x) < 0$
   (c) $a > b$ and $f(x) > 0$
   (d) $a > b$ and $f(x) < 0$
   (e) None of the above

5. Actuary Exam—Part II
   Two men patronize the same barber shop. If they both arrive independently between 3 PM and 4 PM on the same day and both stay for 15 minutes, what is the probability that they are in the barber shop for part or all of the same time?
   (a) $\frac{3}{8}$ (b) $\frac{1}{6}$ (c) $\frac{1}{2}$ (d) $\frac{9}{16}$ (e) $\frac{5}{8}$

6. Actuary Exam—Part II
   In a certain process for enameling copper wire, small bare spots occur at random with an average frequency of 2 such spots per 1000 feet of wire. What is the probability that a 5000 foot roll of this copper wire will contain no more than 1 such bare spot?
   (a) $10e^{-10}$ (b) $11e^{-10}$ (c) $61e^{-10}$ (d) $1 - 50e^{-10}$ (e) $1 - 10e^{-10}$

7. *Actuary Exam—Part II*
   Each day, X arrives at a point A between 8:00 and 9:00 AM, his times of arrival being uniformly distributed. Y arrives independently at A between 8:30 and 9:00 AM, his times of arrival also being uniformly distributed. What is the probability that Y arrives before X?
   (a) $\frac{1}{8}$ (b) $\frac{1}{6}$ (c) $\frac{2}{9}$ (d) $\frac{1}{4}$ (e) $\frac{1}{2}$

Other Books

Batschelet, E., *Introduction to Mathematics for Life Sciences,* Springer-Verlag, New York, 1972.

Chiang, Alpha, *Fundamental Methods of Mathematical Economics,* 2nd ed., McGraw-Hill, New York, 1974.

Ferguson, C. E., *Microeconomic Theory,* 3rd ed., Richard D. Irwin Press, Homewood, Ill., 1972.

Grossman, Stanley, and James Turner, *Mathematics for the Biological Sciences,* Macmillan, New York, 1974.

Mizrahi, M. and M. Sullivan, *Calculus and Analytic Geometry,* Wadsworth Publishing Co., Belmont, CA, 1982.

Springer, C. H., R. E. Herlihy, and R. I. Beggs, *Advanced Methods and Models,* Mathematics for Management Series, Vol. 2, Richard D. Irwin Press, Homewood, Ill., 1965.

Thrall, Robert M., ed., *Some Mathematical Models in Biology,* University of Michigan, Ann Arbor, 1967.

9
Numerical Techniques*

1. Newton's Method of Solving Equations
2. Trapezoidal Rule

*This chapter is from *Calculus and Analytic Geometry* by Abe Mizrahi and Michael Sullivan. Copyright © 1982 by Wadsworth, Inc. Reprinted by permission of Wadsworth Publishing Company, Belmont, California 94002.

1. Newton's Method of Solving Equations
Newton's method will enable us to find, to any desired degree of accuracy, the real solutions of many equations.

Suppose we let $y = f(x)$ denote a function whose derivative f' is continuous, and we wish to find the real solutions of the equation $f(x) = 0$. Graphically, this means that we are to find the x-intercepts of the graph. Now suppose that, from a graph or from tables or by trial calculations, we have found that the equation $f(x) = 0$ has a real solution in some open interval containing the number $x = c_1$. We draw the tangent line to the graph of f at the point $P_1 = (c_1, f(c_1))$ and let P_2 be the point where this tangent line intersects the x-axis (see Figure 1). Then, in

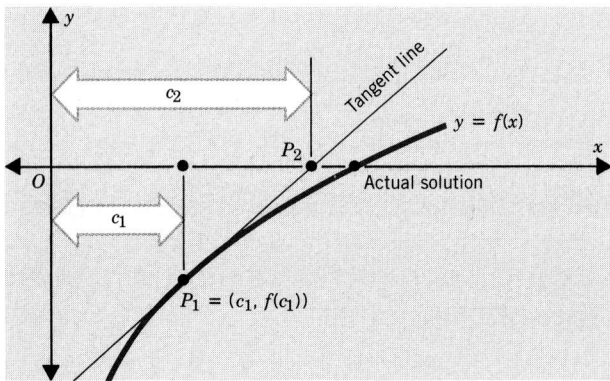

Figure 1

general, if $x = c_1$ is a fair *first approximation* to the required solution of the given equation $f(x) = 0$, then the x-intercept $c_2 = \overline{OP_2}$ of the tangent line will give a better, or *second approximation* to the solution. Graphically, this is the idea behind Newton's method of solving equations.

We may derive a formula for calculating the approximation c_2 as follows: The coordinates of P_1 are $(c_1, f(c_1))$, and the slope of the tangent line to the graph of f at P_1 is $f'(c_1)$. Therefore, the equation of this tangent line at P_1 is

$$y - f(c_1) = f'(c_1)(x - c_1)$$

Putting $y = 0$ in this equation and solving for x in order to find the point of intersection of the tangent line with the x-axis, we obtain

$$x = c_2 = c_1 - \frac{f(c_1)}{f'(c_1)}$$

Hence:

If $x = c_1$ **is a sufficiently close first approximation to a real solution of the equation** $f(x) = 0$, **then the formula**

(1)
$$c_2 = c_1 - \frac{f(c_1)}{f'(c_1)}$$

gives a closer, second approximation to the solution.

We may now use c_2 in formula (1) in place of c_1 and get a third, and probably closer, approximation to the required solution of the equation; namely,

$$c_3 = c_2 - \frac{f(c_2)}{f'(c_2)}$$

The process may be repeated as often as required to give the desired degree of accuracy.

Example 1 Use Newton's method to find a third approximation to the real positive solution r of the equation $x^3 - 2x - 5 = 0$, where $2 \leq r \leq 3$.

Solution Let $f(x) = x^3 - 2x - 5$. We find $f(2) = -1$ and $f(3) = 16$, which indicates that the given equation has a solution between 2 and 3, and considerably nearer 2 than 3. Therefore, we take $c_1 = 2$ and apply Newton's formula. Then, $f(c_1) = f(2) = -1$, and since $f'(x) = 3x^2 - 2$, we have $f'(c_1) = f'(2) = 10$. By formula (1),

(2) $$c_2 = 2 - \frac{(-1)}{10} = 2.1$$

Now we apply formula (1) again, with $c_1 = 2.1$. We find $f(2.1) = 0.061$ and $f'(2.1) = 11.23$. Hence, a third approximation to the solution is

(3) $$c_3 = 2.1 - \frac{0.061}{11.23} = 2.1 - 0.0054 = 2.0946$$

∎

Example 2 Use Newton's method to find a third approximation to the real solution of the equation $e^x + x - 2 = 0$.

Solution To get the first approximation c_1, we note that the real solution of $e^x + x - 2 = 0$ occurs at the intersection of $y = e^x$ and $y = 2 - x$, which is some number between 0 and 1 (see Figure 2). We begin with $c_1 = 0$. (You may want to try to begin with $c_1 = 1$ to see what happens). To use (1), we set $f(x) = e^x + x - 2$. Then, $f'(x) = e^x + 1$, and

$$c_2 = c_1 - \frac{f(c_1)}{f'(c_1)} = 0 - \frac{f(0)}{f'(0)} = -\frac{-1}{2} = 0.5$$

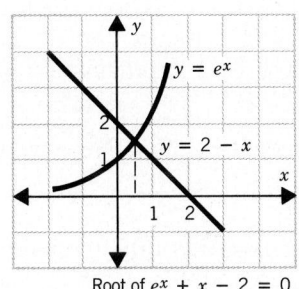

Root of $e^x + x - 2 = 0$

Figure 2

Since

$$f(0.5) = e^{1/2} + \frac{1}{2} - 2 \approx 0.1487$$

↑ Use a calculator

and

$$f'(0.5) = e^{1/2} + 1 \approx 2.6487$$

we find the third approximation to be

$$c_3 = 0.5 - \frac{0.1487}{2.6487} = 0.5 - 0.056 = 0.444$$

∎

Exercise 1
Solutions to Odd-Numbered Problems begin on page 429.

In Problems 1–8 solve by using Newton's method. In each equation find a third approximation to the solution indicated.

1. $x^3 + 3x - 5 = 0$; solution between 1 and 2
2. $x^3 - 4x + 2 = 0$; solution between 1 and 2
3. $2x^3 + 3x^2 + 4x - 1 = 0$; solution between 0 and 1
4. $x^3 - x^2 - 2x + 1 = 0$; solution between 0 and 1
5. $x^3 - 6x - 12 = 0$; solution between 3 and 4
6. $3x^3 + 5x - 40 = 0$; solution between 2 and 3
7. $x^4 - 2x^3 + 21x - 23 = 0$; solution between 1 and 2
8. $x^4 - x^3 + x - 2 = 0$; solution between 1 and 2

In Problems 9 and 10 follow the pattern of Example 2 to find a second approximation to the solution.

9. $x + \ln x - 2 = 0$
10. $e^x + x^3 = 0$

11. The volume of a spherical segment is given by $V = \frac{1}{3}\pi h^2(3R - h)$, where R is the radius of the sphere and h is the height of the segment. If $R = 4$ feet and $V = 12$ cubic feet, find a second approximation to h.
12. A solid wooden sphere of diameter d and specific gravity S sinks in water to a depth h, which is determined by the equation $2x^3 - 3x^2 + S = 0$, where $x = h/d$. Find a second approximation to h for a maple ball of diameter 6 inches for which $S = 0.786$.
13. The equation $5e^{-x} + x - 5 = 0$ arises in the quantum theory of radiation in physics. Find a second approximation to x.

2. Trapezoidal Rule

To integrate the definite integral $\int_a^b f(x)\,dx$, the fundamental formula of calculus requires that we know an antiderivative of the integrand f. When it is not possible to find an antiderivative of the integrand f or when the integrand f is defined by an empirical table of values or by an empirical graph, we turn to numerical

2. TRAPEZOIDAL RULE

techniques to approximate the value of the definite integral. To aid in the following discussion, we shall assume that the integrand f is nonnegative and continuous on the closed interval $[a, b]$.

Most of the methods of approximate integration are based on the fact that a definite integral equals the area under a graph, so that any method of approximating this area will also give an approximation to the integral. One of the most widely used numerical techniques of approximate integration is the *trapezoidal rule*.

Trapezoidal Rule. **If a function f is continuous on the closed interval $[a, b]$, then**

(1) $$\int_a^b f(x)\, dx \approx \left(\frac{1}{2}\right)\left(\frac{b-a}{n}\right)[f(x_0) + 2f(x_1) + 2f(x_2) + \cdots + 2f(x_{n-1}) + f(x_n)]$$

where the closed interval $[a, b]$ has been partitioned into n subintervals $[x_0, x_1]$, $[x_1, x_2], \ldots, [x_{n-1}, x_n]$, each of length $(b - a)/n$.

Error. **The error between the exact value of the integral $\int_a^b f(x)\, dx$ and the approximate value given by the trapezoidal rule may be estimated by the formula**

(2) $$\text{Error} \leq \frac{(b-a)^3 M}{12n^2}$$

where M is the largest value of $|f''(x)|$ on the closed interval $[a, b]$.

Example 1 Use the trapezoidal rule with $n = 4$ to approximate $\int_0^1 \frac{dx}{1+x^2}$. Estimate the error in using this approximation.

Solution We partition the interval $[0, 1]$ into four subintervals of equal length, namely, $[0, \frac{1}{4}]$, $[\frac{1}{4}, \frac{1}{2}]$, $[\frac{1}{2}, \frac{3}{4}]$, $[\frac{3}{4}, 1]$. The corresponding values of f are

$$f(0) = \frac{1}{1+0} = 1 \quad f\left(\frac{1}{4}\right) = \frac{1}{1+(\frac{1}{4})^2} = \frac{16}{17} \approx 0.94117 \quad f\left(\frac{1}{2}\right) = \frac{1}{1+(\frac{1}{2})^2} = \frac{4}{5} = 0.8$$

$$f\left(\frac{3}{4}\right) = \frac{1}{1+(\frac{3}{4})^2} = \frac{16}{25} = 0.64 \quad f(1) = \frac{1}{1+1} = \frac{1}{2} = 0.5$$

It is convenient to set up a table, as shown in Table 1. The sum of the entries in the bottom row of the table is 6.26234, so that by the trapezoidal rule (1) we get

$$\int_0^1 \frac{dx}{1+x^2} \approx \left(\frac{1}{8}\right)(6.26234) \approx 0.78279$$

Table 1

	$x=0$	$x=\frac{1}{4}$	$x=\frac{1}{2}$	$x=\frac{3}{4}$	$x=1$
$f(x) = 1/(1+x^2)$	1	0.94117	0.8	0.64	0.5
Factor	×1	×2	×2	×2	×1
Product	1	1.88234	1.6	1.28	0.5

We now use (2) to estimate the error. For this, we need to find the maximum value of $|f''(x)|$, which in turn requires that we find $f'''(x)$. Some calculations will lead to

$$f(x) = \frac{1}{1+x^2} \qquad f'(x) = \frac{-2x}{(1+x^2)^2} \qquad f''(x) = \frac{2(3x^2-1)}{(1+x^2)^3}$$

$$f'''(x) = 24x(1-x^2)(1+x^2)^{-4}$$

Because $f'''(x) > 0$ for $0 < x < 1$, f'' has no critical numbers in $(0, 1)$. The largest value for $|f''(x)|$ occurs at the endpoint 0. For $M = |f''(0)| = 2$, $b = 1$, $a = 0$, and $n = 4$, an upper estimate to the error is

$$\text{Error} \leq \frac{(1-0)^3(2)}{(12)(4^2)} = \frac{1}{96} \approx 0.0104$$

■

The next example illustrates how the trapezoidal rule is used when only discrete information is known.

Example 2 A tree trunk is 140 feet long. At a distance x feet from one end its sectional area A is given in square feet by Table 2 at intervals of 20 feet:

Table 2

x	0	20	40	60	80	100	120	140
A	120	124	128	130	132	136	144	158

Find the approximate volume* of the tree trunk.

Solution As the footnote indicates the volume is

$$V = \int_0^{140} A \, dx$$

Since $n = 7$, $a = 0$, and $b = 140$, by the trapezoidal rule, we find

$$V \approx \frac{140}{14}[120 + 2(124) + 2(128) + 2(130) + 2(132) + 2(136) + 2(144) + 158]$$

$$= 18{,}660 \text{ cubic feet}$$

■

Proof of the Trapezoidal Rule

The trapezoidal rule is based on the idea of representing a definite integral by an area under a graph and on approximating this area by a collection of trapezoids obtained by replacing the graph by a set of chords. Suppose that we are to evaluate $\int_a^b f(x) \, dx$ approximately. The integral is then equal to the area enclosed

---

*It can be shown that if the cross sectional area A is known, then the volume is equal to the definite integral of A. That is, Volume $= \int_a^b A \, dx$.

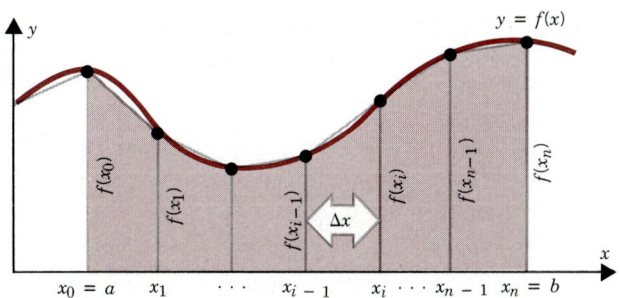

Area of trapezoid in ith subinterval $= \frac{1}{2}[f(x_{i-1}) + f(x_i)]\Delta x$

Figure 3

by the graph of f, the x-axis, and the lines $x = a$ and $x = b$ (see Figure 3). Now, partition the interval $[a, b]$ into n subintervals

$$[a, x_1), [x_1, x_2], \ldots, [x_{i-1}, x_i], \ldots, [x_{n-1}, b]$$

each of length $\Delta x = (b - a)/n$. The ordinates corresponding to $x_0 = a$, $x_1, x_2, \ldots, x_{n-1}, x_n = b$ are $f(x_0), f(x_1), f(x_2), \ldots, f(x_{n-1}), f(x_n)$. When we join consecutive points on the graph by straight line segments (chords), trapezoids are formed and the sum of the areas of the trapezoids is taken as an approximation to the area under the graph. Since the area of a trapezoid is equal to half the sum of the length of the parallel sides times the altitude, we have the following equation for the sum of the areas of the trapezoids:

$$\text{Area} = (\tfrac{1}{2})[f(x_0) + f(x_1)]\Delta x + (\tfrac{1}{2})[f(x_1) + f(x_2)]\Delta x + \cdots + (\tfrac{1}{2})[f(x_{n-1}) + f(x_n)]\Delta x$$
$$= (\tfrac{1}{2})\Delta x[f(x_0) + 2f(x_1) + 2f(x_2) + \cdots + 2f(x_{n-1}) + f(x_n)]$$

Setting $\Delta x = (b - a)/n$, we have formula (1). ∎

Exercise 2
Solutions to Odd-Numbered Problems begin on page 430.

In Problems 1–6 use the trapezoidal rule to approximate each integral.

1. $\displaystyle\int_0^4 x^2\, dx; \quad n = 8$
2. $\displaystyle\int_0^3 x^3\, dx; \quad n = 6$
3. $\displaystyle\int_1^2 \frac{dx}{x}; \quad n = 4$
4. $\displaystyle\int_0^1 \frac{dx}{1 + x}; \quad n = 6$
5. $\displaystyle\int_0^1 e^{x^2}\, dx; \quad n = 4$
6. $\displaystyle\int_1^2 \frac{dx}{x^2}; \quad n = 4$

7. Show that $\int_1^2 dx/x = \ln 2$. Then use the trapezoidal rule with $n = 5$ to approximate $\int_1^2 dx/x$ and hence obtain an approximation to $\ln 2$.
8. In the table, S is the area in square meters of the cross section of a railroad cutting, and x meters is the corresponding distance along the line:

x	0	25	50	75	100	125	150
S	105	118	142	120	110	90	78

Use the trapezoidal rule to calculate the number of cubic meters of earth removed to make the cutting from $x = 0$ to $x = 150$. Do not attempt to compute an error, since a function f for the area is not known.

9. A series of soundings taken across a river channel is given in the table, where x is the distance from one shore and y is the corresponding depth:

x	0	10	20	30	40	50	60	70	80
y	5	10	13.2	15	15.6	12	6	4	0

Draw the section and find its area by the trapezoidal rule.

10. The area of the horizontal section of a reservoir is A square meters at a height x meters from the bottom; corresponding values of x and A are given in the table:

x	0	2.5	5	7.5	10	12.5	15	17.5	20	22.5	25
A	0	2510	3860	4870	5160	5590	5810	6210	6890	7680	8270

Find the volume of water in the reservoir by use of the trapezoidal rule.

11. Use the trapezoidal rule to approximate the area of the pond pictured in the illustration.

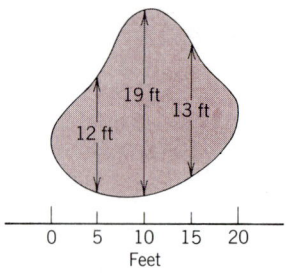

12. The velocity v (in meters per second) of a particle at time t is given by the table:

t	0	0.5	1	1.5	2	2.5	3
v	5.1	5.3	5.6	6.1	6.8	6.7	6.5

Use the trapezoidal rule to approximate the distance traveled from $t = 0$ to $t = 3$.

Appendix
A Review

1. Introduction
2. Exponents
3. Multiplication and Division of Polynomials
4. Factoring
5. Least Common Denominator
6. Geometry Formulas
7. Absolute Value

1. Introduction

Algebra makes use of letters such as x, y, or z to represent a quantity that is *unknown*. The *term* $3x$, for example, means 3 times x, where x is an unknown, or *variable*, quantity. The expression $5x + 2y$ consists of two terms. The term $5x$ is a product, having two *factors*, 5 and x. The number 5 is called the *coefficient* of the term $5x$.

If $x = 3$ and $y = 4$, then $5x + 2y$ has the value

$$5(3) + 2(4) = 15 + 8 = 23$$

Terms that are alike can be added. For example,

$$5x + 6x = 11x$$

Unlike terms cannot be added and are expressed as a formal sum. Thus, $5x$ and $2y$ are unlike terms and cannot be combined.

An algebraic expression containing only one term is often called a *monomial*. For example, $3x, 5y, 8xy$ are monomials.

An algebraic expression having two terms is called a *binomial*. For example, $3x + 2y, x + 5, x^2 + x$ are binomials.

A polynomial is the sum of two or more monomials. For example, $x^2 + x + 3$, $x^3 + 5, x^4 + 3x^2 + x + 1$ are polynomials.

In an expression such as $x - (y + z)$, we use parentheses to indicate that y and z are to be added first, and the sum is then to be subtracted from x. To add or subtract one algebraic expression from another involves the laws of arithmetic. The symbols most often used as grouping symbols are parentheses ( ), brackets [ ], and braces { }. For example, to simplify

$$10x - \{x - [2(x - 1) + 3] - x + 4\},$$

we remove the parentheses first, the brackets second, and finally the braces. Thus,

$$\begin{aligned}10x - \{x - [2x - 2 + 3] - x + 4\} &= 10x - \{x - [2x + 1] - x + 4\} \\ &= 10x - \{x - 2x - 1 - x + 4\} \\ &= 10x - \{-2x + 3\} \\ &= 10x + 2x - 3 \\ &= 12x - 3\end{aligned}$$

2. Exponents

For n any positive integer and x any real number, we define

$$x^1 = x \qquad x^2 = x \cdot x \qquad x^n = \underbrace{x \cdot x \cdot \ldots \cdot x}_{n \text{ times}}$$

2. EXPONENTS

For $x \neq 0$ any real number,

$$x^0 = 1 \qquad x^{-1} = \frac{1}{x} \qquad x^{-2} = \frac{1}{x^2} \qquad x^{-n} = \frac{1}{x^n}$$

Example 1 (a) $2^3 = 2 \cdot 2 \cdot 2 = 8$ (b) $3^{-2} = \frac{1}{3^2} = \frac{1}{9}$

(c) $5x^{-2} = 5\left(\frac{1}{x^2}\right) = \frac{5}{x^2}$ (d) $8^0 = 1$

(e) $(\frac{2}{3})^{-2} = \frac{1}{(\frac{2}{3})^2} = \frac{1}{\frac{4}{9}} = \frac{9}{4}$

∎

In the expression x^n, x is called the *base* and n is the *exponent*.

Laws of Exponents The five laws of exponents are

I. $x^n x^m = x^{n+m}$

In words, when *multiplying* expressions with the *same base,* **retain the base** and *add* **the exponents.**

II. $(x^n)^m = x^{nm}$

In words, when an expression x^n is raised to the power m, retain the base x and *multiply* **the exponents** n **and** m.

III. $(ax)^n = a^n x^n$

In words, if a product is raised to a power, this expression is equal to the product of each factor raised to the same power.

IV. $\frac{x^n}{x^m} = x^{n-m}$

In words, when *dividing* expressions with the *same base,* **retain the base** and *subtract* **the exponents.**

V. $\left(\frac{x}{a}\right)^n = \frac{x^n}{a^n}$

Example 2 (a) $x^2 x^4 = x^{2+4} = x^6$ (b) $x^3 x^{-2} = x^{3+(-2)} = x^1 = x$

(c) $(x^3)^4 = x^{3 \cdot 4} = x^{12}$ (d) $(x^{-1})^4 = x^{-4} = \frac{1}{x^4}$

(e) $(4x)^2 = 4^2 x^2 = 16x^2$ (f) $8x^3 = 2^3 x^3 = (2x)^3$

(g) $(3x)^{-2} = 3^{-2} x^{-2} = \frac{1}{3^2} \frac{1}{x^2} = \frac{1}{9x^2}$

(h) $\frac{x^5}{x^3} = x^{5-3} = x^2$ (i) $\frac{x^3}{x^5} = x^{3-5} = x^{-2} = \frac{1}{x^2}$

(j) $\dfrac{(x+1)^3}{x+1} = (x+1)^2$

(k) $\left(\dfrac{x}{2}\right)^3 = \dfrac{x^3}{8}$

(l) $\dfrac{x^2}{9} = \dfrac{x^2}{3^2} = \left(\dfrac{x}{3}\right)^2$

The laws of exponents as stated above are valid for integer exponents. By properly defining the meaning of $x^{p/q}$, $p, q \neq 0$ integers, they will also be valid for rational number exponents. First, we introduce the notion of a qth root.

The qth root of x, $\sqrt[q]{x}$, is a symbol for the number, which when raised to the power q, equals x. If x is positive, then $\sqrt[q]{x}$ is positive.

Example 3 (a) $\sqrt[3]{8} = 2$ since $2^3 = 8$ (b) $\sqrt[2]{64} = 8$ since $8^2 = 64$

We read $\sqrt[3]{x}$ as the *cube root of x* and $\sqrt[2]{x}$ as the *square root of x*. (Usually, we abbreviate square roots by $\sqrt{}$, dropping the 2.)

No meaning is assigned to even roots of negative numbers since any real number raised to an even power is positive. For example, $\sqrt{4} = 2$, whereas $\sqrt{-4}$ has no meaning in the set of real numbers. Of course, $\sqrt[3]{27} = 3$ while $\sqrt[3]{-64} = -4$ since $(-4)^3 = -64$. Thus, meaning is given to odd roots of negative numbers. Finally, following the usual convention, even roots of positive numbers are *always* positive. Thus, even though $(2)^2 = 4$ and $(-2)^2 = 4$, in computing $\sqrt{4}$, we only take the positive root. That is, $\sqrt{4} = 2$.

We define $x^{p/q}$ as

$$x^{p/q} = (x^{1/q})^p = (\sqrt[q]{x})^p \quad \text{or} \quad x^{p/q} = (x^p)^{1/q} = \sqrt[q]{x^p}$$

With this definition, the laws of exponents are preserved for rational exponents.

Example 4 (a) $\sqrt[3]{x^2} = x^{2/3}$

(b) $\sqrt{8} = \sqrt{4 \cdot 2} = \sqrt{4} \cdot \sqrt{2} = 2\sqrt{2}$

(c) $\sqrt{x} \cdot \sqrt[3]{x} = x^{1/2} \cdot x^{1/3} = x^{1/2 + 1/3} = x^{5/6} = \sqrt[6]{x^5}$

(d) $27^{1/3} = \sqrt[3]{27} = 3$

(e) $(-32)^{1/5} = \sqrt[5]{-32} = -2$

(f) $\sqrt{(x+2)^2} = [(x+2)^{1/2}]^2 = (x+2)^1 = x+2$

Exercise 2
Solutions to Odd-Numbered Problems begin on page 430.

Express each of the following as a single number.

1. 3^3
2. 4^2
3. 2^{-3}
4. 4^{-2}
5. $\left(\tfrac{1}{2}\right)^3$
6. $\left(\tfrac{2}{3}\right)^{-3}$

7. $(343)^{1/3}$ 8. $(81)^{1/4}$ 9. $(-243)^{1/5}$
10. $(-32)^{3/5}$ 11. $2^{-1} + 4^{-1}$ 12. $16^{1/2} + 8^{1/3}$
13. $(3^{-2})(3^4)$ 14. $(9^{-3/2})(9^{1/2})$ 15. $\dfrac{(2^{-2})(2^4)}{2^5}$
16. $\dfrac{(3^{-3/2})(3^{5/2})}{3^{1/2}}$ 17. $(\sqrt{2})(\sqrt[4]{32})$ 18. $(\sqrt[3]{3})(\sqrt{27})$
19. $\sqrt{\tfrac{81}{16}}$ 20. $\sqrt{\tfrac{25}{4}}$

Simplify each of the following expressions and write your answer using only positive exponents.

21. $x^4 x^2$ 22. $\dfrac{(2^3)x^2}{x^5}$ 23. $(ax^2)(3x^3)$

24. $\dfrac{(2x)^3}{(3x)^2}$ 25. $\sqrt{(x+2)^4}$ 26. $\sqrt{x^3} \cdot \sqrt{x}$

27. $\sqrt[3]{(x+1)^2} \cdot \sqrt[3]{(x+1)^4}$ 28. $\left(\dfrac{1}{\sqrt{x}}\right)^{-2}$ 29. $x^{-1} + x^{-2}$

30. $4x^0 + x^{-1}$

3. Multiplication and Division of Polynomials

To multiply two polynomials, we use the *distributive law*,

$$a(b + c) = ab + ac$$

For example, $2(x + 3) = 2x + 2(3) = 2x + 6$. To multiply most polynomials, we need to use the distributive law more than once.

Example 1   (a) $(x + 2)(x + 3) = x(x + 3) + 2(x + 3)$
$= (x^2 + 3x) + (2x + 6) = x^2 + 5x + 6$
(b) $(x - 1)(x^2 - x - 4) = x(x^2 - x - 4) - 1(x^2 - x - 4)$
$= (x^3 - x^2 - 4x) - x^2 + x + 4$
$= x^3 - 2x^2 - 3x + 4$

The division of certain simple expressions requires using laws of exponents and applying the following rule:

$$\frac{a + b}{c} = \frac{a}{c} + \frac{b}{c}$$

Example 2   (a) $\dfrac{x^4 + x^3}{x} = \dfrac{x^4}{x} + \dfrac{x^3}{x} = x^3 + x^2$

340 A REVIEW

(b) $\dfrac{12t^4 - 6t}{3t} = \dfrac{12t^4}{3t} - \dfrac{6t}{3t} = 4t^3 - 2$

(c) $\dfrac{18x^3 + 6x^2 + 4x}{2x} = \dfrac{18x^3}{2x} + \dfrac{6x^2}{2x} + \dfrac{4x}{2x} = 9x^2 + 3x + 2$ ∎

To divide one polynomial by another, follow these steps:

1. **Arrange the terms of each polynomial in descending powers of the variable.**
2. **Divide the first term in the dividend by the first term in the divisor to get the first term of the quotient.**
3. **Multiply the divisor by the term of the quotient thus obtained, and subtract this result from the dividend to obtain a new dividend.**
4. **Repeat this procedure until a remainder is obtained that is zero or of lower degree than the divisor.**

Example 3

$$
\begin{array}{r}
\text{divisor} \nearrow \quad x^2 - 2x + 1 \quad \leftarrow \text{quotient} \\
x - 1 \overline{\smash{)}\, x^3 - 3x^2 + 3x - 1} \quad \leftarrow \text{dividend} \\
\text{subtract} \quad \underline{x^3 - x^2} \\
-2x^2 + 3x \\
\text{subtract} \rightarrow \underline{-2x^2 + 2x} \\
x - 1 \\
\text{subtract} \longrightarrow \underline{x - 1} \\
0 \leftarrow \text{remainder}
\end{array}
$$

In other words,

$$\dfrac{x^3 - 3x^2 + 3x - 1}{x - 1} = x^2 - 2x + 1$$ ∎

Example 4

$$
\begin{array}{r}
x - 2 \\
x + 2 \overline{\smash{)}\, x^2 + 4} \\
\underline{x^2 + 2x} \\
-2x + 4 \\
\underline{-2x - 4} \\
8 \leftarrow \text{remainder}
\end{array}
$$

In other words,

$$\dfrac{x^2 + 4}{x + 2} = x - 2 + \dfrac{8}{x + 2}$$ ∎

Exercise 3
Solutions to Odd-Numbered Problems begin on page 431.

In Problems 1-12 find the product.

1. $3(x + 2)$
2. $4(x - 1)$
3. $x(x^2 - 1)$
4. $x(x^2 + 4)$

5. $(x - 2)(x + 1)$
6. $(x + 3)(x + 2)$
7. $(x - 2)(x - 1)$
8. $(x + 4)(x - 2)$
9. $(x - 2)(x^2 + x + 1)$
10. $(x + 1)(x^2 + 2x + 2)$
11. $(x + 1)(2x^4 - x^2 + 4)$
12. $(x - 2)(3x^3 - x^2 + 1)$

In Problems 13-20 find the quotient.

13. $\dfrac{x^2 + 4x}{x}$
14. $\dfrac{x^3 - x^2}{x}$
15. $\dfrac{3x^3 - 9x}{3x}$
16. $\dfrac{8x^4 + 4x^2}{2x}$
17. $\dfrac{3x^2 + 5x - 6}{x - 4}$
18. $\dfrac{4x^2 - 8x + 9}{2x - 1}$
19. $\dfrac{2x^3 + 3x^2 + 10x + 15}{2x + 3}$
20. $\dfrac{x^4 - x^2 - 12}{x - 2}$

4. Factoring

To factor expressions such as $x^2 - x$ and $3x + 9$, we use the distributive law and write

$$x^2 - x = x(x - 1) \qquad 3x + 9 = 3(x + 3)$$

A further application of the distributive law enables us to find the following useful products:

(I) $(x + a)^2 = x^2 + 2ax + a^2$
(II) $(x + a)(x - a) = x^2 - a^2$
(III) $(x + a)(x + b) = x^2 + (a + b)x + ab$

We arrive at these results by using the distributive law. Thus, in (I), we have

$$(x + a)^2 = (x + a)(x + a) = x(x + a) + a(x + a)$$
$$= xx + xa + ax + aa$$
$$= x^2 + ax + ax + a^2 = x^2 + 2ax + a^2$$

Note that the square of a sum has three terms, and the middle term is of the form $2ax$.

The derivations of (II) and (III) are left as an exercise. See Problem 31 in Exercise 4.

Examples of the above rules follow.

Example 1
(a) $(x + 2)^2 = x^2 + 2(2x) + 2^2 = x^2 + 4x + 4$
(b) $(x - 3)^2 = x^2 - 2(3x) + (-3)^2 = x^2 - 6x + 9$
(c) $(x - 2)(x + 2) = x^2 - (2)^2 = x^2 - 4$
(d) $(x + 1)(x + 3) = x^2 + x + 3x + 3 = x^2 + 4x + 3$
(e) $(x - 2)(x - 3) = x^2 - 2x - 3x + 6 = x^2 - 5x + 6$

Example 2 Factor:
(a) $x^2 + 6x + 9$ (b) $x^2 - 9$ (c) $x^2 + 7x + 12$

Solution (a) From (I),
$$x^2 + 6x + 9 = (x + 3)^2$$
(b) By (II),
$$x^2 - 9 = (x - 3)(x + 3)$$
Formula (II) is sometimes called the *difference of two squares*.
(c) By (III),
$$x^2 + 7x + 12 = (x + 3)(x + 4)$$
We obtain these factors by trial and error, keeping in mind that we want to find a product that equals 12 (i.e., 4 • 3) and sums to 7 (i.e., 4 + 3). ■

Some additional factoring principles are listed below.

(IV) $x^3 - a^3 = (x - a)(x^2 + ax + a^2)$
(V) $x^3 + a^3 = (x + a)(x^2 - ax + a^2)$
(VI) $x^3 + 3ax^2 + 3a^2x + a^3 = (x + a)^3$

Example 3 (a) $x^3 - 8 = (x - 2)(x^2 + 2x + 4)$
(b) $x^3 + 1 = (x + 1)(x^2 - x + 1)$
(c) $x^3 + 6x^2 + 12x + 8 = (x + 2)^3$ ■

Exercise 4
Solutions to Odd-Numbered Problems begin on page 431.

Find the following products:

1. $(x + 1)^2$
2. $(x + 4)(x - 4)$
3. $(x + 6)(x + 1)$
4. $(3x - 7)^2$
5. $(x + \sqrt{2})(x - \sqrt{2})$
6. $(2x + 5)(3x - 1)$
7. $(4x + 1)(x - 3)$
8. $(9x + 2)^2$
9. $(x + 1)^3$
10. $(x - 2)^3$
11. $(x - 1)(x^2 + x + 1)$
12. $(x + 2)(x^2 - 2x + 4)$
13. $(2x - 3)^3$
14. $(2x + 5)^3$

Factor the following expressions:

15. $x^2 - 4$
16. $x^2 - 16$
17. $x^2 + 7x + 6$
18. $x^2 - 7x + 12$
19. $2x^2 - x - 1$
20. $6x^2 + 13x - 5$
21. $3x^2 + 5x - 2$
22. $9x^2 + 12x + 4$
23. $x^3 + 27$

24. $x^3 - 1$
25. $8x^3 + 1$
26. $27x^3 + 1$
27. $x^3 + 3x^2 + 3x + 1$
28. $x^3 - 3x^2 + 3x - 1$
29. Simplify:
    (a) $(x + h)^2 - x^2$
    (b) $(x + h)^3 - x^3$
30. Use $x = 9$, $a = 4$, and $b = 1$ to convince yourself of the following statements:
    (a) $\dfrac{x + a}{x} \neq a$
    (b) $\dfrac{ax + b}{x} \neq a + b$
    (c) $(x + a)^2 \neq x^2 + a^2$
    (d) $\sqrt{x + a} \neq \sqrt{x} + \sqrt{a}$
*31. Verify the product rules (II) and (III) by repeated use of the distributive law.

5. Least Common Denominator

The least common denominator (LCD) of two or more fractions is the smallest number that is exactly divisible by each of the denominators of the fractions.

Example 1 The least common denominator of the fractions $\tfrac{1}{2}$, $\tfrac{1}{3}$, and $\tfrac{3}{4}$ is 12, since 12 is the smallest number that is divisible by 2, 3, and 4. ■

We use the LCD to add fractions.

Example 2
$$\frac{1}{2} + \frac{1}{3} + \frac{3}{4} = \frac{6}{12} + \frac{4}{12} + \frac{9}{12} = \frac{6 + 4 + 9}{12} = \frac{19}{12}$$
■

We add algebraic expressions in the same way.

Example 3
$$\frac{x + 1}{3} + \frac{x - 1}{6} = \frac{2(x + 1)}{6} + \frac{(x - 1)}{6}$$
$$= \frac{2x + 2 + x - 1}{6} = \frac{3x + 1}{6}$$
■

Example 4
$$\frac{1}{\sqrt{x}} + \sqrt{x} = \frac{1}{\sqrt{x}} + \frac{\sqrt{x}}{1} = \frac{1}{\sqrt{x}} + \frac{\sqrt{x} \cdot \sqrt{x}}{\sqrt{x}} = \frac{1 + x}{\sqrt{x}}$$
■

Example 5 Rationalize the numerator:
$$\frac{\sqrt{x + 5} - \sqrt{x}}{5}$$

Solution We rationalize the numerator by multiplying the numerator and the denominator by $\sqrt{x+5} + \sqrt{x}$. Then,

$$\frac{\sqrt{x+5} - \sqrt{x}}{5} = \frac{(\sqrt{x+5} - \sqrt{x})}{5} \frac{(\sqrt{x+5} + \sqrt{x})}{(\sqrt{x+5} + \sqrt{x})}$$

$$= \frac{x+5-x}{5(\sqrt{x+5} + \sqrt{x})} = \frac{5}{5(\sqrt{x+5} + \sqrt{x})}$$

$$= \frac{1}{\sqrt{x+5} + \sqrt{x}}$$ ∎

Exercise 5
Solutions to Odd-Numbered Problems begin on page 431.

In Problems 1–4 perform the indicated operation.

1. $\dfrac{2}{3x-3} + \dfrac{x+1}{x^2}$

2. $\dfrac{5}{3x} + \dfrac{1}{x+1}$

3. $\dfrac{7}{x+3} - \dfrac{7}{x}$

4. $\dfrac{x}{3} + \dfrac{2x-5}{6}$

In Problems 5 and 6 rationalize the numerator of each expression.

5. $\dfrac{\sqrt{2x+5} + \sqrt{2x}}{5}$

6. $\dfrac{\sqrt{x+4} - \sqrt{x}}{2}$

---

6. Geometry Formulas

Pythagorean theorem For a *right triangle* with legs a and b and hypotenuse c, the *Pythagorean theorem* states that

$$c^2 = a^2 + b^2$$

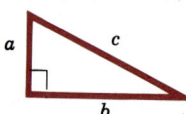

Rectangle For a *rectangle* with base b and altitude h:

$$\text{Area} = A = bh$$
$$\text{Perimeter} = P = 2b + 2h$$

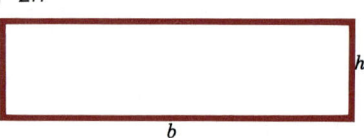

Triangle For a *triangle* with base b and altitude h:

$$\text{Area} = A = \frac{bh}{2}$$

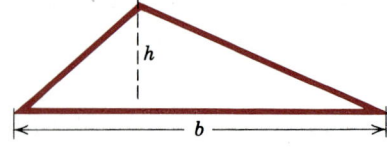

6. GEOMETRY FORMULAS

Trapezoid For a *trapezoid* with bases b_1 and b_2 and altitude h:

$$\text{Area} = A = \frac{b_1 + b_2}{2} h$$

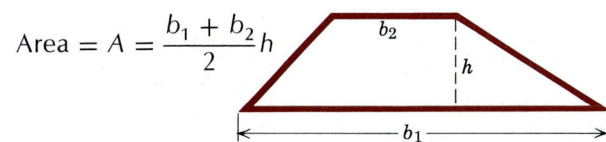

Circle For a *circle* with radius R:

$$\text{Circumference} = C = 2\pi R$$
$$\text{Area} = A = \pi R^2$$

where $\pi \approx 3.14$.

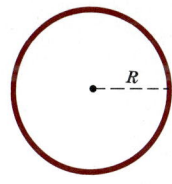

Rectangular Box For a *rectangular box* with length l, width w, and height h:

$$\text{Volume} = V = lwh$$
$$\text{Surface area} = 2lw + 2lh + 2wh$$

If $l = w = h = x$, then $V = x^3$ is the volume of a cube.

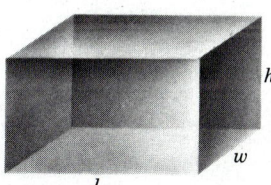

Sphere For a *sphere* with radius R:

$$\text{Volume} = V = \frac{4\pi R^3}{3}$$
$$\text{Surface area} = S = 4\pi R^2$$

where $\pi \approx 3.14$.

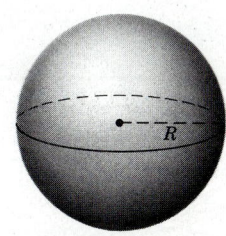

Right Circular Cylinder For a *right circular cylinder* with radius R and height h:

$$\text{Volume} = V = \pi R^2 h$$
$$\text{Surface area} = S = 2\pi R^2 + 2\pi Rh$$

where $\pi \approx 3.14$.

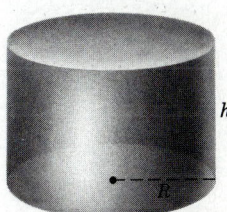

Right Circular Cone For a *right circular cone* with radius R, height h,

$$\text{Volume} = V = \frac{\pi R^2 h}{3}$$

where $\pi \approx 3.14$.

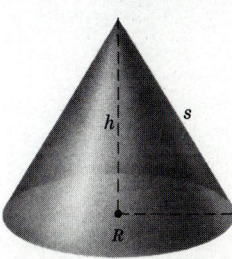

Example 1 The perimeter of a rectangular hall is 34 meters. If its width is 4 meters, what is its length?

Solution
$$2l + 2w = 34$$
$$2l + 2(4) = 34$$
$$2l = 34 - 8 = 26$$
$$l = 13 \text{ meters}$$

■

Exercise 6
Solutions to Odd-Numbered Problems begin on page 431.

Find the length of the missing side of the right triangle in the picture.

1. $a = 3, \ b = 4, \ c = ?$
2. $a = ?, \ b = 6, \ c = 10$
3. $a = 5, \ b = ?, \ c = 13$
4. $a = 5, \ b = 6, \ c = ?$

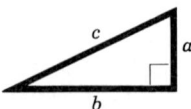

Find the area and the perimeter of a rectangle with base b and altitude h if:

5. $b = 2, \ h = 3$
6. $b = 5, \ h = 2$
7. $b = \frac{1}{2}, \ h = \frac{1}{3}$
8. $b = \frac{3}{4}, \ h = \frac{4}{3}$

Find the area of a triangle with base b and altitude h if:

9. $b = 2, \ h = 1$
10. $b = 3, \ h = 4$
11. $b = \frac{1}{2}, \ h = \frac{3}{4}$
12. $b = \frac{3}{4}, \ h = \frac{4}{3}$

Find the area and circumference of a circle with radius R if (use $\pi \approx 3.14$):

13. $R = 1$
14. $R = 2$
15. $R = \frac{1}{2}$
16. $R = \frac{3}{4}$

Find the volume and surface area of a rectangular box with length l, width w, and height h if:

17. $l = 1, \ w = 1, \ h = 1$
18. $l = 2, \ w = 3, \ h = 4$
19. $l = \frac{1}{2}, \ w = \frac{3}{2}, \ h = \frac{4}{3}$
20. $l = \frac{3}{4}, \ w = \frac{1}{2}, \ h = \frac{5}{3}$

21. Find the volume and surface area of a sphere of radius 2 (use $\pi \approx 3.14$).
22. Find the volume of a right circular cylinder with radius 2 and height 3 (use $\pi \approx 3.14$).
23. Find the volume of a right circular cone with radius 3 and height 4 (use $\pi \approx 3.14$).

7. Absolute Value

We now introduce the concept of *absolute value of a real number*. The absolute value of a number is the magnitude of that number, or the nonnegative value of that number. Thus, the absolute value of 5 is 5; the absolute value of −6 is 6. The definition of absolute value is given below.

Absolute Value The *absolute value of a real number x*, **denoted by the $|x|$, is defined as**

$$x \text{ if } x \geq 0 \quad \text{and} \quad -x \text{ if } x < 0$$

For example, since $-4 < 0$, then the second rule must be used to get

$$|-4| = -(-4) = 4$$

Example 1 (a) $|8| = 8$ (b) $|0| = 0$ (c) $|-15| = 15$ ∎

Example 2 Find all x for which

$$|x + 4| = 13$$

Solution There are two possibilities; either $x + 4 \geq 0$ or else $x + 4 < 0$. In the first instance, we have

$$|x + 4| = x + 4 = 13$$
$$x = 9$$

In the second case, if $x + 4 < 0$, we have

$$|x + 4| = -(x + 4) = 13$$
$$x = -17$$

Thus, the solutions are $x = 9$ and $x = -17$. ∎

Some properties obeyed by absolute value are the following:

(I) The absolute value of *any* real number is always nonnegative; that is,

$$|x| \geq 0$$

(II) The absolute value of the product $|xy|$ of two real numbers equals the product of their absolute values $|x| \cdot |y|$; that is,

$$|xy| = |x| \cdot |y|$$

(III) The absolute value of the sum $|x + y|$ of two real numbers never exceeds the sum of their absolute values $|x| + |y|$; that is,

$$|x + y| \leq |x| + |y|$$

348 A REVIEW

(IV) The square of the absolute value of x equals x^2; that is,
$$|x|^2 = x^2$$

Example 3 The following example serves as an illustration of these four properties.

(a) $|-5| = 5 \geq 0$ (b) $|0| = 0 \geq 0$ (c) $|6| = 6 \geq 0$
(d) $|(-5)(-2)| = |10| = 10$ (e) $|-5| \cdot |-2| = 5 \cdot 2 = 10$
(f) $|-5 + 4| = |-1| = 1 \leq |-5| + |4| = 5 + 4 = 9$
(g) $|-3|^2 = 3^2 = 9$ (h) $(-3)^2 = 9$ ∎

A geometric interpretation of the absolute value of a real number x is that *it measures the distance from the origin 0 to the point x*. This means that the quantity $|x - a|$ measures the distance from a to x.

For example, if $|x - 3| = 6$, then x lies either 6 units to the right or 6 units to the left of 3. See Figure 1.

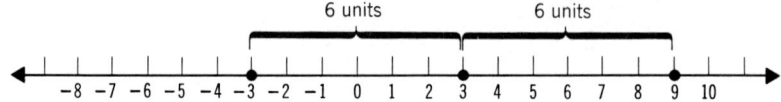

Figure 1

That is, the solutions of $|x - 3| = 6$ are -3 and 9.

If we ask for all x obeying $|x - 3| \leq 6$, we are asking for all points x whose distance from 3 is less than or equal to 6. Clearly, from Figure 1, the solution set is all x from -3 to 9 inclusive. It is convenient to use the notation
$$-3 \leq x \leq 9$$
to symbolize the set of real numbers x from -3 to 9 inclusive.

Example 4 Find all x for which
$$|x| < 7$$

Solution We are asked to find all numbers x, whose distance from the origin 0 is less than 7. From Figure 1, we can see that any x between -7 and 7 satisfies the condition. Thus, the solution is
$$-7 < x < 7$$ ∎

In general, if $|x| \leq a$, $a > 0$, then x must obey
$$-a \leq x \leq a$$
and if $|x| < a$, $a > 0$, then
$$-a < x < a$$

Example 5 Find all x for which

$$|x + 3| < 6$$

Solution Think of $x + 3$ as a single unknown quantity so that by following the same pattern as in Example 4, we obtain

$$-6 < x + 3 < 6$$

If we add -3 to each term in the above inequality, we find

$$-6 + (-3) < (x + 3) + (-3) < 6 + (-3)$$
$$-9 < x < 3$$

Exercise 7
Solutions to Odd-Numbered Problems begin on page 431.

In Problems 1-8 find the solution set and graph the solution.

1. $|x| < 5$
2. $|x| \leq 3$
3. $|x - 3| < 4$
4. $|x| < 6$
5. $|x| \leq 4$
6. $|x + 2| \leq 6$
7. $|2x - 4| + 5 \leq 9$
8. $|3x - 7| \leq 10$

Tables

Table 1
Partial Table of Values of e^x and e^{-x}

x	e^x	e^{-x}	x	e^x	e^{-x}	x	e^x	e^{-x}	x	e^x	e^{-x}
0.0	1.000	1.000	2.5	12.18	0.082	5.0	148.4	0.0067	7.5	1,808.0	0.00055
0.1	1.105	0.905	2.6	13.46	0.074	5.1	164.0	0.0061	7.6	1,998.2	0.00050
0.2	1.221	0.819	2.7	14.88	0.067	5.2	181.3	0.0055	7.7	2,208.3	0.00045
0.3	1.350	0.741	2.8	16.44	0.061	5.3	200.3	0.0050	7.8	2,440.6	0.00041
0.4	1.492	0.670	2.9	18.17	0.055	5.4	221.4	0.0045	7.9	2,697.3	0.00037
0.5	1.649	0.607	3.0	20.09	0.050	5.5	244.7	0.0041	8.0	2,981.0	0.00034
0.6	1.822	0.549	3.1	22.20	0.045	5.6	270.4	0.0037	8.1	3,294.5	0.00030
0.7	2.014	0.497	3.2	24.53	0.041	5.7	298.9	0.0033	8.2	3,641.0	0.00027
0.8	2.226	0.449	3.3	27.11	0.037	5.8	330.3	0.0030	8.3	4,023.9	0.00025
0.9	2.460	0.407	3.4	29.96	0.033	5.9	365.0	0.0027	8.4	4,447.1	0.00022
1.0	2.718	0.368	3.5	33.12	0.030	6.0	403.4	0.0025	8.5	4,914.8	0.00020
1.1	3.004	0.333	3.6	36.60	0.027	6.1	445.9	0.0022	8.6	5,431.7	0.00018
1.2	3.320	0.301	3.7	40.45	0.025	6.2	492.8	0.0020	8.7	6,002.9	0.00017
1.3	3.669	0.273	3.8	44.70	0.022	6.3	544.6	0.0018	8.8	6,634.2	0.00015
1.4	4.055	0.247	3.9	49.40	0.020	6.4	601.8	0.0017	8.9	7,332.0	0.00014
1.5	4.482	0.223	4.0	54.60	0.018	6.5	665.1	0.0015	9.0	8,103.1	0.00012
1.6	4.953	0.202	4.1	60.34	0.017	6.6	735.1	0.0014	9.1	8,955.3	0.00011
1.7	5.474	0.183	4.2	66.69	0.015	6.7	812.4	0.0012	9.2	9,897.1	0.00010
1.8	6.050	0.165	4.3	73.70	0.014	6.8	897.8	0.0011	9.3	10,938	0.00009
1.9	6.686	0.150	4.4	81.45	0.012	6.9	992.3	0.0010	9.4	12,088	0.00008
2.0	7.389	0.135	4.5	90.02	0.011	7.0	1,096.6	0.0009	9.5	13,360	0.00007
2.1	8.166	0.122	4.6	99.48	0.010	7.1	1,212.0	0.0008	9.6	14,765	0.00007
2.2	9.025	0.111	4.7	109.95	0.009	7.2	1,339.4	0.0007	9.7	16,318	0.00006
2.3	9.974	0.100	4.8	121.51	0.008	7.3	1,480.3	0.0007	9.8	18,034	0.00006
2.4	11.023	0.091	4.9	134.29	0.007	7.4	1,636.0	0.0006	9.9	19,930	0.00005

Table 2
Common Logarithms

x	0	1	2	3	4	5	6	7	8	9
1.0	.0000	.0043	.0086	.0128	.0170	.0212	.0253	.0294	.0334	.0374
1.1	.0414	.0453	.0492	.0531	.0569	.0607	.0645	.0682	.0719	.0755
1.2	.0792	.0828	.0864	.0899	.0934	.0969	.1004	.1038	.1072	.1106
1.3	.1139	.1173	.1206	.1239	.1271	.1303	.1355	.1367	.1399	.1430
1.4	.1461	.1492	.1523	.1553	.1584	.1614	.1644	.1673	.1703	.1732
1.5	.1761	.1790	.1818	.1847	.1875	.1903	.1931	.1959	.1987	.2014
1.6	.2041	.2068	.2095	.2122	.2148	.2175	.2201	.2227	.2253	.2279
1.7	.2304	.2330	.2355	.2380	.2405	.2430	.2455	.2480	.2504	.2529
1.8	.2553	.2577	.2601	.2625	.2648	.2672	.2695	.2718	.2742	.2765
1.9	.2788	.2810	.2833	.2856	.2878	.2900	.2923	.2945	.2967	.2989
2.0	.3010	.3032	.3054	.3075	.3096	.3118	.3139	.3160	.3181	.3201
2.1	.3222	.3243	.3263	.3284	.3304	.3324	.3345	.3365	.3385	.3404
2.2	.3424	.3444	.3464	.3483	.3502	.3522	.3541	.3560	.3579	.3598
2.3	.3617	.3636	.3655	.3674	.3692	.3711	.3729	.3747	.3766	.3784
2.4	.3802	.3820	.3838	.3856	.3874	.3892	.3909	.3927	.3945	.3962
2.5	.3979	.3997	.4014	.4031	.4048	.4065	.4082	.4099	.4116	.4133
2.6	.4150	.4166	.4183	.4200	.4216	.4232	.4249	.4265	.4281	.4298
2.7	.4314	.4330	.4346	.4362	.4378	.4393	.4409	.4425	.4440	.4456
2.8	.4472	.4487	.4502	.4518	.4533	.4548	.4564	.4579	.4594	.4609
2.9	.4624	.4639	.4654	.4669	.4683	.4698	.4713	.4728	.4742	.4757
3.0	.4771	.4786	.4800	.4814	.4829	.4843	.4857	.4871	.4886	.4900
3.1	.4914	.4928	.4942	.4955	.4969	.4983	.4997	.5011	.5024	.5038
3.2	.5051	.5065	.5079	.5092	.5105	.5119	.5132	.5145	.5159	.5172
3.3	.5185	.5198	.5211	.5224	.5237	.5250	.5263	.5276	.5289	.5307
3.4	.5315	.5328	.5340	.5353	.5366	.5378	.5391	.5403	.5416	.5428
3.5	.5441	.5453	.5465	.5478	.5490	.5502	.5514	.5527	.5539	.5551
3.6	.5563	.5575	.5587	.5599	.5611	.5623	.5635	.5647	.5658	.5670
3.7	.5682	.5694	.5705	.5717	.5729	.5740	.5752	.5763	.5775	.5786
3.8	.5798	.5809	.5821	.5832	.5843	.5855	.5866	.5877	.5888	.5899
3.9	.5911	.5922	.5933	.5944	.5955	.5966	.5977	.5988	.5999	.6010
4.0	.6021	.6031	.6042	.6053	.6064	.6075	.6085	.6096	.6107	.6117
4.1	.6128	.6138	.6149	.6160	.6170	.6180	.6191	.6201	.6212	.6222
4.2	.6232	.6243	.6253	.6263	.6274	.6284	.6294	.6304	.6314	.6325
4.3	.6335	.6345	.6355	.6365	.6375	.6385	.6395	.6405	.6415	.6425
4.4	.6435	.6444	.6454	.6464	.6474	.6484	.6493	.6503	.6513	.6522
4.5	.6532	.6542	.6551	.6561	.6571	.6580	.6590	.6599	.6609	.6618
4.6	.6628	.6637	.6646	.6656	.6665	.6675	.6684	.6693	.6702	.6712
4.7	.6721	.6730	.6739	.6749	.6758	.6767	.6776	.6785	.6794	.6803
4.8	.6812	.6821	.6830	.6839	.6848	.6857	.6866	.6875	.6884	.6893
4.9	.6902	.6911	.6920	.6928	.6937	.6946	.6955	.6964	.6972	.6981
5.0	.6990	.6998	.7007	.7016	.7024	.7033	.7042	.7050	.7059	.7067
5.1	.7076	.7084	.7093	.7101	.7110	.7118	.7126	.7135	.7143	.7152
5.2	.7160	.7168	.7177	.7185	.7193	.7202	.7210	.7218	.7226	.7235
5.3	.7243	.7251	.7259	.7267	.7275	.7284	.7292	.7300	.7308	.7316
5.4	.7324	.7332	.7340	.7348	.7356	.7364	.7372	.7380	.7388	.7396

Table 2 (continued)

x	0	1	2	3	4	5	6	7	8	9
5.5	.7404	.7412	.7419	.7427	.7435	.7443	.7451	.7459	.7466	.7474
5.6	.7482	.7490	.7497	.7505	.7513	.7520	.7528	.7536	.7543	.7551
5.7	.7559	.7566	.7574	.7582	.7589	.7597	.7604	.7612	.7619	.7627
5.8	.7634	.7642	.7649	.7657	.7664	.7672	.7679	.7686	.7694	.7701
5.9	.7709	.7716	.7723	.7731	.7738	.7745	.7752	.7760	.7767	.7774
6.0	.7782	.7789	.7796	.7803	.7810	.7818	.7825	.7832	.7839	.7846
6.1	.7853	.7860	.7868	.7875	.7882	.7889	.7896	.7903	.7910	.7917
6.2	.7924	.7931	.7938	.7945	.7952	.7959	.7966	.7973	.7980	.7987
6.3	.7993	.8000	.8007	.8014	.8021	.8028	.8035	.8041	.8048	.8055
6.4	.8062	.8069	.8075	.8082	.8089	.8096	.8102	.8109	.8116	.8122
6.5	.8129	.8136	.8142	.8149	.8156	.8162	.8169	.8176	.8182	.8189
6.6	.8195	.8202	.8209	.8215	.8222	.8228	.8235	.8241	.8248	.8254
6.7	.8261	.8267	.8274	.8280	.8287	.8293	.8299	.8306	.8312	.8319
6.8	.8325	.8331	.8338	.8344	.8351	.8357	.8363	.8370	.8376	.8382
6.9	.8388	.8395	.8401	.8407	.8414	.8420	.8426	.8432	.8439	.8445
7.0	.8451	.8457	.8463	.8470	.8476	.8482	.8488	.8494	.8500	.8506
7.1	.8513	.8519	.8525	.8531	.8537	.8543	.8549	.8555	.8561	.8567
7.2	.8573	.8579	.8585	.8591	.8597	.8603	.8609	.8615	.8621	.8627
7.3	.8633	.8639	.8645	.8651	.8657	.8663	.8669	.8675	.8681	.8686
7.4	.8692	.8698	.8704	.8710	.8716	.8722	.8727	.8733	.8739	.8745
7.5	.8751	.8756	.8762	.8768	.8774	.8779	.8785	.8791	.8797	.8802
7.6	.8808	.8814	.8820	.8825	.8831	.8837	.8842	.8848	.8854	.8859
7.7	.8865	.8871	.8876	.8882	.8887	.8893	.8899	.8904	.8910	.8915
7.8	.8921	.8927	.8932	.8938	.8943	.8949	.8954	.8960	.8965	.8971
7.9	.8976	.8982	.8987	.8993	.8998	.9004	.9009	.9015	.9020	.9025
8.0	.9031	.9036	.9042	.9047	.9053	.9058	.9063	.9069	.9074	.9079
8.1	.9085	.9090	.9096	.9101	.9106	.9112	.9117	.9122	.9128	.9133
8.2	.9138	.9143	.9149	.9154	.9159	.9165	.9170	.9175	.9180	.9186
8.3	.9191	.9196	.9201	.9206	.9212	.9217	.9222	.9227	.9232	.9238
8.4	.9243	.9248	.9253	.9258	.9263	.9269	.9274	.9279	.9284	.9289
8.5	.9294	.9299	.9304	.9309	.9315	.9320	.9325	.9330	.9335	.9340
8.6	.9345	.9350	.9355	.9360	.9365	.9370	.9375	.9380	.9385	.9390
8.7	.9395	.9400	.9405	.9410	.9415	.9420	.9425	.9430	.9435	.9440
8.8	.9445	.9450	.9455	.9460	.9465	.9469	.9474	.9479	.9484	.9489
8.9	.9494	.9499	.9504	.9509	.9513	.9518	.9523	.9528	.9533	.9538
9.0	.9542	.9547	.9552	.9557	.9562	.9566	.9571	.9576	.9581	.9586
9.1	.9590	.9595	.9600	.9605	.9609	.9614	.9619	.9624	.9628	.9633
9.2	.9638	.9643	.9647	.9652	.9657	.9661	.9666	.9671	.9675	.9680
9.3	.9685	.9689	.9694	.9699	.9703	.9708	.9713	.9717	.9722	.9727
9.4	.9731	.9736	.9741	.9745	.9750	.9754	.9759	.9763	.9768	.9773
9.5	.9777	.9782	.9786	.9791	.9795	.9800	.9805	.9809	.9814	.9818
9.6	.9823	.9827	.9832	.9836	.9841	.9845	.9850	.9854	.9859	.9863
9.7	.9868	.9872	.9877	.9881	.9886	.9890	.9894	.9899	.9903	.9908
9.8	.9912	.9917	.9921	.9926	.9930	.9934	.9939	.9943	.9948	.9952
9.9	.9956	.9961	.9965	.9969	.9974	.9978	.9983	.9987	.9991	.9996

Table 3
Natural Logarithms

x	ln x	x	ln x	x	ln x
		4.5	1.5041	9.0	2.1972
0.1	−2.3026	4.6	1.5261	9.1	2.2083
0.2	−1.6094	4.7	1.5476	9.2	2.2192
0.3	−1.2040	4.8	1.5686	9.3	2.2300
0.4	−0.9163	4.9	1.5892	9.4	2.2407
0.5	−0.6931	5.0	1.6094	9.5	2.2513
0.6	−0.5108	5.1	1.6292	9.6	2.2618
0.7	−0.3567	5.2	1.6487	9.7	2.2721
0.8	−0.2231	5.3	1.6677	9.8	2.2824
0.9	−0.1054	5.4	1.6864	9.9	2.2925
1.0	0.0000	5.5	1.7047	10	2.3026
1.1	0.0953	5.6	1.7228	11	2.3979
1.2	0.1823	5.7	1.7405	12	2.4849
1.3	0.2624	5.8	1.7579	13	2.5649
1.4	0.3365	5.9	1.7750	14	2.6391
1.5	0.4055	6.0	1.7918	15	2.7081
1.6	0.4700	6.1	1.8083	16	2.7726
1.7	0.5306	6.2	1.8245	17	2.8332
1.8	0.5878	6.3	1.8405	18	2.8904
1.9	0.6419	6.4	1.8563	19	2.9444
2.0	0.6931	6.5	1.8718	20	2.9957
2.1	0.7419	6.6	1.8871	25	3.2189
2.2	0.7885	6.7	1.9021	30	3.4012
2.3	0.8329	6.8	1.9169	35	3.5553
2.4	0.8755	6.9	1.9315	40	3.6889
2.5	0.9163	7.0	1.9459	45	3.8067
2.6	0.9555	7.1	1.9601	50	3.9120
2.7	0.9933	7.2	1.9741	55	4.0073
2.8	1.0296	7.3	1.9879	60	4.0943
2.9	1.0647	7.4	2.0015	65	4.1744
3.0	1.0986	7.5	2.0149	70	4.2485
3.1	1.1314	7.6	2.0281	75	4.3175
3.2	1.1632	7.7	2.0142	80	4.3820
3.3	1.1939	7.8	2.0541	85	4.4427
3.4	1.2238	7.9	2.0669	90	4.4998
3.5	1.2528	8.0	2.0794	95	4.5539
3.6	1.2809	8.1	2.0919	100	4.6052
3.7	1.3083	8.2	2.1041		
3.8	1.3350	8.3	2.1163		
3.9	1.3610	8.4	2.1282		
4.0	1.3863	8.5	2.1401		
4.1	1.4110	8.6	2.1518		
4.2	1.4351	8.7	2.1633		
4.3	1.4586	8.8	2.1748		
4.4	1.4816	8.9	2.1861		

Solutions to Odd-Numbered Problems

CHAPTER 1

Exercise 1 (page 6)

1. $2x + 5 = 7$
   $2x = 7 - 5$
   $2x = 2$
   $x = 1$

3. $6 - x = 0$
   $6 = x$
   $x = 6$

5. $3(2 - x) = 9$
   $2 - x = 3$
   $x = 2 - 3$
   $x = -1$

7. $\frac{4x}{3} + \frac{x}{3} = 5$
   $4x + x = 3 \cdot 5$
   $5x = 15$
   $x = 3$

9. $\frac{3x - 5}{x - 3} = 1$
   $3x - 5 = x - 3$
   $3x - x = -2 + 5$
   $2x = 2$
   $x = 1$

11. $x^2 - x - 12 = 0$
    $(x - 4)(x + 3) = 0$
    $x = 4$ or $x = -3$

13. $x^2 - 5x + 6 = 0$
    $(x - 3)(x - 2) = 0$
    $x = 3$ or $x = 2$

15. $x^2 - x = 0$
    $x(x - 1) = 0$
    $x = 0$ or $x = 1$

17. $2x^2 + 3x - 2 = 0$
    $(2x - 1)(x + 2) = 0$
    $2x - 1 = 0$ or $x + 2 = 0$
    $x = \frac{1}{2}$ or $x = -2$

19. $\frac{x^2 - 9}{x + 5} = 0$
    $x^2 - 9 = (x + 5)(0)$
    $x^2 - 9 = 0$
    $(x + 3)(x - 3) = 0$
    $x = -3$ or $x = 3$

21. $\frac{x^2 - 10x + 6}{x^2 + x - 5} = 0$
    $x^2 - 10x + 6 = (x^2 + x - 5)(0)$
    $x^2 - 10x + 6 = 0$
    Quad. eqn.
    $x = \frac{10 \pm \sqrt{100 - 24}}{2} = \frac{10 \pm \sqrt{76}}{2}$
    $x \approx 9.3589$ or $x \approx .6411$

23. $\frac{1}{x - 1} + \frac{2}{x + 2} = 0$
    $x + 2 + 2(x - 1) = 0(x - 1)(x + 2)$
    $x + 2 + 2x - 2 = 0$
    $3x = 0$
    $x = 0$

25. $\frac{9}{x} - x = 6$
    $9 - x^2 = 6x$
    $x^2 + 6x - 9 = 0$
    $x = \frac{-6 \pm \sqrt{36 + 36}}{2}$
    $x = \frac{-6 \pm 6\sqrt{2}}{2} = -3 \pm 3\sqrt{2}$
    $x \approx 1.2426$ or $x \approx -7.2426$

27. $3x + 5 \leq 2$
    $3x \leq 2 - 5$
    $3x \leq -3$
    $x \leq -1$

29. $3x + 5 \geq 2$
    $3x \geq -3$
    $x \geq -1$

31. $-3x + 5 \leq 2$
    $-3x \leq -3$
    $3x \geq 3$
    $x \geq 1$

33. $6x - 3 \geq 8x + 5$
    $6x - 8x \geq 5 + 3$
    $-2x \geq 8$
    $x \leq -4$

35. $A = (4, 2); B = (6, 2); C = (5, 3); D = (-2, 1); E = (-2, -3); F = (3, -2); G = (6, -2); H = (5, 0)$

37. $\frac{f_2 - a_2}{f_1 - a_1} = \frac{-2 - 2}{3 - 4} = \frac{-4}{-1} = 4$

39. $\frac{a_2 - c_2}{a_1 - c_1} = \frac{2 - 3}{4 - 5} = \frac{-1}{-1} = 1$

358 CHAPTER 1 SOLUTIONS

41.

$y = x - 3$

x	0	3	2	-2	4	-4
y	-3	0	-1	-5	1	-7

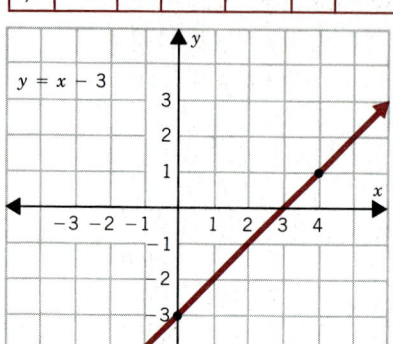

43.

$2x - y = 6$

x	0	3	2	-2	4	-4
y	-6	0	-2	-10	2	-14

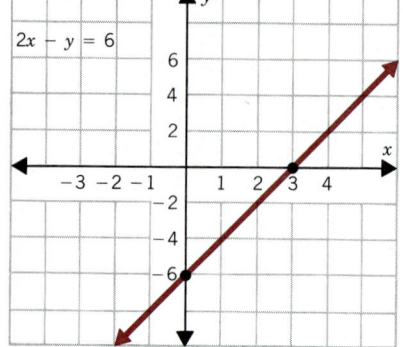

45.

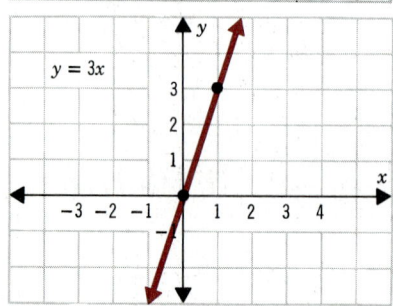

47.

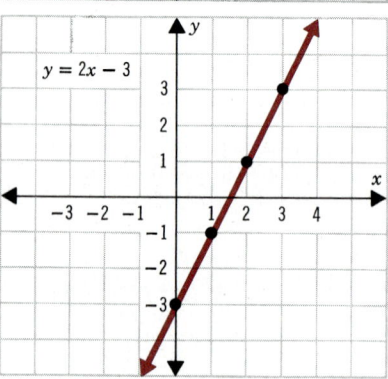

49.

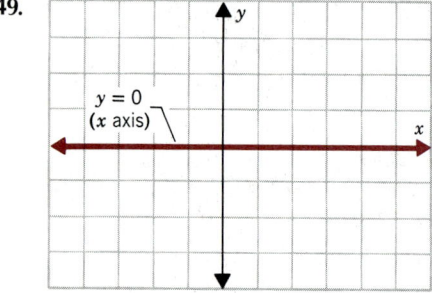

51.

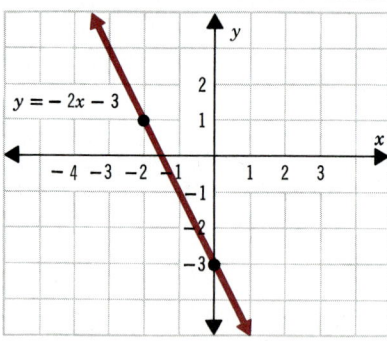

53.

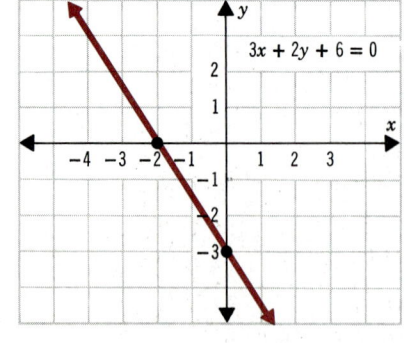

55. The lines are parallel.

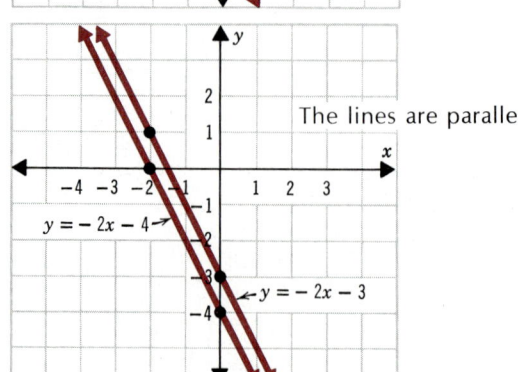

Exercise 2 (page 14)

1. $\dfrac{1-3}{0-2} = \dfrac{-2}{-2} = 1$

3. $\dfrac{-4-0}{-5-(-3)} = \dfrac{-4}{-2} = 2$

5. $\dfrac{4.0-0.3}{1.5-0.1} = \dfrac{3.7}{1.4} = \dfrac{37}{14}$

7. $y - 3 = 2[x - (-2)]$
$y - 3 = 2(x + 2)$
$2x - y + 7 = 0$

9. $y - (-1) = -\tfrac{2}{3}(x - 1)$
$y + 1 = -\tfrac{2}{3}x + \tfrac{2}{3}$
$2x + 3y + 1 = 0$

11. $m = \dfrac{2-3}{-1-1} = \dfrac{-1}{-2} = \dfrac{1}{2}$
$y - 3 = \tfrac{1}{2}(x - 1)$
$x - 2y + 5 = 0$

13. $y = -3x + 3$
$3x + y - 3 = 0$

15. $m = \dfrac{-1-0}{0-2} = \dfrac{-1}{-2} = \dfrac{1}{2}$
$y = \tfrac{1}{2}x - 1$
$x - 2y - 2 = 0$

17. $x - 1 = 0$

19. Slope $\tfrac{3}{2}$, y-intercept -3

21. Slope $-\tfrac{1}{2}$, y-intercept 2

23. Slope undefined, no y-intercept

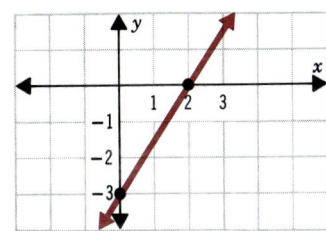
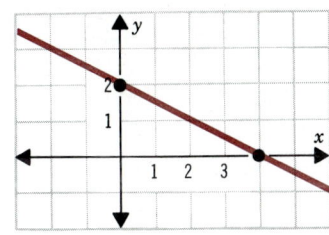
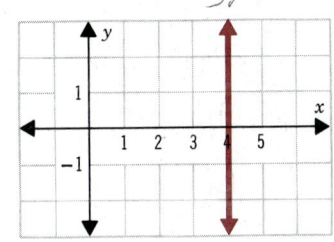

25. $x = 2y$ **27.** $x + y = 2$ **29.** $°F = \tfrac{9}{5}°C + 32$ or $°C = \tfrac{5}{9}(F - 32)$; $\tfrac{5}{9}(70 - 32) = \tfrac{190}{9} = 21.111\ldots$

Exercise 3 (page 21)

1. $m_1 = m_2 = -1$

3. $m_1 = m_2 = \tfrac{2}{3}$

5. $y = 5 - x$
$3x - (5 - x) - 7 = 0$
$3x - 5 + x - 7 = 0$
$4x = 12$
$x = 3$
$y = 5 - 3 = 2$
$(x, y) = (3, 2)$

7. $y = 2 - 3x$
$3x - 2(2 - 3x) + 5 = 0$
$3x - 4 + 6x + 5 = 0$
$9x + 1 = 0$
$x = -\tfrac{1}{9}$
$y = 2 - 3(-\tfrac{1}{9}) = \tfrac{7}{3}$
$(x, y) = (-\tfrac{1}{9}, \tfrac{7}{3})$

9. $\left. \begin{array}{l} 2x - 3y + 4 = 0 \\ 3x + 2y - 7 = 0 \end{array} \right\}$ $\begin{array}{l} 4x - 6y + 8 = 0 \\ \underline{9x + 6y - 21 = 0} \\ 13x - 13 = 0 \\ x = 1 \end{array}$
$2(1) - 3y + 4 = 0$
$y = 2$
$(x, y) = (1, 2)$

11. $\left. \begin{array}{l} 3x - 4y + 8 = 0 \\ 2x + y - 2 = 0 \end{array} \right\}$ $\begin{array}{l} 3x - 4y + 8 = 0 \\ \underline{8x + 4y - 8 = 0} \\ 11x = 0 \\ x = 0 \end{array}$
$3(0) - 4y + 8 = 0$
$y = 2$
$(x, y) = (0, 2)$

13. $\left. \begin{array}{l} -2x + 3y - 7 = 0 \\ 3x + 2y - 9 = 0 \end{array} \right\}$ $\begin{array}{l} -6x + 9y - 21 = 0 \\ \underline{6x + 4y - 18 = 0} \\ 13y - 39 = 0 \\ y = 3 \end{array}$
$3x + 2(3) - 9 = 0$
$x = 1$
$(x, y) = (1, 3)$

15. L: $2x - 3y + 6 = 0$
  $-3y = -2x - 6$
  $y = \tfrac{2}{3}x + 2$

M: $4x - 6y + 7 = 0$
  $-6y = -4x - 7$
  $y = \tfrac{2}{3}x + \tfrac{7}{6}$

No solution, the lines L and M are parallel; same slope; different y-intercepts.

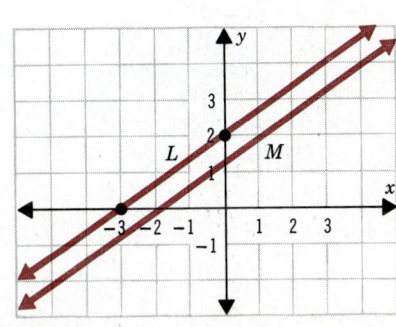

360 CHAPTER 1 SOLUTIONS

17. $L: -2x + 3y + 6 = 0$ $M: 4x - 6y - 12 = 0$
    $\ 3y = 2x - 6$ $\ -6y = -4x + 12$
    $\ y = \frac{2}{3}x - 2$ $\ y = \frac{2}{3}x - 2$
    Infinitely many solutions; the lines L and M are identical; same slope; same y-intercept.

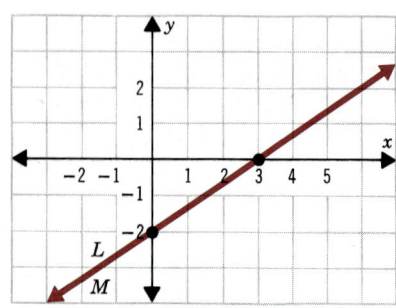

19. $L: 3x - 3y + 10 = 0$ $M: x + y - 2 = 0$
    $\ -3y = -3x - 10$ $\ y = -x + 2$
    $\ y = x + \frac{10}{3}$
    One solution; since the slopes are different, the lines intersect.
    $x_0 + \frac{10}{3} = -x_0 + 2$
    $2x_0 = -\frac{4}{3}$
    $x_0 = -\frac{2}{3}$
    $y_0 = \frac{2}{3} + 2 = \frac{8}{3}$
    L and M intersect at $(-\frac{2}{3}, \frac{8}{3})$.

21. $2x - y = 6$
    $-y = -2x + 6$
    $y = 2x - 6$
    A line parallel to this line must have slope 2:
    $y - 2 = 2(x - 1)$
    $y = 2x$

23. Let x = Number of caramels, y = Number of creams; then $x(0.05) + y(0.10) = 4.0$ and $x + y = 50$:
    $5x + 10y = 400$
    $\underline{5x + 5y = 250}$
    $5y = 150$
    $y = 30$
    $x = 20$
    20 caramels, 30 creams; increase the number of caramels to increase profits.

25. Let x = Bond investment; y = Savings; then $x + y = 50{,}000$ and $0.15x + 0.07y = 6000$:
    $0.15x + 0.07(50{,}000 - x) = 6000$
    $0.15x + 3500 - 0.07x = 6000$
    $0.08x = 2500$
    $x = \$31{,}250$ in bonds
    $y = \$18{,}750$ in savings certificates

27. $x + y = 13$ and $5x + 25y = 165$; solving by substitution
    $5(13 - y) + 25y = 165$
    $65 - 5y + 25y = 165$
    $20y = 100$
    $y = 5$ quarters
    $x = 8$ nickels

29. x = cc of 15% acid; y = cc of 5% acid; then $y = 100 - x$ and $0.15x + 0.05y = (0.08)(100)$:
    $0.15x + 0.05(100 - x) = 8$
    $0.15x + 5 - 0.05x = 8$
    $0.1x = 3$
    $x = 30$ cc of 15% acid
    $y = 70$ cc of 5% acid

CHAPTER 1 SOLUTIONS **361**

31. $x =$ Number of adults; $y =$ Number of children; then $y = 5200 - x$ and $2.75x + 1.50y = 11{,}875$:
$$2.75x + 1.50(5200 - x) = 11{,}875$$
$$1.25x + 7800 = 11{,}875$$
$$1.25x = 4075$$
$$x = 3260 \text{ adults}$$
$$y = 1940 \text{ children}$$

Exercise 5 (page 29)

1. (a) $A = 1000(1 + 0.18t) = 1000 + 180t$ (b) $1000 + 180 \cdot \tfrac{1}{2} = 1090$ (c) 1180 (d) 1360

3. $C = \$10x + \$600$, $R = \$30x$
$$10x + 600 = 30x$$
$$600 = 20x$$
$$30 = x$$

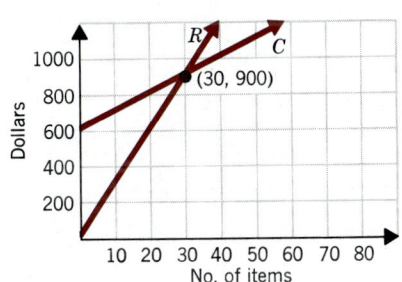

5. $C = \$0.2x + \$50$, $R = \$0.3x$
$$0.2x + 50 = 0.3x$$
$$2x + 500 = 3x$$
$$500 = x$$

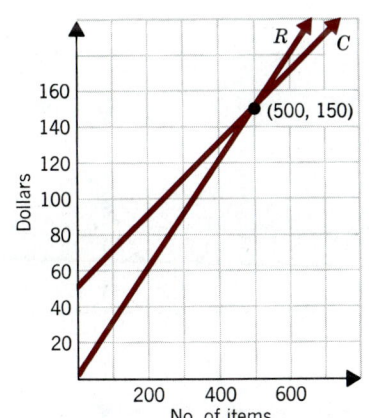

7. $R = \$1x$, $C = 0.75x + 300$
$$x = 0.75x + 300$$
$$0.25x = 300$$
$$x = 1200 \text{ items}$$

9. (a) $\$80{,}000$ (b) $\$95{,}000$ (c) $\$105{,}000$ (d) $\$120{,}000$

11. $S = p + 1$, $D = 3 - p$
$$p + 1 = 3 - p$$
$$2p = 2$$
$$p = \$1$$

13. $S = 20p + 500$, $D = 1000 - 30p$
$$20p + 500 = 1000 - 30p$$
$$50p = 500$$
$$p = \$10$$

15. $S = 0.7p + 0.4$, $D = -0.5p + 1.6$
$0.7p + 0.4 = -0.5p + 1.6$
$1.2p = 1.2$
$p = 1$

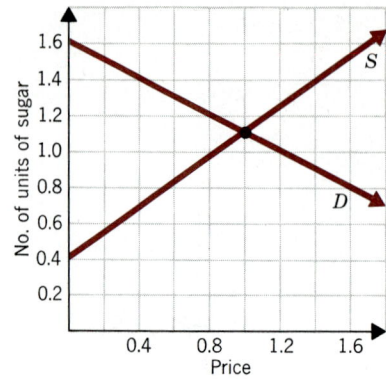

Review Exercises (page 32)

1. $3x + 6 = 2x - 1$
$x = -7$

3. $-2(x + 3) = x + 5$
$-2x - 6 = x + 5$
$-3x = 11$
$x = -\frac{11}{3}$

5. $\frac{4x - 1}{x + 2} = 5$
$4x - 1 = 5(x + 2)$
$4x - 1 = 5x + 10$
$-x = 11$
$x = -11$

7. $2x - 1 \leq 5$
$2x \leq 6$
$x \leq 3$

9. $3x + 7 \geq -2x + 2$
$5x \geq -5$
$x \geq -1$

11.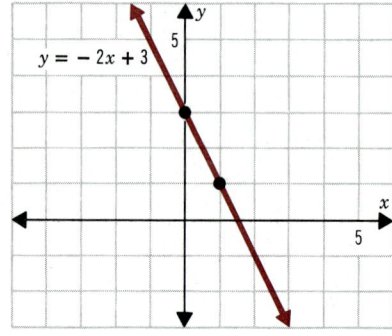

13.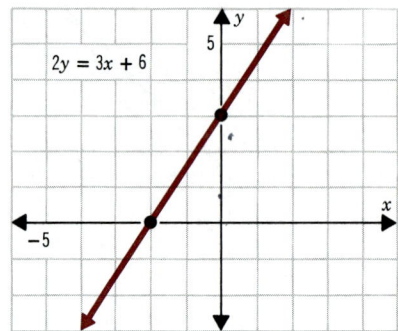

15. Slope $= \frac{4 - 2}{-3 - 1} = \frac{2}{-4} = \frac{-1}{2}$
$y - 2 = -\frac{1}{2}(x - 1)$
$2y - 4 = -x + 1$
$x + 2y - 5 = 0$

17. Slope $= \frac{3 - 0}{-2 - 0} = \frac{-3}{2}$
$y - 0 = -\frac{3}{2}(x - 0)$
$2y = -3x$
$3x + 2y = 0$

19. $y = 2(x + 1)$
$2x - y + 2 = 0$

21. $y - 3 = 1(x - 1)$
$x - y + 2 = 0$

23. $-9x - 2y + 18 = 0$
$2y = -9x + 18$
$y = (-\frac{9}{2})x + 9$
Slope $= -\frac{9}{2}$
y-intercept $= 9$

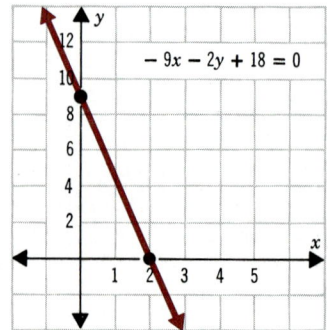

25. $4x + 2y - 9 = 0$
    $2y = -4x + 9$
    $y = -2x + \frac{9}{2}$
    Slope $= -2$
    y-intercept $= \frac{9}{2}$

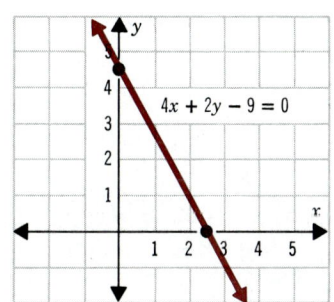

27. $3x - 4y + 12 = 0$ $6x - 8y + 9 = 0$
    $-4y = -3x - 12$ $-8y = -6x - 9$
    $y = \frac{3}{4}x + 3$ $y = \frac{3}{4}x + \frac{9}{8}$
    No solution; the lines are parallel.

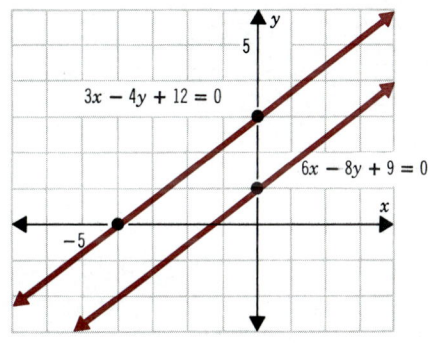

29. $x - y + 2 = 0$ $3x - 4y + 12 = 0$
    $y = x + 2$ $-4y = -3x - 12$
         $y = (\frac{3}{4})x + 3$
    The lines intersect.
    $3x - 4y + 12 = 0$ $3x - 4y + 12 = 0$
    $x - y + 2 = 0$ $\underline{3x - 3y + 6 = 0}$
         $-y + 6 = 0$
         $y = 6$
    $x = 6 - 2 = 4$
    One solution; the lines intersect at $(4, 6)$.

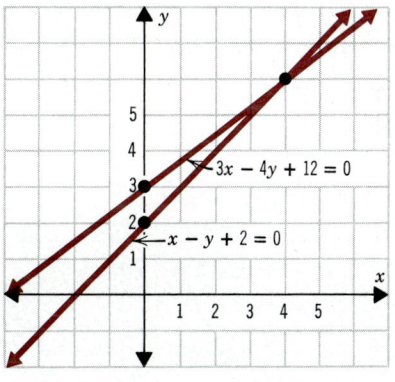

31. $4x + 6y + 12 = 0$ $2x + 3y + 6 = 0$
    $6y = -4x - 12$ $3y = -2x - 6$
    $y = -\frac{2}{3}x - 2$ $y = -\frac{2}{3}x - 2$
    Infinitely many solutions; the lines are identical.

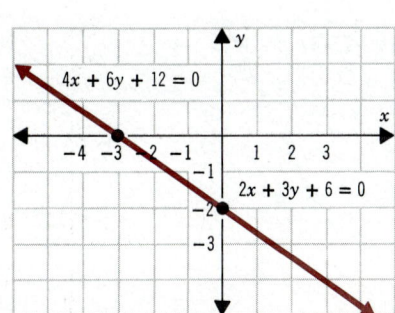

364 CHAPTER 2 SOLUTIONS

33. Let x = Amount in bonds, y = Amount in bank; then $y = 90{,}000 - x$ and $0.16x + 0.06y = 10{,}000$:
$$0.16x + (0.06)(90{,}000 - x) = 10{,}000$$
$$0.16x - 0.06x + 5400 = 10{,}000$$
$$0.10x = 4600$$
$$x = \$46{,}000 \text{ in bonds}$$
$$y = \$44{,}000 \text{ in bank}$$

35. (a)
(b)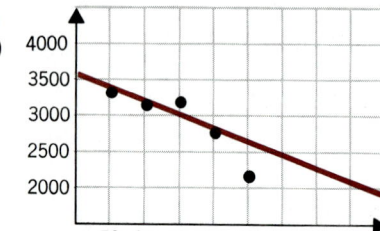

(Using only the last two digits of the date.)

(c) We used the points $(82, 2800)$ and $(79, 3400)$. If you chose two different points, your answer may be different.
$$m = \frac{3400 - 2800}{79 - 82} = \frac{600}{-3} = -200$$
$$y - 2800 = -200(x - 82)$$
(d) $y = -200(84 - 82) + 2800 = 2400$

CHAPTER 2

Exercise 1 (page 45)

1. $3x + 5 \leq 2$
$3x \leq -3$
$x \leq -1$

3. $3x + 5 \geq 2$
$3x \geq -3$
$x \geq -1$

5. $6x - 3 \geq 8x + 5$
$-2x \geq 8$
$x \leq -4$

7. $14x - 21x + 16 \leq 3x - 2$
$-10x \leq -18$
$x \geq 1.8$

9. $x^2 - 5x + 6 \geq 0$
$(x - 3)(x - 2) \geq 0$
$x \geq 3$ or $x \leq 2$

11. $x^2 + 7x < -12$
$x^2 + 7x + 12 < 0$
$(x + 4)(x + 3) < 0$
$-4 < x < -3$

13.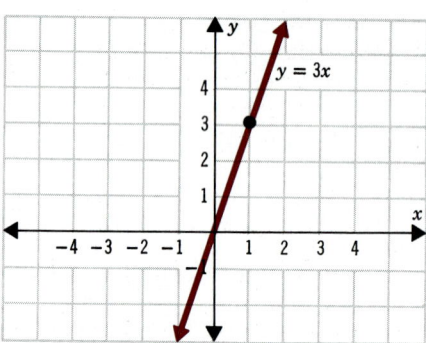

15.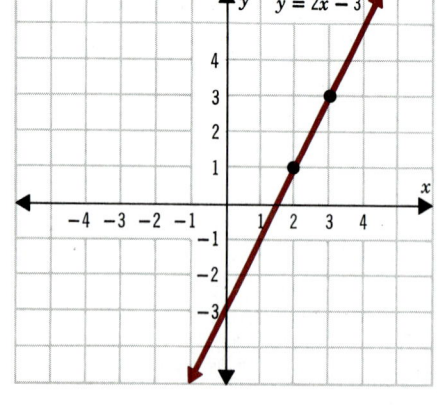

17.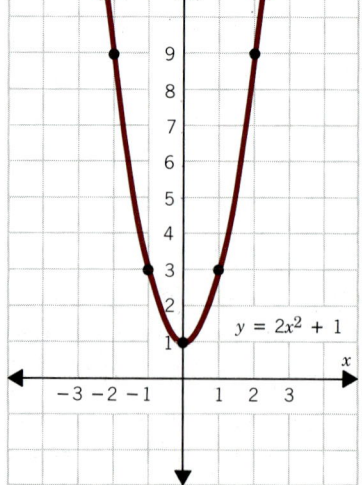

19.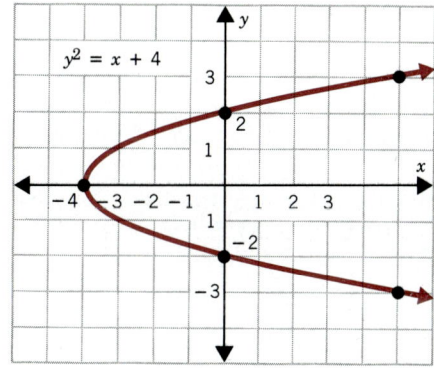

21. $|P_1P_2| = \sqrt{(3-3)^2 + (1+4)^2} = 5$
23. $|P_1P_2| = \sqrt{(-0.2-2)^2 + (-0.4+0.6)^2} = \sqrt{4.88}$
25.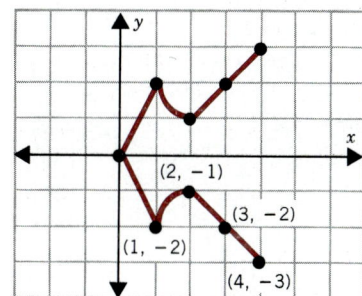

27.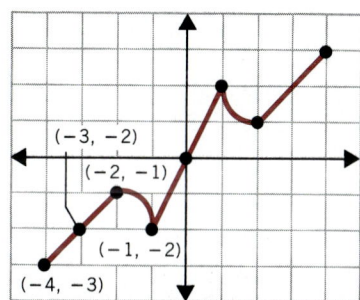

29. $y = 3x^2$; symmetric with respect to the y-axis
31. $xy = 1$; symmetric with respect to the origin
33. $y = x^3 + 1$; no symmetries
35. $y = -x^5 + 3x$; symmetric with respect to the origin
37. The medians of the triangle are the line segments
    (a) from $(0,0)$ to $(4,3)$, length $= \sqrt{16+9} = 5$;
    (b) from $(8,0)$ to $(0,3)$, length $= \sqrt{64+9} = \sqrt{75}$;
    (c) from $(0,6)$ to $(4,0)$, length $= \sqrt{16+36} = \sqrt{52}$

39. $|P_1P_2| = \sqrt{(-4-2)^2 + (1-1)^2} = 6$
    $|P_1P_3| = \sqrt{(-4-2)^2 + (-3-1)^2} = \sqrt{52}$
    $|P_2P_3| = \sqrt{(-4+4)^2 + (-3-1)^2} = 4$
    This is a right triangle; $6^2 + 4^2 = 52$.

41. $|P_1P_2| = \sqrt{(0+2)^2 + (7+1)^2} = \sqrt{68}$
    $|P_1P_3| = \sqrt{(3+2)^2 + (2+1)^2} = \sqrt{34}$
    $|P_2P_3| = \sqrt{(3-0)^2 + (2-7)^2} = \sqrt{34}$
    This is an isosceles right triangle.

Exercise 2 (page 55)

1.
Inputs	Outputs
3	17
1	7
0	2
−4	−18
−2	−8

3. (a) $f(3) = (3)(3) - 2 = 9 - 2 = 7$
   (b) $f(-2) = 3(-2) - 2 = -8$
   (c) $f(0) = 3(0) - 2 = -2$
   (d) $f(x+2) = 3(x+2) - 2 = 3x + 4$
   (e) $f(x+h) = 3(x+h) - 2 = 3x + 3h - 2$
   (f) $f\left(\dfrac{1}{x}\right) = 3\left(\dfrac{1}{x}\right) - 2 = \left(\dfrac{3}{x}\right) - 2$

5. $f(0) = 3$, $f(2) = 4$, since $(0, 3)$ and $(2, 4)$ are points on the graph
7. $f(-4)$ is negative, since the point $(-4, f(-4))$ lies below the x-axis
9. $x = -3, 6, 10$
11. No; $f(-1) < 3 < f(1)$

366 CHAPTER 2 SOLUTIONS

13. Not a function; vertical lines may intersect the graph at two points.
15. Not a function   17. Function   19. Function   21. Function
23. Not a function; if $x = 0$ (for example), then $y^2 = 1$, so y may be either 1 or -1. Thus, $f(0)$ is not uniquely determined.
25. Function; $f(x) = 1 - x^2$ 27. Not a function; $f(1)$ is not uniquely determined.   29. Function
31. $\mathbb{R}$ 33. $\{x \mid x \geq 1\}$ 35. $\mathbb{R}$ 37. $\{x \mid x \neq 2\}$
39. $\{x \mid x > 0\}$ 41. Domain $= \mathbb{R}$ 43. Domain $= \mathbb{R}$

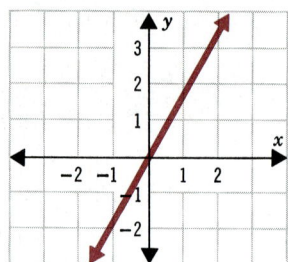
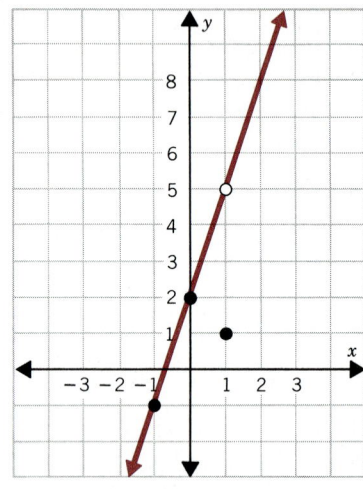

45. Domain $= \{x \mid x < 5\}$ 47. Domain $= \{x \mid x \geq -2\}$

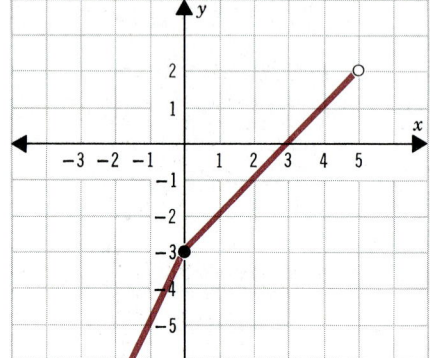
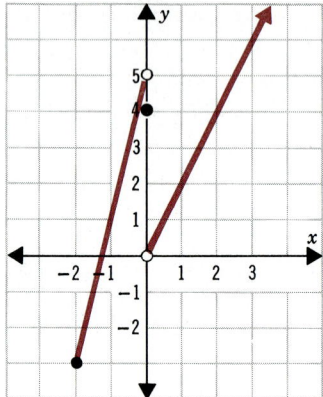

49. No; $f(3) = \dfrac{3 + \frac{1}{2}}{3 - 6} = \dfrac{\frac{7}{2}}{-3} = -\frac{7}{6} \neq 14$

51. Yes; $f(-1) = \dfrac{3}{1 + 1} = \frac{3}{2}$

53. (a) -2 (b) -1 (c) 1 (d) 8
55. (a) Height when $x = 1$ is $H(1) = 20 - 13(1)^2 = 7$ meters
$H(1.1) = 20 - 13(1.1)^2 = 4.27$
$H(1.2) = 20 - 13(1.2)^2 = 1.28$
$H(1.3) = 20 - 13(1.3)^2 = -1.97$ (The rock is embedded in the surface of the planet.)
(b) The rock hits the ground when $H(x) = 0$, that is when
$$0 = 20 - 13x^2$$
$$13x^2 = 20$$
$$x^2 = \tfrac{20}{13}$$
$$x = \sqrt{\tfrac{20}{13}} \approx 1.24 \text{ seconds}$$

57. (a) $C(10) = \dfrac{50}{110-10} = \dfrac{50}{100} = 0.5 = \$500$ (d) $C(70) = \dfrac{350}{110-70} = \dfrac{350}{40} = 8.75 = \$8750$

    (b) $C(30) = \dfrac{150}{110-30} = \dfrac{150}{80} = 1.875 = \$1875$ (e) $C(90) = \dfrac{450}{110-90} = \dfrac{450}{20} = 22.5 = \$22{,}500$

    (c) $C(50) = \dfrac{250}{110-50} = \dfrac{250}{60} \approx 4.16667 = \$4166.67$

59. $R(x) = \$1.50x; \; C(x) = \$150 + 0.60x$
$P(x) = R(x) - C(x) = 1.5x - (150 + 0.6x) = 0.9x - 150$
$P(100) = 0.9(100) - 150 = 90 - 150 = -60$
$P(200) = 30; \; P(300) = 120; \; P(400) = 210; \; P(500) = 300$

Exercise 3 (page 65)

1. $8^{1/3} = 2$ **3.** $\left(\dfrac{1}{27}\right)^{1/3} = \dfrac{1}{3}$ **5.** $27^{2/3} = (27^{1/3})^2 = 9$

7. $27^{-1/3} = \dfrac{1}{27^{1/3}} = \dfrac{1}{3}$ **9.** $8^{-2/3} = \dfrac{1}{8^{2/3}}$

                                                                    $= \dfrac{1}{(8^{1/3})^2} = \dfrac{1}{4}$

11. Opens up

Vertex: $x = -\dfrac{b}{2a} = -\dfrac{1}{4}$
$y = 2(\tfrac{1}{16}) - \tfrac{1}{4} - 3 = -\tfrac{25}{8}$
Intercepts: $x = 0, \; y = -3$
$y = 0, \; 2x^2 + x - 3 = (2x+3)(x-1) = 0$
$x = -\tfrac{3}{2}, \; x = 1$

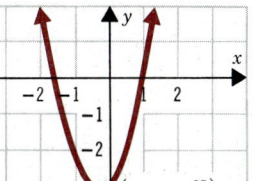
$y = 2x^2 + x - 3$

13. Opens up
Vertex: $x = 0, \; y = -4$
Intercepts: $x = 0, \; y = -4$
           $y = 0, \; x = \pm 2$

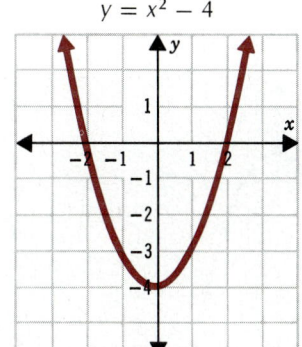
$y = x^2 - 4$

15. Opens up
Vertex: $x = 0, \; y = 1$
Intercepts: $x = 0, \; y = 1$

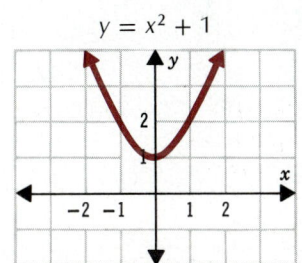
$y = x^2 + 1$

368 CHAPTER 2 SOLUTIONS

17. Opens down
Vertex: $x = 0, y = 1$
Intercepts: $x = 0, y = 1$
$\quad\quad\quad\quad y = 0, x = \pm 1$

$y = -x^2 + 1$

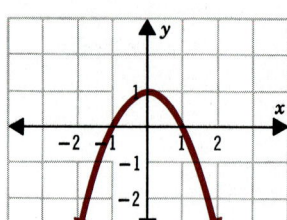

19. Opens up
Vertex: $x = \frac{7}{2}, y = \left(\frac{49}{4}\right) - \left(\frac{49}{2}\right) + 12 = -\frac{1}{4}$
Intercepts: $x = 0, y = 12$
$\quad\quad\quad\quad y = 0, x = 3, x = 4$

$y = x^2 - 7x + 12$

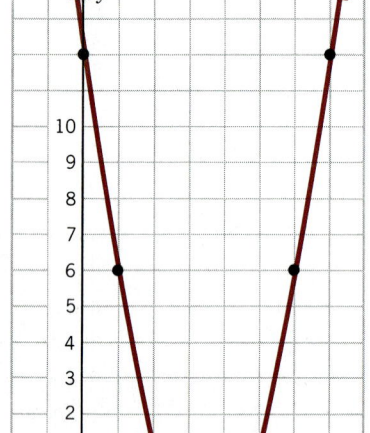

21. Opens down
Vertex: $x = 0, y = 4$
Intercepts: $x = 0, y = 4$
$\quad\quad\quad\quad y = 0, x = \pm 2$

$y = 4 - x^2$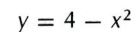

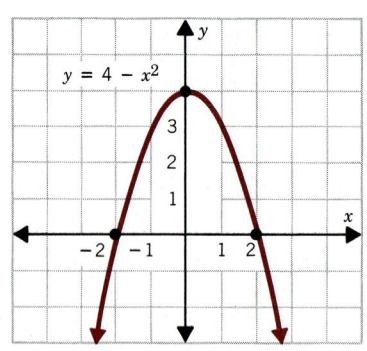

23. $\quad y = x^2 - 15 = 2x$
$\quad\quad x^2 - 15 - 2x = 0$
$\quad\quad (x - 5)(x + 3) = 0$
$\quad\quad x = 5, \quad y = 10$
$\quad\quad x = -3, y = -6$
$\quad\quad (5, 10) \text{ and } (-3, -6)$

25. $x^2 - 2x - 5 = 0$
$\quad x = \dfrac{2 \pm \sqrt{4 + 20}}{2}$
$\quad x = \dfrac{2 \pm 2\sqrt{6}}{2}$
$\quad x = 1 \pm \sqrt{6}$

27. $(\frac{1}{2})x^2 - 3x + 1 = 0$
$\quad x = \dfrac{3 \pm \sqrt{9 - 2}}{2(\frac{1}{2})}$
$\quad x = 3 \pm \sqrt{7}$

29. $10x^2 - 5x + 5 = 0$
(Div. by 5)
$2x^2 - x + 1 = 0$
$x = \dfrac{1 \pm \sqrt{1-8}}{4}$
$x = \dfrac{1 \pm \sqrt{-7}}{4}$ (No real solution)

31. $x + \dfrac{1}{x-6} = 2$
$x(x-6) + 1 = 2(x-6)$
$x^2 - 6x + 1 = 2x - 12$
$x^2 - 8x + 13 = 0$
$x = \dfrac{8 \pm \sqrt{64-52}}{2}$
$x = \dfrac{8 \pm 2\sqrt{3}}{2}$
$x = 4 \pm \sqrt{3}$

33. $\dfrac{3}{x} + x = 5$
$3 + x^2 = 5x$
$x^2 - 5x + 3 = 0$
$x = \dfrac{5 \pm \sqrt{25-12}}{2}$
$x = \dfrac{5 \pm \sqrt{13}}{2}$

35. $1 + \dfrac{8}{x} = -\dfrac{15}{x^2}$
$x^2 + 8x = -15$
$x^2 + 8x + 15 = 0$
$(x+5)(x+3) = 0$
$x = -5, x = -3$

37. $\dfrac{3}{1+x} + \dfrac{2}{1-x} = 6$
$3 - 3x + 2 + 2x = 6 - 6x^2$
$6x^2 - x - 1 = 0$
$x = \dfrac{1 \pm \sqrt{1+24}}{12}$
$x = \dfrac{1 \pm 5}{12}$
$x = \tfrac{1}{2}$ or $x = -\tfrac{1}{3}$

39. $R = C$
$-4p^2 + 4000p = -400p + 960,000$
$4p^2 - 4400p + 960,000 = 0$
$p^2 - 1100p + 240,000 = 0$
$(p - 800)(p - 300) = 0$
$p = 300$ or $p = 800$
To insure a profit:
$300 < p < 800$

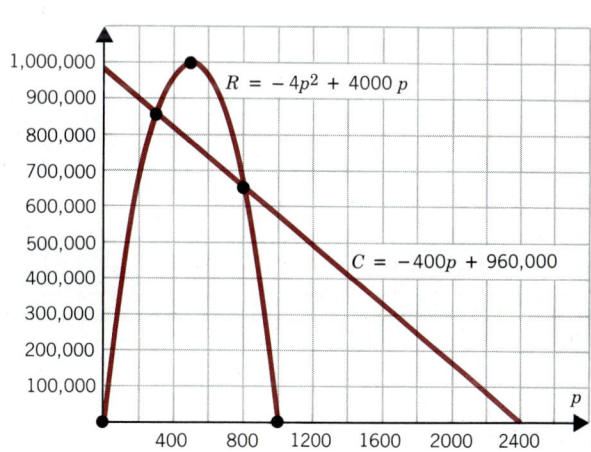

41. $g(x) + h(x) = (3x - 1) + (2 - x^2)$
$= 1 + 3x - x^2$

43. $g(x) \cdot h(x) = (3x-1)(2-x^2) = 6x - 3x^3 - 2 + x^2$
$= -3x^3 + x^2 + 6x - 2$

45. $g(x)/f(x) = \dfrac{3x-1}{x^2-1}$

47. $f(x)/k(x) = \dfrac{x^2-1}{x^{3/2}} = x^{1/2} - x^{-3/2}$

49. $k(x)/p(x) = \dfrac{x^{3/2}}{x^{-6}} = x^{3/2}x^6 = x^{15/2}$

51. $f(x) + g(x) = \dfrac{x}{x-8} + \dfrac{-x}{x+2}$
$= \dfrac{x(x+2)}{(x-8)(x+2)} - \dfrac{x(x-8)}{(x+2)(x-8)}$
$= \dfrac{x^2 + 2x - x^2 + 8x}{(x-8)(x+2)}$
$= \dfrac{10x}{x^2 - 6x - 16}$

53. $f(x) + g(x) = \dfrac{x-4}{x+6} + \dfrac{x-6}{x+4}$
$= \dfrac{(x-4)(x+4)}{(x+6)(x+4)} + \dfrac{(x-6)(x+6)}{(x+4)(x+6)}$
$= \dfrac{x^2 - 16 + x^2 - 36}{(x+6)(x+4)}$
$= \dfrac{2x^2 - 52}{x^2 + 10x + 24}$

370 CHAPTER 2 SOLUTIONS

Exercise 4 (page 73)

1.

x	0.9	0.99	0.999	1.001	1.01	1.1
$f(x) = 2x$	1.8	1.98	1.998	2.002	2.02	2.2

$\lim_{x \to 1^-} f(x) = 2$ $\lim_{x \to 1^+} f(x) = 2$ $\lim_{x \to 1} f(x) = 2$

3.

x	0.1	0.01	0.001	-0.001	-0.01	-0.1
$f(x) = x^2 + 2$	2.01	2.0001	2.000001	2.000001	2.0001	2.01

$\lim_{x \to 0^-} f(x) = 2$ $\lim_{x \to 0^+} f(x) = 2$ $\lim_{x \to 0} f(x) = 2$

5.

x	1.9	1.99	1.999	2.001	2.01	2.1
$f(x) = \dfrac{x^2 - 4}{x - 2}$	3.9	3.99	3.999	4.001	4.01	4.1

$\lim_{x \to 2^-} f(x) = 4$ $\lim_{x \to 2^+} f(x) = 4$ $\lim_{x \to 2} f(x) = 4$

7.

x	-1.1	-1.01	-1.001	-0.999	-0.99	-0.9
$f(x) = \dfrac{x^3 + 1}{x + 1}$	1.11	1.0101	1.001001	.999001	.9901	.91

$\lim_{x \to -1^-} f(x) = 1$ $\lim_{x \to -1^+} f(x) = 1$ $\lim_{x \to -1} f(x) = 1$

9. $\lim_{x \to 2^-} f(x) = \lim_{x \to 2^-}(2x + 5) = 9$

$\lim_{x \to 2^+} f(x) = \lim_{x \to 2^+}(2x + 5) = 9$

$\lim_{x \to 2} f(x) = 9$

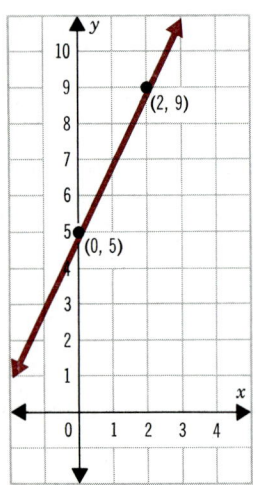

11. $\lim_{x \to 1^-} f(x) = \lim_{x \to 1^-}(3x - 1) = 2$

$\lim_{x \to 1^+} f(x) = \lim_{x \to 1^+} 3x - 1 = 2$

$\lim_{x \to 1} f(x) = 2$

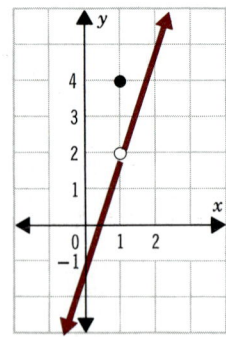

13. $\lim\limits_{x \to 1^-} f(x) = \lim\limits_{x \to 1^-} (3x - 1) = 2$

$\lim\limits_{x \to 1^+} f(x) = \lim\limits_{x \to 1^+} 2x = 2$

$\lim\limits_{x \to 1} f(x) = 2$

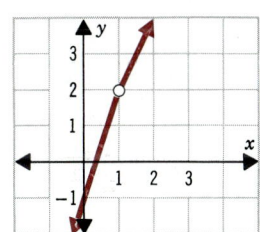

15. $\lim\limits_{x \to 0^-} f(x) = \lim\limits_{x \to 0^-} x^2 = 0$

$\lim\limits_{x \to 0^+} f(x) = \lim\limits_{x \to 0^+} (2x + 1) = 3$

$\lim\limits_{x \to 0} f(x)$ does not exist

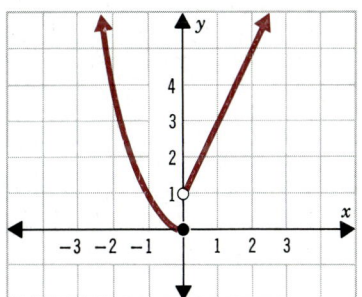

17. (a) $\lim\limits_{x \to 0^+} f(x) = \lim\limits_{x \to 0^+} (2x + 1) = 1$

(b) $\lim\limits_{x \to 1^-} f(x) = \lim\limits_{x \to 1^-} (2x + 1) = 1$

(c) $f(x)$ does not exist for $x < 0$ nor for $x > 1$

Exercise 5 (page 79)

1. $\lim\limits_{x \to 0} 3 = 3$ 3. $\lim\limits_{x \to 0} x = 0$

5. $\lim\limits_{x \to 2}(2x - 1) = \lim\limits_{x \to 2} 2x - \lim\limits_{x \to 2} 1 = \lim\limits_{x \to 2} 2 \cdot \lim\limits_{x \to 2} x - \lim\limits_{x \to 2} 1 = 2 \cdot 2 - 1 = 3$

7. $\lim\limits_{x \to 4}(x^2 + x) = \lim\limits_{x \to 4} x^2 + \lim\limits_{x \to 4} x = 4^2 + 4 = 20$

9. $\lim\limits_{x \to -3}(2x^2 - 1) = \lim\limits_{x \to -3} 2x^2 - \lim\limits_{x \to -3} 1 = \lim\limits_{x \to -3} 2 \cdot \lim\limits_{x \to -3} x^2 - \lim\limits_{x \to -3} 1 = 2(-3)^2 - 1 = 17$

11. $\lim\limits_{x \to 2} \dfrac{x^2 + 4}{x + 3} = \dfrac{\lim\limits_{x \to 2}(x^2 + 4)}{\lim\limits_{x \to 2}(x + 3)} = \dfrac{\lim\limits_{x \to 2} x^2 + \lim\limits_{x \to 2} 4}{\lim\limits_{x \to 2} x + \lim\limits_{x \to 2} 3} = \dfrac{4 + 4}{2 + 3} = \dfrac{8}{5}$

13. $\lim\limits_{x \to 2} \dfrac{x^2 - 4}{x - 2} = \lim\limits_{x \to 2} \dfrac{(x + 2)(x - 2)}{(x - 2)} = \lim\limits_{x \to 2}(x + 2) = 4$

15. $\lim\limits_{x \to -4} \dfrac{x^3 + 64}{x + 4} = \lim\limits_{x \to -4} \dfrac{(x + 4)(x^2 - 4x + 16)}{(x + 4)} = \lim\limits_{x \to -4}(x^2 - 4x + 16) = (-4)^2 + 16 + 16 = 48$

17. $\lim\limits_{x \to 1} \dfrac{\left(\frac{1}{x}\right) - 1}{x - 1} = \lim\limits_{x \to 1} \dfrac{1 - x}{x(x - 1)} = \lim\limits_{x \to 1} \dfrac{-1}{x} = -1$

19. $\lim\limits_{x \to 1} \dfrac{\sqrt{x} - 1}{x - 1} = \lim\limits_{x \to 1} \dfrac{(\sqrt{x} - 1)(\sqrt{x} + 1)}{(x - 1)(\sqrt{x} + 1)} = \lim\limits_{x \to 1} \dfrac{x - 1}{(x - 1)(\sqrt{x} + 1)} = \lim\limits_{x \to 1} \dfrac{1}{\sqrt{x} + 1} = \tfrac{1}{2}$

21. $\lim\limits_{x \to 4} \dfrac{x^2 - 16}{x - 4} = \lim\limits_{x \to 4} \dfrac{(x + 4)(x - 4)}{(x - 4)} = \lim\limits_{x \to 4}(x + 4) = 8$

23. $\lim\limits_{x \to 2} \dfrac{3x^2 + x - 14}{x - 2} = \lim\limits_{x \to 2}(3x + 7) = 13$

25. $\lim\limits_{x \to c}[2f(x)] = 2 \cdot 5 = 10$ 27. $\lim\limits_{x \to c} g(x)^3 = 2^3 = 8$

29. $\lim\limits_{x \to c} \dfrac{x^n - c^n}{x - c} = \lim\limits_{x \to c}(x^{n-1} + cx^{n-2} + \cdots + c^{n-1}) = \underbrace{\dfrac{c^{n-1} + cc^{n-2} + \cdots + c^{n-1}}{n \text{ terms}}} = nc^{n-1}$

Exercise 6 (page 79)

1. $\lim_{x \to 1}(2 - 7x) = 2 - 7(1) = -5$

3. $\lim_{x \to 2}(x^4 - 2x + 1) = (2)^4 - 2(2) + 1 = 13$

5. $\lim_{x \to 3}(x^2 - 2x + 7)/(2 - x) = \dfrac{(3)^2 - 2(3) + 7}{2 - 3} = -10$

7. $\lim_{x \to \frac{1}{2}}(2 - x^2)/(x^3 - 5x + 1) = \dfrac{2 - (\frac{1}{2})^2}{(\frac{1}{2})^3 - 5(\frac{1}{2}) + 1} = \dfrac{2 - \frac{1}{4}}{\frac{1}{8} - \frac{5}{2} + 1} = \frac{7}{4} \div (-\frac{11}{8}) = -\frac{14}{11}$

9. $\lim_{x \to -1}(1 - 6x + x^2)/(x^3 - 2x + 1) = \dfrac{1 - 6(-1) + (-1)^2}{(-1)^3 - 2(-1) + 1} = \dfrac{8}{2} = 4$

11. $\lim_{x \to 2^-} f(x) = \lim_{x \to 2}(2x + 5) = 9$; $\lim_{x \to 2^+} f(x) = \lim_{x \to 2}(4x + 1) = 9$; $f(2) = 2 \cdot 2 + 5 = 9$; $f(x)$ is continuous at $x = 2$

13. $\lim_{x \to 1^-} f(x) = \lim_{x \to 1}(3x - 1) = 2$; $\lim_{x \to 1^+} f(x) = \lim_{x \to 1} 2x = 2$; $f(1) = 4$; $f(x)$ is not continuous at $x = 1$, since $\lim_{x \to 1} f(x) \neq f(1)$

15. $f(x)$ is not continuous at $x = 1$, since $f(1)$ is not defined

17. $\lim_{x \to 0^-} f(x) = \lim_{x \to 0} x^2 = 0$; $\lim_{x \to 0^+} f(x) = \lim_{x \to 0} 2x = 0$; $f(0) = 0$; $f(x)$ is continuous at $x = 0$

19. $f(x)$ is not continuous at $x = 2$, since $f(2)$ is not defined; the discontinuity can be removed by defining $f(2) = 4$, since $\lim_{x \to 2} f(x) = 4$

21. (a) $C(x) = \begin{cases} 1200 & \text{if } x = 0 \\ 0.5x + 1700 & \text{if } 0 < x \leq 10{,}000 \\ 0.5x + 2200 & \text{if } 10{,}000 < x \leq 20{,}000 \\ 0.5x + 2700 & \text{if } 20{,}000 < x \leq 30{,}000 \\ 0.5x + 3200 & \text{if } 30{,}000 < x \leq 40{,}000 \end{cases}$

(b) $D = \{x \mid 0 \leq x \leq 40{,}000\}$

(c) Discontinuous at $x = 0$, $x = 10{,}000$, $x = 20{,}000$, $x = 30{,}000$

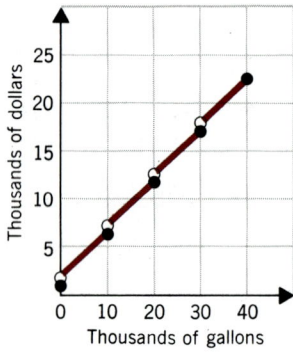

23. Discontinuous at $x = 1, 2, 3, 4, 6,$ and 8

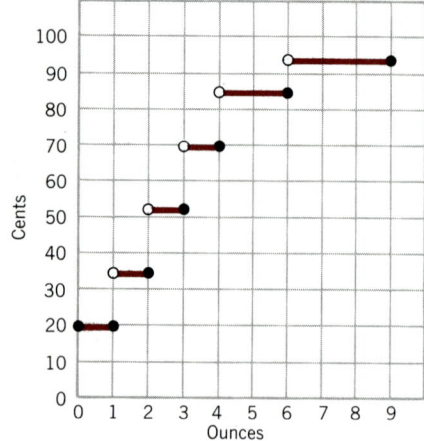

Review Exercises (page 81)
1. Yes   3. Yes   5. No   7. (a) $f(0) = 0$ (b) $f(2+h) = 8 + 6h + h^2$ (c) $f(2) = 8$
9. $y = x^2 + 2$ 11. $y = x^2 + 2x - 8$

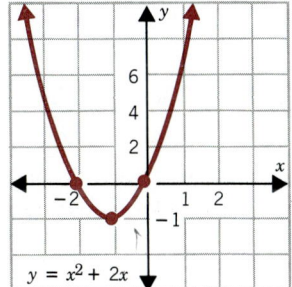

13. $y = x^2 + 2x + 4$ 15. Domain = $\mathbb{R}$

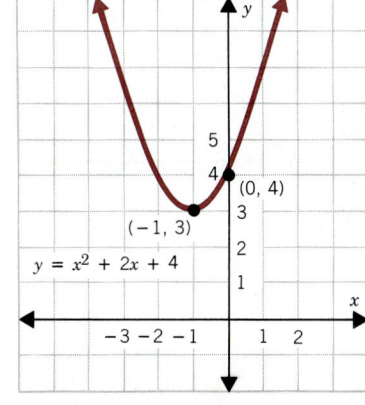

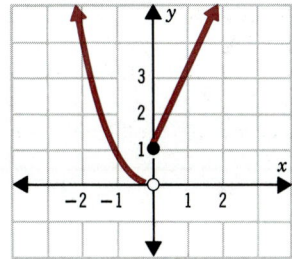

17. $\mathbb{R}$ 19. $\{x \mid x \leq 2\}$ 21. $\{x \mid x \neq 4\}$ 23. $\{x \mid x \neq 2, x \neq -3\}$
25. $\{x \mid x \neq -\frac{1}{3}, x \neq 2\}$ 27. $\{x \mid x > 1 \text{ or } x < -1\}$
29. $\quad y = x^2 - 3 = 2x$
$\quad x^2 - 2x - 3 = 0$
$(x - 3)(x + 1) = 0$
$x = 3, \quad y = 6$
$x = -1, \quad y = -2$
$(3, 6)$ and $(-1, -2)$

31. 5   33. 4   35. 7   37. $\lim\limits_{x \to 1} \dfrac{x^2 - 1}{x - 1} = \lim\limits_{x \to 1}(x + 1) = 2$

39. $\lim\limits_{x \to 1} \dfrac{x^3 - 1}{x - 1} = \lim\limits_{x \to 1}(x^2 + x + 1) = 3$

41. $\lim\limits_{x \to 1} \dfrac{\left(\dfrac{1}{x}\right) - 1}{x - 1} = \lim\limits_{x \to 1} \dfrac{1 - x}{x(x - 1)} = \lim\limits_{x \to 1} \dfrac{-1}{x} = -1$

43. $\dfrac{f(x) - f(4)}{x - 4} = \dfrac{4x - 16}{x - 4} = \dfrac{4(x - 4)}{x - 4}$; $\lim\limits_{x \to 4} \dfrac{f(x) - f(4)}{x - 4} = 4$

45. $\dfrac{f(x) - f(4)}{x - 4} = \dfrac{x^2 + x - 20}{x - 4} = \dfrac{(x + 5)(x - 4)}{x - 4}$; $\lim\limits_{x \to 4} \dfrac{f(x) - f(4)}{x - 4} = 9$

47. Continuous; $\lim\limits_{x \to 0} f(x) = 5 = f(0)$

49. Continuous; $\lim\limits_{x \to 4} f(x) = \lim\limits_{x \to 4}(x + 4) = 8 = f(4)$

51. Not continuous; $\lim\limits_{x \to 2} f(x) = 12$, $f(2) = 4$

CHAPTER 3

Exercise 1 (page 97)

1. (a) $\Delta y = f(4) - f(0) = 18 - (-2) = 20$
  (b) $\Delta y = f(2) - f(-3) = 8 - (-17) = 25$
  (c) $\Delta x = 4 - 0 = 4$; $\dfrac{\Delta y}{\Delta x} = \dfrac{20}{4} = 5$
  (d) $\Delta x = 2 - (-3) = 5$; $\dfrac{\Delta y}{\Delta x} = \dfrac{25}{5} = 5$
  (e) $\dfrac{\Delta y}{\Delta x} = \dfrac{5d - 2 - (5c - 2)}{d - c} = \dfrac{5d - 2 - 5c + 2}{d - c} = \dfrac{5d - 5c}{d - c} = 5$
  The average rate of change is equal to 5 regardless of c and d. The graph is a straight line with slope = 5.

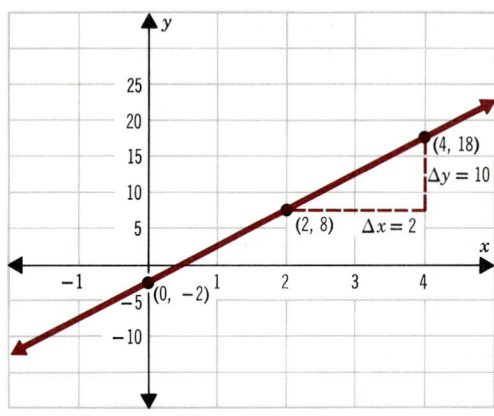

3. (a) $\Delta y = f(1) - f(-1) = 1 - (-1) = 2$ (b) $\Delta y = f(4) - f(0) = 64 - 0 = 64$
  (c) $\Delta x = 1 - (-1) = 2$; $\dfrac{\Delta y}{\Delta x} = \dfrac{2}{2} = 1$ (d) $\Delta x = 4 - 0 = 4$; $\dfrac{\Delta y}{\Delta x} = \dfrac{64}{4} = 16$

5. (a) $\dfrac{\Delta s}{\Delta t} = \dfrac{f(3) - f(2)}{3 - 2} = \dfrac{72 - 36}{1} = 36$ (b) $\dfrac{\Delta s}{\Delta t} = \dfrac{f(2.5) - f(2)}{2.5 - 2} = \dfrac{52.5 - 36}{0.5} = 33$

7. $\dfrac{\Delta y}{\Delta x} = \dfrac{(3x - 5) - (-2)}{x - 1} = \dfrac{3x - 3}{x - 1} = 3$ **9.** $\dfrac{\Delta y}{\Delta x} = \dfrac{(3x^2 - 2) - (1)}{x - 1} = \dfrac{3x^2 - 3}{x - 1} = 3(x + 1)$

11. $\dfrac{\Delta y}{\Delta x} = \dfrac{(x - x^2) - 0}{x - 1} = \dfrac{x(1 - x)}{x - 1} = -x$ **13.** $\dfrac{\Delta y}{\Delta x} = \dfrac{(x^2 + 2x - 2) - 1}{x - 1} = \dfrac{x^2 + 2x - 3}{x - 1} = x + 3$

15. $\dfrac{\Delta y}{\Delta x} = \dfrac{(3x^2 - 2x + 5) - 6}{x - 1} = \dfrac{3x^2 - 2x - 1}{x - 1} = 3x + 1$

CHAPTER 3 SOLUTIONS 375

17. $m = \text{Slope} = \dfrac{\Delta y}{\Delta x} = \dfrac{f(x) - f(0)}{x - 0} = \dfrac{(3x - 2) - (-2)}{x} = \dfrac{3x}{x} = 3$

19. $m = \dfrac{f(x) - f(0)}{x - 0} = \dfrac{(x^2 + 4) - (4)}{x} = x$ 21. $m = \dfrac{f(x) - f(0)}{x - 0} = \dfrac{(x^3 + 2) - 2}{x} = \dfrac{x^3}{x} = x^2$

23. (a) $\dfrac{\Delta s}{\Delta t} = \dfrac{f(3.5) - f(3)}{3.5 - 3} = \dfrac{196 - 144}{0.5} = \dfrac{52}{0.5} = 104$ feet per second

    (b) $\dfrac{\Delta s}{\Delta t} = \dfrac{f(3.1) - f(3)}{3.1 - 3} = \dfrac{153.76 - 144}{0.1} = \dfrac{9.76}{0.1} = 97.6$ feet per second

25. From 0 to 16 hours:
    $\dfrac{\Delta s}{\Delta t} = \dfrac{\sqrt{16} - \sqrt{0}}{16 - 0} = \dfrac{4}{16} = \dfrac{1}{4} = 0.25$ kph
    From 1 to 4 hours:
    $\dfrac{\Delta s}{\Delta t} = \dfrac{\sqrt{4} - \sqrt{1}}{4 - 1} = \dfrac{2 - 1}{4 - 1} = \dfrac{1}{3} = 0.33$ kph
    From 1 to 2 hours:
    $\dfrac{\Delta s}{\Delta t} = \dfrac{\sqrt{2} - \sqrt{1}}{2 - 1} \approx \dfrac{1.41 - 1}{1} = 0.41$ kph

27. From 0 to 50 pounds:
    $\dfrac{\Delta y}{\Delta x} = \dfrac{47.3 - 45}{50 - 0} = \dfrac{2.3}{50} = 0.046$
    From 10 to 40 pounds:
    $\dfrac{\Delta y}{\Delta x} = \dfrac{47.6 - 46.2}{40 - 10} = \dfrac{1.4}{30} = 0.047$
    From 10 to 30 pounds:
    $\dfrac{\Delta y}{\Delta x} = \dfrac{48.2 - 46.2}{30 - 10} = \dfrac{2}{20} = 0.1$

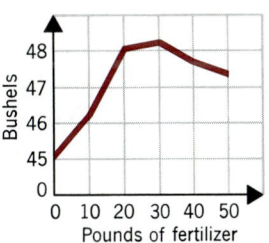

29. $\dfrac{\Delta M}{\Delta t} = \dfrac{[4.5 - (0.03)(2)^2] - [4.5 - (0.03)(0)^2]}{2 - 0}$

    $= \dfrac{4.38 - 4.5}{2} = -0.06$

The mass is decreasing at an average rate of 0.06 gram per hour for the first 2 hours.

Exercise 2 (page 104)

1. $\dfrac{f(x) - f(1)}{x - 1} = \dfrac{2x + 3 - 5}{x - 1} = \dfrac{2x - 2}{x - 1} = 2; \quad f'(1) = \lim\limits_{x \to 1} \dfrac{f(x) - f(1)}{x - 1} = 2$

3. $\dfrac{f(x) - f(0)}{x - 0} = \dfrac{(x^2 - 2) + 2}{x} = \dfrac{x^2}{x} = x; \quad f'(0) = \lim\limits_{x \to 0} \dfrac{f(x) - f(0)}{x - 0} = \lim\limits_{x \to 0} x = 0$

5. $\dfrac{f(x) - f(-1)}{x - (-1)} = \dfrac{(3x^2 + x + 5) - 7}{x + 1} = \dfrac{3x^2 + x - 2}{x + 1} = 3x - 2; \quad f'(-1) = \lim\limits_{x \to -1} \dfrac{f(x) - f(-1)}{x - (-1)} = -5$

7. $\dfrac{f(x) - f(4)}{x - 4} = \dfrac{\left(\frac{1}{x}\right) - \left(\frac{1}{4}\right)}{x - 4} = \dfrac{\frac{(4 - x)}{4x}}{x - 4} = \dfrac{(4 - x)}{4x(x - 4)} = -\dfrac{1}{4x}; \quad f'(4) = \lim\limits_{x \to 4}\left(-\dfrac{1}{4x}\right) = -\dfrac{1}{16}$

9. $\dfrac{f(x + \Delta x) - f(x)}{\Delta x} = \dfrac{2(x + \Delta x) + 3 - (2x + 3)}{\Delta x} = \dfrac{2x + 2\Delta x + 3 - 2x - 3}{\Delta x} = \dfrac{2\Delta x}{\Delta x} = 2; \quad f'(x) = 2$

11. $\dfrac{f(x + \Delta x) - f(x)}{\Delta x} = \dfrac{(x + \Delta x)^2 - 2 - (x^2 - 2)}{\Delta x} = \dfrac{x^2 + 2x\,\Delta x + (\Delta x)^2 - 2 - x^2 + 2}{\Delta x} = \dfrac{2x\,\Delta x + (\Delta x)^2}{\Delta x}$

    $= 2x + \Delta x; \quad f'(x) = \lim\limits_{\Delta x \to 0}(2x + \Delta x) = 2x$

376 CHAPTER 3 SOLUTIONS

13. $\dfrac{f(x + \Delta x) - f(x)}{\Delta x} = \dfrac{3(x + \Delta x)^2 + (x + \Delta x) + 5 - (3x^2 + x + 5)}{\Delta x}$

$= \dfrac{3x^2 + 6x \Delta x + 3(\Delta x)^2 + x + \Delta x + 5 - 3x^2 - x - 5}{\Delta x}$

$= \dfrac{6x \Delta x + 3(\Delta x)^2 + \Delta x}{\Delta x} = 6x + 1 + 3\Delta x; \quad f'(x) = \lim_{\Delta x \to 0}(6x + 1 + 3\Delta x) = 6x + 1$

15. $\dfrac{f(x + \Delta x) - f(x)}{\Delta x} = \dfrac{5 - 5}{\Delta x} = 0; \quad f'(x) = \lim_{\Delta x \to 0} 0 = 0$

17. $\dfrac{f(x + \Delta x) - f(x)}{\Delta x} = \dfrac{\dfrac{1}{x + \Delta x} - \dfrac{1}{x}}{\Delta x} = \dfrac{\dfrac{x - (x + \Delta x)}{(x + \Delta x)x}}{\Delta x}$

$= \dfrac{x - x - \Delta x}{(x + \Delta x)x \cdot \Delta x} = -\dfrac{1}{(x + \Delta x)x}; \quad f'(x) = \lim_{\Delta x \to 0} -\dfrac{1}{(x + \Delta x)x} = -\dfrac{1}{x^2}$

19. $\dfrac{f(x + \Delta x) - f(x)}{\Delta x} = \dfrac{m(x + \Delta x) + b - mx - b}{\Delta x} = \dfrac{mx + m\Delta x + b - mx - b}{\Delta x} = \dfrac{m \Delta x}{\Delta x} = m = f'(x)$

21. $s'(0) = \lim_{t \to 0} \dfrac{s(t) - s(0)}{t - 0} = \lim_{t \to 0} \dfrac{3t^2 + 4t}{t} = \lim_{t \to 0}(3t + 4) = 4$

$s'(2) = \lim_{t \to 2} \dfrac{s(t) - s(2)}{t - 2} = \lim_{t \to 2} \dfrac{3t^2 + 4t - 20}{t - 2} = \lim_{t \to 2}(3t + 10) = 16$

$s'(t) = \lim_{\Delta t \to 0} \dfrac{s(t + \Delta t) - s(t)}{\Delta t} = \lim_{\Delta t \to 0} \dfrac{3(t + \Delta t)^2 + 4(t + \Delta t) - 3t^2 - 4t}{\Delta t}$

$= \lim_{\Delta t \to 0} \dfrac{6t \Delta t + (\Delta t)^2 + 4 \Delta t}{\Delta t} = 6t + 4$

23. $s'(t) = 100 - 32t; \quad s'(0) = 100$ feet per second, $s'(1) = 100 - 32 = 68$ feet per second, $s'(4) = 100 - 128 = -28$ feet per second;
$s(t) = 0$ when $t = 0$ or $16t = 100$ ($t = 6.25$ seconds);
$s'(t) = 0$ when $32t = 100$ ($t = 3.125$ seconds)

Exercise 3 (page 108)

1. $f'(3) = 6$
$y - 9 = 6(x - 3)$
$y = 6x - 9$

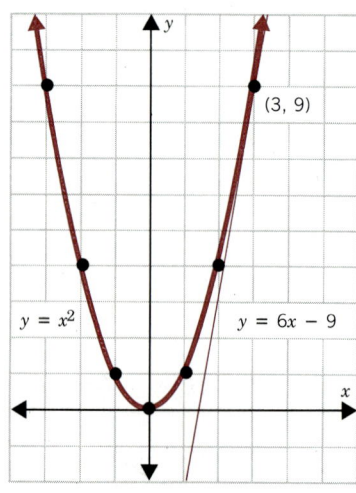

3. $f'(1) = 4$
    $y - 4 = 4(x - 1)$
    $y = 4x$

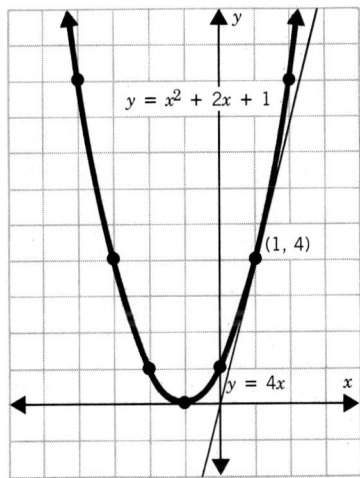

5. $f'(1) = -1$
    $y - 1 = -(x - 1)$
    $y = -x + 2$

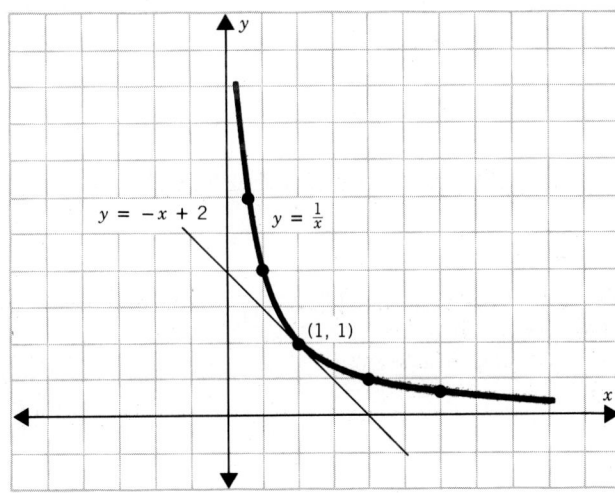

7. No; $f'(1) = 2$, so the equation of the tangent line is $y - 1 = 2(x - 1)$ or $y = 2x - 1$, which does not pass through $(2, 5)$.
9. Let (a, a^2) be a point on the graph. The tangent line at such a point is $y - a^2 = 2a(x - a)$ or $y = 2ax - a^2$. If $(1, 0)$ is a point on this line, $a = 2$. Thus, the point on the graph is $(2, 4)$.
11. (a) $\Delta A = \pi(R + \Delta R)^2 - \pi R^2 = 2\pi R \, \Delta R + \pi(\Delta R)^2$
    (b) $\Delta C = 2\pi(R + \Delta R) - 2\pi R = 2\pi \, \Delta R$
    (c) $\dfrac{\Delta A}{\Delta R} = 2\pi R + \pi \Delta R$ (d) $\dfrac{\Delta C}{\Delta R} = 2\pi$ (e) $\lim\limits_{\Delta R \to 0} \dfrac{\Delta C}{\Delta R} = 2\pi$
13. (a) $A = .11 \, m^{2/3}$
    $A = (.11)(64)^{2/3} = (.11)(16) = 1.76$ square meters
    (b) $A = .112 \, m^{2/3}$
    $\dfrac{dA}{dm} = (.112)(\tfrac{2}{3}m^{-1/3}) \approx .07467 m^{-1/3}$

Exercise 4 (page 111)

1. $f'(x) = 0$ 3. $f'(x) = 4x^{4-1} = 4x^3$ 5. $f'(x) = 5x^4$ 7. $f'(x) = -\dfrac{2}{x^3}$ 9. $f'(x) = \tfrac{2}{3}x^{-1/3}$

11. $f'(x) = \dfrac{-3}{x^4}$ 13. $f'(x) = 1.3x^{0.3}$ 15. $f'(x) = \sqrt{5}\,x^{\sqrt{5}-1}$ 17. $m = f'(1) = 4$

19. $m = f'(4) = \frac{1}{4}$ **21.** $m = f'(-8) = -\frac{1}{3}(-8)^{-4/3} = -\frac{1}{48}$

23. $\dfrac{d}{dx}(x^5) = 5x^4$ **25.** $\dfrac{d}{dx}(x^{-1/3}) = -\frac{1}{3}x^{-4/3}$ **27.** $\dfrac{d}{dx}(\sqrt[3]{x}) = \dfrac{d}{dx}(x^{1/3}) = \frac{1}{3}x^{-2/3} = \dfrac{1}{3\sqrt[3]{x^2}}$

29. STEP 1: $f(2) = \frac{1}{2}$ STEP 2: $f(x) - f(2) = \dfrac{1}{x} - \dfrac{1}{2} = \dfrac{2-x}{2x}$

STEP 3: $\dfrac{f(x) - f(2)}{x - 2} = \dfrac{2-x}{2x(x-2)} = \dfrac{-1}{2x}$ STEP 4: $\lim\limits_{x \to 2} \dfrac{f(x) - f(2)}{x - 2} = \lim\limits_{x \to 2} \dfrac{-1}{2x} = -\frac{1}{4}$

31. $s(t) = t^{3/2}$; $s'(t) = \frac{3}{2}t^{1/2}$; $s'(1) = \frac{3}{2} \cdot 1 = 1.5$ feet per second

If $s(t) = 8 = t^{3/2}$, then $t^3 = 64$ and $t = 4$ is the time required for the child to strike the ground.
$s'(4) = \frac{3}{2}\sqrt{4} = 3$ feet per second

Exercise 5 (page 115)

1. $f'(x) = 45x^{14}$ **3.** $f'(x) = 3$ **5.** $f'(x) = 2x + 3$ **7.** $f'(x) = 40x^4 - 5$ **9.** $f'(x) = \frac{4}{3}x^3 - 3$

11. $f'(x) = 3\pi x^2 + 3x$ **13.** $f'(x) = \frac{1}{3}(5x^4) = \frac{5}{3}x^4$ **15.** $f'(x) = 2ax + b$

17. $f'(x) = (3x^2 - 5)(2) + (2x + 1)(6x) = 6x^2 - 10 + 12x^2 + 6x = 18x^2 + 6x - 10$

19. $f'(t) = (2t^5 - t)(3t^2) + (t^3 + 1)(10t^4 - 1) = 6t^7 - 3t^3 + 10t^7 - t^3 + 10t^4 - 1$
$= 16t^7 + 10t^4 - 4t^3 - 1$

21. $f'(t) = -3t^{-4}$ **23.** $f'(x) = -40x^{-5} - 6x^{-3} = \dfrac{-40}{x^5} - \dfrac{6}{x^3}$ **25.** $f'(s) = \dfrac{(s+1)2 - 2s(1)}{(s+1)^2} = \dfrac{2}{(s+1)^2}$

27. $f'(x) = \dfrac{(3x+4)8x - (4x^2 - 2)3}{(3x+4)^2} = \dfrac{24x^2 + 32x - 12x^2 + 6}{(3x+4)^2} = \dfrac{12x^2 + 32x + 6}{(3x+4)^2}$

29. $f'(t) = 3 - \dfrac{1}{3t^2}$ **31.** $f'(u) = \dfrac{(1+2u)(-2) - (1-2u)(2)}{(1+2u)^2} = \dfrac{-2 - 4u - 2 + 4u}{(1+2u)^2} = \dfrac{-4}{(1+2u)^2}$

33. $f'(x) = 9x^2 + \dfrac{2}{3x^3}$ **35.** $f'(t) = -\dfrac{1}{t^2} + \dfrac{2}{t^3} - \dfrac{3}{t^4}$ **37.** $f'(w) = \dfrac{-3w^2}{(w^3 - 1)^2}$

39. $f'(x) = 3x^2 + 3$; $f'(0) = 3 = m$; $y + 1 = 3(x - 0)$ or $y = 3x - 1$

41. $f'(x) = \dfrac{(x+1)3x^2 - x^3(1)}{(x+1)^2} = \dfrac{2x^3 + 3x^2}{(x+1)^2}$; $f'(1) = \dfrac{2+3}{2^2} = \frac{5}{4}$; $y - \frac{1}{2} = \frac{5}{4}(x - 1)$

43. $f'(x) = 6x - 12 = 0$ when $x = 2$ **45.** $f'(x) = 3x^2 - 3 = 0$ when $x = \pm 1$

47. $f'(x) = \dfrac{(x+1)2x - x^2}{(x+1)^2} = \dfrac{x^2 + 2x}{(x+1)^2} = 0$ when $x = -2$ or 0

49. (a) $f'(x) = 2x(3x - 2) + x^2(3) = x(9x - 4)$
(b) $f(x) = 3x^3 - 2x^2$; $f'(x) = 9x^2 - 4x = x(9x - 4)$

51. $\sqrt{3}$ **53.** $2\pi R$ **55.** $4\pi R^2$

57. $\dfrac{ds}{dt} = 3t^2 - 1$; $\dfrac{ds}{dt} = -1$ meter per second when $t = 0$; $\dfrac{ds}{dt} = 74$ meters per second when $t = 5$

59. Instantaneous rate of change of satisfaction with respect to reward is
$S'(r) = \dfrac{a(g - r) - ar(-1)}{(g - r)^2} = \dfrac{ag}{(g - r)^2}$

61. Rate at any time t is $8 - 2t^{-1/2}$; hence, at $t = 4$, Rate $= 8 - \dfrac{2}{\sqrt{4}} = 7$ liters per hour

63. $F'(x) = \dfrac{d}{dx}[f(x) - g(x)] = \dfrac{d}{dx}\{f(x) + [-g(x)]\}$
$= \dfrac{d}{dx}f(x) + \dfrac{d}{dx}[-g(x)] = f'(x) + [-g'(x)] = f'(x) - g'(x)$

65. $\dfrac{d}{dx}[f(x)g(x)h(x)] = \dfrac{d}{dx}\{f(x)[g(x)h(x)]\} = f'(x)[g(x)h(x)] + f(x)\dfrac{d}{dx}[g(x)h(x)]$
$= f'(x)g(x)h(x) + f(x)[g(x)h'(x) + g'(x)h(x)]$
$= g(x)h(x)f'(x) + f(x)g(x)h'(x) + h(x)f(x)g'(x)$

$\dfrac{d}{dx}[f(x)]^3 = \dfrac{d}{dx}[f(x)f(x)f(x)] = 3f(x)f(x)f'(x) = 3[f(x)]^2 f'(x)$

67. $\dfrac{dy}{dx} = (x-1)(x^2+5)(3x^2) + (x-1)(x^3-1)(2x) + (x^2+5)(x^3-1)$

Exercise 6 (page 123)

1. $f'(x) = 2(3x+5)(3) = 6(3x+5)$ **3.** $f'(x) = -3(6x-5)^{-4}(6) = -18(6x-5)^{-4}$

5. $f'(x) = 4(x^2+5)^3(2x) = 8x(x^2+5)^3$ **7.** $f'(t) = 7(t^5 - t^2 + t)^6(5t^4 - 2t + 1)$

9. $f'(x) = 3\left(x - \dfrac{1}{x}\right)^2\left(1 + \dfrac{1}{x^2}\right)$

11. $f'(z) = \dfrac{(z+1)3(3z-1)^2(3) - (3z-1)^3}{(z+1)^2} = \dfrac{(3z-1)^2[9(z+1) - (3z-1)]}{(z+1)^2} = \dfrac{2(3z-1)^2[3z+5]}{(z+1)^2}$

13. $f'(x) = \dfrac{(x^3-1)2(x^2+1)(2x) - (x^2+1)^2(3x^2)}{(x^3-1)^2} = \dfrac{x(x^2+1)(x^3-3x-4)}{(x^3-1)^2}$

15. $f'(x) = \tfrac{1}{2}(x^2+1)^{-1/2}(2x) = \dfrac{x}{\sqrt{x^2+1}}$

17. $f'(x) = \tfrac{1}{3}(3x-1)^{-2/3}(3) = (3x-1)^{-2/3}$

19. $f'(x) = x[\tfrac{1}{2}(x^2+1)^{-1/2}(2x)] + (x^2+1)^{1/2}(1) = \dfrac{x^2}{\sqrt{x^2+1}} + \sqrt{x^2+1}$

$= \dfrac{x^2}{\sqrt{x^2+1}} + \dfrac{x^2+1}{\sqrt{x^2+1}} = \dfrac{2x^2+1}{\sqrt{x^2+1}}$

21. $f'(x) = x^2[\tfrac{1}{2}(3x+1)^{-1/2}(3)] + (3x+1)^{1/2}(2x) = \dfrac{3x^2}{2\sqrt{3x+1}} + 2x\sqrt{3x+1}$

$= \dfrac{3x^2}{2\sqrt{3x+1}} + \dfrac{4x(3x+1)}{2\sqrt{3x+1}} = \dfrac{15x^2+4x}{2\sqrt{3x+1}}$

23. $f'(x) = 2\left(\dfrac{3x-1}{3x+1}\right)\left[\dfrac{3(3x+1) - (3x-1)3}{(3x+1)^2}\right] = \dfrac{12(3x-1)}{(3x+1)^3}$

25. $y = u^5,\ u = x^3 + 1$

$\dfrac{dy}{dx} = \dfrac{dy}{du}\dfrac{du}{dx}$

$= (5u^4)(3x^2)$
$= 5(x^3+1)^4(3x^2)$
$= 15x^2(x^3+1)^4$

27. $y = \dfrac{u}{u+1},\ u = x^2 + 1$

$\dfrac{dy}{dx} = \dfrac{dy}{du}\dfrac{du}{dx}$

$= \left[\dfrac{(u+1) - u}{(u+1)^2}\right][2x]$

$= \left(\dfrac{1}{(u+1)^2}\right)2x$

$= \dfrac{2x}{((x^2+1)+1)^2}$

$= \dfrac{2x}{(x^2+2)^2}$

29. $y = (u+1)^2,\ u = \dfrac{1}{x}$

$\dfrac{dy}{dx} = \dfrac{dy}{du}\dfrac{du}{dx}$

$= 2(u+1)\left(-\dfrac{1}{x^2}\right)$

$= \dfrac{-2}{x^2}\left(\dfrac{1}{x}+1\right)$

$= -\dfrac{2}{x^3} - \dfrac{2}{x^2}$

31. $y = (u^3-1)^5,\ u = x^{-2}$

$\dfrac{dy}{dx} = \dfrac{dy}{du}\dfrac{du}{dx}$

$= [5(u^3-1)^4(3u^2)][-2x^{-3}]$
$= -30x^{-3}u^2(u^3-1)^4$
$= -30x^{-3}(x^{-2})^2((x^{-2})^3 - 1)^4$
$= -30x^{-7}\left(\dfrac{1}{x^6} - 1\right)^4$
$= -30x^{-31}(1-x^6)^4$

33. $y = \dfrac{1}{u^3 + 1}$, $u = \dfrac{1}{x^3 + 1}$

$\dfrac{dy}{dx} = \dfrac{dy}{du} \cdot \dfrac{du}{dx}$

$= [-(u^3 + 1)^{-2}(3u^2)][-(x^3 + 1)^{-2}3x^2]$

$= \dfrac{9u^2 x^2}{(u^3 + 1)^2 (x^3 + 1)^2}$

$= \dfrac{\dfrac{9x^2}{(x^3+1)^2}}{\left(\dfrac{1}{(x^3+1)^3} + 1\right)^2 (x^3 + 1)^2}$

$= \dfrac{9x^2 (x^3+1)^2}{\left(\dfrac{1}{(x^3+1)^3} + 1\right)^2 (x^3+1)^6}$

$= \dfrac{9x^2(x^3+1)^2}{(1 + (x^3+1)^3)^2}$

35. (a) $y = u^2$, $u = x^3 + 1$

$\dfrac{dy}{dx} = \dfrac{dy}{du} \cdot \dfrac{du}{dx}$
$= (2u)(3x^2)$
$= 6x^2(x^3 + 1)$
$= 6x^5 + 6x^2$

(b) $y' = 2(x^3 + 1)(3x^2)$
$= 6x^5 + 6x^2$

(c) $y = x^6 + 2x^3 + 1$
$y' = 6x^5 + 6x^2$

(d) all equal

37. $f'(x) = \sqrt{1 - x^2} - x[\tfrac{1}{2}(1 - x^2)^{-1/2}(2x)] = \dfrac{1 - 2x^2}{\sqrt{1 - x^2}}$; $f'(x) = 0$ when $x = \dfrac{\pm 1}{\sqrt{2}}$;

$f'(x)$ does not exist at $x = \pm 1$

39. $s(t) = \sqrt{(52 - 8t)^2 + (12t)^2} = 4\sqrt{13t^2 - 52t + 169}$

$s'(t) = \tfrac{1}{2} \cdot 4(13t^2 - 52t + 169)^{-1/2}(26t - 52) = \dfrac{52(t - 2)}{\sqrt{13(t^2 - 4t + 13)}}$

$s'(1) = \dfrac{52(1 - 2)}{\sqrt{13(1 - 4 + 13)}} = -\dfrac{52}{\sqrt{130}} = -4.56$ kph

$s'(4) = \dfrac{52(4 - 2)}{\sqrt{13(16 - 16 + 13)}} = \dfrac{104}{\sqrt{13^2}} = \dfrac{104}{13} = 8$ kph

Since $8 > 0$, the ships are receding at 10 P.M.

$s'(t) = 0 \Rightarrow t - 2 = 0$
$t = 2$

Hence, they are closest at 8 P.M.

41. $A'(t) = 3(t^{1/4} + 3)^2 \cdot \tfrac{1}{4} t^{-3/4} = \dfrac{3(t^{1/4} + 3)^2}{4\sqrt[4]{t^3}}$; $A'(16) = \dfrac{3(16^{1/4} + 3)^2}{4\sqrt[4]{16^3}} = \dfrac{3(5)^2}{32} = \tfrac{75}{32} = 2.34375$

Exercise 7 (page 129)

1. $f'(x) = 2$; $f''(x) = 0$
3. $f'(x) = 6x + 1$; $f''(x) = 6$
5. $f'(x) = 1 - \dfrac{1}{x^2}$; $f''(x) = 0 - (-2x^{-3}) = \dfrac{2}{x^3}$
7. $f'(t) = \dfrac{(t + 1) - t}{(t + 1)^2} = \dfrac{1}{(t + 1)^2}$; $f''(t) = -2(t + 1)^{-3} = \dfrac{-2}{(t + 1)^3}$
9. $f'(x) = \dfrac{2x(x + 1) - x^2}{(x + 1)^2} = \dfrac{x(x + 2)}{(x + 1)^2}$;

$f''(x) = \dfrac{(x + 1)^2[(x + 2) + x] - x(x + 2)[2(x + 1)]}{(x + 1)^4} = \dfrac{2(x + 1)^2 - 2x(x + 2)}{(x + 1)^3} = \dfrac{2}{(x + 1)^3}$

11. $f'(w) = 3(w^2 + 3)^2 2w = 6w(w^2 + 3)^2$;
$f''(w) = 6(w^2 + 3)^2 + 12w(w^2 + 3)(2w) = 6(w^2 + 3)(5w^2 + 3)$

13. $f'(x) = \dfrac{1}{2\sqrt{x}}$; $f''(x) = \dfrac{-1}{4x^{3/2}}$

15. $f'(z) = \frac{3}{2}(3z^2 + 1)^{1/2}(6z) = 9z(3z^2 + 1)^{1/2}$
$f''(z) = 9(3z^2 + 1)^{1/2} + 9z[\frac{1}{2}(3z^2 + 1)^{-1/2}(6z)] = \dfrac{9(6z^2 + 1)}{\sqrt{3z^2 + 1}}$

17. $f^{(4)}(x) = 0$, since the degree of the polynomial is <4

19. $\dfrac{d^{20}}{dx^{20}}(8x^{19} - 2x^{14} + 2x^5) = 0$

21. $\dfrac{d^8}{dx^8}(\frac{1}{8}x^8 - \frac{1}{7}x^7 + x^5 - x^3) = 8!(\frac{1}{8}) = 7! = 5040$

23. $v = \dfrac{ds}{dt} = 32t + 20;\quad a = \dfrac{dv}{dt} = 32$

25. $v = 9.8t + 4;\quad a = 9.8$

27. $f'(x) = n(2x + 3)^{n-1}(2)$
$f''(x) = n(n-1)(2x+3)^{n-2}(2)^2$
$f'''(x) = n(n-1)(n-2)(2x+3)^{n-3}(2)^3$
$\vdots$
$f^{(n)}(x) = 2^n n!$

29. $f'(x) = x^2 g'(x) + 2xg(x)$
$f''(x) = x^2 g''(x) + 2xg'(x) + 2xg'(x) + 2g(x) = x^2 g''(x) + 4xg'(x) + 2g(x)$

31. $s'(t) = 80 - 32t$
(a) $s'(2) = 16$ feet per second
(b) $s'(t) = 0$ when $t = \frac{80}{32} = 2.5$ seconds
(c) $s(2.5) = 6 + 80(2.5) - 16(2.5)^2 = 106$ feet
(d) $s''(t) = -32$ feet per second per second
(e) $s(t) = 0$ when $t = \dfrac{10 \pm \sqrt{106}}{4} = \dfrac{10 + \sqrt{106}}{4}$ since t must be greater than 0; $\dfrac{10 + \sqrt{106}}{4} \approx 5.07$ seconds
(f) $s'\left(\dfrac{10 + \sqrt{106}}{4}\right) = 80 - 32\left(\dfrac{10 + \sqrt{106}}{4}\right) = -8\sqrt{106}$ feet per second
(g) Since $s(0) = 6$ and the maximum height is 106, the ball travels 100 feet to a height of 106 feet and then falls to the ground: $100 + 106 = 206$ feet.

33. $s'(t) = -3(2-t)^2(-1) = 3(2-t)^2$;
$s'(1) = 3$ meters per second; $s''(t) = -6(2-t) = 6t - 12$

Exercise 8 (page 134)

1. f is continuous at c.
3. f is not differentiable at c.
5. $\lim\limits_{x \to 1^-} \dfrac{f(x) - f(1)}{x - 1} = \lim\limits_{x \to 1^-} \dfrac{2x + 3 - 5}{x - 1} = 2;\quad \lim\limits_{x \to 1^+} \dfrac{f(x) - f(1)}{x - 1} = \lim\limits_{x \to 1^+} \dfrac{x^2 + 4 - 5}{x - 1} = 2;\quad f'(1) = 2$

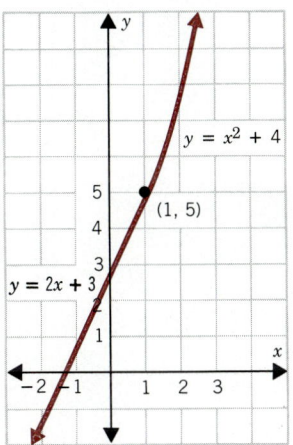

7. $\lim\limits_{x\to 1/2^-} \dfrac{f(x)-f(\frac{1}{2})}{x-\frac{1}{2}} = \lim\limits_{x\to 1/2^-} \dfrac{(-4+2x)-(-3)}{x-\frac{1}{2}} = \lim\limits_{x\to 1/2^-} \dfrac{2x-1}{x-\frac{1}{2}} = 2;$

$\lim\limits_{x\to 1/2^+} \dfrac{f(x)-f(\frac{1}{2})}{x-\frac{1}{2}} = \lim\limits_{x\to 1/2^+} \dfrac{4x^2-4-(-3)}{x-\frac{1}{2}} = \lim\limits_{x\to 1/2^+} \dfrac{4x^2-1}{x-\frac{1}{2}} = 4;$ $f'(\frac{1}{2})$ does not exist

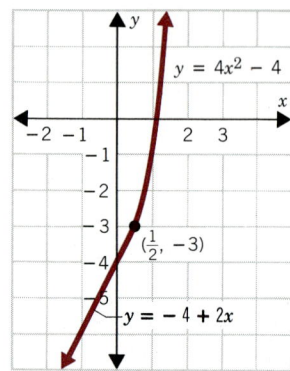

9. $\lim\limits_{x\to 2^-} \dfrac{f(x)-f(2)}{x-2} = \lim\limits_{x\to 2^-} \dfrac{4-x^2}{x-2} = -4;$ $\lim\limits_{x\to 2^+} \dfrac{f(x)-f(2)}{x-2} = \lim\limits_{x\to 2^+} \dfrac{x^2-4}{x-2} = 4;$ $f'(2)$ does not exist

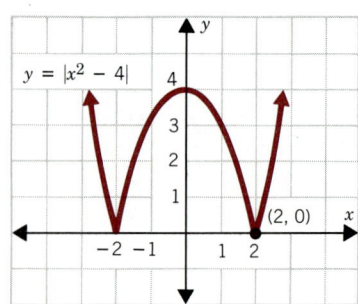

11. $\lim\limits_{x\to -1^-} \dfrac{f(x)-f(-1)}{x-(-1)} = \lim\limits_{x\to -1^-} \dfrac{(2x^2+1)-0}{x+1} = \lim\limits_{x\to -1^-} \dfrac{2x^2+1}{x+1}$ does not exist; $f'(-1)$ does not exist

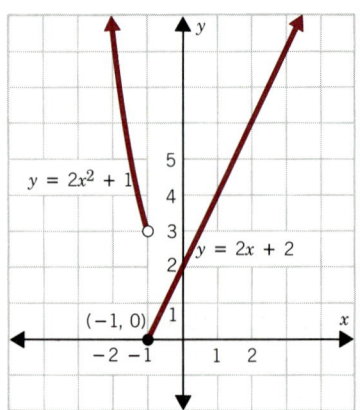

13. (a) $\lim_{x \to 0^-} f(x) = \lim_{x \to 0^-} x^3 = 0$;
$\lim_{x \to 0^+} f(x) = \lim_{x \to 0^+} x^2 = 0$;
$f(0) = 0$
Continuous at 0

(b) $\lim_{x \to 0^-} \frac{f(x) - f(0)}{x - 0} = \lim_{x \to 0^-} \frac{x^3}{x} = 0$;
$\lim_{x \to 0^+} \frac{f(x) - f(0)}{x - 0} = \lim_{x \to 0^+} \frac{x^2}{x} = 0$;
$f'(0) = 0$

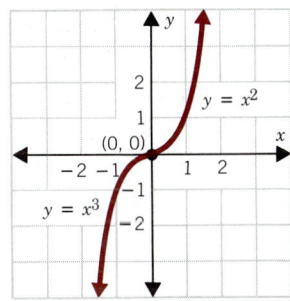

This works because the graphs of the (continuous) functions $f(x) = x^2$ and $f(x) = x^3$ both pass through the same point at $x = 0$ and have the same tangent line at that point.

15. Velocity just before impact = $\lim_{t \to 5^-} \frac{t^3 - 125}{t - 5} = 75$ feet per second; Velocity just after impact = 0

The formulas quoted are not completely accurate, in the following sense: Assuming that they accurately describe the motion of the vehicle before and after impact, the velocity during impact will decrease very rapidly (but not instantaneously) from 75 feet per second to 0. During some short interval of time (which it might be possible to record with a high-speed camera) the velocity will be less than 75 but greater than 0 feet per second.

Review Exercises (page 136)

1. $\frac{\Delta y}{\Delta x} = \frac{f(1) - f(0)}{1 - 0} = \frac{3 - 1}{1} = 2$ 3. $\frac{\Delta y}{\Delta x} = \frac{f(1) - f(0)}{1 - 0} = \frac{0 - (-3)}{1 - 0} = 3$

5. $f'(0) = \lim_{x \to 0} \frac{f(x) - f(0)}{x - 0} = \lim_{x \to 0} \frac{2x^2 + 1 - 1}{x} = \lim_{x \to 0} 2x = 0$

7. $f'(-1) = \lim_{x \to -1} \frac{f(x) - f(-1)}{x + 1} = \lim_{x \to -1} \frac{(x^2 + 2x - 3) - (-4)}{x + 1} = \lim_{x \to -1} \frac{(x + 1)^2}{x + 1} = \lim_{x \to -1} (x + 1) = 0$

9. $f'(x) = 12x^3 - 4x + 5$ 11. $f'(x) = 4x^{-1/2} = \frac{4}{\sqrt{x}}$ 13. $g'(x) = \frac{-4}{x^2}$

15. $f'(t) = 3(t^2 + 1)^2(2t) = 6t(t^2 + 1)^2$ 17. $f'(x) = \frac{-3}{x^2} - \frac{1}{3}$

19. $f'(x) = (3x^2 + x + 1)(12x^2 - 1) + (4x^3 - x + 2)(6x + 1) = 60x^4 + 16x^3 + 3x^2 + 10x + 1$

21. $g'(t) = \frac{(t^2 + 1) - t(2t)}{(t^2 + 1)^2} = \frac{1 - t^2}{(t^2 + 1)^2}$ 23. $f'(w) = \frac{2}{5}(\frac{5}{2}w^{3/2}) - 2(\frac{3}{2})w^{1/2} = w^{3/2} - 3w^{1/2}$

25. $f'(x) = 3(3x - 2)^2(3) + 10(3x - 2)(3) = 9(3x - 2)^2 + 30(3x - 2) = 3(3x - 2)(9x + 4)$

27. $f(x) = (x + x^{1/2})^{1/2}$ 29. $f'(t) = 2t\sqrt{t - 1} + \frac{t^2}{2}(t - 1)^{-1/2}$

$f'(x) = \frac{1}{2}(x + x^{1/2})^{-1/2}(1 + \frac{1}{2}x^{-1/2})$

31. $f'(x) = 12x^3 - 4x \Rightarrow f'(1) = 8$ 33. $f'(x) = \frac{2}{3}x^{-1/3} \Rightarrow f'(1) = \frac{2}{3}$
$f''(x) = 36x^2 - 4 \Rightarrow f''(1) = 32$ $f''(x) = -\frac{2}{9}x^{-4/3} \Rightarrow f''(1) = -\frac{2}{9}$

35. $f'(x) = -\frac{3}{x^2} \Rightarrow f'(1) = -3$

$f''(x) = \frac{6}{x^3} \Rightarrow f''(1) = 6$

37. $f'(x) = 6x^2 + 6x - 12$; $f'(x) = 0$ when $x^2 + x - 2 = 0 \Rightarrow (x + 2)(x - 1) = 0 \Rightarrow x = -2, 1$
$f''(x) = 12x + 6$; $f''(-2) = -18$; $f''(1) = 18$

39. $f'(x) = \frac{3}{2}(x^2 - 1)^{1/2}(2x) = 3x(x^2 - 1)^{1/2}$; $f'(x) = 0$ when $x = \pm 1$ (0 is not in the domain of f)
$f''(x) = 3(x^2 - 1)^{1/2} + \frac{3x}{2}(x^2 - 1)^{-1/2}(2x) = 3(x^2 - 1)^{-1/2}[(x^2 - 1) + x^2] = 3(x^2 - 1)^{-1/2}(2x^2 - 1)$;
$f''(x)$ does not exist for $x = \pm 1$

384 CHAPTER 4 SOLUTIONS

41. $\dfrac{dP}{dV} = \dfrac{d}{dV}(2V^{-1}) = -2V^{-2} = \dfrac{-2}{V^2}$

43. $\dfrac{dy}{dx} = \dfrac{d}{dx}(kx^2(x^3 + 450x - 3500))$
$= \dfrac{d}{dx}(kx^5 + 450kx^3 - 3500kx^2)$
$= 5kx^4 + 1350kx^2 - 7000kx$
$= 5kx(x^3 + 270x - 1400)$

45. $\dfrac{dA}{dT} = \dfrac{d}{dT}(100 + 0.5T)$
$= .05$
.05 Is called monthly interest payment.

47. (a) $v(0) = 5100$
$v(3) = \dfrac{5000}{4} + 100 = 1350$
$\dfrac{v(3) - v(0)}{3} = \dfrac{1350 - 5100}{3} = \dfrac{-3750}{3} = -1250$
$1250 is lost per year on the average for the first 3 years.

(b) $\dfrac{dv}{dt} = \dfrac{-5000}{(1+t)^2}$

(c) $\dfrac{dv}{dt}$ when $t = 1 = \dfrac{-5000}{4} = -1250$

(d) $\dfrac{dv}{dt}$ when $t = 3 = \dfrac{-5000}{16} = -312.50$

CHAPTER 4

Exercise 1 (page 149)

1. $f(x)$ is increasing for $x > 1$
$f(x)$ is decreasing for $x < 1$
Critical number is $x = 1$.

3. $f(x)$ is increasing for $x < 0$ or $x > 1$
$f(x)$ is decreasing for $0 < x < 1$
Critical numbers are $x = 0, 1$.

5. $f(x)$ is increasing for $x > \frac{1}{2}$
$f(x)$ is decreasing for $x < \frac{1}{2}$
Critical numbers are $x = 0, \frac{1}{2}$.

7. $f'(x) = 2x - 4 = 2(x - 2)$
Critical number: $x = 2$
$x < 2 \Rightarrow f'(x) < 0 \Rightarrow f(x)$ is decreasing
$x > 2 \Rightarrow f'(x) > 0 \Rightarrow f(x)$ is increasing
$(2, 2)$ is a relative minimum

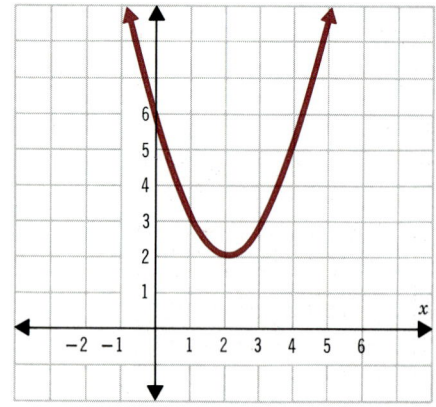

CHAPTER 4 SOLUTIONS 385

9. $f'(x) = 3x^2 - 18x = 3x(x - 6)$
   Critical numbers: $x = 0, 6$
   $x < 0 \Rightarrow f'(x) > 0 \Rightarrow f(x)$ is increasing
   $0 < x < 6 \Rightarrow f'(x) < 0 \Rightarrow f(x)$ is decreasing
   $x > 6 \Rightarrow f'(x) > 0 \Rightarrow f(x)$ is increasing
   $(0, -27)$ is a relative maximum
   $(6, -135)$ is a relative minimum

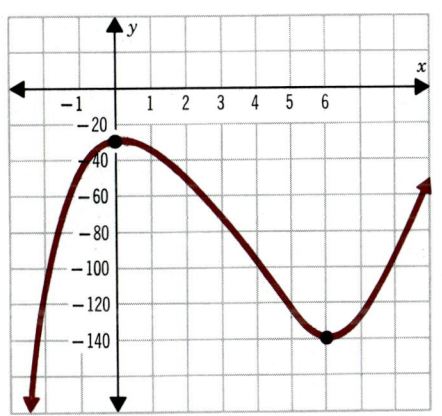

11. $f'(x) = 3x^2 - 18x + 27 = 3(x^2 - 6x + 9) = 3(x - 3)^2$
    Critical number: $x = 3$
    $x < 3 \Rightarrow f'(x) > 0 \Rightarrow f(x)$ is increasing
    $x > 3 \Rightarrow f'(x) > 0 \Rightarrow f(x)$ is increasing
    $(3, 0)$ is neither a relative maximum nor a relative minimum

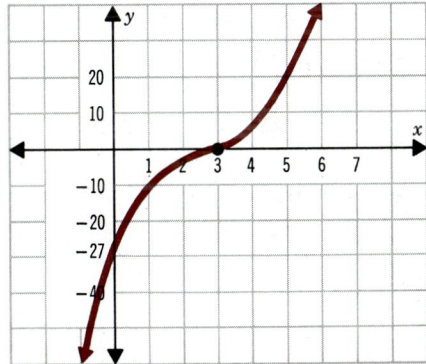

13. $f'(x) = 4x^3 - 4x = 4x(x^2 - 1) = 4x(x + 1)(x - 1)$
    Critical numbers: $0, -1, 1$
    $x < -1 \Rightarrow f'(x) < 0 \Rightarrow f(x)$ is decreasing
    $-1 < x < 0 \Rightarrow f'(x) > 0 \Rightarrow f(x)$ is increasing
    $0 < x < 1 \Rightarrow f'(x) < 0 \Rightarrow f(x)$ is decreasing
    $x > 1 \Rightarrow f'(x) > 0 \Rightarrow f(x)$ is increasing
    $(0, 1)$ is a relative maximum
    $(-1, 0)$ is a relative minimum
    $(1, 0)$ is a relative minimum

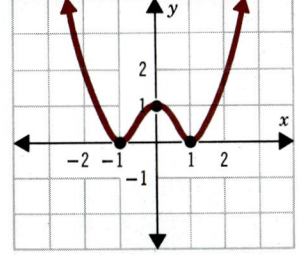

15. $f'(x) = 2x - 8 = 2(x - 4)$
    $f''(x) = 2$
    Critical number: $x = 4$
    $f''(4) = 2 > 0 \Rightarrow (4, -9)$ is a relative minimum

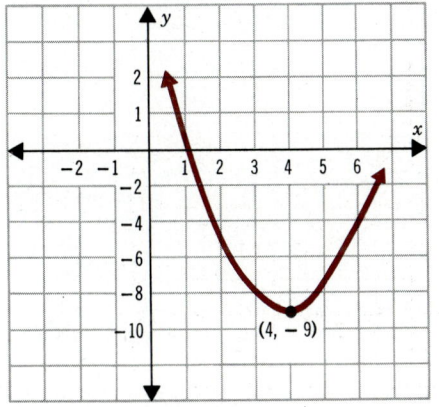

386 CHAPTER 4 SOLUTIONS

17. $f'(x) = 3x^2 - 12x = 3x(x - 4)$
$f''(x) = 6x - 12$
Critical numbers: $x = 0, 4$
$f''(0) = -12 < 0 \Rightarrow (0, 1)$ is a relative maximum
$f''(4) = 12 > 0 \Rightarrow (4, -31)$ is a relative minimum

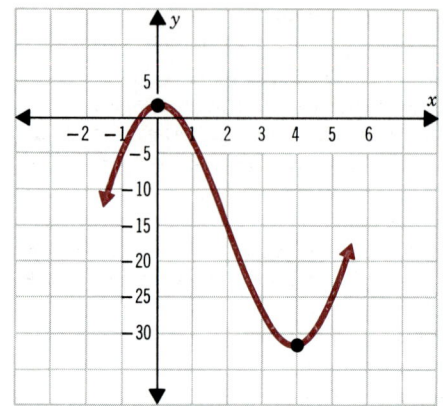

19. $f'(x) = 3x^2 - 18x - 21 = 3(x^2 - 6x - 7) = 3(x - 7)(x + 1)$
$f''(x) = 6x - 18$
Critical numbers: $x = 7, -1$
$f''(7) = 24 > 0 \Rightarrow (7, -243)$ is a relative minimum
$f''(-1) = -24 < 0 \Rightarrow (-1, 13)$ is a relative maximum

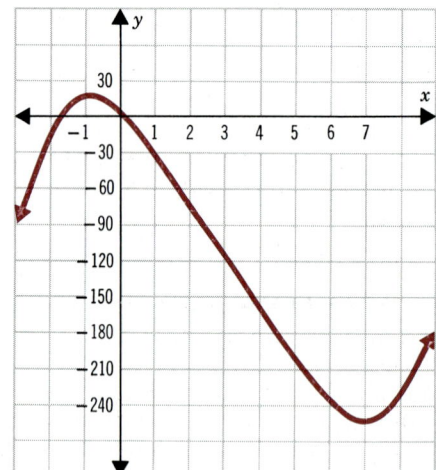

21. $f'(x) = 4x^3 - 12x = 4x(x^2 - 3)$
$f''(x) = 12x^2 - 12$
Critical numbers: $x = 0, \sqrt{3}, -\sqrt{3}$
$f''(0) = -12 \Rightarrow (0, 1)$ is a relative maximum
$f''(\sqrt{3}) = 24 > 0 \Rightarrow (\sqrt{3}, -8)$ is a relative minimum
$f''(-\sqrt{3}) = 24 > 0 \Rightarrow (-\sqrt{3}, -8)$ is a relative minimum

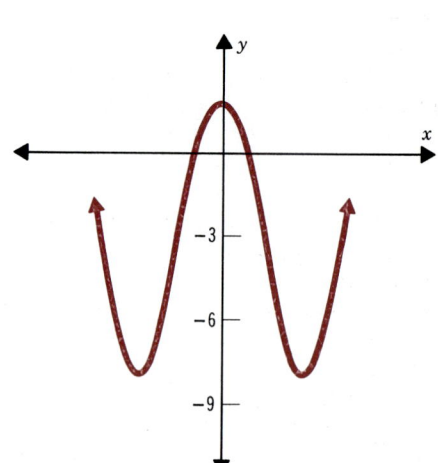

23. $f'(x) = -4x + 4 = -4(x - 1)$
$f''(x) = -4$
Critical number: $x = 1$
$f''(1) = -4 < 0 \Rightarrow (1, -3)$ is a relative maximum

25. $f'(x) = 6x^2 + 6x = 6x(x+1)$
    $f''(x) = 12x + 6$
    Critical numbers: $x = 0, -1$
    $f''(0) = 6 > 0 \Rightarrow (0, 4)$ is a relative minimum
    $f''(-1) = -6 < 0 \Rightarrow (-1, 5)$ is a relative maximum

27. $f'(x) = 3x^2 + 12x + 12 = 3(x^2 + 4x + 4) = 3(x+2)^2$
    $f''(x) = 6x + 12$
    Critical number: $x = -2$
    $f''(-2) = 0$; hence, the second-derivative test does not apply
    Using the first-derivative test, we find that $(-2, -7)$ is neither a relative maximum nor a relative minimum.

29. $f'(x) = 12x^3 - 36x^2 = 12x^2(x - 3)$
    $f''(x) = 36x^2 - 72x$
    Critical numbers: $x = 0, 3$
    $f''(3) > 0 \Rightarrow (3, -76)$ is a relative minimum
    For $x < 0$, $f'(x) < 0 \Rightarrow f(x)$ is decreasing.
    For $0 < x < 3$, $f'(x) < 0 \Rightarrow f(x)$ is decreasing.
    Hence, $(0, 5)$ is neither a relative maximum nor a relative minimum

31. $f'(x) = 4x^3 + 12x^2 + 12x = 4x(x^2 + 3x + 3)$
    $f''(x) = 12x^2 + 24x + 12$
    Critical number: $x = 0$
    $f''(0) = 12 > 0 \Rightarrow (0, 0)$ is a relative minimum

33. $f'(x) = 3x^2 + 6x - 9 = 3(x^2 + 2x - 3) = 3(x+3)(x-1)$
    $f''(x) = 6x + 6$
    Critical numbers: $x = -3, 1$
    $f''(1) = 12 > 0 \Rightarrow (1, -4)$ is a relative minimum
    $f''(-3) = -12 < 0 \Rightarrow (-3, 28)$ is a relative maximum

35. $f'(x) = 3x^2 + 2x = x(3x + 2)$
    $f''(x) = 6x + 2$
    Critical numbers: $x = 0, -\frac{2}{3}$
    $f''(0) = 2 > 0 \Rightarrow (0, 0)$ is a relative minimum
    $f''(-\frac{2}{3}) < 0 \Rightarrow (-\frac{2}{3}, \frac{4}{27})$ is a relative maximum

37. $f'(x) = 4x^3 + 2x = 2x(2x^2 + 1)$
    $f''(x) = 12x^2 + 2$
    Critical number: $x = 0$
    $f''(0) = 2 > 0 \Rightarrow (0, 0)$ is a relative minimum

39. $f'(x) = \frac{2}{3}x^{-1/3} + \frac{1}{3}x^{-2/3} = \frac{1}{3}x^{-2/3}(2x^{1/3} + 1)$
    $f''(x) = -\frac{2}{9}x^{-4/3} - \frac{2}{9}x^{-5/3} = -\frac{2}{9}x^{-5/3}(x^{1/3} + 1)$
    Critical number: $x = -\frac{1}{8}$
    $f''(-\frac{1}{8}) > 0 \Rightarrow (-\frac{1}{8}, -\frac{1}{4})$ is a relative minimum

41. $f'(x) = \frac{1}{3}x^{-1/3}(5x - 20) = \frac{5}{3}x^{-1/3}(x - 4)$
    $f''(x) = \frac{10}{9}x^{-4/3}(x + 2)$
    Critical number: $x = 4$
    $f''(4) > 0 \Rightarrow (4, f(4)) = (4, -12\sqrt[3]{2})$ is a relative minimum

43. $f(x) = \frac{8}{3}x^{-1/3}(x^2 - 2)$
    $f''(x) = \frac{8}{9}x^{-4/3}(5x^2 + 2)$
    Critical numbers: $x = \pm\sqrt{2}$
    $f''(\sqrt{2}) > 0 \Rightarrow (\sqrt{2}, f(\sqrt{2})) = (\sqrt{2}, -6\sqrt[3]{2})$ is a relative minimum
    $f''(-\sqrt{2}) > 0 \Rightarrow (-\sqrt{2}, f(-\sqrt{2})) = (-\sqrt{2}, -6\sqrt[3]{2})$ is a relative minimum

45. $s(t) = \sqrt{(12t)^2 + (30 - 6t)^2} = 6\sqrt{5t^2 - 10t + 25}$
    $s'(t) = \frac{1}{2} \cdot 6(10t - 10)(5t^2 - 10t + 25)^{-1/2} = \frac{30(t-1)}{\sqrt{5t^2 - 10t + 25}}$
    $s(t) = 0 \Rightarrow t = 1$ hour (11 A.M.)
    $s(1) = 6\sqrt{20} \approx 26.8$ kilometers

47. Time with no wind: $t = \dfrac{d}{r} = \dfrac{3000}{500} = 6$ hours

(a) $t = \dfrac{3000}{500 + 25} \approx 5$ hours 43 minutes; Time saved = 17 minutes

(b) $t = \dfrac{3000}{500 - 50} = 6$ hours 40 minutes; Time lost = 40 minutes

(c) $C(500) = 100 + \dfrac{500}{10} + \dfrac{36,000}{500} = \$222$

(d) $C(525) = 100 + \dfrac{525}{10} + \dfrac{36,000}{525} \approx \$221.07$

(e) $C(450) = 100 + \dfrac{450}{10} + \dfrac{36,000}{450} = \$225$

(f) $C'(x) = \dfrac{1}{10} - \dfrac{36,000}{x^2}$; $C'(x) = 0 \Rightarrow x^2 = 360,000$

$x = 600$ miles per hour

(g) $C(600) = 100 + \dfrac{600}{10} + \dfrac{36,000}{600} = \$220$

49. (a) $f'(c) = 4c - 2 = 0 \Rightarrow c = \tfrac{1}{2}$
$f(\tfrac{1}{2}) = 2(\tfrac{1}{2})^2 - 2(\tfrac{1}{2}) = -\tfrac{1}{2}$
Hence, $(c, f(c)) = (\tfrac{1}{2}, -\tfrac{1}{2})$

(b) $f'(c) = 4c^3 = 0 \Leftrightarrow c = 0$
$f(0) = -1$
Hence, $(c, f(c)) = (0, -1)$

(c) $f'(c) = 4c^3 - 4c = 0 \Rightarrow c = 0, c = 1,$ or $c = -1$
$f(0) = -8$
$f(1) = 1 - 2 - 8 = -9$
$f(-1) = 1 - 2 - 8 = -9$
Hence, there are three such points, $(0, -8)$, $(1, -9)$, and $(-1, -9)$ in the interval $[-2, 2]$.

Exercise 2 (page 154)

1. $f'(x) = 2x + 2 = 2(x + 1)$
Critical number: $x = -1$
$f(-1) = -1$
$f(-3) = 3$
$f(3) = 15$
-1 is absolute minimum
15 is absolute maximum

3. $f'(x) = -6 - 2x = -2(3 + x)$
There are no critical numbers in the domain of $f(x)$.
$f(0) = 1$
$f(4) = 1 - 24 - 16 = -39$
1 is absolute maximum
-39 is absolute minimum

5. $f'(x) = 3x^2 - 6x = 3x(x - 2)$
Critical number: $x = 2$
$f(2) = -4$
$f(1) = -2$
$f(4) = 16$
-4 is absolute minimum
16 is absolute maximum

7. $f'(x) = 4x^3 - 4x = 4x(x^2 - 1)$
$f(0) = 1$
$f(1) = 0$
1 is absolute maximum
0 is absolute minimum

9. $f'(x) = \tfrac{2}{3}x^{-1/3}$
There are no critical numbers in the domain of $f(x)$.
$f'(x)$ does not exist at $x = 0$
$f(-1) = 1$
$f(1) = 1$
$f(0) = 0$
0 is absolute minimum
1 is absolute maximum

11. $f'(x) = \dfrac{1}{\sqrt{x}}$
No critical numbers
$f(1) = 2$
$f(4) = 4$
4 is absolute maximum
2 is absolute minimum

13. $f'(x) = \dfrac{1 - 2x^2}{\sqrt{1 - x^2}}$

   Critical numbers: $x = \dfrac{1}{\sqrt{2}}, \dfrac{-1}{\sqrt{2}}$

   $f(-1) = 0 = f(1)$

   $f\left(\dfrac{1}{\sqrt{2}}\right) = \dfrac{1}{\sqrt{2}}\sqrt{\tfrac{1}{2}} = \tfrac{1}{2}$

   $f\left(\dfrac{-1}{\sqrt{2}}\right) = -\tfrac{1}{2}$

   $-\tfrac{1}{2}$ is absolute minimum
   $\tfrac{1}{2}$ is absolute maximum

15. $f'(x) = \dfrac{x(x - 2)}{(x - 1)^2}$

   $f(0) = 0$
   $f(-1) = -\tfrac{1}{2}$
   $f(\tfrac{1}{2}) = -\tfrac{1}{2}$
   0 is absolute maximum
   $-\tfrac{1}{2}$ is absolute minimum

17. $f'(x) = \dfrac{2(x + 2)(4x - 1)}{3(x - 1)^{1/3}}$

   Critical numbers: $x = -2, x = \tfrac{1}{4}$
   $f'(x)$ does not exist at $x = 1$
   $f(-2) = 0$
   $f(\tfrac{1}{4}) = \tfrac{81}{32}\dfrac{\sqrt[3]{9}}{\sqrt[3]{2}}$
   $f(1) = 0$
   $f(-4) = 4\sqrt[3]{25}$
   $f(5) = 98\sqrt[3]{2}$
   0 is absolute minimum
   $98\sqrt[3]{2}$ is absolute maximum

19. $f'(x) = \dfrac{11 - 2x}{3(x - 4)^{2/3}(x - 1)^2}$

   Critical number: $x = \tfrac{11}{2}$
   $f'(x)$ does not exist at $x = 4$
   $f(\tfrac{11}{2}) = \dfrac{2\sqrt[3]{\tfrac{3}{2}}}{9}$
   $f(4) = 0$
   $f(2) = \sqrt[3]{-2}$
   $f(12) = \tfrac{2}{11}$
   $\sqrt[3]{-2}$ is absolute minimum
   $\dfrac{2\sqrt[3]{\tfrac{3}{2}}}{9}$ is absolute maximum

21. $C'(x) = 1.60\left(-\dfrac{1600}{x^2} + 1\right) = 1.60\left(\dfrac{x^2 - 1600}{x^2}\right)$

   Critical number in [10, 75] is $x = 40$
   $C(40) = \$128$
   $C(10) = \$272$
   $C(75) = \$154.13$
   The most economical speed is 40 mph

23. $P'(x) = 1000 - 50x$
   $P'(x) = 1000 - 50x > 0$ for $x < 20$
   $P(x)$ is increasing on (0, 20)
   Critical number: $x = 20$
   $P(20) = 10{,}000$
   $P(0) = 0$
   $P(40) = 0$
   A selling price of 20 yields maximum profit.

Exercise 3 (page 162)
1. $y =$ Side along highway; $x =$ Other side
   Perimeter $= 2x + y = 3000$
   Hence, $y = 3000 - 2x$
   Area $= A = xy = 3000x - 2x^2$
   $A'(x) = 3000 - 4x = 0 \Rightarrow x = 750$
   $A''(750) = -4 < 0 \Rightarrow x = 750$ maximizes area
   Thus, $A(750) = 1{,}125{,}000$ square meters is the largest area that can be enclosed.
3. Area $= A = xy$; $2x + 2y = 200$
   Hence, $y = 100 - x$
   $A(x) = x(100 - x) = 100x - x^2$
   $A'(x) = 100 - 2x = 0 \Rightarrow x = 50$
   $A'(50) = -2 < 0 \Rightarrow x = 50$ maximizes area
   The maximum area that can be enclosed is $A(50) = 100(50) - (50)^2 = 2500$ square meters.
   The dimensions are 50 meters by 50 meters.

390 CHAPTER 4 SOLUTIONS

5. A = Area = xy
   Perimeter = $3y + 2x = 30{,}000$
   Thus, $y = \frac{1}{3}(30{,}000 - 2x)$ so that $A = \frac{30{,}000}{3}x - \frac{2}{3}x^2$.

   $A'(x) = \frac{30{,}000}{3} - \frac{4}{3}x = 0 \Rightarrow x = 7500$

   $A''(7500) = -\frac{4}{3} < 0 \Rightarrow x = 7500$ meters maximizes area

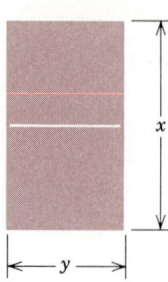

   Thus, the largest area that can be enclosed is $A(7500) = \frac{30{,}000}{3}(7500) - \frac{2}{3}(7500)^2 = 37{,}500{,}000$ square meters. The dimensions are 7500 by 5000 meters.

7. Area = $A = xy$
   Perimeter = $L = 2x + 2y$
   Thus, $y = \frac{L - 2x}{2}$ so that $A(x) = \frac{x}{2}(L - 2x) = \frac{L}{2}x - x^2$.

   $A'(x) = \frac{L}{2} - 2x = 0 \Rightarrow x = \frac{L}{4}$

   $A''\left(\frac{L}{4}\right) = -2 < 0 \Rightarrow x = \frac{L}{4}$ maximizes area

   The dimensions are $\frac{L}{4}$ by $\frac{L}{4}$.

9. $V(x) = x(24 - 2x)^2 = 4x^3 - 96x^2 + 576x$
   $V'(x) = 12x^2 - 192x + 576$
   $= 12(x^2 - 16x + 48)$
   $= 12(x - 12)(x - 4) = 0 \Rightarrow x = 12, 4$
   $V''(x) = 24x - 192;\quad V''(12) > 0;\quad V''(4) < 0$

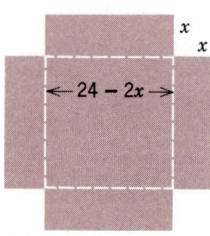

   Hence the dimension $x = 4$ maximizes volume and $4 \times 16 \times 16$ are the dimensions of the box.

11. Volume = $V = x^2 y = 8000$ or $y = \frac{8000}{x^2}$
    Material used = $A = 4xy + 2x^2$
    Hence, $A(x) = \frac{32{,}000}{x} + 2x^2$

    $A'(x) = -\frac{32{,}000}{x^2} + 4x = \frac{4x^3 - 32{,}000}{x^2} = 0 \Rightarrow x = \sqrt[3]{8000} = 20$

    $A''(x) = -\frac{64{,}000}{x^3} + 4;\quad A''(20) > 0$

    Thus, the dimensions $x = 20$ minimizes the amount of material used. Note that $y = \frac{8000}{x^2} = \frac{8000}{400} = 20$ when $x = 20$. (The box is a cube.)

13. Volume $= \pi R^2 h = 10$
    Cost $= C = (2.00)2\pi R^2 + (1.50)2\pi Rh$
    $h = \dfrac{10}{\pi R^2}$ so that $C(R) = 4\pi R^2 + \dfrac{30}{R}$

    $C'(R) = 8\pi R - \dfrac{30}{R^2} = \dfrac{8\pi R^3 - 30}{R^2} = 0 \Rightarrow R = \sqrt[3]{\dfrac{15}{4\pi}} \approx 1.09$ meters

    $C''(R) = 8\pi + \dfrac{60}{R^3};\quad C''(1.09) > 0 \Rightarrow R = 1.09$ meters minimizes cost

    Hence, $R = 1.09$ meters and $h = \dfrac{10}{\pi(1.09)^2} = 2.67$ meters are the dimensions of the container that minimize the cost.

15. For an increase of x members, $R(x) = (60 + x)(200 - 2x) = 12{,}000 + 80x - 2x^2$.
    $R'(x) = 80 - 4x = 0 \Rightarrow x = 20;\ R''(20) = -4$, so $x = 20$ maximizes revenue
    Hence, 80 members gives a maximum revenue.

17. $T =$ Time of trip along $AC +$ Time of trip along $CB = \dfrac{\text{Distance } AC}{2.5} + \dfrac{\text{Distance } CB}{4}$

    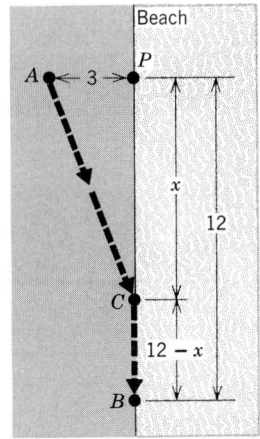

    Distance $AC = \sqrt{x^2 + 9}$
    Distance $CB = 12 - x$

    Hence, $T(x) = \dfrac{1}{2.5}(x^2 + 9)^{1/2} + \dfrac{1}{4}(12 - x)$

    $T'(x) = \dfrac{x}{2.5}(x^2 + 9)^{-1/2} - \dfrac{1}{4} = 0 \Rightarrow x \approx 2.4$

    Now the restrictions are x are $0 \le x \le 12$.
    $T(0) = 4.2$
    $T(12) = 4.95$
    $T(2.4) = 3.94$

    Hence, if the woman rows to point C, 2.4 km downshore from P, and walks the rest of the way, the time will be kept to a minimum.

19. $V = \pi R^2 h$
    Surface area $= S = 2\pi R^2 + 2\pi Rh$

    $h = \dfrac{V}{\pi R^2}$ so that $S(R) = 2\pi R^2 + \dfrac{2V}{R}$

    $S'(R) = 4\pi R - \dfrac{2V}{R^2} = \dfrac{4\pi R^3 - 2V}{R^2}$

    $S'(R) = 0 \Rightarrow R = \sqrt[3]{\dfrac{V}{2\pi}}$

    Now, $h = \dfrac{V}{\pi R^2} = \dfrac{V}{\pi \left(\dfrac{V}{2\pi}\right)^{2/3}} = 2\dfrac{V^{1/3}}{\pi^{1/3} 2^{1/3}} = 2\sqrt[3]{\dfrac{V}{2\pi}} = 2R.$

21. Perimeter $= 22 = 2y + x + \frac{\pi x}{2} = 2y + \frac{(2+\pi)x}{2}$

Area $= A = xy + \frac{1}{2}\pi\left(\frac{x}{2}\right)^2 = xy + \frac{\pi}{8}x^2$

Now, $y = \frac{44-(2+\pi)x}{4}$; hence, $A(x) = x\left[\frac{44-(2+\pi)x}{4}\right] + \frac{\pi}{8}x^2 = 11x - \frac{1}{8}(4+\pi)x^2$.

$A'(x) = 11 - \frac{1}{4}(4+\pi)x = 0 \Rightarrow x = \frac{44}{4+\pi}$

$A''(x) = -\frac{1}{4}(4+\pi) \Rightarrow A''\left(\frac{44}{4+\pi}\right) < 0 \Rightarrow x = \frac{44}{4+\pi}$ maximizes area

Hence, a window of dimensions $x = \frac{44}{(4+\pi)}$ and $y = \frac{22}{(4+\pi)}$ feet will maximize the amount of sunlight passing through.

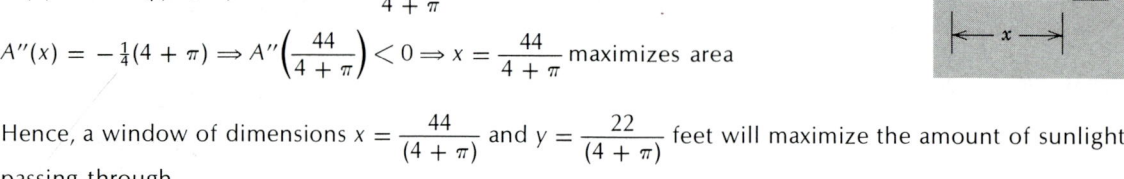

Exercise 4 (page 169)

1. $f'(x) = 2x - 2$; $f''(x) = 2$
$f''(x)$ is never zero
Hence, there are no inflection points.
Since $f''(x) > 0$, $f(x)$ is concave up.

3. $f'(x) = 3x^2 - 18x$;
$f''(x) = 6x - 18 = 0 \Rightarrow x = 3$; $f(3) = -52$
$x < 3 \Rightarrow f''(x) < 0 \Rightarrow$ concave down
$x > 3 \Rightarrow f''(x) > 0 \Rightarrow$ concave up
Hence, $(3, -52)$ is an inflection point.

5. $f'(x) = 4x^3 - 12x^2$;
$f''(x) = 12x^2 - 24x = 12x(x-2) = 0 \Rightarrow x = 0, 2$
$x < 0 \Rightarrow f''(x) > 0 \Rightarrow$ concave up
$0 < x < 2 \Rightarrow f''(x) < 0 \Rightarrow$ concave down
$x > 2 \Rightarrow f''(x) > 0 \Rightarrow$ concave up
$f(0) = 10$; $f(2) = -6$
Hence, $(0, 10)$ and $(2, -6)$ are inflection points.

7. $f'(x) = 1 - \frac{1}{x^2}$; $f''(x) = \frac{2}{x^3}$
There are no inflection points.
$x < 0 \Rightarrow f''(x) < 0 \Rightarrow$ concave down
$x > 0 \Rightarrow f''(x) > 0 \Rightarrow$ concave up

9. $f'(x) = x^{-2/3} + 2$; $f''(x) = -\frac{2}{3}x^{-5/3} = \frac{-2}{3x\sqrt[3]{x^2}}$
$f''(x)$ does not exist at $x = 0$; $f(0) = 0$
$x < 0 \Rightarrow f''(x) > 0 \Rightarrow$ concave up
$x > 0 \Rightarrow f''(x) < 0 \Rightarrow$ concave down
Hence, $(0, 0)$ is an inflection point.

11. $f'(x) = 6x^2 - 12x + 6 = 6(x-1)^2 = 0 \Rightarrow x = 1$
Thus, $x = 1$ is a critical number.
$f''(x) = 12(x-1) = 0 \Rightarrow x = 1$
$x < 1 \Rightarrow f''(x) < 0 \Rightarrow$ concave down
$x > 1 \Rightarrow f''(x) > 0 \Rightarrow$ concave up
$f(1) = -1$
Hence, $(1, -1)$ is an inflection point.

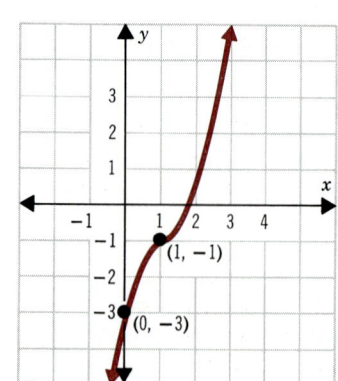

CHAPTER 4 SOLUTIONS 393

13. $f'(x) = -6x^2 + 30x - 36 = -6(x-3)(x-2) = 0 \Rightarrow x = 3, 2$
$f(3) = -20; \quad f(2) = -21$
$f''(x) = -12x + 30 = -6(2x - 5)$
$f''(3) = -6 < 0 \Rightarrow (3, -20)$ is a relative maximum
$f''(2) = 6 > 0 \Rightarrow (2, -21)$ is a relative minimum
$f''(x) = 0 \Rightarrow x = \frac{5}{2}$
$x < \frac{5}{2} \Rightarrow f''(x) > 0 \Rightarrow$ concave up
$x > \frac{5}{2} \Rightarrow f''(x) < 0 \Rightarrow$ concave down
Hence, $(\frac{5}{2}, -\frac{82}{4})$ is an inflection point.

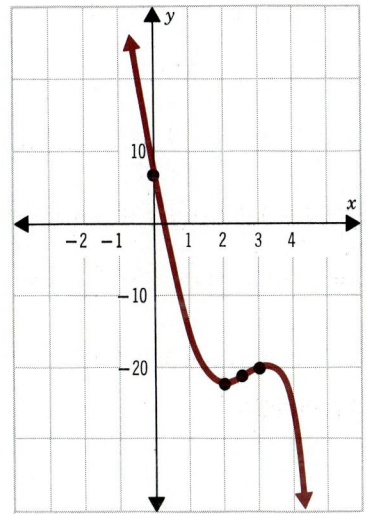

15. $f'(x) = 4x^3 - 4 = 4(x^3 - 1) = 0 \Rightarrow x = 1; \quad f(1) = -3$
$f''(x) = 12x^2$
$f''(1) = 12 > 0 \Rightarrow (1, -3)$ is a relative minimum
$f''(x) > 0$ for each $x \neq 0$; therefore, $f(x)$ is concave up

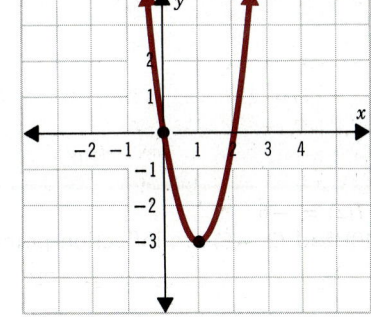

17. $f'(x) = 24x^5 - 24x^3 = 24x^3(x^2 - 1) = 0 \Rightarrow x = 0, +1, -1$
Hence, $x = 0, +1, -1$ are critical numbers.
$f''(x) = 120x^4 - 72x^2 = 24x^2(5x^2 - 3)$
$f''(1) > 0 \Rightarrow (1, -2)$ is a relative minimum
$f''(-1) > 0 \Rightarrow (-1, -2)$ is a relative minimum
$f''(0) = 0 \Rightarrow$ no information available
By first-derivative test, $(0, 0)$ is a relative maximum.
For inflection points:
$f''(x) = 0 \Rightarrow x = 0, \sqrt{\frac{3}{5}}, -\sqrt{\frac{3}{5}}$
$x > \sqrt{\frac{3}{5}} \Rightarrow f''(x) > 0 \Rightarrow$ concave up
$0 < x < \sqrt{\frac{3}{5}} \Rightarrow f''(x) < 0 \Rightarrow$ concave down
$-\sqrt{\frac{3}{5}} < x < 0 \Rightarrow f''(x) < 0 \Rightarrow$ concave down
$x < -\sqrt{\frac{3}{5}} \Rightarrow f''(x) > 0 \Rightarrow$ concave up
Hence, $(-\sqrt{\frac{3}{5}}, -1.3)$ and $(\sqrt{\frac{3}{5}}, -1.3)$ are inflection points.

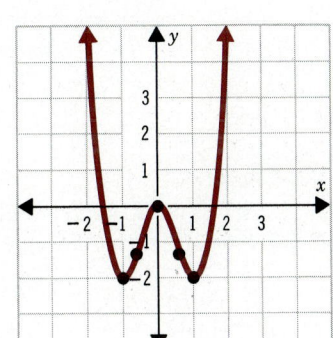

394 CHAPTER 4 SOLUTIONS

19. $f'(x) = 15x^4 - 15x^2 = 15x^2(x^2 - 1) = 0 \Rightarrow x = 0, +1, -1;$ $f(0) = 0, f(1) = -2, f(-1) = 2$
$f''(x) = 60x^3 - 30x = 30x(2x^2 - 1)$
$f''(0) = 0 \Rightarrow$ no information available
$f''(1) > 0 \Rightarrow (1, -2)$ is a relative minimum
$f''(-1) < 0 \Rightarrow (-1, 2)$ is a relative maximum
By first-derivative test, $(0, 0)$ is neither a relative maximum nor a relative minimum.

$f''(x) = 30x(2x^2 - 1) = 0 \Rightarrow x = 0, +\frac{1}{\sqrt{2}}, -\frac{1}{\sqrt{2}};$ $f\left(\pm\frac{1}{\sqrt{2}}\right) = \mp\frac{7}{4\sqrt{2}}$

$0 < x < \frac{1}{\sqrt{2}} \Rightarrow f''(x) < 0 \Rightarrow f''(x)$ is concave down

$x > \frac{1}{\sqrt{2}} \Rightarrow f''(x) > 0 \Rightarrow f(x)$ is concave up

$-\frac{1}{\sqrt{2}} < x < 0 \Rightarrow f''(x) > 0 \Rightarrow f(x)$ is concave up

$x < -\sqrt{\frac{1}{2}} \Rightarrow f''(x) < 0 \Rightarrow f(x)$ is concave down

Thus, $(0, 0)$, $\left(\frac{1}{\sqrt{2}}, -\frac{7}{4\sqrt{2}}\right)$, and $\left(-\frac{1}{\sqrt{2}}, \frac{7}{4\sqrt{2}}\right)$ are inflection points.

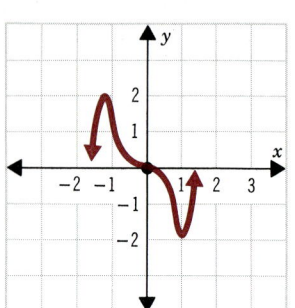

21. $f'(x) = \frac{2}{3}x^{-1/3} + \frac{1}{3}x^{-2/3} = \frac{1}{3}x^{-2/3}(2x^{1/3} + 1)$
$f'(x) = 0 \Rightarrow x = -\frac{1}{8};$ $f(-\frac{1}{8}) = -\frac{1}{4}$
$f''(x) = -\frac{2}{9}x^{-4/3} - \frac{2}{9}x^{-5/3} = -\frac{2}{9}x^{-5/3}(x^{1/3} + 1)$
$f''(-\frac{1}{8}) > 0 \Rightarrow (-\frac{1}{8}, -\frac{1}{4})$ is a relative minimum
$f''(x) = 0 \Rightarrow x = -1;$ $f(-1) = 0$
$f''(x)$ does not exist when $x = 0;$ $f(0) = 0$
$x < -1 \Rightarrow f''(x) < 0 \Rightarrow f(x)$ is concave down
$-1 < x < 0 \Rightarrow f''(x) > 0 \Rightarrow f(x)$ is concave up
$0 < x \Rightarrow f''(x) < 0 \Rightarrow f(x)$ is concave down
Hence, $(-1, 0)$ and $(0, 0)$ are inflection points.
There is a vertical tangent at $(0, 0)$.

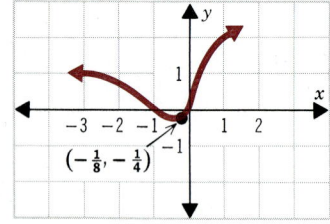

23. $f'(x) = \frac{5}{3}x^{2/3} - \frac{20}{3}x^{-1/3} = \frac{5(x - 4)}{3x^{1/3}} = 0 \Rightarrow x = 4;$

$f(4) = -6(16^{1/3})$

$f''(x) = \frac{10}{9}x^{-1/3} + \frac{20}{9}x^{-4/3} = \frac{10(x + 2)}{9x^{4/3}} = 0 \Rightarrow x = -2;$

$f(-2) = -12(4^{1/3})$
$f''(4) > 0 \Rightarrow (4, -6\sqrt[3]{16})$ is a relative minimum
$x < -2 \Rightarrow f''(x) < 0 \Rightarrow$ concave down
$-2 < x < 0 \Rightarrow f''(x) > 0 \Rightarrow$ concave up
$x > 0 \Rightarrow f''(x) > 0 \Rightarrow$ concave up
Hence, $(-2, -12\sqrt[3]{4})$ is an inflection point.

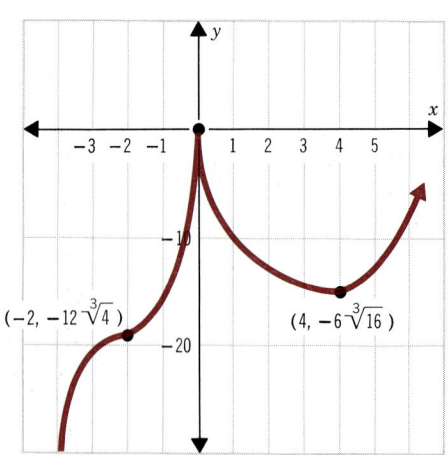

25. $f'(x) = \frac{8}{3}x^{5/3} - \frac{16}{3}x^{-1/3} = \frac{8}{3}x^{-1/3}(x^2 - 2) = 0 \Rightarrow x = \pm\sqrt{2}$
$f(\pm\sqrt{2}) = -6(2^{1/3})$
$f''(x) = \frac{40}{9}x^{2/3} + \frac{16}{9}x^{-4/3} = \frac{8}{9}x^{-4/3}(5x^2 + 2)$
$f''(\sqrt{2}) > 0 \Rightarrow (\sqrt{2}, -6\sqrt[3]{2})$ is a relative minimum
$f''(-\sqrt{2}) > 0 \Rightarrow (-\sqrt{2}, -6\sqrt[3]{2})$ is a relative minimum
$f'(0)$ and $f''(0)$ do not exist; $f(0) = 0$
There is a vertical tangent line at $(0, 0)$.
$f''(x) > 0$ for all $x \neq 0 \Rightarrow$
the graph is concave up everywhere except at $(0, 0)$.

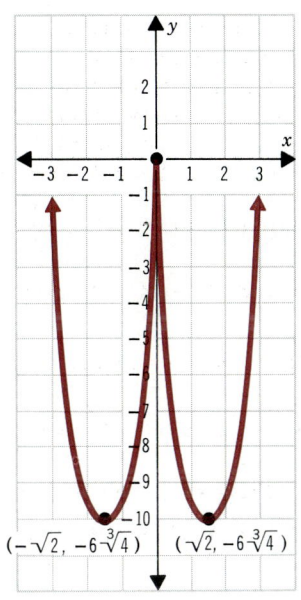

27. $f(0) = 10; \quad f(6) = 15; \quad f(9) = 10; \quad f(10) = 0$
Relative maximum at $(6, 15)$
Relative minimum at $(10, 0)$
Inflection point at $(9, 10)$
y-intercept at $(0, 10)$

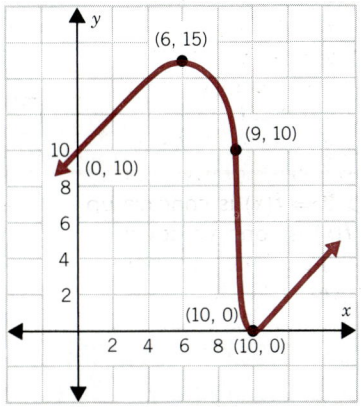

29. $f(1) = 5; \quad f(2) = 3; \quad f(3) = 1$
Relative maximum at $(1, 5)$
Relative minimum at $(3, 1)$
Inflection point at $(2, 3)$

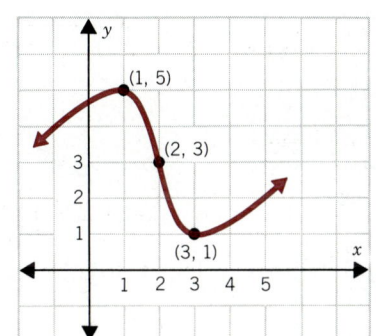

396 CHAPTER 4 SOLUTIONS

31. $C'(x) = 0.003x^2 - 0.6x + 30$
    $C''(x) = 0.006x - 0.6 = 0$
    $$x = 100$$
    For $x < 100$, $C''(x) < 0 \Rightarrow$ concave down
    For $x > 100$, $C''(x) > 0 \Rightarrow$ concave up
    Hence, $(100, 1042)$ is an inflection point.
    Although the cost function $C(x)$ is increasing throughout its domain (since $C'(x) > 0$ for all $x \neq 0$), the cost rate, $C'(x)$, changes from decreasing to increasing at $x = 100$.
33. $f''(x) = 2a \neq 0$; thus, it is impossible for $f(x)$ to have an inflection point; the graph is either always concave up (if $a > 0$) or always concave down (if $a < 0$)

Exercise 5 (page 175)

1. $\lim\limits_{x \to +\infty} \dfrac{x^3 + x^2 + 2x - 1}{x^3 + x + 1} = \lim\limits_{x \to +\infty} \dfrac{1 + \dfrac{1}{x} + \dfrac{2}{x^2} - \dfrac{1}{x^3}}{1 + \dfrac{1}{x^2} + \dfrac{1}{x^3}} = 1$

3. $\lim\limits_{x \to +\infty} \dfrac{2x + 4}{x - 1} = \lim\limits_{x \to +\infty} \dfrac{2 + \dfrac{4}{x}}{1 - \dfrac{1}{x}} = 2$

5. $\lim\limits_{x \to +\infty} \dfrac{3x^2 - 1}{x^2 + 4} = \lim\limits_{x \to +\infty} \dfrac{3 - \dfrac{1}{x^2}}{1 + \dfrac{4}{x^2}} = 3$

7. $\lim\limits_{x \to -\infty} \dfrac{5x^3 - 1}{x^2 + 1} = \lim\limits_{x \to -\infty} \dfrac{5x - \dfrac{1}{x^2}}{1 + \dfrac{1}{x^2}} = -\infty$

9. $\lim\limits_{x \to +\infty} \left(3 + \dfrac{1}{x}\right) = 3 = \lim\limits_{x \to -\infty} \left(3 + \dfrac{1}{x}\right)$; $y = 3$ is a horizontal asymptote
    $\lim\limits_{x \to 0^-} \left(3 + \dfrac{1}{x}\right) = -\infty$; $\lim\limits_{x \to 0^+} \left(3 + \dfrac{1}{x}\right) = +\infty$; $x = 0$ is a vertical asymptote

11. $\lim\limits_{x \to +\infty} \dfrac{2}{(x - 1)^2} = 0 = \lim\limits_{x \to -\infty} \dfrac{2}{(x - 1)^2}$; $y = 0$ is a horizontal asymptote
    $\lim\limits_{x \to 1} \dfrac{2}{(x - 1)^2} = +\infty$; $x = 1$ is a vertical asymptote

13. $\lim\limits_{x \to +\infty} \dfrac{x^2}{x^2 - 4} = 1 = \lim\limits_{x \to -\infty} \dfrac{x^2}{x^2 - 4}$; $y = 1$ is a horizontal asymptote
    $\lim\limits_{x \to -2^-} \dfrac{x^2}{x^2 - 4} = +\infty = \lim\limits_{x \to 2^+} \dfrac{x^2}{x^2 - 4}$; $\lim\limits_{x \to -2^+} \dfrac{x^2}{x^2 - 4} = -\infty = \lim\limits_{x \to 2^-} \dfrac{x^2}{x^2 - 4}$;
    $x = 2$ and $x = -2$ are vertical asymptotes

15. $f(x) = \dfrac{2}{x^2 - 4}$; $f'(x) = -\dfrac{4x}{(x^2 + 4)^2}$; $f''(x) = \dfrac{6x^2 + 8}{(x^2 - 4)^3}$
    y-intercept: $(0, -\tfrac{1}{2})$
    No x-intercept, since $f(x)$ is never 0
    Symmetry about y-axis
    Horizontal asymptote: $y = 0$
    Vertical asymptotes: $x = 2, x = -2$
    Relative maximum at $(0, -\tfrac{1}{2})$
    Concave up on $(-\infty, -2)$ and $(2, +\infty)$
    Concave down on $(-2, 2)$
    No inflection point

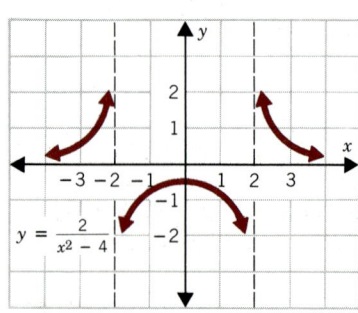

17. y-intercept: $(0, -1)$
x-intercept: $(\frac{1}{2}, 0)$
Horizontal asymptote: $y = 2$
Vertical asymptote: $x = -1$
No relative extrema, since $f'(x)$ is never 0
No inflection points
Concave up and increasing on $(-\infty, -1)$
Concave down and increasing on $(-1, +\infty)$
$f(x) > 2$ for all $x < -1$
$f(x) < 2$ for all $x > -1$

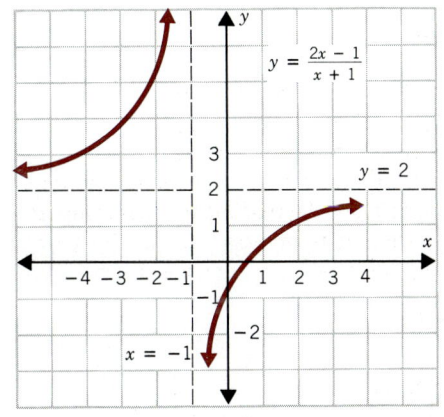

19. The only intercept is $(0, 0)$.
Symmetric about the origin
Horizontal asymptote: $y = 0$
Relative maximum at $(1, \frac{1}{2})$
Relative minimum at $(-1, -\frac{1}{2})$
Inflection points at $(-\sqrt{3}, -\frac{\sqrt{3}}{4})$, $(0, 0)$, and $(\sqrt{3}, \frac{\sqrt{3}}{4})$
Concave down on $(-\infty, -\sqrt{3}]$ and $[0, \sqrt{3}]$;
Concave up on $[-\sqrt{3}, 0]$ and $[\sqrt{3}, +\infty)$.

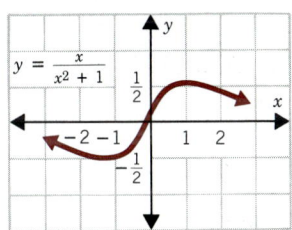

21. y-intercept: $(0, -\frac{1}{2})$
No x-intercept
Horizontal asymptote: $y = 0$
Vertical asymptotes: $x = -4, x = 4$
Relative maximum at $(0, -\frac{1}{2})$
Concave up on $(-\infty, -4)$ and $(4, +\infty)$
Concave down on $(-4, 4)$
No inflection point

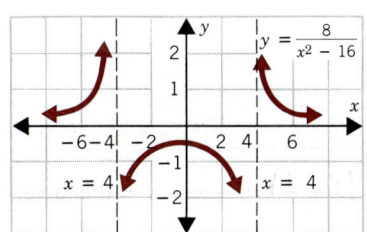

23.

Exercise 6 (page 181)
1. $C(x) = 2x + 5$; $R(x) = 8x - x^2$
   (a) $MR = R'(x) = 8 - 2x = 2(4 - x)$
   (b) $R'(3) = 2(4 - 3) = 2$ million dollars; $R'(4) = 0$ dollars
   (c) $MC = C'(x) = 2$ million dollars
   (d) 5 million dollars
   (e) $VC = 2x$; $VC(4) = 8$ million dollars
   (f) $R(x) = C(x) \Rightarrow 2x + 5 = 8x - x^2 \Rightarrow x^2 - 6x + 5 = 0 \Rightarrow (x - 5)(x - 1) = 0 \Rightarrow x = 5, 1$

(g) Profit $= P(x) = R(x) - C(x) = 6x - x^2 - 5$
(h) $P'(x) = 6 - 2x = 0 \Rightarrow x = 3$; thus, producing 3 thousand units yields maximum profit
(i) $P(3) = 18 - 9 - 5 = 4$ million dollars
(j) $MR(3) = R'(3) = 2$ million dollars
(k) $R(3) = 24 - 9 = 15$ million dollars
(l) $2(3) = 6$ million dollars

3. $R(x) = xp(x) = 20x - 0.03x^2$
$C(x) = 3 + 0.02x$
Profit $= P(x) = R(x) - C(x) = 20x - 0.03x^2 - (3 + 0.02x) = 19.98x - 0.03x^2 - 3$
$P'(x) = 19.98 - 0.06x = 0 \Rightarrow x = 333$ units

5. $C(x) = 200 + 35x + 0.02x^2$
Price function $p(x)$ is constant at $39. Hence,
$R(x) = xp(x) = 39x$
Profit $= P(x) = R(x) - C(x) = 39x - (200 + 35x + 0.02x^2) = 4x - 200 - 0.02x^2$
$P'(x) = 4 - 0.04x = 0 \Rightarrow x = 100$ tons

7. (a) The price function is a linear relationship, so that
$$p(x) - 20 = \frac{0.50}{-50}(x - 500)$$
$$p(x) = -0.01x + 25$$
(b) Profit $= P(x) = R(x) - C(x) = xp(x) - C(x) = -0.01x^2 + 25x - (4200 + 5.10x + 0.0001x^2)$
$= -0.0101x^2 + 19.9x - 4200$
$P'(x) = -0.0202x + 19.9 = 0 \Rightarrow x = 985$ articles
(c) $p(985) = (-0.01)(985) + 25 = \$15.15$

9. (a) C represents the cost function
R represents the revenue function
x represents the number of units
(b) 300 is the fixed cost and has no effect on the marginal cost since $MC = \frac{dC}{dx}$ and the derivative of a constant is 0
(c) $MC = \frac{d}{dx}(300 + 5x) = 5$ variable cost to produce each unit
$MR = \frac{d}{dx}(4x) = 4$ revenue received per item (selling price)
(d) For each additional unit produced the marginal profit is $MR - MC$, or -1; that is, a loss of 1 occurs for each unit that is produced.
(e) Straight lines, positive slope
(f) This graph represents a losing situation since the graph of R will always lie below C and the vertical distance (loss) is increasing for increasing values of x.

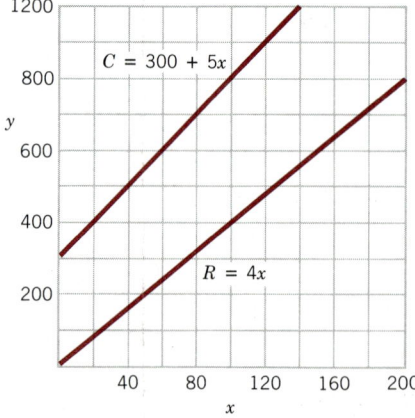

CHAPTER 4 SOLUTIONS 399

11. $A(x) = \dfrac{C(x)}{x} \Rightarrow A'(x) = \dfrac{xC'(x) - C(x)}{x^2}$

$A'(x) = 0 \Rightarrow xC'(x) = C(x) \Rightarrow C' = \dfrac{C(x)}{x}$

$A''(x) = \dfrac{x^2 \cdot xC'' - 2x(xC' - C)}{x^4}$; when $C' = \dfrac{C}{x}$ and $C'' > 0$, then

$A'' = \dfrac{C''}{x} > 0 \Rightarrow A(x)$ is a minimum

Exercise 7 (page 186)

1. $t = 18 - 3x^2$ or $x = \dfrac{\sqrt{18 - t}}{\sqrt{3}}$

$R(x) = xt = 18x - 3x^3$
$R'(x) = 18 - 9x^2 = 0 \Rightarrow x = \sqrt{2} \approx 1.41$
$R(1.41) = 16.97$ (maximum revenue)
$t = 18 - 3(1.41)^2 = 12\%$

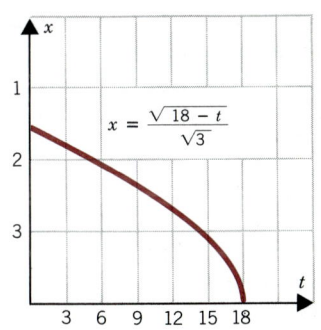

2. $C(t) = \dfrac{30k}{t} + \dfrac{450}{t} = \dfrac{30t^{4/3}}{t} + \dfrac{450}{t} = 30t^{1/3} + \dfrac{450}{t}$

$C'(t) = 10t^{-2/3} - \dfrac{450}{t^2} = \dfrac{10t^{4/3} - 450}{t^2}$

$C'(t) = 0 \Rightarrow t^{4/3} = 45 \Rightarrow t = 45^{3/4} \approx 17.37$ years

3. Let $x = $ Number of lawnmowers in each order.
Then, $C(x) = $ Storage cost + Order cost $= 10x + 20\left(\dfrac{50}{x}\right) = 10x + \dfrac{1000}{x}$

$C'(x) = 10 - \dfrac{1000}{x^2} = 0 \Rightarrow x^2 = 100 \Rightarrow x = 10$

$C''(x) = \dfrac{2000}{x^3}$, so $C''(10) > 0$

Hence, 10 lawnmowers should be ordered and 5 orders should be placed to minimize the cost.

4. $C(x) = $ (Number of orders per quarter)(delivery cost per order) + $10x$

$= \left(\dfrac{2500}{x}\right)90 + 10x = \dfrac{225{,}000}{x} + 10x$

$C'(x) = \dfrac{-225{,}000}{x^2} + 10$

$C''(x) = \dfrac{450{,}000}{x^3} > 0$ for $x > 0$ Hence minimum occurs at the critical value.

$C'(x) = 0 \Rightarrow \dfrac{-225{,}000}{x^2} + 10 = 0$

$x^2 = \dfrac{225{,}000}{10} = 22{,}500$

$x = 150$

In order to minimize total cost 150 cases should be ordered at a time.

5. Set-up cost $S = k_1 M$, $k_1 > 0$
Operating cost $O = \dfrac{k_2}{M}$, $k_2 > 0$

Total cost $T = k_1 M + \dfrac{k_2}{M}$

$\dfrac{dT}{dM} = k_1 - \dfrac{k_2}{M^2}$

$\dfrac{d^2T}{dM^2} = \dfrac{2k_2}{M^3} > 0$ since $M > 0$. Hence minimum occurs at the critical value

$k_1 - \dfrac{k_2}{M^2} = 0$ multiply by M

$k_1 M - \dfrac{k_2}{M} = 0$

$k_1 M = \dfrac{k_2}{M}$

Set-up cost = Operating cost

Review Exercises (page 188)

1.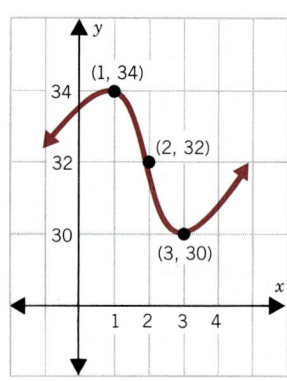

3. $f'(x) = 3x^2 - 6x + 3 = 0 \Rightarrow (x-1)^2 = 0 \Rightarrow x = 1$
Critical number: $x = 1$; $f(1) = 0$
$f'(x) > 0$ for every $x \neq 1 \Rightarrow f$ is increasing for all x
$f''(x) = 6x - 6 = 6(x-1) = 0 \Rightarrow x = 1$
$x < 1 \Rightarrow f''(x) < 0 \Rightarrow$ concave down
$x > 1 \Rightarrow f''(x) > 0 \Rightarrow$ concave up
Thus, $(1, 0)$ is an inflection point.

5. $f'(x) = 4x^3 - 4x = 4x(x^2 - 1)$
    $f'(x) = 0 \Rightarrow x = 0, 1, -1$
    Critical points: $(0, 0)$, $(1, -1)$, and $(-1, -1)$
    $x < -1 \Rightarrow f'(x) < 0 \Rightarrow f(x)$ is decreasing
    $-1 < x < 0 \Rightarrow f'(x) > 0 \Rightarrow f(x)$ is increasing
    $0 < x < 1 \Rightarrow f'(x) < 0 \Rightarrow f(x)$ is decreasing
    $x > 1 \Rightarrow f'(x) > 0 \Rightarrow f(x)$ is increasing
    $(-1, -1)$ is a relative minimum
    $(0, 0)$ is a relative maximum
    $(1, -1)$ is a relative minimum
    $f''(x) = 12x^2 - 4 = 4(3x^2 - 1) = 0 \Rightarrow x = \pm \dfrac{1}{\sqrt{3}}$; $f\left(\dfrac{1}{\sqrt{3}}\right) = -\dfrac{5}{9} = f\left(-\dfrac{1}{\sqrt{3}}\right)$

    $x < -\dfrac{1}{\sqrt{3}} \Rightarrow f''(x) > 0 \Rightarrow$ concave up

    $-\dfrac{1}{\sqrt{3}} < x < +\dfrac{1}{\sqrt{3}} \Rightarrow f''(x) < 0 \Rightarrow$ concave down

    $x > \dfrac{1}{\sqrt{3}} \Rightarrow f''(x) > 0 \Rightarrow$ concave up

    Hence, $\left(\dfrac{1}{\sqrt{3}}, -\dfrac{5}{9}\right)\left(-\dfrac{1}{\sqrt{3}}, -\dfrac{5}{9}\right)$ are inflection points.

7. $f'(x) = 5x^4 - 5 = 5(x^4 - 1)$
    $f'(x) = 0 \Rightarrow x = 1, -1$
    Thus, $(1, -4)$ and $(-1, 4)$ are critical points.
    $x < -1 \Rightarrow f'(x) > 0 \Rightarrow f(x)$ is increasing
    $-1 < x < 1 \Rightarrow f'(x) < 0 \Rightarrow f(x)$ is decreasing
    $x > 1 \Rightarrow f'(x) > 0 \Rightarrow f(x)$ is increasing
    Thus, $(1, -4)$ is a relative minimum
    and $(-1, 4)$ is a relative maximum.
    $f''(x) = 20x^3$; $f''(x) = 0 \Rightarrow x = 0$
    $x < 0 \Rightarrow f''(x) < 0 \Rightarrow f(x)$ is concave down
    $x > 0 \Rightarrow f''(x) > 0 \Rightarrow f(x)$ is concave up
    Thus, $(0, 0)$ is an inflection point.

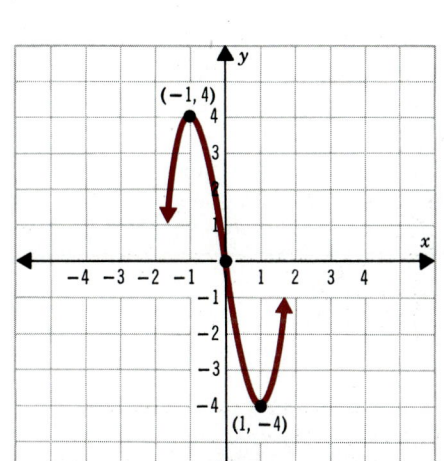

9. $f'(x) = 4x^3 - 12x^2 + 8x = 4x(x^2 - 3x + 2) = 4x(x - 1)(x - 2)$
    $f'(x) = 0 \Rightarrow x = 0, x = 1, x = 2$
    Critical points: $(0, 0)$, $(1, 1)$, $(2, 0)$
    $x < 0 \Rightarrow f'(x) < 0 \Rightarrow f(x)$ is decreasing
    $0 < x < 1 \Rightarrow f'(x) > 0 \Rightarrow f(x)$ is increasing
    $1 < x < 2 \Rightarrow f'(x) < 0 \Rightarrow f(x)$ is decreasing
    $x > 2 \Rightarrow f'(x) > 0 \Rightarrow f(x)$ is increasing
    Thus, $(0, 0)$ is a relative minimum; $(1, 1)$ is a relative maximum;
    and $(2, 0)$ is a relative minimum.
    $f''(x) = 12x^2 - 24x + 8 = 4(3x^2 - 6x + 2)$
    $f''(x) = 0 \Rightarrow x = 1 \pm \dfrac{\sqrt{3}}{3}$

    Inflection points at $x = 1 + \dfrac{\sqrt{3}}{3}$ and $x = 1 - \dfrac{\sqrt{3}}{3}$

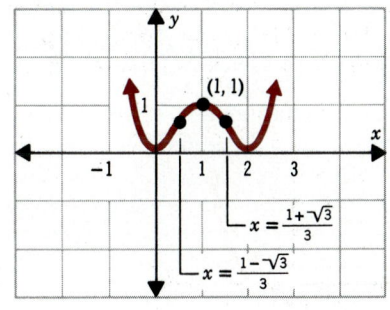

11. $f'(x) = \frac{4}{3}x^{1/3} + \frac{4}{3}x^{-2/3} = \frac{4}{3}x^{-2/3}(x+1)$
$f'(x) = 0 \Rightarrow x = -1$ is a critical number
$x < -1 \Rightarrow f'(x) < 0 \Rightarrow f(x)$ is increasing
$-1 < x < 0 \Rightarrow f'(x) > 0 \Rightarrow f(x)$ is increasing
$x > 0 \Rightarrow f'(x) > 0 \Rightarrow f(x)$ is increasing
Thus, $(-1, -3)$ is a relative minimum.
$f''(x) = \frac{4}{9}x^{-2/3} - \frac{8}{9}x^{-5/3} = \frac{4}{9}x^{-5/3}(x-2)$
$f''(x) = 0 \Rightarrow x = 2$
$0 < x < 2 \Rightarrow f''(x) < 0 \Rightarrow f(x)$ is concave down
$x > 2 \Rightarrow f''(x) > 0 \Rightarrow f(x)$ is concave up
$x < 0 \Rightarrow f''(x) > 0 \Rightarrow f(x)$ is concave up
Hence, $(2, f(2)) = (2, 7.56)$ and $(0,0)$ are inflection points.
At $(0,0)$ there is a vertical tangent line.

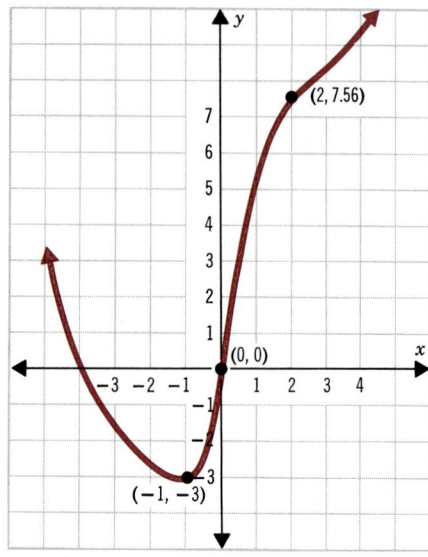

13. $f'(x) = \dfrac{2(x^2+1) - 2x(2x)}{(x^2+1)^2} = \dfrac{2(1-x^2)}{(x^2+1)^2}$
$f'(x) = 0 \Rightarrow x = 1, -1$ are critical numbers
$x < -1 \Rightarrow f'(x) < 0 \Rightarrow f(x)$ is decreasing
$-1 < x < 1 \Rightarrow f'(x) > 0 \Rightarrow f(x)$ is increasing
$x > 1 \Rightarrow f'(x) < 0 \Rightarrow f(x)$ is decreasing
Hence, $(1,1)$ is a relative maximum and $(-1,-1)$ is a relative minimum.
$f''(x) = \dfrac{(x^2+1)^2(-4x) - (2-2x^2)4x(x^2+1)}{(x^2+1)^4} = \dfrac{4x(x^2-3)}{(x^2+1)^3}$
$f''(x) = 0 \Rightarrow x = 0, \sqrt{3}, -\sqrt{3}$
$x < -\sqrt{3} \Rightarrow f''(x) < 0 \Rightarrow$ concave down
$-\sqrt{3} < x < 0 \Rightarrow f''(x) > 0 \Rightarrow$ concave up
$0 < x < \sqrt{3} \Rightarrow f''(x) < 0 \Rightarrow$ concave down
$x > \sqrt{3} \Rightarrow f''(x) > 0 \Rightarrow$ concave up
Hence, $(0,0)$, $\left(\sqrt{3}, \dfrac{\sqrt{3}}{2}\right)$, and $\left(-\sqrt{3}, -\dfrac{\sqrt{3}}{2}\right)$ are inflection points.

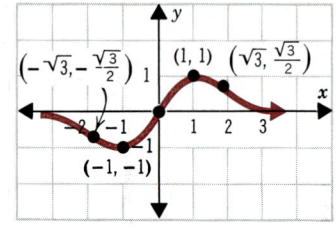

15. $f'(x) = 3x^2 - 6x + 3 = 3(x^2 - 2x + 1) = 3(x-1)^2 = 0 \Rightarrow x = 1$
$f(1) = 0;\quad f(0) = -1$
Absolute maximum $= 0$; absolute minimum $= -1$

17. $f'(x) = 4x^3 - 4x = 4x(x^2 - 1) = 0 \Rightarrow x = 0, 1, -1$
$f(-1) = -1;\quad f(1) = -1;\quad f(0) = 0$
Absolute minimum $= -1$; absolute maximum $= 0$

19. $f'(x) = \frac{4}{3}x^{1/3} + \frac{4}{3}x^{-2/3} = \frac{4}{3}x^{-2/3}(x+1) = 0 \Rightarrow x = -1$
$f'(x)$ does not exist at $x = 0$
$f(0) = 0;\quad f(-1) = -3;\quad f(1) = 5$
Absolute maximum $= 5$; absolute minimum $= -3$

21. $P'(x) = 120 - 6x = 0$ when $x = 20$ thousand dollars

23. Statement (e) must be true. (If we also assume that $f'(x)$ exists everywhere on some open interval containing $[-1, 3]$, then (a), (b), and (d) must also be true.)

CHAPTER 5 SOLUTIONS 403

25. $f'(x) = \dfrac{5}{x^2 + 1} = 5(x^2 + 1)^{-1}$
$f''(x) = -5(x^2 + 1)^{-2} 2x$
$= \dfrac{-10x}{(x^2 + 1)^2}$
$f''(x) > 0$ if $x < 0$
$f''(x) < 0$ if $x > 0$
Hence 0 is an inflection point.
$f'(x)$ exists everywhere and $\dfrac{5}{x^2 + 1} \neq 0$ for all x.
$f'(x) > 0$ for all x

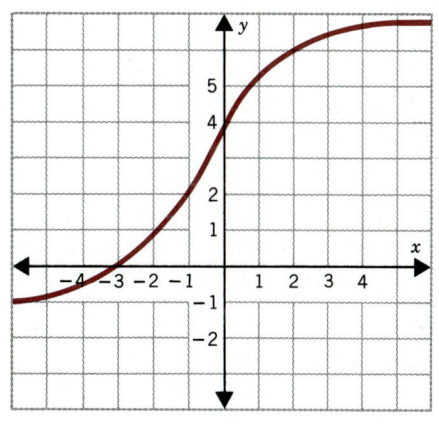

27. (a) A loss is incurred for $0 < x < a$ and $x > d$.
 (b) A profit is incurred for $a < x < d$.
 (c) Marginal cost is increasing for $x > b$.
     Marginal cost is decreasing for $0 < x < b$.

29. $R = (1500 + 200x)(300 - 15x)$
$R' = 37,500 - 6000x$
$x = 6.25$, price to maximize revenue should be
$300 - (6.25)(15) = 206.25$

CHAPTER 5

Exercise 1 (page 199)

1. $27^{2/3} = 3^2 = 9$ 3. $9^{3/2} = 3^3 = 27$ 5. $16^{-1/2} = \dfrac{1}{16^{1/2}} = \dfrac{1}{4}$

7. $27^{-2/3} = \dfrac{1}{27^{2/3}} = \dfrac{1}{9}$ 9. $(\tfrac{1}{9})^{1/2} = \tfrac{1}{3}$

11. $(\tfrac{1}{8})^{-1/3} = 8^{1/3} = 2$ 13. $(9^{1/3})(3)^{1/3} = (3^{2/3})(3^{1/3}) = 3$
15. $[(8^{-1})(8^{1/3})]^3 = [8^{-2/3}]^3 = 8^{-2} = \tfrac{1}{8^2} = \tfrac{1}{64}$
17. $2^4 = 16$ 19. $2^{1/3} \approx 1.260$ 21. $2^{0.1} \approx 1.072$
23. $f(x) = 3^x$ 25. $f(x) = (\tfrac{1}{3})^x$ 27. $f(x) = 4^{0.5x}$

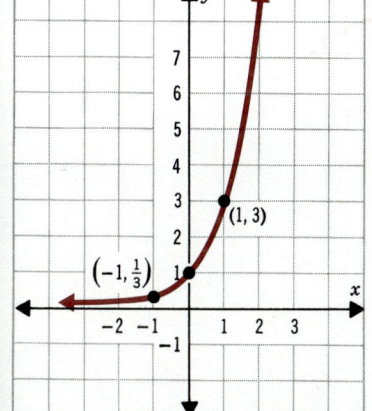 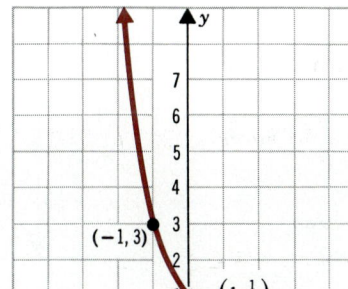 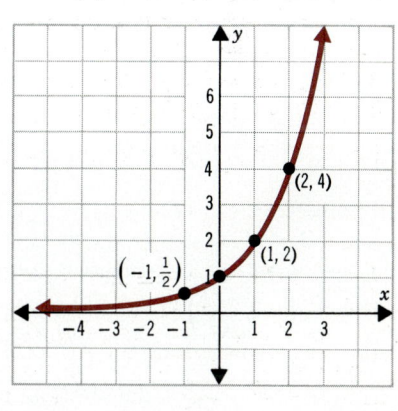

29. (b) $1 - e^{-1} = 0.632$; thus, 63.2% of the potential customers have responded after 10 days
    (d) $R(t) = 10,000,000(1 - e^{-0.1t})(0.70)$
        $P(t) = 7,000,000(1 - e^{-0.1t}) - 30,000 - 5000t$
        $P(28) = 7,000,000(1 - e^{-2.8}) - 30,000 - 140,000 = \$6,404,329$
31. Let x = Time in years and $P(x)$ denote earnings after x years. If P_0 denotes earnings at time $x = 0$, then
    $P(x) = (1 + 0.10)^x P_0 = (1.10)^x P_0$.

404 CHAPTER 5 SOLUTIONS

33. First year: $P(1) = 10{,}000 + 25{,}000(\frac{1}{4})^{0.5}$
$= 10{,}000 + 25{,}000(\frac{1}{2}) = \$22{,}500$
Third year: $P(3) = 10{,}000 + 25{,}000(\frac{1}{4})^{1.5}$
$= 10{,}000 + 25{,}000(0.125) = \$13{,}125$
Fifth year: $P(5) = 10{,}000 + 25{,}000(\frac{1}{4})^{2.5}$
$= 10{,}000 + 25{,}000(0.03125) = \$10{,}781.25$

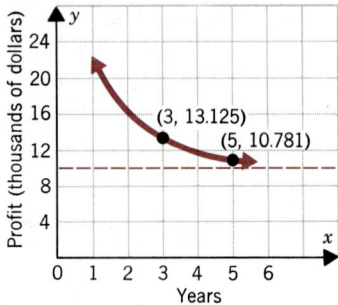

35. (a) $f(x+1) = a^{x+1} = a^x a = af(x)$
(b) $f(x+1) - f(x) = a^{x+1} - a^x = a^x a - a^x = a^x(a-1) = (a-1)f(x)$
(c) $f(x+h) = a^{x+h} = a^x a^h = a^h f(x)$

Exercise 2 (page 207)

1. $f'(x) = 5e^x$ **3.** $f'(x) = e^{5x}\dfrac{d}{dx}(5x) = 5e^{5x}$ **5.** $f'(x) = 8e^{-x/2}\left[\dfrac{d}{dx}\left(-\dfrac{1}{2}x\right)\right] = -4e^{-x/2}$

7. $f'(x) = e^x + xe^x = e^x(1+x)$ **9.** $f'(x) = e^{x^2} \cdot 2x = 2xe^{x^2}$

11. $f'(x) = e^{\sqrt{x}}\dfrac{d}{dx}(\sqrt{x}) = \dfrac{1}{2\sqrt{x}}e^{\sqrt{x}}$ **13.** $f(x) = (e^x)^{1/2} = e^{x/2}$; hence, $f'(x) = \frac{1}{2}e^{x/2}$

15. $f'(x) = 0 - e^x = e^x$ **17.** $f'(x) = \dfrac{e^x - e^{-x}}{2}$

19. $f'(x) = -3e^{-3x} - 3$
21. $f'(x) = e^x(3e^{3x} + e^{-x}) + (e^{3x} - e^{-x})e^x = e^x(3e^{3x} + e^{-x} + e^{3x} - e^{-x}) = e^x(4e^{3x}) = 4e^{4x}$

23. $f'(x) = (4x+1)e^{2x^2+x+1}$ **25.** $f'(x) = \dfrac{xe^x - e^x}{x^2} = \dfrac{e^x(x-1)}{x^2}$

27. $f'(x) = \left(1 + \dfrac{1}{x^2}\right)e^{x-(1/x)}$ **29.** $f'(x) = \dfrac{e^x}{2\sqrt{1+e^x}}$

31. $y' = ae^{ax}$; $y'' = a^2 e^{ax}$; $y''' = a^3 e^{ax}$ **33.** $y'' - 4y = 2 \cdot 2e^{2x} - 4e^{2x} = 0$

35. $f'(x) = e^x + xe^x = e^x(x+1)$
Critical number: $x = -1$
$f''(x) = 2e^x + xe^x = e^x(2+x)$
$f''(-1) > 0 \Rightarrow \left(-1, \dfrac{-1}{e}\right)$ is a relative minimum
$f''(x) = 0 \Rightarrow x = -2$
$x < -2 \Rightarrow f''(x) < 0 \Rightarrow f(x)$ is concave down
$x > -2 \Rightarrow f''(x) > 0 \Rightarrow f(x)$ is concave up
Hence, $\left(-2, \dfrac{-2}{e^2}\right)$ is an inflection point.

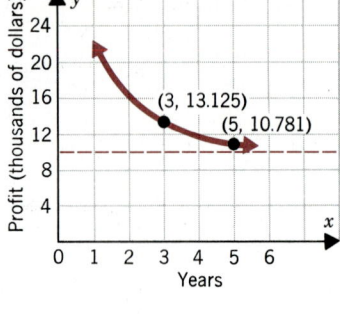

37. (a) $f'(x) = e^x > 0$ for all x
$f(-10) = \dfrac{1}{e^{10}}$ is the absolute minimum
$f(10) = e^{10}$ is the absolute maximum
(b) $f'(x) = -e^{-x} < 0$ for all x
$f(10) = \dfrac{1}{e^{10}}$ is the absolute minimum
$f(-10) = e^{10}$ is the absolute maximum

39. Compounded continuously: $A = Pe^{rt} = 1000e^{(0.08)2} = 1000e^{0.16} \approx 1000(1.1735) = \$1173.50$
Compounded quarterly: $A = P(1+r)^t = 1000\left(1 + \dfrac{0.085}{4}\right)^4 = 1000(1.02125)^8 \approx \$1183.20$
The investment at 8.5% compounded quarterly is better.

41. Compounded continuously: $A = Pe^{rt}$
$$2000 = Pe^{(0.1)3}$$
$$P = \frac{2000}{e^{0.3}} \approx \$1481.59$$

Compounded quarterly: $P = \dfrac{A}{(1+r)^t} = \dfrac{2000}{\left(1+\dfrac{0.1}{4}\right)^{12}} \approx \$1487.11$

43. Continuously compounded, $P = P_0 e^{0.08t}$ where t is time measured in years. If $P = 2P_0$ then $e^{0.08t} = 2$. From Table 3, $0.08t \approx .7$ or $t \approx 8.75$ years = 105 months.

Compounded quarterly, $P = P_0\left(1 + \dfrac{.08}{4}\right)^{4t}$. If $P = 2P_0$ then $(1.02)^{4t} = 2$, which implies that $4t \approx 36$ or $t \approx 9$ years = 108 months.

45. $E'(x) = -75{,}000(\ln 0.15)0.15^x - 500$
$E''(x) = -75{,}000(\ln 0.15)^2 0.15^x < 0$ for all x
Thus, the maximum value for $E(x)$ occurs when
$-75{,}000(\ln 0.15)0.15^x - 500 = 0$
$$(150 \ln 0.15)0.15^x = -1$$
$$150 \ln \left(\tfrac{3}{20}\right) 0.15^x = -1$$
$$150(\ln \tfrac{20}{3})0.15^x = 1$$
$$0.15^x = \frac{1}{150 \ln \left(\tfrac{20}{3}\right)}$$
$$x \ln 0.15 = \ln\left[\frac{1}{150 \ln \left(\tfrac{20}{3}\right)}\right]$$
$$= -\ln[150 \ln (\tfrac{20}{3})]$$
$$x = -\frac{\ln[150 \ln (\tfrac{20}{3})]}{\ln 0.15}$$
$$= -\frac{\ln (150) + \ln (\ln 20 - \ln 3)}{\ln 0.15}$$
$$x \approx 3.98$$

Exercise 3 (page 214)

1. $f(x) = \log_5 x$ **3.** $f(x) = \log_{1/3} x$

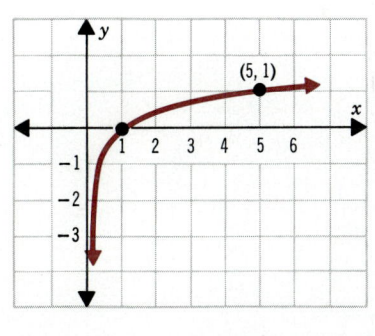
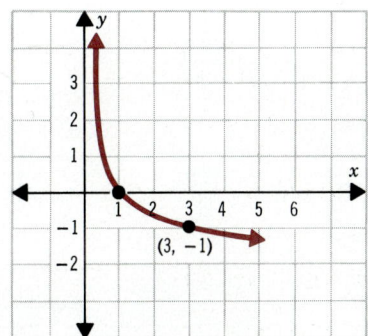

5. $9 = 3^2$ **7.** $\tfrac{1}{81} = 3^{-4}$ **9.** $P = a^Q$ **11.** $3 = \log_{10}(1000)$ **13.** $\tfrac{1}{2} = \log_a 3$
15. $\log_2 32 = \log_2 2^5 = 5 \log_2 2 = 5$ **17.** $\log_{10} 10^{-3} = -3 \log_{10} 10 = -3$
19. $\log_2 24 - \log_2 12 = \log_2 \tfrac{24}{12} = \log_2 2 = 1$
21. $\log_{10} 12 = \log_{10} 3 + \log_{10} 2^2 = (0.4771) + 2(0.3010) = 1.079$
23. $\log_{10} 7.5 = \log_{10} \tfrac{15}{2} = \log_{10} 15 - \log_{10} 2 = \log_{10} 5 + \log_{10} 3 - \log_{10} 2$
$= (0.6990) + (0.4771) - (0.3010) = 0.8751$
25. $\log_{10} 36 = \log_{10} 2^2 + \log_{10} 3^2 = 2(\log_{10} 2 + \log_{10} 3) = 1.5562$

406 CHAPTER 5 SOLUTIONS

27. $\log_{10} \frac{1}{8} = \log_{10} 2^{-3} = -3 \log_{10} 2 = -0.9030$

29. $\ln 3 + \ln x = \ln(3x)$

31. $\frac{1}{2} \ln 16 = \ln 16^{1/2} = \ln 4$

33. $\ln[(x + 1)(x + 2)(x + 3)]$

35. $f'(x) = \frac{6}{x}$

37. $f'(x) = \frac{3}{3x} = \frac{1}{x}$

39. $f'(x) = 8 \frac{\frac{1}{2}}{\frac{x}{2}} = \frac{8}{x}$

41. $f'(x) = \ln x + 1$

43. $f'(x) = \frac{2x}{x^2} = \frac{2}{x}$

45. $f(x) = \frac{1}{2} \ln x \Rightarrow f'(x) = \frac{1}{2x}$

47. $f(x) = (\ln x)^{1/2} \Rightarrow f'(x) = \frac{1}{2}(\ln x)^{-1/2} \frac{1}{x} = \frac{1}{2x(\ln x)^{1/2}}$

49. $f'(x) = -\frac{1}{x^2} \ln x + \frac{1}{x^2} = \frac{1}{x^2}(1 - \ln x)$

51. $f'(x) = e^{\ln x} \frac{d}{dx}(\ln x) = \frac{1}{x} e^{\ln x} = 1$

53. $f'(x) = x\left(\frac{2x}{x^2 + 4}\right) + \ln(x^2 + 4) = \frac{2x^2}{x^2 + 4} + \ln(x^2 + 4)$

55. $f(x) = \frac{1}{2} x \ln(x^2 + 1); \quad f'(x) = \frac{1}{2} x \left(\frac{2x}{x^2 + 1}\right) + \frac{1}{2} \ln(x^2 + 1) = \frac{x^2}{x^2 + 1} + \frac{\ln(x^2 + 1)}{2}$

57. $\ln|f(x)| = 2\ln(x^2 + 1) + 4\ln|2x^3 - 1|$

$\frac{f'(x)}{f(x)} = \frac{4x}{x^2 + 1} + \frac{24x^2}{2x^3 - 1}$

$f'(x) = (x^2 + 1)^2(2x^3 - 1)^4\left(\frac{4x}{x^2 + 1} + \frac{24x^2}{2x^3 - 1}\right)$

$= 4x(x^2 + 1)^2(2x^3 - 1)^4\left(\frac{1}{x^2 + 1} + \frac{6x}{2x^3 - 1}\right)$

$= 4x(x^2 + 1)^2(2x^3 - 1)^4 \frac{(2x^3 - 1 + 6x^3 + 6x)}{(x^2 + 1)(2x^3 - 1)}$

$= 4x(x^2 + 1)(2x^3 - 1)^3(8x^3 + 6x - 1)$

59. $\ln|f(x)| = \ln|x^3 + 1| + \ln|x - 1| + \ln(x^4 + 5)$

$\frac{f'(x)}{f(x)} = \frac{3x^2}{x^3 + 1} + \frac{1}{x - 1} + \frac{4x^3}{x^4 + 5}$

$f'(x) = (x^3 + 1)(x - 1)(x^4 + 5)\left(\frac{3x^2}{x^3 + 1} + \frac{1}{x - 1} + \frac{4x^3}{x^4 + 5}\right)$

61. $\ln|f(x)| = 2\ln|x| + \ln|x^3 + 1| - \frac{1}{2}\ln(x^2 + 1)$

$\frac{f'(x)}{f(x)} = \frac{2}{x} + \frac{3x^2}{x^3 + 1} - \frac{x}{x^2 + 1}$

$f'(x) = \frac{x^2(x^3 + 1)}{\sqrt{x^2 + 1}}\left(\frac{2}{x} + \frac{3x^2}{x^3 + 1} - \frac{x}{x^2 + 1}\right)$

63. $f'(x) = 2^x \ln 2$

65. $f'(x) = \frac{1}{x \ln 2}$

67. $f(x) = x \ln x, \ x > 0$
$f'(x) = \ln x + 1$
Critical number: $x = e^{-1} = \frac{1}{e}; \quad f(e^{-1}) = -e^{-1}$

$f''(x) = \frac{1}{x}$

$f''(e^{-1}) > 0 \Rightarrow \left(\frac{1}{e}, -\frac{1}{e}\right)$ is a relative minimum

$f''(x) = \frac{1}{x} > 0$ for all $x > 0; \Rightarrow f(x)$ is concave up for all $x > 0$

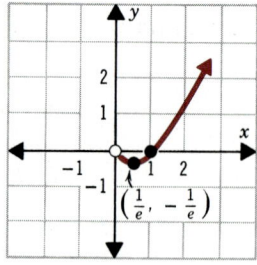

69. $y' = ae^{ax}; \quad y'' = a^2 e^{ax}; \quad y''' = a^3 e^{ax}$

71. $MC = C'(x) = \frac{1}{x + 1}$

73. $P(x) = R(x) - C(x) = 0.1x - \ln(x+1) - 20$
    $P'(x) = 0.1 - \dfrac{1}{x+1} = 0$ when $x + 1 = 10$ or $x = 9$

    $P''(x) = \dfrac{1}{(x+1)^2} > 0$ for all x
    Thus, $P(x)$ has no maximum value.

75. (a) $f'(x) = ae^{ax}$; $f''(x) = a^2 e^{ax}$; ...; $f^{(n)}(x) = a^n e^{ax}$
    (b) $f'(x) = \dfrac{1}{x}$; $f''(x) = -\dfrac{1}{x^2}$; ...; $f^{(n)}(x) = -\dfrac{(-1)^n}{x^n}$

77. (a) $\tfrac{1}{2}A = A\, 2^{-kt}$
    $2^{-1}A = A\, 2^{-kt}$
    $2^{-1} = 2^{-kt}$
    $-1 = -kt$
    $t = \dfrac{1}{k}$

    (b) $\dfrac{dc}{dt} = A(\ln 2)2^{-kt}(-k)$
    $= -(kA \ln 2)2^{-kt}$

Exercise 4 (page 219)
1. (a) $S(0) = 2000 - 1000e^0 = 1000$
   (b) $S'(t) = (-1000)(-.4)e^{-.4t}$
       $S'(t) = 400e^{-.4t}$
       $S'(5) = 400e^{-2} \approx 54.134$
   (c) $\lim\limits_{t \to \infty} (2000 - 1000e^{-.4t}) = 2000 - 1000(0) = 2000$

3. $f(x) = \dfrac{20{,}000}{1 + 50e^{-x}}$

   $f'(x) = \dfrac{20{,}000(50e^{-x})}{(1 + 50e^{-x})^2} = \dfrac{(10^6)e^{-x}}{(1 + 50e^{-x})^2}$
   $f'(x) > 0$ for all $x \ge 0 \Rightarrow f$ is increasing
   $f''(x) = \dfrac{-(1 + 50e^{-x})^2(10^6)e^{-x} - (10^6)e^{-x}2(1 + 50e^{-x})(-50e^{-x})}{(1 + 50e^{-x})^4} = \dfrac{-(10^6)e^{-x}(1 - 50e^{-x})}{(1 + 50e^{-x})^3}$
   $f''(x) = 0 \Rightarrow e^x = 50 \Rightarrow x = \ln 50 \approx 3.9$
   $x > 3.9 \Rightarrow f''(x) < 0 \Rightarrow$ concave down
   $x < 3.9 \Rightarrow f''(x) > 0 \Rightarrow$ concave up
   Hence, $(3.9, 10{,}000)$ is an inflection point.
   Therefore, in 3.9 months,
   the sales rate is a maximum.

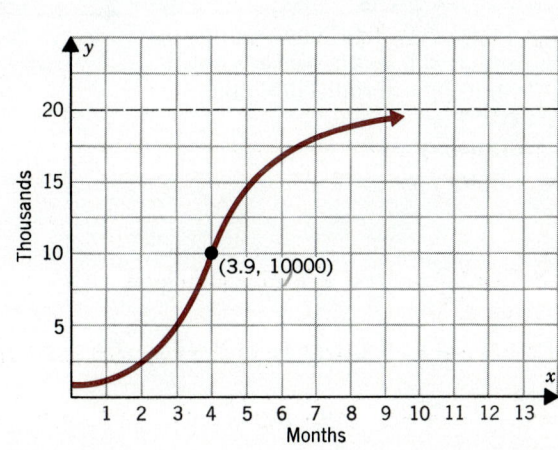

5. The rate of spreading of the flu is
$$N'(t) = \frac{99{,}990{,}000e^{-t}}{(1 + 9999e^{-t})^2}$$
which reaches a maximum when
$$N''(t) = 99{,}990{,}000e^{-t}\left[\frac{9999e^{-t} - 1}{(1 + 9999e^{-t})^3}\right] = 0$$
$9999e^{-t} = 1$
$e^t = 9999$
$t \approx 9.21$ days

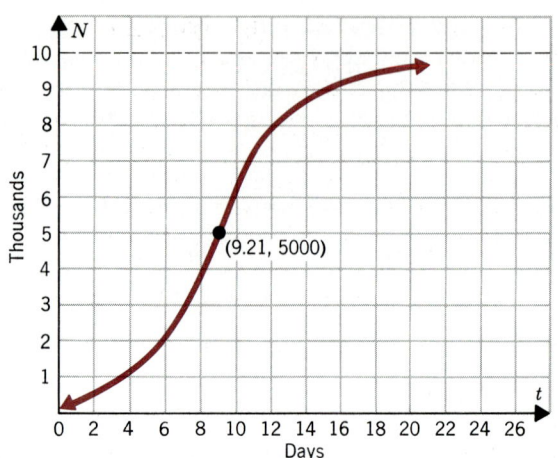

Review Exercises (page 221)

1. $f'(x) = 20e^{5x}$ **3.** $f'(x) = \dfrac{15}{x}$ **5.** $f'(x) = 4xe^{2x^2+5}$ **7.** $f'(x) = \dfrac{4x}{2x^2 + 5}$

9. $\ln|f(x)| = 2\ln(x^2 + 1) + 3\ln|x^2 - 1|$
$$\frac{f'(x)}{f(x)} = \frac{4x}{x^2 + 1} + \frac{6x}{x^2 - 1}$$
$$f'(x) = (x^2 + 1)^2(x^2 - 1)^3\left(\frac{4x}{x^2 + 1} + \frac{6x}{x^2 - 1}\right) = 2x(x^2 + 1)(x^2 - 1)^2(5x^2 + 1)$$

11. $20{,}000 = 10{,}000e^{.01x}$ where x is measured in months
$e^{.01x} = 2$; $.01x = \ln 2$; $x = 100\ln 2 \approx 69.3$ months

13. $f(x) = \dfrac{2000}{1 + 4e^{-x}}$

$f(0) = \dfrac{2000}{5} = 400$

y-intercept: $(0, 400)$

$f'(x) = \dfrac{8000e^{-x}}{(1 + 4e^{-x})^2}$

$f'(x) > 0$ for all x; hence, $f(x)$ is increasing

$f''(x) = \dfrac{-8000e^{-x}[1 - 4e^{-x}]}{(1 + 4e^{-x})^3} = 0 \Rightarrow$

$e^x = 4 \Rightarrow x = \ln 4 \approx 1.38$

$x < 1.38 \Rightarrow f''(x) > 0 \Rightarrow f$ is concave up
$x > 1.38 \Rightarrow f''(x) < 0 \Rightarrow f$ is concave down
Hence, $(\ln 4, 1000)$ is an inflection point.

15. $S(10) = 4000e^{-3} = 199$ sales
$S'(t) = -(0.3)(4000)e^{-0.3t}$
$S'(10) = -1200e^{-3} = -59.7$

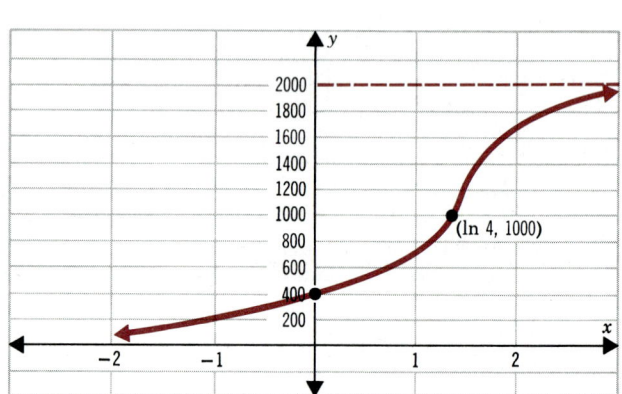

CHAPTER 6

Exercise 1 (page 226)

1. $f(2, 1) = 2^2 + 1 = 5$ **3.** $f(2, 1) = \sqrt{2 \cdot 1} = \sqrt{2}$ **5.** $f(2, 1) = \dfrac{1}{2(2) + 1} = \dfrac{1}{5}$

7. $f(2, 1) = \dfrac{4 - 1}{2 - 1} = \dfrac{3}{1} = 3$ **9.** $f(2, 1) = 0$

CHAPTER 6 SOLUTIONS 409

11. (a) $f(1,0) = 3 + 0 + 0 = 3$ (b) $f(0,1) = 0 + 2 + 0 = 2$ (c) $f(2,1) = 6 + 2 + 2 = 10$
    (d) $f(x + \Delta x, y) = 3(x + \Delta x) + 2y + (x + \Delta x)y = 3x + 2y + xy + (3 + y)\Delta x$
    (e) $f(x, y + \Delta y) = 3x + 2(y + \Delta y) + x(y + \Delta y) = 3x + 2y + xy + (2 + x)\Delta y$
13. (a) $f(0,0) = 0 + 0 = 0$ (b) $f(0,1) = 0 + 0 = 0$ (c) $f(a^2, t^2) = \sqrt{a^2 t^2} + a^2 = at + a^2$
    (d) $f(x + \Delta x, y) = \sqrt{(x + \Delta x)y} + x + \Delta x = \sqrt{xy + y\Delta x} + x + \Delta x$
    (e) $f(x, y + \Delta y) = \sqrt{x(y + \Delta y)} + x = \sqrt{xy + x\Delta y} + x$
15. A plane parallel to the xz-plane passing through $y = 3$ 17. The yz-plane
19. A line in the xz-plane parallel to the z-axis and passing through the point $(1,0,0)$
21. (a) $f(x + \Delta x, y) = 3(x + \Delta x) + 4y = 3x + 4y + 3\Delta x$
    (b) $f(x + \Delta x, y) - f(x, y) = 3x + 4y + 3\Delta x - (3x + 4y) = 3\Delta x$
    (c) $\dfrac{f(x + \Delta x, y) - f(x, y)}{\Delta x} = \dfrac{3\Delta x}{\Delta x} = 3$
    (d) $\lim\limits_{\Delta x \to 0} \dfrac{f(x + \Delta x, y) - f(x, y)}{\Delta x} = \lim\limits_{\Delta x \to 0} 3 = 3$
23. $C(R, u) = \$300(\text{Area of top} + \text{Area of bottom}) + 500(\text{Area of side})$
    $= \$300(2)(\pi R^2) + 500 \cdot 2\pi R \cdot h$
    $= 600\pi R^2 + 1000\pi R h$

Exercise 2 (page 231)

1. $f_x = -2y - 24x$ $f_y = 3y^2 - 2x + 2y$ 3. $f_x = e^y + ye^x + 1$ $f_y = xe^y + e^x$
   $f_{xx} = -24$ $f_{yy} = 6y + 2$ $f_{xx} = ye^x$ $f_{yy} = xe^y$
   $f_{xy} = -2$ $f_{yx} = -2$ $f_{xy} = e^y + e^x$ $f_{yx} = e^y + e^x$

5. $f_x = \dfrac{1}{y}$ $f_y = -\dfrac{x}{y^2}$
   $f_{xx} = 0$ $f_{yy} = \dfrac{2x}{y^3}$
   $f_{xy} = -\dfrac{1}{y^2}$ $f_{yx} = -\dfrac{1}{y^2}$

7. $f(x, y) = \ln(x^2 + y^2)$
   $f_x = \dfrac{2x}{x^2 + y^2}$
   $f_{xx} = \dfrac{2(x^2 + y^2) - 2x(2x)}{(x^2 + y^2)^2} = \dfrac{2y^2 - 2x^2}{(x^2 + y^2)^2}$
   $f_{xy} = \dfrac{-2y(2x)}{(x^2 + y^2)^2} = \dfrac{-4xy}{(x^2 + y^2)^2}$
   $f_y = \dfrac{2y}{x^2 + y^2}$
   $f_{yy} = \dfrac{2(x^2 + y^2) - 2y(2y)}{(x^2 + y^2)^2} = \dfrac{2x^2 - 2y^2}{(x^2 + y^2)^2}$
   $f_{yx} = \dfrac{-2x(2y)}{(x^2 + y^2)^2} = \dfrac{-4xy}{(x^2 + y^2)^2}$

9. $f(x, y) = \dfrac{10 - x + 2y}{xy}$
   $f_x = \dfrac{xy(-1) - (10 - x + 2y)(y)}{(xy)^2} = \dfrac{-(10 + 2y)}{x^2 y}$
   $f_y = \dfrac{2(xy) - (10 - x + 2y)(x)}{(xy)^2} = \dfrac{x - 10}{xy^2}$
   $f_{xy} = \dfrac{-2(x^2 y) + (10 + 2y)x^2}{x^4 y^2} = \dfrac{10}{x^2 y^2}$
   $f_{yx} = \dfrac{xy^2 - (x - 10)y^2}{x^2 y^4} = \dfrac{10}{x^2 y^2}$
   $f_{xx} = \dfrac{2(10 + 2y)}{x^3 y}$
   $f_{yy} = \dfrac{-2(x - 10)}{xy^3}$

11. $f(x, y) = x^3 + y^2$
    $f_x = 3x^2$ $f_{xy} = 0$
    $f_y = 2y$ $f_{yx} = 0$
13. $f(x, y) = 3x^4 y^2 + 7x^2 y$
    $f_x = 12x^3 y^2 + 14xy$ $f_{xy} = 24x^3 y + 14x$
    $f_y = 6x^4 y + 7x^2$ $f_{yx} = 24x^3 y + 14x$

15. $f(x, y) = \dfrac{y}{x^2}$

$f_x = -\dfrac{2y}{x^3}$ $f_{xy} = \dfrac{-2}{x^3}$

$f_y = \dfrac{1}{x^2}$ $f_{yx} = \dfrac{-2}{x^3}$

17. Slope of tangent line is

$\dfrac{\partial z}{\partial x} = \tfrac{1}{4} \cdot 5 \cdot \tfrac{1}{2}(16 - x^2)^{-1/2}(-2x) = \dfrac{-5x}{4\sqrt{16 - x^2}}$

At $\left(2, 3, 5\dfrac{\sqrt{3}}{2}\right)$, the value of $\dfrac{\partial z}{\partial x}$ is

$m = \dfrac{-5(2)}{4\sqrt{16 - 4}} = \dfrac{-5}{2 \cdot 2\sqrt{3}} = \dfrac{-5}{4\sqrt{3}}$

19. $\dfrac{\partial z}{\partial x} = -50$ means that when the cost of margarine is held fixed and the cost of butter per pound increases by \$1, the demand for butter decreases by 50 pounds.

$\dfrac{\partial z}{\partial y} = 90$ means that when the cost of butter is held fixed and the cost of margarine per pound increases by \$1, the demand for butter increases by 90 pounds.

Exercise 3 (page 235)

1. $f_x = 6x - 2y = 0$ when $y = 3x$; $f_y = -2x + 2y = 0$ when $x = y$
$x = y = 3x \Rightarrow x = y = 0$; thus, $(0, 0)$ is a critical point
$f_{xx} = 6 > 0$; $f_{yy} = 2$; $f_{xy} = -2$
$f_{xx}f_{yy} = 12 > 4 = f_{xy}^2$ (for all x, y)
Thus, there is a relative minimum at $(0, 0)$.
$f(0, 0) = 0$

3. $f_x = 2x - 3 = 0$ when $x = \tfrac{3}{2}$; $f_y = 2y = 0$ when $y = 0$
Thus, $(\tfrac{3}{2}, 0)$ is a critical point.
$f_{xx} = 2$; $f_{yy} = 2$; $f_{xy} = 0$
$f_{xx}f_{yy} = 4 > f_{xy}^2$
Since $f_{xx}(\tfrac{3}{2}, 0) > 0$, there is a relative minimum at $(\tfrac{3}{2}, 0)$.
$f(\tfrac{3}{2}, 0) = 9.75$

5. $f_x = 2x + 4 = 0$ when $x = -2$; $f_y = -2y + 8 = 0$ when $y = 4$
Thus, $(-2, 4)$ is a critical point.
$f_{xx} = 2$; $f_{yy} = -2$; $f_{xy} = 0$
$f_{xx}f_{yy} = -4 < 0 = f_{xy}^2$
There is a saddle point at $(-2, 4)$. Since this is the only critical point, there is no relative maximum or minimum.

7. $f_x = 2x - 4$; $f_y = 8y + 8$; $f_x = f_y = 0$ when $x = 2$ and $y = -1$
Thus $(2, 1)$ is a critical point.
$f_{xx} = 2 > 0$; $f_{yy} = 8 > 0$; $f_{xy} = 0$; $f_{xx}f_{yy} = 16 > f_{xy}^2$
Thus there is a relative minimum at $(2, -1)$.
$f(2, -1) = -9$

9. $f_x = 2x + y - 6$; $f_y = 2y + x$
$f_x = f_y = 0$ when $x = 4$ and $y = -2$
Thus, there is a critical point at $(4, -2)$.
$f_{xx} = 2 > 0$; $f_{yy} = 2$; $f_{xy} = 1$
$f_{xx}f_{yy} = 4 > 1 = f_{xy}^2$
Thus, there is a relative minimum at $(4, -2)$.
$f(4, -2) = -6$

11. $f_x = 2x + y$; $f_y = -2y + x$
$f_x = f_y = 0$ when $x = y = 0$
Thus, $(0, 0)$ is a critical point.
$f_{xx} = 2$; $f_{yy} = -2$; $f_{xy} = 1$
$f_{xx}f_{yy} = -4 < 1 = f_{xy}^2$
There is a saddle point at $(0, 0)$.
No relative maximum or minimum exists.

13. $f_x = 3x^2 - 6y$; $f_y = -6x + 3y^2$; $f_x = f_y = 0$ when $x = y = 0$ or $x = y = 2$. Thus $(0,0)$ and $(2,2)$ are critical points. $f_{xx} = 6x$; $f_{yy} = 6y$; $f_{xy} = 6$; $f_{xx}(0,0) = 0 = f_{yy}(0,0)$; $f_{xx}f_{yy} = 0 < f_{xy}^2$ at $(0,0)$. There is a saddle point at $(0,0)$. $f_{xx}f_{yy} = 144 > f_{xy}^2$ at $(2,2)$. Since $f_{xx}(2,2) > 0$, there is a relative minimum at $(2,2)$. $f(2,2) = -8$.

15. This function has no relative extrema. $f_x = 3x^2 + 2xy$; $f_y = x^2 + 2y$; $f_x = 0 = f_y$ when $x = y = 0$ or $x = 3$ and $y = -\frac{9}{2}$; the critical points are $(0,0)$ and $(3, -\frac{9}{2})$. Since $f_x(x,0) > 0$ for all $x \neq 0$, $f(0,0)$ is neither a relative maximum or a relative minimum. At $(3, -\frac{9}{2})$, the function has a saddle point, since $f_{xx}f_{yy} = 18 < 36 = f_{xy}^2$.

17. $f_x = \dfrac{-y}{(x+y)^2}$; $f_y = \dfrac{x}{(x+y)^2}$; $f_x = 0$ if and only if $y = 0$ and $x \neq 0$; $f_y = 0$ if and only if $x = 0$ and $y \neq 0$. There are no critical points, hence no relative extrema.

19. Revenue $= R(x,y) = xp + yq = x(12 - x) + y(8 - y) = 12x + 8y - x^2 - y^2$
    Cost $= C(x,y) = x^2 + 2xy + 3y^2$
    Profit $= P(x,y) = R(x,y) - C(x,y) = 12x + 8y - 2x^2 - 4y^2 - 2xy$
    $P_x = 12 - 4x - 2y$; $P_y = 8 - 8y - 2x$
    $P_x = P_y = 0$ when $x = \frac{20}{7}$ and $y = \frac{2}{7}$
    Hence, $(\frac{20}{7}, \frac{2}{7})$ is a critical point.
    $P_{xx} = -4$; $P_{yy} = -8$; $P_{xy} = -2$
    $P_{xx}(\frac{20}{7}, \frac{2}{7}) P_{yy}(\frac{20}{7}, \frac{2}{7}) = 32 > 4 = [P_{xy}(\frac{20}{7}, \frac{2}{7})]^2$; $P_{xx}(\frac{20}{7}, \frac{2}{7}) = -4 < 0$
    There is a relative maximum at $x = \frac{20}{7}$, $y = \frac{2}{7}$; the prices are $p = \frac{64}{7}$, $q = \frac{54}{7}$, and the profit is $P(\frac{20}{7}, \frac{2}{7}) = \frac{128}{7} \approx 18.28$ thousand dollars.

21. $z_x = 2y - 4x - 8$; $z_y = 2x - 2y + 6$
    $z_x = z_y = 0$ when $x = -1$ and $y = 2$
    $z_{xx} = -4$; $z_{yy} = -2$; $z_{xy} = 2$
    $z_{xx} z_{yy} = 8 > 4 = z_{xy}^2$
    Since $z_{xx}(-1, 2) < 0$, there is a relative maximum at $(-1, 2)$. Since this is the only relative maximum, the height of the mountain must be $f(-1, 2) = 14$ thousand feet.

23. $R_x = 2x(a - x)t^2 e^{-t} - x^2 t^2 e^{-t} = xt^2 e^{-t}(2a - 3x)$; $R_t = x^2(a - x)(2te^{-t} - t^2 e^{-t}) = x^2(a - x)te^{-t}(2 - t)$
    If $R_x = 0$, then $x = 0$, $t = 0$, or $x = \dfrac{2a}{3}$.
    If $R_t = 0$, then $x = 0$, $x = a$, $t = 0$, or $t = 2$.
    If $x = 0$, $t = 0$, or $x = a$, then $R(x,t) = 0$. Thus, the maximum reaction must occur for $x = \dfrac{2a}{3}$ units administered at $t = 2$ hours after the drug has been administered.

Exercise 4 (page 240)

1. $F(x, y, \lambda) = 3x + 4y + \lambda(x^2 + y^2 - 9)$
   $\dfrac{\partial F}{\partial x} = 3 + 2\lambda x = 0$; $\dfrac{\partial F}{\partial y} = 4 + 2\lambda y = 0$; $\dfrac{\partial F}{\partial \lambda} = x^2 + y^2 - 9 = 0$; $\lambda = -\dfrac{3}{2x}$
   Hence, $4 + 2y\left(-\dfrac{3}{2x}\right) = 0 \Rightarrow 4x - 3y = 0 \Rightarrow x = \frac{3}{4}y$; $(\frac{3}{4}y)^2 + y^2 = \frac{25}{16}y^2 = 9 \Rightarrow y = \pm\frac{12}{5}$; $x = \pm\frac{9}{5}$
   $f(-\frac{9}{5}, -\frac{12}{5}) = -15$
   The maximum value of $f(x,y)$ is $f(\frac{9}{5}, \frac{12}{5}) = 3(\frac{9}{5}) + 4(\frac{12}{5}) = \frac{75}{5} = 15$.

3. $F(x, y, \lambda) = 12xy - 3y^2 - x^2 + \lambda(x + y - 16)$
   $\dfrac{\partial F}{\partial x} = 12y - 2x + \lambda = 0$; $\dfrac{\partial F}{\partial y} = 12x - 6y + \lambda = 0$; $\dfrac{\partial F}{\partial \lambda} = x + y - 16 = 0$;
   $\lambda = 2x - 12y = -12x + 6y \Rightarrow 7x - 9y = 0$
   Using $x + y = 16$, we obtain $x = 9$ and $y = 7$.
   The maximum value of $f(x,y)$ is $f(9,7) = 528$.

5. $F(x, y, \lambda) = x^2 + y^2 + \lambda(x + y - 1)$
   $\dfrac{\partial F}{\partial x} = 2x + \lambda = 0$; $\dfrac{\partial F}{\partial y} = 2y + \lambda = 0$; $\dfrac{\partial F}{\partial \lambda} = x + y - 1 = 0$; $\lambda = -2x = -2y \Rightarrow -x + y = 0$;
   Using $x + y = 1$, we obtain $y = \frac{1}{2} = x$.
   Hence, the minimum value is $f(\frac{1}{2}, \frac{1}{2}) = \frac{1}{4} + \frac{1}{4} = \frac{1}{2}$

412 CHAPTER 6 SOLUTIONS

7. $F(x, y, \lambda) = 5x^2 + 6y^2 - xy + \lambda(x + 2y - 24)$
$\frac{\partial F}{\partial x} = 10x - y + \lambda = 0; \quad \frac{\partial F}{\partial y} = 12y - x + 2\lambda = 0; \quad \frac{\partial F}{\partial \lambda} = x + 2y - 24 = 0;$
$2\lambda = 2y - 20x = -12y + x \Rightarrow 2y - 3x = 0 \Rightarrow y = \frac{3x}{2}$

If we substitute $x + 2\left(\frac{3x}{2}\right) = 24 \Rightarrow x = 6$, we obtain $y = 9$.
The minimum value of $f(x, y)$ is $f(6, 9) = 612$.

9. $F(x, y, \lambda) = 18x^2 + 9y^2 + \lambda(x + y - 54)$
$\frac{\partial F}{\partial x} = 36x + \lambda = 0; \quad \frac{\partial F}{\partial y} = 18y + \lambda = 0; \quad \frac{\partial F}{\partial \lambda} = x + y - 54 = 0;$
$\lambda = -36x = -18y \Rightarrow 18y - 36x = 0 \Rightarrow y - 2x = 0$
Using $x + y - 54 = 0$, we get $3x = 54 \Rightarrow x = 18, y = 36$.
The minimum cost is $C(18, 36) = 18 \cdot 18 \cdot 18 + 9 \cdot 36 \cdot 36 = 17{,}496$.

11. $F(x, y, \lambda) = x^2 + 3xy - 6x + \lambda(x + y - 40)$
$\frac{\partial F}{\partial x} = 2x + 3y - 6 + \lambda = 0; \quad \frac{\partial F}{\partial y} = 3x + \lambda = 0; \quad \frac{\partial F}{\partial \lambda} = x + y - 40 = 0;$
$\lambda = -3x = -2x - 3y + 6 \Rightarrow -x + 3y = 6$
Using $x + y = 40$, we obtain $y = \frac{23}{2}$ and $x = \frac{57}{2}$.
Maximum production occurs when $x = \frac{57}{2}$ and $y = \frac{23}{2}$.

Exercise 5 (page 245)

1. $\quad x_1^2 + x_2^2 + x_3^2 + x_4^2 = 0 + 1 + 4 + 9 = 14$
$\quad x_1 + x_2 + x_3 + x_4 = 0 + 1 + 2 + 3 = 6$
$x_1y_1 + x_2y_2 + x_3y_3 + x_4y_4 = 0 + 4 + 4 - 3 = 5$
$\quad\quad\quad\quad n = 4$
$y_1 + y_2 + y_3 + y_4 = 7 + 4 + 2 + (-1) = 12$
$\quad\quad 14m + 6b = 5$
$\quad\quad 6m + 4b = 12$
$m = -2.6 \quad b = 6.9 \quad y = -2.6x + 6.9$

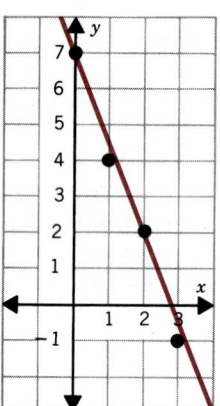

3. $\quad x_1^2 + x_2^2 + x_3^2 + x_4^2 = 0 + 1 + 9 + 36 = 46$
$\quad x_1 + x_2 + x_3 + x_4 = 0 + 1 + 3 + 6 = 10$
$x_1y_1 + x_2y_2 + x_3y_3 + x_4y_4 = 0 + 2 + 18 + 48 = 68$
$\quad\quad y_1 + y_2 + y_3 + y_4 = 1 + 2 + 6 + 8 = 17$
$46m + 10b = 68$
$10m + 4b = 17$
$m = \frac{17}{14}, \quad b = \frac{17}{14} \quad y = \frac{17}{14}x + \frac{17}{14}$

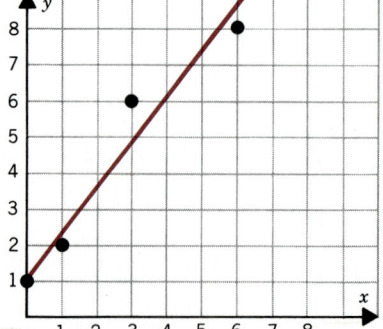

CHAPTER 6 SOLUTIONS 413

5. $x_1^2 + x_2^2 + x_3^2 + x_4^2 = 1 + 4 + 9 + 25 = 39$
$x_1 + x_2 + x_3 + x_4 = 1 + 2 + 3 + 5 = 11$
$x_1y_1 + x_2y_2 + x_3y_3 + x_4y_4 = 0 + 2 + 9 + 20 = 31$
$y_1 + y_2 + y_3 + y_4 = 0 + 1 + 3 + 4 = 8$
$39m + 11b = 31$
$11m + 4b = 8$
$m = \frac{36}{35}, \quad b = -\frac{29}{35}$
$y = \frac{36}{35}x - \frac{29}{35}$

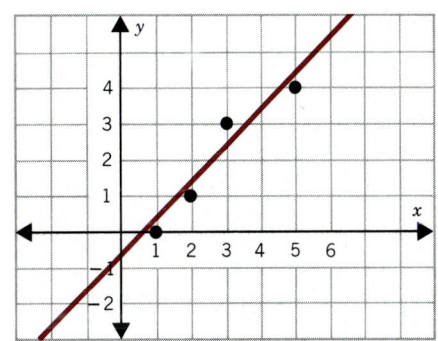

7. $x_1^2 + x_2^2 + x_3^2 + x_4^2 = 1 + 4 + 9 + 36 = 50$
$x_1 + x_2 + x_3 + x_4 = 1 + 2 + 3 + 6 = 12$
$x_1y_1 + x_2y_2 + x_3y_3 + x_4y_4 = 0 + 2 + 12 + 36 = 50$
$y_1 + y_2 + y_3 + y_4 = 0 + 1 + 4 + 6 = 11$
$50m + 12b = 50$
$12m + 4b = 11$
$m = \frac{17}{14}, \quad b = -\frac{25}{28}, \quad y = \frac{17}{14}x - \frac{25}{28}$

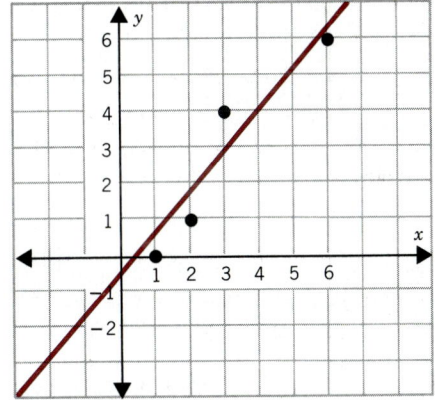

9. $x_1^2 + x_2^2 + x_3^2 + x_4^2 + x_5^2 = 1 + 9 + 25 + 36 + 64 = 135$
$x_1 + x_2 + x_3 + x_4 + x_5 = 1 + 3 + 5 + 6 + 8 = 23$
$x_1y_1 + x_2y_2 + x_3y_3 + x_4y_4 + x_5y_5 = 1 + 12 + 15 + 30 + 64 = 122$
$y_1 + y_2 + y_3 + y_4 + y_5 = 1 + 4 + 3 + 5 + 8 = 21$
$135m + 23b = 122$
$23m + 5b = 21$
$m = \frac{127}{146}, \quad b = \frac{29}{146}$
$y = \frac{127}{146}x + \frac{29}{146}$
$y = .870x + .199$

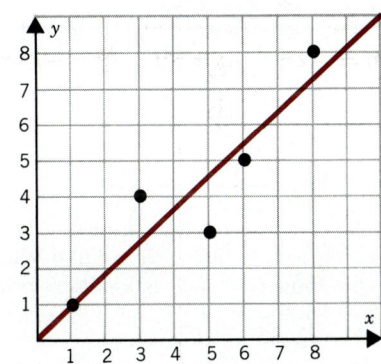

11. (a) $x_1^2 + x_2^2 + x_3^2 + x_4^2 = 9 + 16 + 36 + 9 = 70$
$x_1 + x_2 + x_3 + x_4 = 3 + 4 + 6 + 3 = 16$
$x_1y_1 + x_2y_2 + x_3y_3 + x_4y_4 = 63 + 108 + 216 + 69 = 456$
$y_1 + y_2 + y_3 + y_4 = 21 + 27 + 36 + 23 = 107$
$70m + 16b = 456$
$16m + 4b = 107$
$m = \frac{14}{3}, \quad b = \frac{97}{12}$
$y = \frac{14}{3}x + \frac{97}{12}$
$y = 4.667x + 8.0833$

(b) $(\frac{14}{3})3 + \frac{97}{12} \approx 22.0833$, 22 cars

13.

x	y	x²	xy
1.5	3.1	2.25	4.65
0.8	1.9	.64	1.52
2.6	4.2	6.76	10.92
1.0	2.3	1.0	2.3
0.6	1.2	.36	.72
2.8	4.9	7.84	13.72
1.2	2.8	1.44	3.36
0.9	2.1	.81	1.89
0.4	1.4	.16	.56
1.3	2.4	1.69	3.12
1.2	2.4	1.44	2.88
2.0	3.8	4.00	7.6
Totals 16.3	32.5	28.39	53.24

$$28.39m + 16.3b = 53.24$$
$$16.3m + 12b = 32.5$$
$$m = 1.455 \quad b = .732$$
$$y = 1.455x + .732$$

Review Exercises (page 247)

1. $z = f(x, y) = x^2 + xy$
   (a) $f(x + \Delta x, y) = (x + \Delta x)^2 + (x + \Delta x)y = x^2 + 2x\,\Delta x + (\Delta x)^2 + xy + y\,\Delta x$
   (b) $f(x + \Delta x, y) - f(x, y) = 2x\,\Delta x + (\Delta x)^2 + y\,\Delta x$
   (c) $\dfrac{f(x + \Delta x, y) - f(x, y)}{\Delta x} = 2x + \Delta x + y$
   (d) $\lim\limits_{\Delta x \to 0} \dfrac{f(x + \Delta x, y) - f(x, y)}{\Delta x} = \lim\limits_{\Delta x \to 0}(2x + \Delta x + y) = 2x + y$

3. (a) $f_x = y - 6 - 2x; \quad f_y = x - 2y$
   $f_x = f_y = 0 \Rightarrow x = -4$ and $y = -2$
   Thus, $(-4, -2)$ is a critical point.
   $f_{xx} = -2; \quad f_{yy} = -2; \quad f_{xy} = 1$
   Since $f_{xx}f_{yy} = 4 > 1 = f_{xy}^2$ (for all x, y) and $f_{xx}(-4, -2) < 0$, there is a relative maximum at $(-4, -2)$.
   (b) $f_x = 2x + 2; \quad f_y = 2y + 4$
   $f_x = f_y = 0 \Rightarrow x = -1, y = -2$
   $f_{xx} = 2; \quad f_{yy} = 2; \quad f_{xy} = 0$
   $f_{xx}f_{yy} = 4 > 0 = f_{xy}^2$ (for all x, y)
   Since $f_{xx}(-1, -2) > 0$, there is a relative minimum at $(-1, -2)$.
   (c) $f_x = 2 - 2x = 0$ when $x = 1; \quad f_y = 4 - 2y = 0$ when $y = 2$
   $f_{xx} = -2; \quad f_{yy} = -2; \quad f_{xy} = 0$
   $f_{xx}f_{yy} > f_{xy}^2$ (for all x, y)
   Since $f_{xx}(1, 2) < 0$, there is a relative maximum at $(1, 2)$.
   (d) $f_x = y = 0$ when $y = 0; \quad f_y = x = 0$ when $x = 0$
   $f_{xx} = f_{yy} = 0; \quad f_{xy} = 1$
   Since $f_{xx}f_{yy} < f_{xy}^2$ for all x, y, there is a saddle point at $(0, 0)$.

(e) $f_x = 2x = 0$ when $x = 0$; $f_y = -9 + 3y^2 = 0$ when $y = \pm\sqrt{3}$
$f_{xx} = 2$; $f_{yy} = 6y$; $f_{xy} = 0$
$f_{xx}(0, \sqrt{3})f_{yy}(0, \sqrt{3}) = 12\sqrt{3} > [f_{xy}(0, \sqrt{3})]^2$
Since $f_{xx}(0, \sqrt{3}) > 0$, there is a relative minimum at $(0, \sqrt{3})$.
$f_{xx}(0, -\sqrt{3})f_{yy}(0, -\sqrt{3}) = -12\sqrt{3} < [f_{xy}(0, -\sqrt{3})]^2$
Thus, there is a saddle point at $(0, -\sqrt{3})$.

5. (a) $F(x, y, \lambda) = 5x^2 - 3y^2 + xy + \lambda(2x - y - 20)$
$\dfrac{\partial F}{\partial x} = 10x + y + 2\lambda = 0$; $\dfrac{\partial F}{\partial y} = -6y + x - \lambda = 0$; $\dfrac{\partial F}{\partial \lambda} = 2x - y - 20 = 0$
Solving $\lambda = x - 6y$, $10x + y + 2\lambda = 0$, $2x - y = 20$, gives $x = 22$, $y = 24$.
The maximum value is $f(22, 24) = 1220$.

(b) $F(x, y, \lambda) = x\sqrt{y} + \lambda(2x + y - 3000)$
$\dfrac{\partial F}{\partial x} = \sqrt{y} + 2\lambda = 0$; $\dfrac{\partial F}{\partial y} = \dfrac{x}{2\sqrt{y}} + \lambda = 0$; $\dfrac{\partial F}{\partial \lambda} = 2x + y - 3000 = 0$
Solving the equations (eliminating λ) yields $x = y = 1000$.
The maximum value is $f(1000, 1000) = 10{,}000\sqrt{10}$.

7. $P(x, y) = R(x, y) - C(x, y)$
$= (2000x - 2x^2 + 100y - y^2 + xy) - (x^2 + 200x + y^2 + 100y - xy)$
$= 1800x - 3x^2 - 2y^2 + 2xy$
$P_x = 1800 - 6x + 2y$; $P_y = -4y + 2x$
$P_x = P_y = 0 \Rightarrow x = 360$, $y = 180$
Thus, $(360, 180)$ is a critical point.
$P_{xx} = -6$; $P_{yy} = -4$; $P_{xy} = 2$; $P_{xx}(360, 180) = -6 < 0$
$P_{xx}(360, 180)P_{yy}(360, 180) = 24 > 4 = [P_{xy}(360, 180)]^2$
The profit will be maximized by 360 units of x and 180 units of y.

CHAPTER 7

Exercise 1 (page 254)

1. $\int 3\,dx = 3x + K$ 3. $\int x\,dx = \tfrac{1}{2}x^2 + K$ 5. $\int x^{1/3}\,dx = \tfrac{3}{4}x^{4/3} + K$

7. $\int x^{-2}\,dx = -x^{-1} + K = \dfrac{-1}{x} + K$ 9. $\int x^{-1/2}\,dx = 2x^{1/2} + K$

11. $\int (x^2 + 2e^x)\,dx = \int x^2\,dx + 2\int e^x\,dx = \tfrac{1}{3}x^3 + 2e^x + K$

13. $\int \left(\dfrac{x-1}{x}\right)dx = \int \left(1 - \dfrac{1}{x}\right)dx = \int dx - \int \dfrac{1}{x}\,dx = x - \ln x + K$

15. $\int \left(\dfrac{3\sqrt{x}+1}{\sqrt{x}}\right)dx = \int (3 + x^{-1/2})\,dx = \int 3\,dx + \int x^{-1/2}\,dx = 3x + 2x^{1/2} + K$

17. $\int \dfrac{x^2 - 4}{x + 2}\,dx = \int \dfrac{(x+2)(x-2)}{(x+2)}\,dx = \int (x - 2)\,dx = \tfrac{1}{2}x^2 - 2x + K$

19. $\int x(x - 1)\,dx = \int (x^2 - x)\,dx = \int x^2\,dx - \int x\,dx = \tfrac{1}{3}x^3 - \tfrac{1}{2}x^2 + K$

21. $\int \dfrac{3x^5 + 2}{x}\,dx = \int \left(3x^4 + \dfrac{2}{x}\right)dx = 3\int x^4\,dx + 2\int \dfrac{1}{x}\,dx = \tfrac{3}{5}x^5 + 2\ln x + K$

23. $\int \dfrac{4e^x + e^{2x}}{e^x}\,dx = \int (4 + e^x)\,dx = 4x + e^x + K$

25. (a) $\int (x\sqrt{x})\,dx = \int x^{3/2}\,dx = \tfrac{2}{5}x^{5/2} + K$; $\int x\,dx = \tfrac{1}{2}x^2 + K_1$; $\int \sqrt{x}\,dx = \tfrac{2}{3}x^{3/2} + K_2$

$\int x\,dx \cdot \int \sqrt{x}\,dx = (\tfrac{1}{2}x^2 + K_1)(\tfrac{2}{3}x^{3/2} + K_2) = \tfrac{1}{3}x^{7/2} + \tfrac{2}{3}x^{3/2}K_1 + \tfrac{1}{2}x^2 K_2 + K_1 K_2$

$\int (x\sqrt{x})\,dx \neq \int x\,dx \cdot \int \sqrt{x}\,dx$

(b) $\int x(x^2 + 1)\,dx = \int (x^3 + x)\,dx = \tfrac{1}{4}x^4 + \tfrac{1}{2}x^2 + K$

$x\int (x^2 + 1)\,dx = x[\tfrac{1}{3}x^3 + x + K_1] = \tfrac{1}{3}x^4 + x^2 + K_1 x$

$\int x(x^2 + 1)\,dx \neq x\int (x^2 + 1)\,dx$

(c) $\int \dfrac{x^2 - 1}{x - 1}\,dx = \tfrac{1}{2}x^2 + x + K;\quad \int (x^2 - 1)\,dx = \tfrac{1}{3}x^3 - x + K_1;\quad \int (x - 1)\,dx = \tfrac{1}{2}x^2 - x + K_2$

$\dfrac{\int (x^2 - 1)\,dx}{\int (x - 1)\,dx} = \dfrac{\tfrac{1}{3}x^3 - x + K_1}{\tfrac{1}{2}x^2 - x + K_2} \neq \tfrac{1}{2}x^2 + x + K$

Exercise 2 (page 259)

1. Let $u = 2x + 1$, $du = 2\,dx$ or $\dfrac{du}{2} = dx$. Then

$\int (2x + 1)^5\,dx = \int u^5 \dfrac{du}{2} = \tfrac{1}{2}\int u^5\,du = \tfrac{1}{12}u^6 + K = \tfrac{1}{12}(2x + 1)^6 + K$

3. Let $u = 2x - 3$, $du = 2\,dx$ or $\dfrac{du}{2} = dx$. Then

$\int e^{2x-3}\,dx = \tfrac{1}{2}\int e^u\,du = \tfrac{1}{2}e^u + K = \tfrac{1}{2}e^{2x-3} + K$

5. Let $u = -2x + 3$, $du = -2\,dx$ or $\dfrac{du}{-2} = dx$. Then

$\int (-2x + 3)^{-2}\,dx = -\tfrac{1}{2}\int u^{-2}\,du = \tfrac{1}{2}u^{-1} + K = \tfrac{1}{2}(-2x + 3)^{-1} + K$

7. $u = x^2 + 4$, $du = 2x\,dx$; $\int (x^2 + 4)^2 x\,dx = \tfrac{1}{2}\int u^2\,du = \tfrac{1}{6}u^3 + K = \tfrac{1}{6}(x^2 + 4)^3 + K$

9. $u = x^3 + 1$, $du = 3x^2\,dx$; $\int e^{x^3+1} x^2\,dx = \tfrac{1}{3}\int e^u\,du = \tfrac{1}{3}e^u + K = \tfrac{1}{3}e^{x^3+1} + K$

11. $\int (e^x + e^{-x})\,dx = \int e^x\,dx + \int e^{-x}\,dx = e^x - e^{-x} + K$

13. $u = x^3 + 2$, $du = 3x^2\,dx$; $\int (x^3 + 2)^6 x^2\,dx = \tfrac{1}{3}\int u^6\,du = \tfrac{1}{21}u^7 + K = \tfrac{1}{21}(x^3 + 2)^7 + K$

15. $u = 1 + x^2$, $du = 2x\,dx$; $\int \dfrac{x\,dx}{\sqrt[3]{1 + x^2}} = \tfrac{1}{2}\int u^{-1/3}\,du = \tfrac{3}{4}u^{2/3} + K = \tfrac{3}{4}(1 + x^2)^{2/3} + K$

17. $u = x + 3$, $du = dx$; $\int x\sqrt{x + 3}\,dx = \int (u - 3)u^{1/2}\,du = \int (u^{3/2} - 3u^{1/2})\,du = \tfrac{2}{5}u^{5/2} - 2u^{3/2} + K$
    $= \tfrac{2}{5}(x + 3)^{5/2} - 2(x + 3)^{3/2} + K$

19. $u = e^x + 1$, $du = e^x\,dx$; $\int \dfrac{e^x\,dx}{e^x + 1} = \int \dfrac{du}{u} = \ln u + K = \ln(e^x + 1) + K$

21. $u = \sqrt{x} = x^{1/2}$, $du = \tfrac{1}{2}x^{-1/2}\,dx = \dfrac{dx}{2\sqrt{x}}$; $\int \dfrac{e^{\sqrt{x}}\,dx}{\sqrt{x}} = 2\int e^u\,du = 2e^u + K = 2e^{\sqrt{x}} + K$

23. $u = x^{1/3} - 1$, $du = \tfrac{1}{3}x^{-2/3}\,dx$; $\int \dfrac{(x^{1/3} - 1)^6\,dx}{x^{2/3}} = 3\int u^6\,du = \tfrac{3}{7}u^7 + K = \tfrac{3}{7}(x^{1/3} - 1)^7 + K$

25. $u = x^2 + 2x + 3$, $du = 2(x + 1)\,dx$; $\int \dfrac{(x + 1)\,dx}{(x^2 + 2x + 3)^2} = \tfrac{1}{2}\int \dfrac{du}{u^2} = -\tfrac{1}{2}u^{-1} + K = \dfrac{-1}{2(x^2 + 2x + 3)} + K$

27. $u = \sqrt{x} + 1$, $du = \dfrac{1}{2\sqrt{x}}\,dx$; $\int \dfrac{dx}{\sqrt{x}(1 + \sqrt{x})^4} = 2\int \dfrac{du}{u^4} = -\tfrac{2}{3}u^{-3} + K = -\tfrac{2}{3}(\sqrt{x} + 1)^{-3} + K$

29. $u = 2x + 3$, $du = 2\,dx$; $\int \dfrac{dx}{2x + 3} = \tfrac{1}{2}\int \dfrac{du}{u} = \tfrac{1}{2}\ln u + K = \tfrac{1}{2}\ln(2x + 3) + K$

31. $u = 4x^2 + 1, du = 8x\, dx;\quad \int \frac{x\, dx}{4x^2 + 1} = \frac{1}{8}\int \frac{du}{u} = \frac{1}{8}\ln u + K = \frac{1}{8}\ln(4x^2 + 1) + K$

33. $u = x^2 + 2x + 2, du = 2(x + 1)\, dx;\quad \int \frac{(x + 1)\, dx}{x^2 + 2x + 2} = \frac{1}{2}\int \frac{du}{u} = \frac{1}{2}\ln u + K = \frac{1}{2}\ln(x^2 + 2x + 2) + K$

35. Let $u = ax + b, du = a\, dx$. Then

$$\int (ax + b)^n\, dx = \frac{1}{a}\int u^n\, du = \frac{1}{a}\left(\frac{1}{n + 1}u^{n+1} + K\right) = \frac{1}{a(n + 1)}(ax + b)^{n+1} + K$$

Exercise 3 (page 263)

1. Let $u = x, du = dx, dv = e^{2x}\, dx, v = \frac{1}{2}e^{2x}$. Then
$$\int xe^{2x}\, dx = \frac{1}{2}xe^{2x} - \frac{1}{2}\int e^{2x}\, dx = \frac{1}{2}xe^{2x} - \frac{1}{4}e^{2x} + K$$

3. Let $u = x^2, du = 2x\, dx, dv = e^{-x}\, dx, v = -e^{-x}$. Then
$$\int x^2 e^{-x}\, dx = -x^2 e^{-x} + 2\int xe^{-x}\, dx$$
Integrating again by parts, let $u = x, du = dx, dv = e^{-x}\, dx, v = -e^{-x}$. Then
$$\int xe^{-x}\, dx = -xe^{-x} + \int e^{-x}\, dx = -xe^{-x} - e^{-x} + K$$
$$\int x^2 e^{-x}\, dx = -x^2 e^{-x} - 2e^{-x}(x + 1) + K$$

5. Let $u = \ln x, du = \frac{dx}{x}, dv = \sqrt{x}\, dx, v = \frac{2}{3}x^{3/2}$. Then
$$\int \sqrt{x}\ln x\, dx = \frac{2}{3}x^{3/2}\ln x - \frac{2}{3}\int x^{1/2}\, dx = \frac{2}{3}x^{3/2}\ln x - \frac{4}{9}x^{3/2} + K$$

7. Let $u = (\ln x)^2, du = \frac{2}{x}\ln x\, dx, dv = dx, v = x$. Then
$$\int (\ln x)^2\, dx = x(\ln x)^2 - 2\int \ln x\, dx$$
We must evaluate $\int \ln x\, dx$ by parts. Using $u = \ln x, du = \frac{dx}{x}, dv = dx, v = x$, we have
$$\int \ln x\, dx = x\ln x - \int dx = x\ln x - x + K$$
$$\int (\ln x)^2\, dx = x(\ln x)^2 - 2(x\ln x - x) + K$$

9. Let $u = \ln 3x, du = \frac{1}{x}dx, dv = x^2\, dx, v = \frac{1}{3}x^3$. Then
$$\int x^2 \ln 3x\, dx = \frac{1}{3}x^3 \ln 3x - \frac{1}{3}\int x^2\, dx = \frac{1}{3}x^3 \ln 3x - \frac{1}{9}x^3 + K$$

11. Let $u = (\ln x)^2, du = 2\frac{\ln x}{x}dx, dv = x^2\, dx, v = \frac{x^3}{3}$. Then

$$\int x^2(\ln x)^2\, dx = \frac{x^3(\ln x)^2}{3} - \frac{2}{3}\int x^2 \ln x\, dx$$

Integrating again by parts, let $u = \ln x, du = \frac{1}{x}dx, dv = x^2\, dx, v = \frac{x^3}{3}$. Then

$$\int x^2 \ln x\, dx = \frac{x^3 \ln x}{3} - \frac{1}{3}\int x^2\, dx = \frac{x^3 \ln x}{3} - \frac{x^3}{9} + K$$

Thus, $\int x^2(\ln x)^2\, dx = \frac{x^3(\ln x)^2}{3} - \frac{2}{3}\left(\frac{x^3 \ln x}{3} - \frac{x^3}{9}\right) + K$

$$= \frac{x^3(\ln x)^2}{3} - \frac{2x^3 \ln x}{9} + \frac{2x^3}{27} + K$$

418 CHAPTER 7 SOLUTIONS

13. Let $u = \ln x$, $du = \frac{1}{x}dx$. Then
$$\int \frac{\ln x}{x}dx = \int u\, du = \frac{u^2}{2} + K$$

Exercise 4 (page 267)
1. The general solution is $y = F(x) = \frac{1}{3}x^3 - x + K$.
   The particular solution subject to the condition $F(0) = 0$ is $F(x) = \frac{1}{3}x^3 - x$.
3. General solution: $y = F(x) = \frac{1}{3}x^3 - \frac{1}{2}x^2 + K$
   Particular solution using $F(3) = 3$,
   $3 = \frac{1}{3}(3)^3 - \frac{1}{2}(3)^2 + K \Rightarrow K = -\frac{3}{2}$
   $F(x) = \frac{1}{3}x^3 - \frac{1}{2}x^2 - \frac{3}{2}$
5. General solution: $y = F(x) = \frac{1}{4}x^4 - \frac{1}{2}x^2 + 2x + K$
   Particular solution using $F(-2) = 1$,
   $1 = F(-2) = \frac{1}{4}(-2)^4 - \frac{1}{2}(-2)^2 - 4 + K \Rightarrow K = 3$
   $F(x) = \frac{1}{4}x^4 - \frac{1}{2}x^2 + 2x + 3$
7. General solution: $F(x) = e^x + K$
   Particular solution using $F(0) = 4$ is $F(x) = e^x + 3$.
9. General solution: $y = F(x) = \frac{1}{2}x^2 + x + \ln x + K$
   Particular solution using $F(1) = 0$ is $y = F(x) = \frac{1}{2}x^2 + x + \ln x - \frac{3}{2}$.
11. The equation is $\frac{dA}{dt} = kA$ and the general solution is $A = A_0 e^{kt}$.
    $t = 0 \Rightarrow A_0 = 100$
    $t = 5 \Rightarrow 150 = 100e^{5k} \Rightarrow k = 0.08$
    Hence, $A(t) = 100e^{0.08t}$.
    $t = 60$ min (1 hour) $\Rightarrow A(60) = 100e^{(0.08)60} = 100e^{4.8} \approx 12{,}151$ bacteria
    $t = 90$ min $\Rightarrow A(90) = 100e^{(0.08)90} \approx 133{,}943$ bacteria
    $1{,}000{,}000 = 100e^{(0.08)t} \Rightarrow t = 115$ minutes
13. $A(t) = A_0 e^{kt}$
    $t = 0 \Rightarrow A_0 = 8$
    $t = 1690 \Rightarrow \frac{1}{2}A = Ae^{k(1690)} \Rightarrow k = -0.0004$
    Thus, $A(t) = 8e^{(-0.0004)t}$.
    $t = 100 \Rightarrow A(100) = 8e^{-0.04} \approx 8(0.9608) = 7.686$ grams
15. $A(t) = A_0 e^{kt}$, where A_0 is the initial amount of carbon-14
    $t = 5600 \Rightarrow \frac{1}{2}A = Ae^{k(5600)} \Rightarrow k = -0.00012$
    Thus, $A = A_0 e^{-(0.00012)t}$.
    If the amount of carbon-14 is 30% of its initial amount A_0, $0.3A_0 = A_0 e^{-(0.00012)t} \Rightarrow t = 9727$ years, so the charcoal is about 9727 years old.
17. $A(t) = 1500e^{kt}$; let t be measured in days
    $A(1) = 1500e^k = 2500$
    $e^k = \frac{25}{15} = \frac{5}{3}$
    $A(t) = 1500(\frac{5}{3})^t$
    $A(3) = 1500(\frac{5}{3})^3 \approx 6{,}944$
19. $A'(t) = 3000e^{2t/5}$
    $A(t) = (\frac{5}{2})3000e^{2t/5} + K = 7500e^{2t/5} + K$
    $A(0) = 7500 + K = 7500;\quad K = 0$
    $A(5) = 7500e^{10/5} = 7500e^2 \approx 7500(7.3890) \approx 55{,}418$
21. (a) $A(t) = A_0 e^{kt}$
    $A_0 = 10{,}000$
    $A(t_1) = 20{,}000 = 10{,}000e^{kt_1}$
    $e^{kt_1} = 2$
    $A(t_1 + 10) = 10{,}000e^{k(t_1+10)} = 10{,}000e^{kt_1}e^{10k} = 20{,}000e^{10k} = 100{,}000$
    $e^{10k} = 5$
    $e^k = 5^{1/10};\quad k = \frac{1}{10}\ln 5 \approx 0.16094$
    $A(t) = 10{,}000e^{kt} = 10{,}000 \cdot 5^{t/10}$

(b) $A(20) = 10,000 \cdot 5^{20/10} = 250,000$
(c) $e^{kt_1} = 2$
$kt_1 = \ln 2$
$t_1 = \dfrac{\ln 2}{k} \approx \dfrac{0.6931}{0.16094} \approx 4.307$ minutes

23. $A(t) = A_0 e^{-\alpha t}$
If h is the half-life of radioactive beryllium, then
$A(h) = A_0 e^{-\alpha h} = \dfrac{A_0}{2}$
$e^{-\alpha h} = \tfrac{1}{2}$
$e^{\alpha h} = 2$
$\alpha h = \ln 2$
$h = \dfrac{\ln 2}{\alpha} \approx \dfrac{0.6931}{1.5 \times 10^{-7}} \approx 0.4621 \times 10^7 = 4,621,000$ years

Exercise 5 (page 272)

1. $R(x) = \int 600\, dx = 600x + K;$ since $R(0) = 0, R(x) = 600x$

3. $R(x) = \int (20x + 5)\, dx = 10x^2 + 5x + K;$ since $R(0) = 0, R(x) = 10x^2 + 5x$

5. $C(x) = \int (14x - 2800)\, dx = 7x^2 - 2800x + K$
$C(0) = 4300 \Rightarrow C(x) = 7x^2 - 2800x + 4300$
$C'(x) = 14x - 2800 = 0 \Rightarrow x = 200$
Thus, cost is a minimum when $x = 200$.

7. $C(x) = \int (20x - 8000)\, dx = 10x^2 - 8000x + K$
$C(0) = 500 \Rightarrow C(x) = 10x^2 - 8000x + 500$
$C'(x) = 20x - 8000 = 0 \Rightarrow x = 400$
Thus, cost is a minimum when $x = 400$.

9. $C(x) = \int (1000 - 20x + x^2)\, dx = 1000x - 10x^2 + \tfrac{1}{3}x^3 + K$
$C(0) = 9000 \Rightarrow C(x) = \tfrac{1}{3}x^3 - 10x^2 + 1000x + 9000$

11. (a) $C(x) = \int (MC)\, dx = \int (16x - 1591)\, dx = 8x^2 - 1591x + K$
$C(0) = 1800 \Rightarrow C(x) = 8x^2 - 1591x + 1800$
(b) $R(x) = 9x$
(c) Profit $= P(x) = R(x) - C(x) = -8x^2 + 1600x - 1800$
(d) $P'(x) = -16x + 1600 = 0 \Rightarrow x = 100$
Thus, the maximum profit that can be obtained in 1 day is
$P(100) = -8(100)^2 + 160,000 - 1800 = \$78,200$

Exercise 6 (page 274)

1. $\displaystyle\int x(3x+5)^2\, dx \underset{\text{Form 1}}{=} \dfrac{(3x+5)^3}{9}\left(\dfrac{3x+5}{4} - \dfrac{5}{3}\right) + K$

3. $\displaystyle\int \dfrac{\sqrt{x+2}}{x^2}\, dx \underset{\text{Form 6}}{=} 2\sqrt{x+2} + 2\int \dfrac{1}{x\sqrt{x+2}}\, dx$
$\underset{\text{Form 5}}{=} 2\sqrt{x+2} + 2\left[\dfrac{1}{\sqrt{2}} \ln\left|\dfrac{\sqrt{x+2} - \sqrt{2}}{\sqrt{x+2} + \sqrt{2}}\right|\right] + K$

5. $\displaystyle\int \dfrac{1}{x\sqrt{2x - x^2}}\, dx \underset{\text{Form 18}}{=} -\sqrt{\dfrac{2-x}{x}} + K$

420 CHAPTER 7 SOLUTIONS

7. $\displaystyle\int x^3 e^{-x}\,dx \underset{\text{Form 19}}{=} (-1)x^3 e^{-x} + 3\int x^2 e^{-x}\,dx$

$\underset{\text{Form 19}}{=} -x^3 e^{-x} + 3[-x^2 e^{-x} + 2\int xe^{-x}\,dx]$

$\underset{\text{Form 19}}{=} -x^3 e^{-x} - 3x^2 e^{-x} + 6[-xe^{-x} + \int e^{-x}\,dx]$

$= -x^3 e^{-x} - 3x^2 e^{-x} - 6xe^{-x} - 6e^{-x} + K$

Review Exercises (page 275)

1. $\displaystyle\int (x^3 - 3x + 1)\,dx = \int x^3\,dx - 3\int x\,dx + \int dx = \tfrac{1}{4}x^4 - \tfrac{3}{2}x^2 + x + K$

3. $\displaystyle\int (x^{1/3} - 4x^{1/2})\,dx = \int x^{1/3}\,dx - 4\int x^{1/2}\,dx = \tfrac{3}{4}x^{4/3} - \tfrac{8}{3}x^{3/2} + K$

5. $\displaystyle\int (1 + e^{-x})\,dx = \int dx + \int e^{-x}\,dx = x - e^{-x} + K$

7. Let $u = x^2 - 1$, $du = 2x\,dx$. Then

$\displaystyle\int x\sqrt{x^2 - 1}\,dx = \tfrac{1}{2}\int u^{1/2}\,du = \tfrac{1}{3}u^{3/2} + K = \tfrac{1}{3}(x^2 - 1)^{3/2} + K$

9. Let $u = 3x - 2$, $du = 3\,dx$. Then

$\displaystyle\int \sqrt{3x - 2}\,dx = \tfrac{1}{3}\int u^{1/2}\,du = \tfrac{2}{9}u^{3/2} + K = \tfrac{2}{9}(3x - 2)^{3/2} + K$

11. Let $u = x^4 + 1$, $du = 4x^3\,dx$. Then

$\displaystyle\int \frac{x^3\,dx}{(x^4 + 1)^{3/2}} = \tfrac{1}{4}\int \frac{du}{u^{3/2}} = -\tfrac{1}{2}u^{-1/2} + K = -\tfrac{1}{2}(x^4 + 1)^{-1/2} + K$

13. Let $u = x^2 + 1$, $du = 2x\,dx$. Then

$\displaystyle\int \frac{5x\,dx}{x^2 + 1} = \tfrac{5}{2}\int \frac{du}{u} = \tfrac{5}{2}\ln u + K = \tfrac{5}{2}\ln(x^2 + 1) + K$

15. Integrate by parts. Let $u = x$, $du = dx$, $dv = e^{x/2}\,dx$, $v = 2e^{x/2}$. Then

$\displaystyle\int xe^{x/2}\,dx = 2xe^{x/2} - 2\int e^{x/2}\,dx = 2xe^{x/2} - 4e^{x/2} + K$

17. Let $u = x^3$, $du = 3x^2\,dx$. Then

$\displaystyle\int x^2 e^{x^3}\,dx = \tfrac{1}{3}\int e^u\,du = \tfrac{1}{3}e^u + K = \tfrac{1}{3}e^{x^3} + K$

19. The general solution is $y = F(x) = x + \ln x + K$; $F(1) = 1 + K = 1$.
    The particular solution is $y = x + \ln x$.

21. The general solution is $y = F(x) = \tfrac{1}{3}(x^2 + 1)^{3/2} + K$; $F(0) = \tfrac{1}{3} + K = 6$.
    The particular solution is $y = \tfrac{1}{3}(x^2 + 1)^{3/2} + \tfrac{17}{3}$.

23. $f(x) = 2x^{1/2} + K$; $f(1) = 2 + K = 1$
    The particular solution is $f(x) = 2x^{1/2} - 1$.

25. $A(t) = A_0 e^{kt}$
    $t = 0 \Rightarrow A_0 = 2000$
    $t = 2 \Rightarrow A(2) = 6000 = 2000\,e^{k(2)} \Rightarrow k = 0.549$
    Hence, $A(t) = 2000e^{(0.549)t}$.
    $A(4.5) = 2000e^{(0.549)(4.5)} = 23{,}656$ bacteria

27. $A(t) = A_0 e^{kt}$
    Half-life is 1000 years so that $\tfrac{1}{2}A_0 = A_0 e^{k(1000)} \Rightarrow k = -0.00069$
    Hence, $A(t) = A_0 e^{-(0.00069)t}$
    $0.2A_0 = A_0 e^{-(0.00069)t} \Rightarrow t = 2332$ years

29. $A(t) = A_0 e^{kt}, \quad A_0 = 6 \times 10^{10}$
$\frac{1}{2} A_0 = A_0 e^{k(5600)}$
$\frac{1}{2} = e^{k(5600)}$
$K = -.000124$
$A(t) = 6 \times 10^{10} e^{-.000124t}$
$2.8 \times 10^{10} = 6 \times 10^{10} e^{-.000124t}$
$\frac{2.8}{6} = e^{-.000124t}$
$t \approx 6146$ years

CHAPTER 8

Exercise 1 (page 282)

1. $\int_1^2 (3x - 1)\, dx = \frac{3}{2}(x^2 |_1^2) - (x|_1^2) = \frac{3}{2}(4 - 1) - (2 - 1) = \frac{3}{2}(3) - 1 = \frac{7}{2}$

3. $\int_0^1 (3x^2 + e^x)\, dx = x_3|_0^1 + e^x|_0^1 = (1 - 0) + (e - 1) = e$

5. $\int_0^1 \sqrt{u}\, du = \frac{2}{3} u^{3/2}|_0^1 = \frac{2}{3} - 0 = \frac{2}{3}$

7. $\int_0^1 (t^2 - t^{3/2})\, dt = \int_0^1 t^2\, dt - \int_0^1 t^{3/2}\, dt = \frac{1}{3} t^3|_0^1 - \frac{2}{5} t^{5/2}|_0^1 = (\frac{1}{3} - 0) - (\frac{2}{5} - 0) = -\frac{1}{15}$

9. $\int_{-2}^3 (x - 1)(x + 3)\, dx = \int_{-2}^3 (x^2 + 2x - 3)\, dx = (\frac{1}{3} x^3 + x^2 - 3x)|_{-2}^3 = 9 - \frac{22}{3} = \frac{5}{3}$

11. $\int_1^2 \frac{x^2 - 1}{x^4}\, dx = \int_1^2 (x^{-2} - x^{-4})\, dx = \int_1^2 x^{-2}\, dx - \int_1^2 x^{-4}\, dx$
$= -x^{-1}|_1^2 - (-\frac{1}{3} x^{-3}|_1^2) = \frac{1}{2} - \frac{7}{24} = \frac{5}{24}$

13. $\int_1^4 (\sqrt[5]{t^2} + \frac{1}{t})\, dt = \int_1^4 t^{2/5}\, dt + \int_1^4 \frac{1}{t}\, dt = t^{7/5}|_1^4 + \ln t|_1^4 = \frac{20}{7} \sqrt[5]{16} - \frac{5}{7} + \ln 4$

15. $\int_1^4 \frac{x + 1}{\sqrt{x}}\, dx = \int_1^4 (x^{1/2} + x^{-1/2})\, dx = \int_1^4 x^{1/2}\, dx + \int_1^4 x^{-1/2}\, dx = \frac{2x^{3/2}}{3}\Big|_1^4 + 2x^{1/2}|_1^4$
$= (\frac{16}{3} - \frac{2}{3}) + (4 - 2) = \frac{20}{3}$

17. $\int_3^3 (5x^4 + 1)^{3/2}\, dx = 0$

19. $\int_{-1}^1 (x + 1)^2\, dx = \frac{1}{3}(x + 1)^3|_{-1}^1 = \frac{8}{3} - 0 = \frac{8}{3}$

21. $\int_1^e (x - \frac{1}{x})\, dx = \int_1^e x\, dx - \int_1^e \frac{1}{x}\, dx = \frac{1}{2} x^2|_1^e - \ln x|_1^e = \frac{1}{2}(e^2 - 1) - (1 - 0) = \frac{1}{2}(e^2 - 3)$

23. $\int_0^1 e^{-x}\, dx = -e^{-x}|_0^1 = -e^{-1} + e^0 = -\frac{1}{e} + 1$

25. $\int_1^3 \frac{dx}{x + 1} = \ln(x + 1)|_1^3 = \ln 4 - \ln 2 = \ln \frac{4}{2} = \ln 2$

27. $\int_0^1 \frac{\sqrt{x}}{x^{3/2} + 1}\, dx = \frac{2}{3} \ln(x^{3/2} + 1)|_0^1 = \frac{2}{3} \ln 2$

29. $\int_1^3 xe^{2x}\, dx = \frac{1}{4} e^{2x}(2x - 1)|_1^3 = \frac{1}{4}(5e^6 - e^2) = \frac{e^2}{4}(5e^4 - 1)$

422 CHAPTER 8 SOLUTIONS

31. $\int_{1}^{2} xe^{-3x}\, dx = -\frac{1}{3}(xe^{-3x} + \frac{1}{3}e^{-3x}|_{1}^{2}) = -\frac{1}{3}[(2e^{-6} + \frac{1}{3}e^{-6}) - (e^{-3} + \frac{1}{3}e^{-3})]$
$= -\frac{1}{9}(7e^{-6} - 4e^{-3}) = -\frac{1}{9}e^{-3}(7e^{-3} - 4)$

33. $\int_{1}^{5} \ln x\, dx = x(\ln x - 1)|_{1}^{5} = 5(\ln 5 - 1) - (-1) = 5\ln 5 - 4$

35. (a) $\int_{-1}^{1} x^2\, dx = \frac{1}{3}x^3|_{-1}^{1} = \frac{1}{3} - (-\frac{1}{3}) = \frac{2}{3}$; $2\int_{0}^{1} x^2\, dx = \frac{2x^3}{3}\Big|_{0}^{1} = \frac{2}{3}$

Hence, $\int_{-1}^{1} x^2\, dx = 2\int_{0}^{1} x^2\, dx$.

(b) $\int_{-1}^{1}(x^4 + x^2)\, dx = \left(\frac{x^5}{5} + \frac{x^3}{3}\right)\Big|_{-1}^{1} = \frac{8}{15} - (-\frac{8}{15}) = \frac{16}{15}$

$2\int_{0}^{1}(x^4 + x^2)\, dx = 2\left(\frac{x^5}{5} + \frac{x^3}{5}\right)\Big|_{0}^{1} = 2(\frac{8}{15}) = \frac{16}{15}$

Exercise 2 (page 293)

1.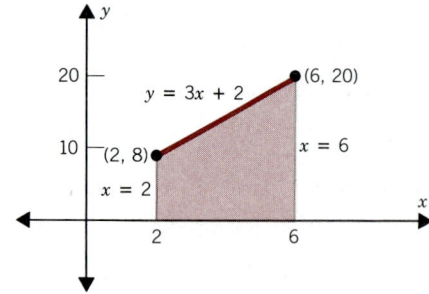

Area = $\int_{2}^{6}(3x + 2)\, dx = (\frac{3}{2}x^2 + 2x)|_{2}^{6}$
$= 66 - 10 = 56$

3.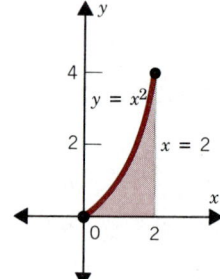

Area = $\int_{0}^{2} x^2\, dx = \frac{x^3}{3}\Big|_{0}^{2} = \frac{8}{3}$

5.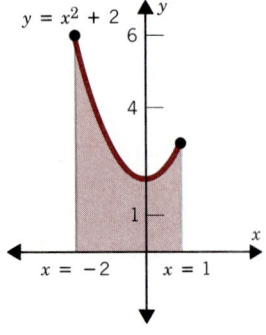

Area = $\int_{-2}^{1}(x^2 + 2)\, dx = \left(\frac{x^3}{3} + 2x\right)\Big|_{-2}^{1}$
$= (\frac{1}{3} + 2) - (-\frac{8}{3} - 4) = 9$

7.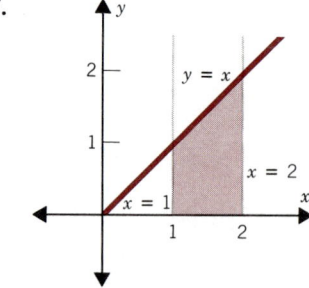

Area = $\int_{1}^{2} x\, dx = \frac{x^2}{2}\Big|_{1}^{2} = 1.5$

9.

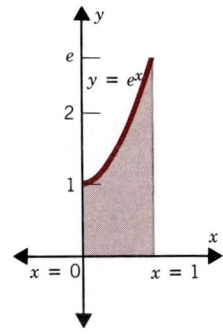

Area $= \int_0^1 e^x \, dx = e^x \big|_0^1 = e - 1$

11.

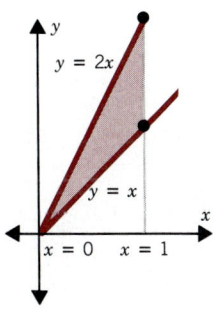

Area $= \int_0^1 (2x - x) \, dx = \int_0^1 x \, dx = \frac{1}{2}$

13.

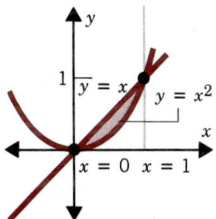

Points of intersection: $y = x = x^2$
when $x = 0 = y$ or $x = 1 = y$

Area $= \int_0^1 (x - x^2) \, dx = \frac{x^2}{2}\big|_0^1 - \frac{x^3}{3}\big|_0^1$

$= \frac{1}{2} - \frac{1}{3} = \frac{1}{6}$

15.

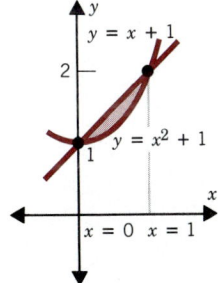

Points of intersection: $y = x + 1 = x^2 + 1$
when $x = 0, y = 1$ or $x = 1, y = 2$

Area $= \int_0^1 [(x + 1) - (x^2 + 1)] \, dx$

$= \int_0^1 (x - x^2) \, dx = \frac{1}{6}$

17.

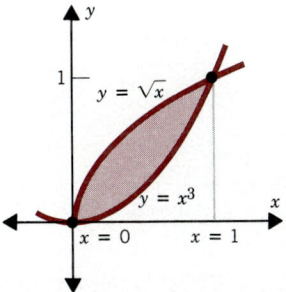

Points of intersection: $y = \sqrt{x} = x^3$
when $x = 0 = y$ and $x = 1 = y$

Area $= \int_0^1 (x^{1/2} - x^3) \, dx = \left(\frac{2}{3}x^{3/2} - \frac{x^4}{4}\right)\big|_0^1$

$= \frac{2}{3} - \frac{1}{4} - 0 = \frac{5}{12}$

19.

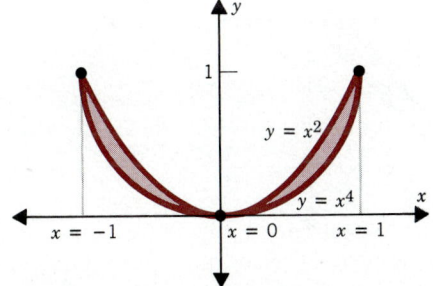

Points of intersection: $y = x^2 = x^4$
when $x = 0 = y$, $x = 1 = y$, and $x = -1$,
$y = 1$

Area $= \int_{-1}^1 (x^2 - x^4) \, dx = 2 \int_0^1 (x^2 - x^4) \, dx$

$= \left(\frac{2}{3}x^3 - \frac{2}{5}x^5\right)\big|_0^1 = \frac{4}{15}$

21.

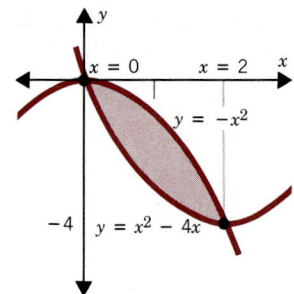

Points of intersection: $y = x^2 - 4x = -x^2$
when $2x^2 - 4x = 0$
$2x(x - 2) = 0$
$x = 0 = y$ and $x = 2, y = -4$

Area $\int_0^2 [-x^2 - (x^2 - 4x)]\, dx$

$= \int_0^2 (-2x^2 + 4x)\, dx$

$= (-\frac{2}{3}x^3 + 2x^2)\big|_0^2 = -\frac{16}{3} + 8 - 0 = \frac{8}{3}$

23.

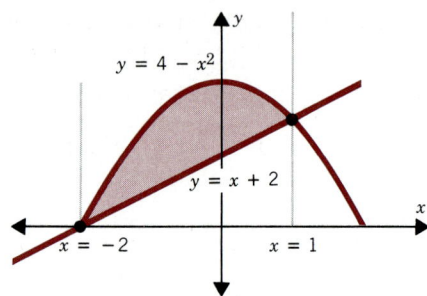

Points of intersection: $y = 4 - x^2 = x + 2$
when $x^2 + x - 2 = 0$
$(x + 2)(x - 1) = 0$
$x = -2, y = 0$ and $x = 1, y = 3$

Area $= \int_{-2}^1 [4 - x^2 - (x + 2)]\, dx$

$= \int_{-2}^1 (2 - x^2 - x)\, dx$

$= 2x\big|_{-2}^1 - \frac{x^3}{3}\big|_{-2}^1 - \frac{x^2}{2}\big|_{-2}^1$

$= 6 - 3 + \frac{3}{2} = 4\frac{1}{2}$

25.

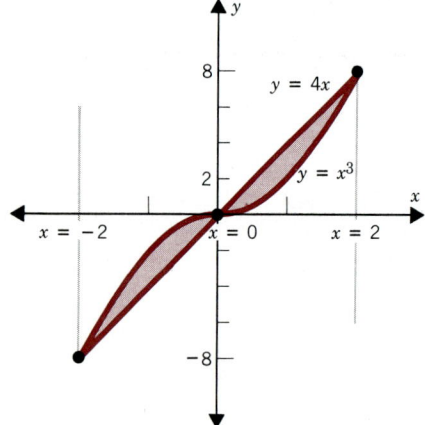

Points of intersection: $y = 4x = x^3$
when $4x - x^3 = 0$
$x(4 - x^2) = 0$
$x = 0 = y, x = 2, y = 8$, and $x = -2, y = -8$

Area $= \int_{-2}^0 (x^3 - 4x)\, dx + \int_0^2 (4x - x^3)\, dx$

$= \frac{x^4}{4}\big|_{-2}^0 - 2x^2\big|_{-2}^0 + 2x^2\big|_0^2 - \frac{x^4}{4}\big|_0^2$

$= -4 - (-8) + 8 - 4 = 8$

27.

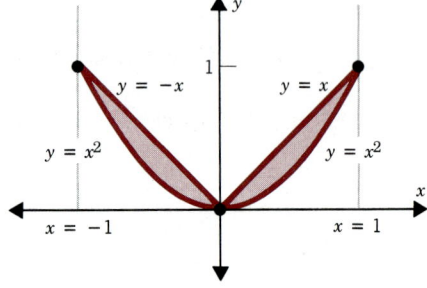

Points of intersection: $y = x = x^2$
when $x = y = 0$ and $x = y = 1$;
$y = -x = x^2$ when $x = y = 0$ and $x = -1$,
$y = 1$; $y = x = -x$ when $x = y = 0$

Area $= \int_{-1}^0 (-x - x^2)\, dx + \int_0^1 (x - x^2)\, dx$

$= 2\int_0^1 (x - x^2)\, dx$

$= 2\left(\frac{x^2}{2} - \frac{x^3}{3}\right)\big|_0^1$

$= 2(\frac{1}{2} - \frac{1}{3}) - 0 = \frac{1}{3}$

CHAPTER 8 SOLUTIONS 425

29. The upper and lower sides of the parallelogram are the lines $y = -x + 2$ and $y = -x - \frac{1}{4}$. Thus, $D = (-2, -\frac{7}{4})$ and $A = (1, -\frac{5}{4})$. The area of the parallelogram is $(4 - \frac{7}{4})(1 + 2) = \frac{27}{4}$. The area of the shaded region is $\int_{-2}^{1} [(-x + 2) - x^2] \, dx = \frac{9}{2}$

31. (a) x^2 (b) $\sqrt{x^2 - 2}$ (c) $\sqrt{x^2 + 2x}$

Exercise 3 (page 299)

1. If $a = 81$ and $b = 100$, then
$$VC = \int_{80}^{100} e^{-0.01x} \, dx = -\frac{1}{0.01} e^{-0.01x} \Big|_{80}^{100} = -100(e^{-1} - e^{-0.8}) \approx -100(0.3679 - 0.4493) = \$8.14$$

If $a = 801$ and $b = 1000$, then
$$VC = \int_{800}^{1000} e^{-0.01x} \, dx = -100(e^{-0.01x}) \Big|_{800}^{1000} = -100(e^{-10} - e^{-8}) \approx -100(0.00005 - 0.00034) = \$0.029$$

3. $T = \int_{0}^{4} 300{,}000 e^{0.1t} \, dt = 3{,}000{,}000 e^{0.1t} \Big|_{0}^{4} = 3 \times 10^6 (e^{0.4} - 1) \approx 3 \times 10^6 (1.4918) = \$4{,}475{,}400$

5. (a) $A = \int_{0}^{10} 50 e^{0.03t} \, dt = \frac{50}{0.03} e^{0.03t} \Big|_{0}^{10} = \frac{5000}{3} (e^{0.3} - 1) \approx 583.2$ trillion cubic feet

(b) $\int_{0}^{T} 50 e^{0.03t} \, dt = \frac{50}{0.03} (e^{0.03T} - 1) = 2200$

$$e^{0.03T} - 1 = \frac{2200(0.03)}{50}$$
$$e^{0.03T} - 1 = 1.32$$
$$e^{0.03T} = 2.32$$
$$0.03T = \ln 2.32$$
$$T = \frac{\ln 2.32}{0.03} \approx 301 \text{ years}$$

7. $A = \int_{0}^{4} P e^{rt} \, dt = 500 \int_{0}^{4} e^{0.055t} \, dt = \frac{500}{0.055} e^{0.055t} \Big|_{0}^{4} = \frac{500}{0.055} (e^{0.22} - 1) \approx \frac{500}{0.055} (0.2461) \approx \$2237.27$

9. $A = 6000 = \int_{0}^{3} P e^{0.06t} \, dt = \frac{P}{0.06} e^{0.06t} \Big|_{0}^{3} = \frac{P}{0.06} (e^{0.18} - 1)$

$P = \frac{(6000)(0.06)}{e^{0.18} - 1} \approx \frac{360}{0.1972} \approx \$1825.56$

Exercise 4 (page 304)

1. $\int_{0}^{1} 3 \, dx = 3x \Big|_{0}^{1} = 3$; $\lim_{\Delta \to 0} [3 \Delta x_1 + 3 \Delta x_2 + \cdots + 3 \Delta x_n]$
$= \lim_{\Delta \to 0} [3(\Delta x_1 + \Delta x_2 + \cdots + \Delta x_n)] = \lim_{\Delta \to 0} (3 \cdot 1) = 3$

3. $\int_{0}^{1} x^2 \, dx = \frac{x^3}{3} \Big|_{0}^{1} = \frac{1}{3}$; $\lim_{\Delta \to 0} (X_1^2 \Delta x_1 + X_2^2 \Delta x_2 + \cdots + X_n^2 \Delta x_n)$

$= \lim_{n \to +\infty} \left(\frac{1^2}{n^2} \cdot \frac{1}{n} + \frac{2^2}{n^2} \cdot \frac{1}{n} + \cdots + \frac{n^2}{n^2} \cdot \frac{1}{n} \right)$

$= \lim_{n \to +\infty} \left[\frac{1}{n^3} (1^2 + 2^2 + 3^2 + \cdots + n^2) \right] = \lim_{n \to +\infty} \frac{n(n+1)(2n+1)}{6n^3}$

$= \lim_{n \to +\infty} \frac{1 \left(1 + \frac{1}{n}\right)\left(2 + \frac{1}{n}\right)}{6} = \frac{2}{6} = \frac{1}{3}$

5. $\left(\frac{1}{X_1} \Delta x_1 + \frac{1}{X_2} \Delta x_2 + \frac{1}{X_3} \Delta x_3 + \frac{1}{X_4} \Delta x_4 \right) = \left(\frac{1}{1.125} + \frac{1}{1.375} + \frac{1}{1.625} + \frac{1}{1.875} \right) 0.25 \approx 0.6912$

Exercise 5 (page 307)

1. $AV = \dfrac{1}{1-0}\displaystyle\int_0^1 x^2\,dx = 1\left(\dfrac{x^3}{3}\right)\Big|_0^1 = \dfrac{1}{3}$

3. $AV = \dfrac{1}{1-(-1)}\displaystyle\int_{-1}^1 (1-x^2)\,dx = \dfrac{1}{2}\left(x-\dfrac{x^3}{3}\right)\Big|_{-1}^1 = \dfrac{2}{3}$

5. $AV = \dfrac{1}{5-1}\displaystyle\int_1^5 3x\,dx = \dfrac{1}{4}\left(\dfrac{3x^2}{2}\right)\Big|_1^5 = \dfrac{3}{8}(25-1) = 9$

7. $AV = \dfrac{1}{4}\displaystyle\int_{-2}^2 (-5x^4 + 4x - 10)\,dx = \dfrac{1}{4}(-x^5 + 2x^2 - 10x)\Big|_{-2}^2 = -26$

9. $AV = \dfrac{1}{1-0}\displaystyle\int_0^1 e^x\,dx = e - 1$

11. $AV = \dfrac{1}{20}\displaystyle\int_0^{20} (5\times 10^9)e^{0.03t}\,dt = \left(\dfrac{10^9}{4}\right)\dfrac{e^{0.03t}}{0.03}\Big|_0^{20} = \left(\dfrac{10^9}{0.12}\right)(e^{0.6}-1) \approx 6.85 \times 10^9$

13. $AV = \dfrac{1}{3}\displaystyle\int_0^3 25x\,dx = \dfrac{25}{6}x^2\Big|_0^3 = 37.5°C$

15. $AV = \dfrac{1}{8}\displaystyle\int_0^8 3t\,dt = \dfrac{3}{16}t^2\Big|_0^8 = 12$ meters per second

Exercise 6 (page 313)

1. $\displaystyle\int_0^2 \tfrac{1}{2}\,dx = \tfrac{1}{2}x\Big|_0^2 = \tfrac{1}{2}(2-0) = 1;\ f(x) \geq 0$ on $[0, 2]$

3. $\displaystyle\int_0^1 2x\,dx = x^2\Big|_0^1 = 1;\ f(x) = 2x \geq 0$ on $[0, 1]$

5. $\displaystyle\int_0^5 \tfrac{3}{250}(10x - x^2)\,dx = \tfrac{3}{250}\left(5x^2 - \dfrac{x^3}{3}\right)\Big|_0^5 = \tfrac{3}{250}(125 - \tfrac{125}{3}) = \tfrac{3}{250}(\tfrac{250}{3}) = 1;$
   $f(x) = \tfrac{3}{250}(10x - x^2) = \tfrac{3}{250}x(10 - x) \geq 0$ on $[0, 5]$

7. $\displaystyle\int_1^e \dfrac{1}{x}\,dx = \ln x\Big|_1^e = \ln e - \ln 1 = 1;\ f(x) = \dfrac{1}{x} \geq 0$ on $[1, e]$

9. $\displaystyle\int_0^3 k\,dx = kx\Big|_0^3 = 3k = 1 \Rightarrow k = \tfrac{1}{3}$ 11. $\displaystyle\int_0^2 kx\,dx = \dfrac{kx^2}{2}\Big|_0^2 = k\cdot 4 = 1 \Rightarrow k = \tfrac{1}{4}$

13. $\displaystyle\int_0^5 k(10x - x^2)\,dx = k\left(5x^2 - \dfrac{x^3}{3}\right)\Big|_0^5 = k(125 - \tfrac{125}{3}) = k(\tfrac{250}{3}) = 1 \Rightarrow k = \tfrac{3}{250}$

15. $\displaystyle\int_1^2 k\dfrac{1}{x}\,dx = k\ln x\Big|_1^2 = k(\ln 2) = 1 \Rightarrow k = \dfrac{1}{\ln 2}$

17. $P(1 \leq x \leq 3) = \displaystyle\int_1^3 \tfrac{1}{5}\,dx = \tfrac{1}{5}x\Big|_1^3 = \tfrac{1}{5}(3-1) = \tfrac{2}{5}$

19. $P(T \geq 6) = 1 - P(T < 6) = 1 - \displaystyle\int_0^6 0.5e^{-0.5t}\,dt = 1 + e^{-0.5t}\Big|_0^6 = 1 + (e^{-3} - 1) = e^{-3} = 0.050$

21. $P(T < 5) = \displaystyle\int_0^5 0.4e^{-0.4t}\,dt = \dfrac{0.4e^{-0.4t}}{-0.4}\Big|_0^5 = 1 - e^{-2} = 0.865$

23. $P(x \leq 10) = \dfrac{1}{4500}\displaystyle\int_0^{10} (30x - x^2)\,dx = \tfrac{1}{4500}(15x^2 - \tfrac{1}{3}x^3)\Big|_0^{10}$
    $= \tfrac{1}{4500}(1500 - \tfrac{1000}{3}) = \tfrac{1}{4500}(\tfrac{3500}{3}) = 0.2593$

    $P(x \geq 20) = \tfrac{1}{4500}\displaystyle\int_{20}^{30} (30x - x^2)\,dx = \tfrac{1}{4500}(15x^2 - \tfrac{1}{3}x^3)\Big|_{20}^{30}$
    $= \tfrac{1}{4500}[13500 - 9000 - (6000 - \tfrac{8000}{3})] = \tfrac{1}{4500}(\tfrac{3500}{3}) = 0.2593$

    $P(10 \leq x \leq 20) = 1 - P(x \leq 10) - P(x \geq 20) = 0.4814$

CHAPTER 8 SOLUTIONS

Exercise 7 (page 322)

1. $\int_{35}^{60} 1000x^{-0.5}\, dx = 2000x^{1/2}\big|_{35}^{60} = 2000(7.746 - 5.916) = 3660$ hours

3. The direct labor required approaches a constant value independent of the number of units produced.

5. Equilibrium: $p = D(x) = S(x)$
$$-0.4x + 15 = 0.8x + 0.5$$
$$-1.2x = -14.5$$
$$x = \tfrac{145}{12} = x^*$$
$p^* = D(x^*) = -0.4(\tfrac{145}{12}) + 15 = \tfrac{61}{6}$
$CS = \int_0^{x^*} D(x)\, dx - x^*p^* = \int_0^{145/12}(-0.4x + 15)\, dx - \tfrac{8845}{72}$
$= (-0.2x^2 + 15x)\big|_0^{145/12} - \tfrac{8845}{72} \approx \$29.20$
$PS = x^*p^* - \int_0^{x^*} S(x)\, dx = \tfrac{8845}{72} - \int_0^{145/12}(0.8x + 0.5)\, dx$
$= \tfrac{8845}{72} - (0.4x^2 + 0.5x)\big|_0^{145/12} \approx \$58.40$

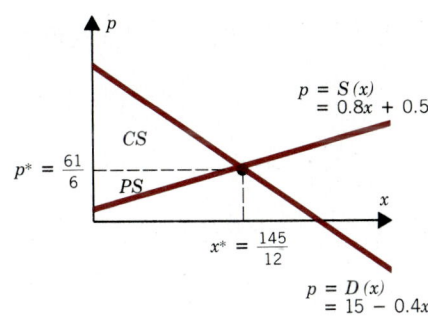

7. $R'(t) = C'(t)$
$19 - t^{1/2} = 3 + 3t^{1/2}$
$4t^{1/2} = 16$
$t^{1/2} = 4$
$t^* = 16$

The operation should continue for 16 years.

$P(t^*) = \int_0^{16}[R'(t) - C'(t)]\, dt = \int_0^{16}[(19 - t^{1/2}) - (3 + 3t^{1/2})]\, dt$
$= \int_0^{16}(16 - 4t^{1/2})\, dt = \left(16t - \tfrac{4t^{3/2}}{\tfrac{3}{2}}\right)\bigg|_0^{16} = 256 - \tfrac{512}{3} = 85.33$ million dollars

Review Exercises (page 323)

1. $\int_2^4 x^2\, dx = \tfrac{1}{3}x^3\big|_2^4 = \tfrac{64}{3} - \tfrac{8}{3} = \tfrac{56}{3}$

3. $\int_{-1}^1 (x^2 + x - 1)\, dx = \tfrac{1}{3}x^3 + \tfrac{1}{2}x^2 - x\big|_{-1}^1 = (\tfrac{1}{3} + \tfrac{1}{2} - 1) - (-\tfrac{1}{3} + \tfrac{1}{2} + 1) = \tfrac{2}{3} - 2 = -\tfrac{4}{3}$

5. $\int_0^4 e^x\, dx = e^x\big|_0^4 = e^4 - e^0 = e^4 - 1$

7. $\int_0^2 x^2\sqrt{x^3 + 1}\, dx = \tfrac{2}{9}(x^3 + 1)^{3/2}\big|_0^2 = \tfrac{54}{9} - \tfrac{2}{9} = \tfrac{52}{9}$

9. $A = \int_0^5 2000e^{0.06t}\, dt = \tfrac{2000}{0.06}e^{0.06t}\bigg|_0^5 = \tfrac{2000}{0.06}(e^{0.3} - e^0) = \$11{,}663$
20% of $100,000 = \$20{,}000 > \$11{,}663$ (not enough)

11. $A = \int_0^{20} 4.2e^{0.1t}\, dt = \tfrac{4.2}{0.1}e^{0.1t}\bigg|_0^{20} = 42(e^2 - 1) = 268$ billion short tons

13. Points of intersection: $x^3 = x^2 = y$ when $x^3 - x^2 = 0$
$x^2(x - 1) = 0$
$x = 0$ or $x = 1$
Thus, the curves intersect at $(0, 0)$ and $(1, 1)$.
For $0 \leq x \leq 1$, $f(x) \geq g(x) \geq 0$:
$A = \int_0^1 (x^2 - x^3)\, dx = \left(\tfrac{x^3}{3} - \tfrac{x^4}{4}\right)\bigg|_0^1 = \tfrac{1}{3} - \tfrac{1}{4} = \tfrac{1}{12}$

15. Points of intersection: $x^2 = \sqrt{x} = y$ when $x^4 = x$
$$x(x^3 - 1) = 0$$
$$x = 0 \text{ or } x = 1$$
Thus, the curves intersect at $(0, 0)$ and $(1, 1)$.
For $0 \leq x \leq 1$, $g(x) \geq f(x) \geq 0$:
$$A = \int_0^1 (x^{1/2} - x^2)\, dx = \left(\tfrac{2}{3}x^{3/2} - \tfrac{x^3}{3}\right)\Big|_0^1 = \tfrac{2}{3} - \tfrac{1}{3} = \tfrac{1}{3}$$

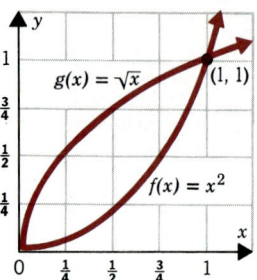

17. Note that $f(x) \geq 0$ for all $x \geq 0$. Hence,
$$A = \int_0^2 \frac{2x}{x^2 + 1}\, dx = \ln(x^2 + 1)\Big|_0^2 = \ln 5 - \ln 1 = \ln 5$$

19. Equilibrium point: $p = D(x) = S(x)$
$$12 - \frac{x}{50} = \frac{x}{20} + 5$$
$$x^* = 100$$
$$p^* = D(100) = 12 - 2 = 10$$
$$CS = \int_0^{100}\left(12 - \frac{x}{50}\right) dx - (10)(100) = \left(12x - \frac{x^2}{100}\Big|_0^{100}\right) - 1000 = \$100$$
$$PS = 1000 - \int_0^{100}\left(\frac{x}{20} + 5\right) dx = 1000 - \left(\frac{x^2}{40} + 5x\Big|_0^{100}\right) = 1000 - (250 + 500) = \$250$$

21. $P(20 \leq x \leq 40) = \dfrac{3}{6.88}\,10^{-5}\displaystyle\int_{20}^{40}(-x^2 + 200x - 5000)\, dx$
$$= \frac{3}{6.88}(10^{-5})(-\tfrac{1}{3}x^3 + 100x^2 - 5000x)\Big|_{20}^{40}$$
$$= \frac{3}{6.88}(10^{-5})\left(\frac{4000}{3}\right) = .0058$$
$P(20 \leq x \leq 60) = \dfrac{3}{6.88}(10^{-5})(-\tfrac{1}{3}x^3 + 100x^2 - 5000x)\Big|_{20}^{60}$
$$= \frac{3}{6.88}(10^{-5})\left(\frac{152{,}000}{3}\right) = .2209$$

23. (a) $P(x \leq 1) = \displaystyle\int_0^1 \tfrac{1}{2}x\, dx = \dfrac{x^2}{4}\Big|_0^1 = \tfrac{1}{4}$

(b) $P(1 \leq x \leq 1.5) = \displaystyle\int_1^{1.5} \tfrac{1}{2}x\, dx = \dfrac{x^2}{4}\Big|_1^{1.5} = \dfrac{(1.5)^2 - 1}{4} = .3125$

(c) $P(1.5 \leq x) = 1 - p(x \leq 1.5) = 1 - (\tfrac{1}{4} + 0.3125)$
$$= .4375$$

25. (a) Uniform density function: $f(x) = \tfrac{1}{2}$ $(0 \leq x \leq 2)$

(b) $P(x \geq 1) = \displaystyle\int_1^2 \tfrac{1}{2}\, dx = \tfrac{1}{2}$

(c) $P(x \leq 1) = 1 - \tfrac{1}{2} = \tfrac{1}{2}$

CHAPTER 9

Exercise 1 (page 330)

1. $f(x) = x^3 + 3x - 5$, $f'(x) = 3x^2 + 3$
   $f(1) = 1 + 3 - 5 = -1$
   $f(2) = 8 + 6 - 5 = 9$
   Let $c_1 = +1$.

   $c_2 = +1 - \dfrac{(-1)}{3+3} = +1 + \tfrac{1}{6} = \tfrac{7}{6}$

   $f(c_2) = (\tfrac{7}{6})^3 + 3(\tfrac{7}{6}) - 5 = \tfrac{19}{216}$, $f'(c_2) = 3(\tfrac{7}{6})^2 + 3 = \tfrac{85}{12}$

   $c_3 = \tfrac{7}{6} - \dfrac{19/216}{85/12} = \tfrac{7}{6} - \tfrac{19}{1530} = \tfrac{883}{765} \approx 1.15425$

3. $f(x) = 2x^3 + 3x^2 + 4x - 1$, $f'(x) = 6x^2 + 6x + 4$
   $f(0) = -1$, $f(1) = 2 + 3 + 4 - 1 = 8$
   Let $c_1 = 0$; $f'(0) = 4$.

   $c_2 = 0 - \dfrac{(-1)}{4} = \tfrac{1}{4}$, $f(\tfrac{1}{4}) = 2(\tfrac{1}{4})^3 + 3(\tfrac{1}{4})^2 + 4(\tfrac{1}{4}) - 1 = \tfrac{7}{32}$

   $f'(\tfrac{1}{4}) = 6(\tfrac{1}{4})^2 + 6(\tfrac{1}{4}) + 4 = \tfrac{47}{8}$

   $c_3 = \tfrac{1}{4} - \dfrac{7/32}{47/8}$

   $c_3 = \tfrac{5}{23} \approx .21739$

5. $f(x) = x^3 - 6x - 12$, $f'(x) = 3x^2 - 6$
   $f(3) = 27 - 18 - 12 = -3$, $f(4) = 64 - 24 - 12 = 28$
   Let $c_1 = 3$; $f'(3) = 27 - 6 = 21$.

   $c_2 = 3 - \dfrac{(-3)}{21} = 3\tfrac{1}{7}$, $f(3\tfrac{1}{7}) = \tfrac{64}{343}$, $f'(3\tfrac{1}{7}) = 23\tfrac{31}{49}$

   $c_3 = 3\tfrac{1}{7} - \dfrac{\tfrac{64}{343}}{23\tfrac{31}{49}} = 3.134962$

7. $f(x) = x^4 - 2x^3 + 21x - 23$, $f'(x) = 4x^3 - 6x^2 + 21$
   $f(1) = 1 - 2 + 21 - 23 = -3$, $f(2) = 16 - 16 + 42 - 23 = 19$
   Let $c_1 = 1$; $f'(1) = 4 - 6 + 21 = 19$.

   $c_2 = 1 - \dfrac{(-3)}{19} = 1\tfrac{3}{19}$

   $c_3 = 1\tfrac{3}{19} - \dfrac{f(1\tfrac{3}{19})}{f'(1\tfrac{3}{19})} \approx 1.1574515$

9. $f(x) = x + \ln x - 2$, $f'(x) = 1 + \dfrac{1}{x}$
   $f(1) = -1$
   $f(2) = \ln 2 \approx .693147$
   Let $c_1 = 1$.

   $c_2 = 1 - \dfrac{f(1)}{f'(1)} = 1 - \dfrac{(-1)}{2} = 1\tfrac{1}{2}$

11. $V = \tfrac{1}{3}\pi h^2(3R - h)$
    $12 = \tfrac{1}{3}\pi h^2(12 - h)$, $4\pi h^2 - \tfrac{1}{3}\pi h^3 - 12 = 0$
    $f(h) = -\tfrac{1}{3}\pi h^3 + 4\pi h^2 - 12$, $f'(h) = -\pi h^2 + 8\pi$
    $f(1) = -.4808$, $f(2) = 29.4071$
    Let $c_1 = 1$.

    $c_2 = 1 - \dfrac{-.4808}{7\pi} \approx 1.02186$

13. $f(x) = 5e^{-x} + x - 5$, $f'(x) = -5e^{-x} + 1$
$f(4) = -.9084$
$f(5) = .0337$
$c_2 = 5 - \dfrac{f(5)}{f'(5)} = 5 - \dfrac{5e^{-5}}{1 - 5e^{-5}} \approx 4.965$

Exercise 2 (page 333)

1. $\displaystyle\int_0^4 x^2\,dx;\quad n = 8$

$\displaystyle\int_0^4 x^2\,dx \approx \tfrac{1}{2}\left(\dfrac{4-0}{8}\right)[f(0) + 2f(\tfrac{1}{8}) + 2f(\tfrac{1}{4}) + 2f(\tfrac{3}{8}) + 2f(\tfrac{1}{2}) + 2f(\tfrac{5}{8}) + 2f(\tfrac{3}{4}) + 2f(\tfrac{7}{8}) + f(1)]$
$= \tfrac{1}{4}(0 + \tfrac{1}{32} + \tfrac{1}{8} + \tfrac{9}{32} + \tfrac{1}{2} + \tfrac{25}{32} + \tfrac{9}{8} + \tfrac{49}{32} + 1)$
$= \tfrac{1}{4}(\tfrac{43}{8}) = \tfrac{43}{32} = 1.34375$

3. $\displaystyle\int_1^2 \dfrac{dx}{x};\quad n = 4$

$\displaystyle\int_1^2 \dfrac{dx}{x} \approx \tfrac{1}{2}\left(\dfrac{2-1}{4}\right)[f(1) + 2f(\tfrac{5}{4}) + 2f(\tfrac{3}{2}) + 2f(\tfrac{7}{4}) + f(2)]$
$= \tfrac{1}{8}[1 + \tfrac{8}{5} + \tfrac{4}{3} + \tfrac{8}{7} + \tfrac{1}{2}] = \tfrac{317}{560} = .56607$

5. $\displaystyle\int_0^1 e^{x^2}\,dx;\quad n = 4$

$\displaystyle\int_0^1 e^{x^2}\,dx \approx \tfrac{1}{2}\left(\dfrac{1-0}{4}\right)(f(0) + 2f(\tfrac{1}{4}) + 2f(\tfrac{1}{2}) + 2f(\tfrac{3}{4}) + f(1))$
$\approx \tfrac{1}{8}(1 + 2e^{1/16} + 2e^{1/4} + 2e^{9/16} + e)$
≈ 1.49068

7. $\displaystyle\int_1^2 \dfrac{dx}{x} = \ln|x|\Big|_1^2 = \ln 2 - \ln 1 = \ln 2$

$\displaystyle\int_1^2 \dfrac{dx}{x} \approx \tfrac{1}{2}\left(\dfrac{2-1}{5}\right)(f(1) + 2f(\tfrac{6}{5}) + 2f(\tfrac{7}{5}) + 2f(\tfrac{8}{5}) + 2f(\tfrac{9}{5}) + f(2))$
$= \tfrac{1}{10}(1 + \tfrac{10}{6} + \tfrac{10}{7} + \tfrac{10}{8} + \tfrac{10}{9} + \tfrac{1}{2})$
$\approx .695635$

9. $n = 8$
$b = 80$
$a = 0$

Area $\approx (\tfrac{1}{2})\left(\dfrac{80-0}{8}\right)(5 + 2(10) + 2(13.2) + 2(15) + 2(15.6) + 2(12) + 2(6) + 2(4) + 0)$
$= \tfrac{1}{20}(156.6) = 7.83$

11. $a = 0$
$b = 20$
$n = 4$

Area $\approx \tfrac{1}{2}\left(\dfrac{20-0}{4}\right)(0 + 2(12) + 2(19) + 2(13) + 0)$
$= \tfrac{5}{2}(88) = 220$

APPENDIX

Exercise 2 (page 338)

1. $3^3 = 27$ **3.** $2^{-3} = \dfrac{1}{2^3} = \tfrac{1}{8}$ **5.** $(\tfrac{1}{2})^3 = \tfrac{1}{8}$ **7.** $(343)^{1/3} = 7$, since $7^3 = 343$
9. $(-243)^{1/5} = -3$, since $(-3)^5 = -243$ **11.** $2^{-1} + 4^{-1} = \tfrac{1}{2} + \tfrac{1}{4} = \tfrac{3}{4}$
13. $(3^{-2})(3^4) = 3^{-2+4} = 3^2 = 9$ **15.** $\dfrac{(2^{-2})(2^4)}{2^5} = 2^{-2+4-5} = 2^{-3} = \tfrac{1}{8}$

APPENDIX SOLUTIONS 431

17. $(\sqrt{2})(\sqrt[4]{32}) = 2^{1/2} \cdot 2^{5/4} = 2^{7/4}$ 19. $\sqrt{\frac{81}{16}} = \frac{9}{4}$
21. x^6 23. $3ax^5$ 25. $(x+2)^2$ 27. $(x+1)^2$ 29. $\frac{1}{x} + \frac{1}{x^2} = \frac{x+1}{x^2}$

Exercise 3 (page 340)
1. $3x + 6$ 3. $x^3 - x$ 5. $x^2 - x - 2$ 7. $x^2 - 3x + 2$ 9. $x^3 - x^2 - x - 2$
11. $2x^5 + 2x^4 - x^3 - x^2 + 4x + 4$ 13. $x + 4$ 15. $x^2 - 3$ 17. $3x + 17 + \frac{62}{x-4}$ 19. $x^2 + 5$

Exercise 4 (page 342)
1. $(x+1)^2 = x^2 + 2x + 1$ 3. $(x+6)(x+1) = x^2 + 7x + 6$ 5. $(x+\sqrt{2})(x-\sqrt{2}) = x^2 - 2$
7. $(4x+1)(x-3) = 4x^2 - 11x - 3$ 9. $(x+1)^3 = x^3 + 3x^2 + 3x + 1$
11. $(x-1)(x^2+x+1) = x^3 - 1$ 13. $(2x-3)^3 = 8x^3 - 36x^2 + 54x - 27$
15. $(x-2)(x+2)$ 17. $(x+6)(x+1)$ 19. $(2x+1)(x-1)$
21. $(3x-1)(x+2)$ 23. $(x+3)(x^2-3x+9)$ 25. $(2x+1)(4x^2-2x+1)$ 27. $(x+1)^3$
29. (a) $(x+h)^2 - x^2 = 2xh + h^2$ (b) $(x+h)^3 - x^3 = 3hx^2 + 3h^2x + h^3$
31. (II) $(x+a)(x-a) = x(x-a) + a(x-a) = x^2 - ax + ax - a^2 = x^2 - a^2$
    (III) $(x+a)(x+b) = x(x+b) + a(x+b) = x^2 + bx + ax + ab = x^2 + (a+b)x + ab$

Exercise 5 (page 344)
1. $\frac{2}{3x-3} + \frac{x+1}{x^2} = \frac{2}{3(x-1)} + \frac{x+1}{x^2} = \frac{2x^2 + 3(x-1)(x+1)}{3x^2(x-1)} = \frac{2x^2 + 3x^2 - 3}{3x^2(x-1)} = \frac{5x^2 - 3}{3x^2(x-1)}$
3. $\frac{7}{x+3} - \frac{7}{x} = \frac{7x - 7(x+3)}{x(x+3)} = \frac{7x - 7x - 21}{x(x+3)} = \frac{-21}{x(x+3)}$
5. $\frac{\sqrt{2x+5} + \sqrt{2x}}{5} \cdot \frac{\sqrt{2x+5} - \sqrt{2x}}{\sqrt{2x+5} - \sqrt{2x}} = \frac{2x+5-2x}{5(\sqrt{2x+5} - \sqrt{2x})} = \frac{5}{5(\sqrt{2x+5} - \sqrt{2x})} = \frac{1}{\sqrt{2x+5} - \sqrt{2x}}$

Exercise 6 (page 346)
1. $c = \sqrt{3^2 + 4^2} = 5$
3. $b = \sqrt{13^2 - 5^2} = 12$
5. $A = 6$; $P = 2 \cdot 5 = 10$
7. $A = \frac{1}{6}$; $P = 2(\frac{1}{2} + \frac{1}{3}) = \frac{5}{3}$
9. $A = \frac{1}{2} \cdot 2 \cdot 1 = 1$
11. $A = \frac{1}{2} \cdot \frac{1}{2} \cdot \frac{3}{4} = \frac{3}{16}$
13. $A \approx 3.14$; $C \approx 6.28$
15. $A = \pi(\frac{1}{2})^2 \approx 0.785$; $C = 2\pi(\frac{1}{2}) \approx 3.14$
17. $V = 1$; $A = 6$
19. $V = \frac{1}{2} \cdot \frac{3}{2} \cdot \frac{4}{3} = 1$; $A = 2(\frac{1}{2} \cdot \frac{3}{2} + \frac{1}{2} \cdot \frac{4}{3} + \frac{3}{2} \cdot \frac{4}{3}) = \frac{41}{6}$
21. $V = \frac{4}{3}\pi \cdot 8 = \frac{32\pi}{3} \approx 33.5$; $S = 16\pi \approx 50.3$
23. $V = \frac{\pi \cdot 9 \cdot 4}{3} = 12\pi \approx 37.7$

Exercise 7 (page 349)
1. $-5 < x < 5$

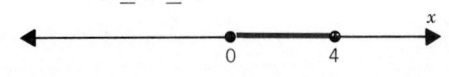

3. $-1 < x < 7$

5. $-4 \leq x \leq 4$

7. $|2x - 4| + 5 \leq 9$
   $|2x - 4| \leq 4$
   $-4 \leq 2x - 4 \leq 4$
   $0 \leq 2x \leq 8$
   $0 \leq x \leq 4$

Answers to Mathematical Questions

CHAPTER 1 (Page 33)

1. b 2. d 3. d 4. b 5. b 6. b 7. d 8. c 9. c 10. c 11. b 12. b

CHAPTER 2 (Page 90)

1. b

CHAPTER 4 (Page 191)

1. b 2. d 3. a 4. c

CHAPTER 5 (Page 222)

1. e 2. a 3. a 4. e

CHAPTER 8 (Page 324)

1. b 2. d 3. c 4. e 5. b 6. b 7. d

Answers to True-False Questions

CHAPTER 1 (Page 31)

1. F **2.** T **3.** T **4.** F **5.** T

CHAPTER 2 (Page 87)

1. T **2.** F **3.** T **4.** F **5.** T **6.** F **7.** F **8.** T **9.** T **10.** F

CHAPTER 3 (Page 135)

1. T **2.** F **3.** T **4.** T **5.** F

CHAPTER 4 (Page 188)

1. T **2.** F **3.** F **4.** F **5.** T

CHAPTER 5 (Page 220)

1. T **2.** F **3.** T **4.** T **5.** F

Answers to Fill-in-the-Blanks Questions

CHAPTER 1 (Page 31)
1. abscissa, ordinate 2. intercepts 3. undefined, zero 4. negative 5. parallel

CHAPTER 2 (Page 88)
1. parabola 2. independent, dependent 3. vertex 4. one 5. rational 6. limit
7. $\angle$ = 8. not have a limit 9. continuous 10. $\neq$

CHAPTER 3 (Page 136)
1. tangent 2. velocity 3. power rule 4. acceleration 5. zero

CHAPTER 4 (Page 188)
1. decreasing 2. decreasing, increasing 3. concave upward 4. horizontal 5. zero

CHAPTER 5 (Page 221)
1. exponential 2. 32 3. $2xe^{x^2}$ 4. e 5. natural logarithm

Index

Abscissa, 4
Absolute:
  maximum, 152
  minimum, 152
  test for, 153
Absolute value, 347
  function, 51
Acceleration, 127
  of gravity, 128
Annuity, 298
Antiderivative, definition of, 250
Area, properties of, 284
Asymptote:
  horizontal, 172
  vertical, 172
Average cost, 182
Average rate of change, 95
Average revenue, 182
Average velocity, 96
  value, 306

Boundary condition, 264
Break-even point, 25

Cartesian coordinates, 4
Chain rule, 119
  proof, 121
Change, average rate of, 95
Closed interval, 44
Compound interest, 205
  formula, 205
  table, 709
Concave down, 165
Concave up, 165
Concavity, test for, 166
Continuous function, 82
Coordinate, 2
  rectangular, 4
Cost average, 182
  function, 178
  marginal, 178, 230
  variable, 295
Critical number, 144
Critical point, 144, 233
Curve, learning, 315
  decay, 198
  growth, 196
  logistic, 218
  modified growth, 217
  smooth, 132
Cylinder, 345

Decay curve, 198
Decreasing function, 141
  test for, 141

Definite integral, 278
Demand equation, 28, 177
  function, 177
Demand level, 654
Density function, 310
  beta, 314
  exponential,
  probability, 309
  uniform, 311
Dependent variable, 48
Derivative, of a^x, 214
  of constant, 109
  definition of, 102
  of difference, 112
  of e^x, 201
  first test, 145
  geometric interpretation, 105
  of $\ln x$, 211
  Leibniz notation, 109
  of $\log a^x$, 214
  logarithmic, 213
  one-sided, 131
  partial, 228
  power rule, 119
  prime notation, 109
  of product, 113
  of quotient, 114
  second, 125
  second test, 147
  of sum, 112
  of x^n, 110
Descartes, Rene, 4
Difference quotient, 96
Differentiable function, 103
Differential equation, 263
  solution of, 263
Discontinuous function, 82
Discriminant, 60
Distance, 43
Distributive law, 339
Domain, 47

e, number, 198, 204
Effective rate of interest, 207
Equation, 2
  demand, 28, 177
  differential, 611
  of line, 12
  linear, 7
  point-slope form, 10
  slope intercept form, 13
  supply, 28
  systems, 17
  tangent line, 106
Exponent, 194, 336

437

fractional, 338
laws of, 194
Exponential function, 195

Function, 47
  absolute value, 51
  average value, 306
  beta density, 314
  composite, 478
  continuous, 82
  cost, 178
  decreasing, 141
  demand, 177
  density, 309
  derivative of, 102
  differentiable, 103
  discontinuous, 82
  even, 283
  exponential, 195
  exponential density, 312
  increasing, 141
  limit of, 67
  linear, 59
  logarithm, 209
  nondifferentiable, 130
  odd, 283
  polynomial, 63
  power, 64
  price, 177
  probability, 308
  probability density, 310
  profit, 179
  quadratic, 59
  rational, 64
  revenue, 178
  smooth, 132
  square root, 52
  step, 84
  two variables, 224
  uniform density, 311
Fundamental formula of integral calculus, 279, 293

Galileo, 128
General equation of a line, 12
Geometry formulas, 344
Goodness of fit, 242
Graph, 5
  computer, 226
Growth curve, 196
  modified, 217

Half-life, 266

Identical lines, 15
Increasing function, 141
  test for, 141

Indefinite integral, 251
Independent variable, 48
Inequalities, linear, 38
Inequality, 3
  solution, 3
Infinite limits, 171
Infinity, limit at, 171
Inflection point, 165
  test for, 166
Integral, definite, 278
  of difference, 253, 280
  indefinite, 251
  table, 273
  of sum, 253, 280
Integrand, 252
Integration, lower limit of, 278
  by parts, 260
  by substitution, 255
  upper limit of, 278
Intercepts, 40
Interest, 23
  compound, 205
  continuous, 206, 265
  effective rate of, 207
  rate of, 23
  simple, 24
Intersecting lines, 16
Interval, closed, 44
  end point, 44
  open, 44

Lagrange multiplier, 237
Learning curve, 315
Leibniz notation, 109
Limit, 70
  algebraic technique, 75
  of constant, 75
  definition of, 70
  of difference, 76
  of $f(x) = x$, 76
  of $f(x) = x^n$, 78
  graphical approach, 68
  infinite, 171
  at infinity, 171
  from left, 68
  numerical approach, 72
  of product, 77
  of quotient, 78
  from right, 69
  of sum, 76
  least squares method of, 241
Line:
  best fit, 242
  general form, 12
  identical, 15
  intercepts of, 12
  intersecting, 16
  parallel, 15

INDEX

point-slope form, 10
secant, 96
slope of, 8
slope-intercept form, 13
straight, 7
tangent, 106
vertical, 9
Linear equation, 7
  function, 59
Linear inequality, 38
Linear regression, 241
  inequality, 38
Logarithm function, 209
  change of base, 213
  common, 210
  derivative of, 211
  natural, 211
  properties of, 209
Logarithmic differentiation, 213
Logistic curve, 218

Marginal cost, 178, 230
Marginal revenue, 178, 316
Market price, 29
Maximum:
  absolute, 152
  relative, 143, 232
Mean Value Theorem, 151
  for integrals, 295
Median, 46
Minimum:
  absolute, 152
  relative, 144, 232
Model:
  average rate measure of synchrony, 321
  consumer's surplus, 316
  inventory cost, 185
  learning curve, 315
  profit over time, 319
  tax revenue, 182
  trade-in-time, 184
  response of the body to drug, 186
  Newton's Method, 328

Norm of partition, 303
Numbers, e, 198, 204
  critical, 144, 300

Open interval, 44
Ordered pair, 4, 49
  triple, 224
Ordinate, 4
Origin, 2

Parabola, 61
Parallel lines, 15

Partial derivative, 228
  geometric interpretation, 229
Partition, 303
Plane, 225
Point:
  break-even, 25
  critical, 144, 233
Point slope equation, 10
Polynomial function, 63
  degree of, 64
Power function, 64
Power rule, 119
Principal, 24
Probability, 308
  density function, 310
Pythagorean theorem, 344

Quadrant, 4
Quadratic formula, 60
Quadratic function, 59

Range, 47
Rate of change, 107
  average, 95
Rate of interest, 24
  effective, 207
Rational function, 64
Rectangular coordinates, 4
Regression analysis, 240
Relative maximum, 143, 232
  test for, 145, 147, 232
Relative minimum, 144, 232
  test for, 145, 147, 232
Revenue function, 178
  average, 182
  marginal, 179
Riemann integral, 303
Riemann sum, 303
Rolle's theorem, 150

Saddle point, 233
  test for, 233
Secant line, 96
Simple interest, 24
Slope intercept form of line, 13
Slope of line, 8
Solution, 2
  of tangent line, 106
  undefined, 8
  differential equation, 263
  particular, 264
Standard normal curve, 361
  step function, 84
Supply equation, 28
Surface, 226
Symmetry, 41
System of equations, 17
  solution of, 18

Table:
   common logarithms, 353
   e^x, e^{-x}
   natural logarithms, 355
Tangent line, 106
   equation, 106
   slope of, 106
Test:
   for absolute maximum, 153
   for concavity, 166
   for decreasing function, 141
   first derivative, 145
   for increasing function, 141
   for inflection point, 166
   relative maximum, 233
   relative minimum, 233
   saddle point, 233
   second derivative, 147
Trapezoid, 345
Trapezoidal Rule, 331
   error, 331
   proof of, 332

Variable, 2, 336
   continuous, 54
   dependent, 48
   discrete, 54
   independent, 48
Velocity, 101
   average, 96
Vertex of parabola, 61
Vertical line, 8
   equation of, 9

x-axis, 4
x-intercept, 12, 40

y-axis, 4
y-intercept, 12, 40

z-axis, 224
Zero of function, 99

METRIC CONVERSION

LENGTH
1 inch (in.) = 2.540 centimeters (cm)

1 foot (ft) = 30.48 centimeters (cm)

1 yard (yd) = 91.44 centimeters (cm) = 0.914 meters (m)

1 mile (mi) = 1609 meters (m) = 1.61 kilometers (km)

AREA
1 square inch (in.^{2}) = 6.452 square centimeters (cm^2)

1 square foot (ft^2) = 0.093 square meters (m^2)

VOLUME
1 cubic inch = 16.39 cubic centimeters (cc) = 0.0164 liters (l)

1 cubic foot = 0.028 cubic meters (m^3)

LIQUID VOLUME
1 quart (qt) = 0.946 liters (l)

1 gallon (gal) = 3.78 liters (l)

WEIGHT
1 pound (lb) = 453.6 grams (gm)

1 ton = 907.2 kilograms (kg)

TEMPERATURE
C = Celsius, F = Fahrenheit

$C = \frac{5}{9}(F - 32)$